Temperate Nuts

Mohammad Maqbool Mir • Munib Ur Rehman •
Umar Iqbal • Shabir Ahmad Mir
Editors

Temperate Nuts

Springer

Editors
Mohammad Maqbool Mir
Division of Fruit Science, FOH
Sher-e-Kashmir University
of Agricultural Sciences
and Technology of Kashmir
Srinagar, Jammu and Kashmir, India

Munib Ur Rehman
Division of Fruit Science, FOH
Sher-e-Kashmir University
of Agricultural Sciences
and Technology of Kashmir
Srinagar, Jammu and Kashmir, India

Umar Iqbal
Division of Fruit Science, FOH
Sher-e-Kashmir University
of Agricultural Sciences
and Technology of Kashmir
Srinagar, Jammu and Kashmir, India

Shabir Ahmad Mir
Department of Food Science & Technology
Government College For Women
Srinagar, Jammu and Kashmir, India

ISBN 978-981-19-9496-8 ISBN 978-981-19-9497-5 (eBook)
https://doi.org/10.1007/978-981-19-9497-5

This Springer imprint is published by the registered company Springer Nature Singapore Pte Ltd.
The registered company address is: 152 Beach Road, #21-01/04 Gateway East, Singapore 189721, Singapore

Contents

Editors and Contributors

About the Editors

Mohammad Maqbool Mir PhD, is an associate professor cum senior scientist in the Division of Fruit Science, Sher-e-Kashmir University of Agricultural Sciences and Technology of Kashmir, Shalimar, Srinagar. He is associated with many externally aided projects and also with research groups working on canopy architectural management, production technology, and standardization of production protocols for different temperate fruits and nuts. He has supervised/co-supervised several MSc and PhD scholars and is associated with many academic and professional societies. To his credit, Dr. Maqbool has published more than 68 scientific papers in different reputed journals at national and international level and 30 other popular articles, book chapters, and extension bulletins, and he has edited 3 books.

Munib Ur Rehman works as an associate professor in the Division of Fruit Science, Faculty of Horticulture, Sher-e-Kashmir University of Agricultural Sciences and Technology of Kashmir, Shalimar, Srinagar. He obtained his master's degree and PhD from Sam Higginbottom University of Agriculture, Technology and Sciences, Allahabad. He has been associated with three externally aided projects as PI and also been associated with research group working on crop production management and standardization of production protocols for temperate fruit and nut crops. He has guided two MSc students besides postgraduate teaching. Munib has published more than 46 research publications in different reputed journals at national and international level along with 10 other review articles and 12 book chapters, and has edited 1 book and extension bulletins.

Umar Iqbal PhD, is an assistant professor cum junior scientist in the Division of Fruit Science, Sher-e-Kashmir University of Agricultural Sciences and Technology of Kashmir, India. He has concluded satisfactorily two externally funded projects by HTM and MIDH as principal investigator and being Co-PI in another externally funded project by Potash Research Institute and International Potash Institute.

Dr. Umar has supervised/co-supervised several MSc and PhD students in fruit science and has published more than 44 papers in Indian and foreign journals. Dr. Umar has contributed chapters in more than four books and edited two books.

Shabir Ahmad Mir obtained his PhD in food technology from Pondicherry University, Puducherry, India. He is presently an assistant professor at the Government College for Women, M. A. Road, Srinagar, India. He received the best PhD Thesis Award for outstanding research in the field of food technology. He has organized several conferences and workshops in food science and technology. Dr. Mir has published numerous international papers and book chapters, and he has edited ten books.

Contributors

Shabnam Ahad Division of Fruit Science, FOH, Sher-e-Kashmir University of Agricultural Sciences and Technology of Kashmir, Srinagar, Jammu and Kashmir, India

M. Amin Mir Ambri Apple Research Centre, Shopian, Sher-e-Kashmir University of Agricultural Sciences and Technology of Kashmir, Srinagar, Jammu and Kashmir, India

Syed Andleeba Division of Soil Science and Agricultural Chemistry, FOA, Sher-e-Kashmir University of Agricultural Sciences and Technology of Kashmir, Wadura, Jammu and Kashmir, India

Tashi Angmo Division of Fruit Science, FOH, Sher-e-Kashmir University of Agricultural Sciences and Technology of Kashmir, Srinagar, Jammu and Kashmir, India

Mohammad Mehdi Arab Department of Horticulture, College of Aburaihan, University of Tehran, Tehran, Iran

T. J. Archana Division of Food Science and Postharvest Technology, ICAR-Indian Agriculture Research Institute, New Delhi, India

Hojjat Ataee Department of Horticultural Sciences, Gorgan University of Agricultural Sciences and Natural Resources, Gorgan, Iran

Zahoor Ahmad Baba Division of Basic Sciences and Humanities, Sher-e-Kashmir University of Agricultural Sciences and Technology of Kashmir, Wadura, Jammu and Kashmir, India

Hujjat Ul Baligah Division of Soil Science and Agricultural Chemistry, FOA, Sher-e-Kashmir University of Agricultural Sciences and Technology of Kashmir, Wadura, Jammu and Kashmir, India

Shafat Ahmad Banday Krishi Vigyan Kendra, Sher-e-Kashmir University of Agricultural Sciences and Technology of Kashmir, Ganderbal, Jammu and Kashmir, India

Ajay K. Banyal Department of Fruit Science, YSPUH&F College of Horticulture and Forestry, Neri, Hamirpur, Himachal Pradesh, India

Sanjeev K. Banyal Department of Fruit Science, YSPUH&F College of Horticulture and Forestry, Neri, Hamirpur, Himachal Pradesh, India

Tabish Jehan Been Division of Fruit Science, FOH, Sher-e-Kashmir University of Agricultural Sciences and Technology of Kashmir, Srinagar, Jammu and Kashmir, India

Khalid M. Bhat Division of Fruit Science, FOH, Sher-e-Kashmir University of Agricultural Sciences and Technology of Kashmir, Srinagar, Jammu and Kashmir, India

M. A. Bhat Division of Plant Pathology, FOH, Sher-e-Kashmir University of Agricultural Sciences and Technology of Kashmir, Srinagar, Jammu and Kashmir, India

Rifat Bhat Division of Fruit Science, FOH, Sher-e-Kashmir University of Agricultural Sciences and Technology of Kashmir, Srinagar, Jammu and Kashmir, India

Shahzad Bhat Division of Fruit Science, FOH, Sher-e-Kashmir University of Agricultural Sciences and Technology of Kashmir, Srinagar, Jammu and Kashmir, India

Z. A. Bhat Division of Plant Pathology, FOH, Sher-e-Kashmir University of Agricultural Sciences and Technology of Kashmir, Srinagar, Jammu and Kashmir, India

Irfan A. Bisati Division of Fruit Science, FOH, Sher-e-Kashmir University of Agricultural Sciences and Technology of Kashmir, Srinagar, Jammu and Kashmir, India

M. H. Chesti Division of Soil Science and Agricultural Chemistry, FOA, Sher-e-Kashmir University of Agricultural Sciences and Technology of Kashmir, Wadura, Jammu and Kashmir, India

M. P. Ellampirai Department of Horticulture, School of Life Sciences, Central University of Tamil Nadu, Thiruvarur, TN, India

Saqib Farooq Department of Food Technology, Islamic University of Science and Technology, Awantipora, Jammu and Kashmir, India

Angela Feechan School of Agriculture and Food Science, University College Dublin, Dublin, Ireland
Institute for Life and Earth Sciences, School of Energy, Geoscience, Infrastructure and Society, Heriot-Watt University, Edinburgh, UK

Nazir Ahmad Ganai Division of Fruit Science, FOH, Sher-e-Kashmir University of Agricultural Sciences and Technology of Kashmir, Srinagar, Jammu and Kashmir, India

S. Gouthami Division of Food Science and Postharvest Technology, ICAR-Indian Agricultural Research Institute, New Delhi, India

Yasmeen Gull Department of Physics, GDC, Kulgam, Jammu and Kashmir, India

Asaad Habibi Department of Horticulture, College of Aburaihan, University of Tehran, Tehran, Iran

Mehvish Hanief Division of Fruit Science, FOH, Sher-e-Kashmir University of Agricultural Sciences and Technology of Kashmir, Srinagar, Jammu and Kashmir, India

G. I. Hassan Research Centre for Residue and Quality Analysis, FOH, Sher-e-Kashmir University of Agricultural Sciences and Technology of Kashmir, Srinagar, Jammu and Kashmir, India

Umar Iqbal Division of Fruit Science, FOH, Sher-e-Kashmir University of Agricultural Sciences and Technology of Kashmir, Srinagar, Jammu and Kashmir, India

Sehrish Jan Division of Fruit Science, FOH, Sher-e-Kashmir University of Agricultural Sciences and Technology of Kashmir, Srinagar, Jammu and Kashmir, India

Sonika Jaryal Division of Fruit Science, FOH, Sher-e-Kashmir University of Agricultural Sciences and Technology of Kashmir, Srinagar, Jammu and Kashmir, India

Aroosa Khalil Division of Fruit Science, FOH, Sher-e-Kashmir University of Agricultural Sciences and Technology of Kashmir, Srinagar, Jammu and Kashmir, India

Inayat M. Khan Division of Soil Science and Agricultural Chemistry, FOA, Sher-e-Kashmir University of Agricultural Sciences and Technology of Kashmir, Wadura, Jammu and Kashmir, India

Saba Q. Khan Division of Fruit Science, FOH, Sher-e-Kashmir University of Agricultural Sciences and Technology of Kashmir, Srinagar, Jammu and Kashmir, India

K. Rama Krishna Department of Horticulture, School of Life Sciences, Central University of Tamil Nadu, Thiruvarur, TN, India

Amit Kumar Division of Fruit Science, FOH, Sher-e-Kashmir University of Agricultural Sciences and Technology of Kashmir, Srinagar, Jammu and Kashmir, India

Sunil Kumar ICAR-National Research Centre on Litchi, Muzaffarpur, India

Zaffar Mahdi Division of Basic Science & Humanities, FOH, Sher-e-Kashmir University of Agricultural Sciences and Technology of Kashmir, Srinagar, Jammu and Kashmir, India

Romana Mahmood Division of Fruit Science, FOH, Sher-e-Kashmir University of Agricultural Sciences and Technology of Kashmir, Srinagar, Jammu and Kashmir, India

A. R. Malik Division of Fruit Science, FOH, Sher-e-Kashmir University of Agricultural Sciences and Technology of Kashmir, Srinagar, Jammu and Kashmir, India

Naveen Kumar Maurya Division of Fruits and Horticultural Technology, ICAR-Indian Agricultural Research Institute, New Delhi, India

Nirmal Kumar Meena Division of Food Science and Postharvest Technology, ICAR-Indian Agricultural Research Institute, New Delhi, India
Agriculture University, Kota, Rajasthan, India

Zaffar Mehdi Division of Basic Science & Humanities, FOH, Sher-e-Kashmir University of Agricultural Sciences and Technology of Kashmir, Srinagar, Jammu and Kashmir, India

M. Menaka Division of Food Science and Postharvest Technology, ICAR-Indian Agricultural Research Institute, New Delhi, India

Mahrukh Mir Division of Fruit Science, FOH, Sher-e-Kashmir University of Agricultural Sciences and Technology of Kashmir, Srinagar, Jammu and Kashmir, India

Mohammad Maqbool Mir Division of Fruit Science, FOH, Sher-e-Kashmir University of Agricultural Sciences and Technology of Kashmir, Srinagar, Jammu and Kashmir, India

Shabir Ahmad Mir Department of Food Science and Technology, Government College for Women, Srinagar, Jammu and Kashmir, India

Shakeel A. Mir Division of Soil Science and Agricultural Chemistry, FOA, Sher-e-Kashmir University of Agricultural Sciences and Technology of Kashmir, Wadura, Jammu and Kashmir, India

Amir Mirzadi Gohari Department of Plant Protection, University College of Agriculture and Natural Resources, University of Tehran, Karaj, Iran

Munib Ur Rehman Division of Fruit Science, FOH, Sher-e-Kashmir University of Agricultural Sciences and Technology of Kashmir, Srinagar, Jammu and Kashmir, India

Safura Nabi Division of Fruit Science, FOH, Sher-e-Kashmir University of Agricultural Sciences and Technology of Kashmir, Srinagar, Jammu and Kashmir, India

Nowsheen Nazir Division of Fruit Science, FOH, Sher-e-Kashmir University of Agricultural Sciences and Technology of Kashmir, Srinagar, Jammu and Kashmir, India

Jamasb Nozari Department of Plant Protection, College of Agriculture and Natural Resources, University of Tehran, Tehran, Iran

Gepu Nyorak Division of Fruit Science, FOH, Sher-e-Kashmir University of Agricultural Sciences and Technology of Kashmir, Srinagar, Jammu and Kashmir, India

Ashaq H. Pandit Division of Fruit Science, FOH, Sher-e-Kashmir University of Agricultural Sciences and Technology of Kashmir, Srinagar, Jammu and Kashmir, India

Mir Uzma Parveze Division of Fruit Science, FOH, Sher-e-Kashmir University of Agricultural Sciences and Technology of Kashmir, Srinagar, Jammu and Kashmir, India

Uday Raj Patial Department of Fruit Science, PAU, Ludhiana, Punjab, India

Satyabrata Pradhan Division of Fruits and Horticultural Technology, ICAR-Indian Agricultural Research Institute, New Delhi, India

K. Prasad Dr. Rajendra Prasad Central Agriculture University, Samastipur, Bihar, India

Tabasum N. Qadri S.P. College, Srinagar, Jammu and Kashmir, India

Bashir Ahmad Rather Mountain Research Centre for Field Crops (MRCFC), Sher-e-Kashmir University of Agricultural Sciences and Technology of Kashmir, Khudwani, Jammu and Kashmir, India

Gh. Hassan Rather Ambri Apple Research Centre, Shopian, Sher-e-Kashmir University of Agricultural Sciences and Technology of Kashmir, Srinagar, Jammu and Kashmir, India

Saadat Sarikhani Department of Horticulture, College of Aburaihan, University of Tehran, Tehran, Iran

Satpal Division of Fruit Science, FOH, Sher-e-Kashmir University of Agricultural Sciences and Technology of Kashmir, Srinagar, Jammu and Kashmir, India

Manzoor Ahamd Shah Department of Food Science and Technology, Government Degree College for Women, Anantnag, Anantnag, Jammu and Kashmir, India

Mahender K. Sharma Division of Fruit Science, FOH, Sher-e-Kashmir University of Agricultural Sciences and Technology of Kashmir, Srinagar, Jammu and Kashmir, India

Nirmal Sharma Division of Fruit Science, FOH, SKUAST-Jammu, Jammu, Jammu and Kashmir, India

Shivali Sharma Division of Fruit Science, FOH, Sher-e-Kashmir University of Agricultural Sciences and Technology of Kashmir, Srinagar, Jammu and Kashmir, India

Vishal Sharma School of Agricultural Biotechnology, Punjab Agricultural University, Ludhiana, Punjab, India

Abdollatif Sheikhi Department of Horticulture, Faculty of Agriculture, Vali-e-Asr University of Rafsanjan, Rafsanjan, Kerman, Iran

Gurupkar Singh Sidhu School of Agricultural Biotechnology, Punjab Agricultural University, Ludhiana, Punjab, India

Shamim A. Simnani Division of Fruit Science, FOH, Sher-e-Kashmir University of Agricultural Sciences and Technology of Kashmir, Srinagar, Jammu and Kashmir, India

Amarjeet S. Sindouri Division of Fruit Science, FOH, Sher-e-Kashmir University of Agricultural Sciences and Technology of Kashmir, Srinagar, Jammu and Kashmir, India

Ajit Kumar Singh Division of Food Science and Postharvest Technology, ICAR-Indian Agricultural Research Institute, New Delhi, India

Jagveer Singh School of Agricultural Biotechnology, Punjab Agricultural University, Ludhiana, Punjab, India
Department of Fruit Science, College of Horticulture & Forestry, Acharya Narendra Deva University of Agriculture & Technology, Ayodhya, India

Kappat Valiyapeediyekkal Sunooj Department of Food Science and Technology, Pondicherry University, Puducherry, India

Anamika Thakur Division of Food Science and Postharvest Technology, ICAR-Indian Agricultural Research Institute, New Delhi, India

Kourosh Vahdati Department of Horticulture, College of Aburaihan, University of Tehran, Tehran, Iran

B. R. Vinod Division of Food Science and Postharvest Technology, ICAR-Indian Agricultural Research Institute, New Delhi, India

Tajamul F. Wani Division of Fruit Science, FOH, Sher-e-Kashmir University of Agricultural Sciences and Technology of Kashmir, Srinagar, Jammu and Kashmir, India

Ashok Yadav ICAR-Central Agroforestry Research Institute, Jhansi, India

Iftisam Yaseen Division of Fruit Science, FOH, Sher-e-Kashmir University of Agricultural Sciences and Technology of Kashmir, Srinagar, Jammu and Kashmir, India

Chapter 1
Global Scenario of Temperate Nuts

Amit Kumar, Nirmal Sharma, Shivali Sharma, Gepu Nyorak, Satpal, Sonika Jaryal, Mohammad Maqbool Mir, Umar Iqbal, Amarjeet S. Sindouri, and Khalid M. Bhat

Abstract The major temperate nut crops grown in different parts of the world include almond, walnut, pecan nut, chestnut, hazelnut and pistachio nut. The geographical features and agro-climatic conditions prevailing in the entire United States of America, China, Turkey and Iran are conducive for the cultivation of temperate nut crops. A continuous increase was observed in the area, production and export of all the temperate nuts crops in the last 5–6 years. In 2020, the United States of America is the leading producer of almond and pistachio nuts, while China is the leading producing country of walnut and chestnut. Mexico and Turkey were the leading producers of pecan nut and hazelnut, respectively. The United States of America (almond, walnut, pistachio nut, pecan nut), Spain (almond), Mexico (pecan nut), Turkey (hazelnut), China (walnut, chestnut, pistachio nut), Italy (hazelnut, chestnut) and Iran (pistachio nut) are the exporting countries of temperate nut crops. All the European countries are importers of all major temperate nut crops.

Keywords Global · Temperate · Nut · Walnut

1 Introduction

Nut crops are a group of botanically unrelated crops belonging to different botanical families. They are grouped together because the fruit type is a nut (generally) and harvesting, postharvest processing, nutritional value and marketing characteristics are relatively similar. A nut is a fruit consisting of a hard or tough shell protecting a kernel which is usually considered as edible part. Based on usage or in gastronomic sense, a wide variety of dry seeds are called nuts, but in botanical way, nut infers

A. Kumar (✉) · S. Sharma · G. Nyorak · Satpal · S. Jaryal · M. M. Mir · U. Iqbal · A. S. Sindouri · K. M. Bhat
Division of Fruit Science, FOH, Sher-e-Kashmir University of Agricultural Sciences and Technology of Kashmir, Srinagar, Jammu and Kashmir, India

N. Sharma
Division of Fruit Science, FOH, SKUAST-Jammu, Jammu, Jammu and Kashmir, India

M. M. Mir et al. (eds.), *Temperate Nuts*,
https://doi.org/10.1007/978-981-19-9497-5_1

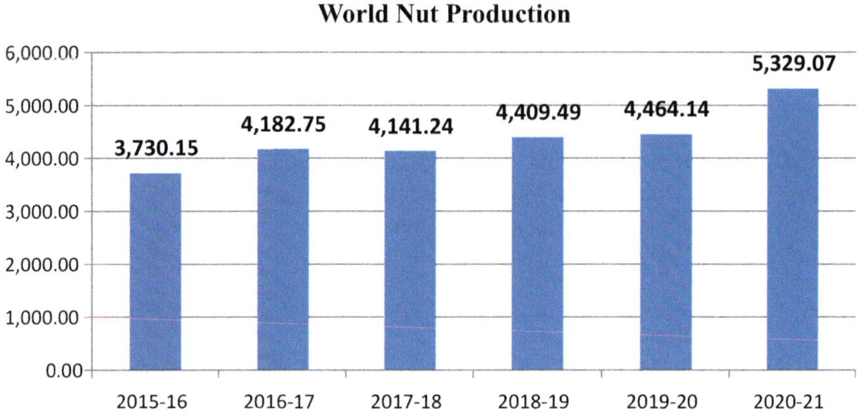

Fig. 1.1 Nut production scenario worldwide during the last 6 years. (Source: Statistical Year Book 2020–2021)

kernel or seed imbedded in hard cover, i.e. shell. Nuts are often wind pollinated and generally produce a lot of pollen, often in specialised male inflorescence, to ensure that the female flowers get fertilised. Nuts are monoecious (walnut and pecan nut) and dioecious (pistachio) in nature, however protandrous and protogynous also exist which are referred as heterodichogamy. Walnut and pecan nut are closely related as both members belongs to family Juglandaceae, however, they are not cross or graft-compatible and are not similar as walnut yields much higher than pecans which is due to propensity to bear nuts from lateral buds on 1-year-old wood in walnut and not just on terminal buds. Majority of the seeds free themselves from the shell naturally upon maturity, except hazelnuts and chestnuts, which have hard shell walls and originate from a compound ovary. According to climatic conditions walnut, pecan nut, hazelnut, chestnut and pistachio nut are temperate nuts and, cashew nut and macadamia nut are tropical nuts.

Over the last decade, global nut production has steadily increased, reaching over 5.3 million MT during 2020–2021 which was 15.0% higher than that produced during 2019–2020 and 65% higher than a decade earlier. Between 2011–2012 and 2020–2021, global nut production increased at an average annual rate of about 212,400 MT per year (Fig. 1.1).

Almonds and walnuts are the most productive nut crops, accounting for 31% and 19% of the global share, respectively, followed by pistachios (19%), hazelnuts (10%), chestnut (5%), pecan nut (2%) and other nut crops accounted for the remaining 14%. When analysing the annual growth rates of various tree nut crops over the last 10 years, it can be seen that the walnut, pecan nut, almond and pistachio crops had the most important linear increments with the following annual growth rates:

- Walnuts: 59,082 MT/year
- Pecans: 7396 MT/year

- Almonds: 53,885 MT/year
- Pistachios: 42,426 MT/year

According to the production data from the last 5 years, the USA led the world nut production with an average share of 40%. The most produced crops were almonds (59% of country production), pistachios (21%) and walnuts (15%). Turkey came in second with 11% of the global share, with hazelnuts and pistachios accounting for 64% and 31% of the country's production, respectively.

Almonds and walnuts accounted for estimated half of total tree nut consumption worldwide in 2019 which was 30% and 20%, respectively, followed by cashews and pistachios which accounted for 18% and 15%, respectively. Following Europe as the leading consumer, Asia and North America are the second and third largest consuming regions with similar market shares.

2 Almond

Worldwide, the production of almond in the year 2020 was 41.40 lakh MT from an area of 21.62 lakh hectare with a productivity of 1.91. During that year, production was 16.23% more than the previous year and 55.15% more in the last years. A regular expansion in the area of almond cultivation was observed in the last 6 years from 17.77 lakh hectares in 2015 to 21.62 lakh hectare in 2020 (Fig. 1.2).

In 2020, the USA was the leading producer of almond in the world which produced 19.36 lakh MT from an area of 4.77 lakh hectare, followed by Spain which produces 3.40 lakh MT from an area of 6.87 lakh hectare (Fig. 1.3).

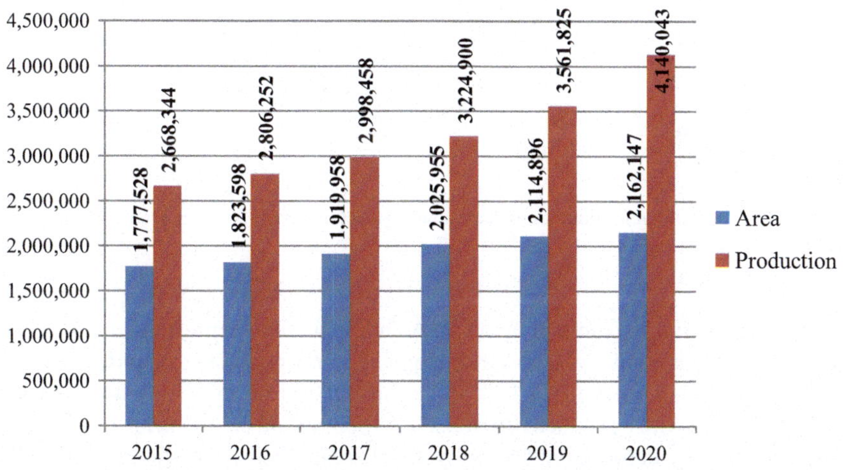

Fig. 1.2 Area and production of almond worldwide during the last 6 years. (Source: FAOSTAT n. d. statistical database)

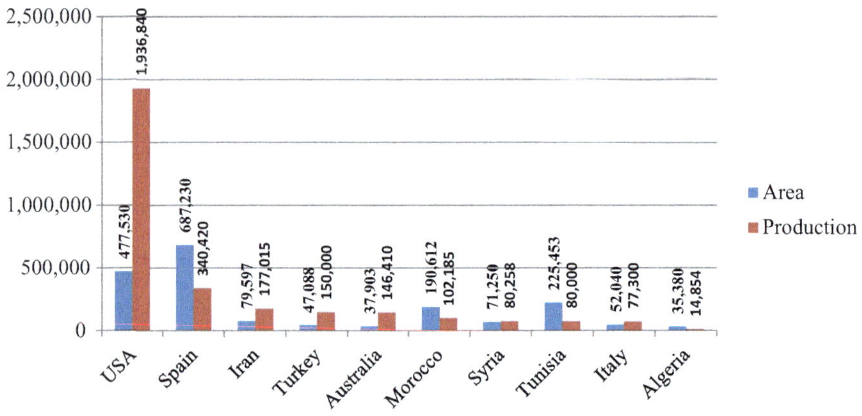

Fig. 1.3 Area and production of almond in different countries during 2020. (Source: FAOSTAT n.d. statistical database)

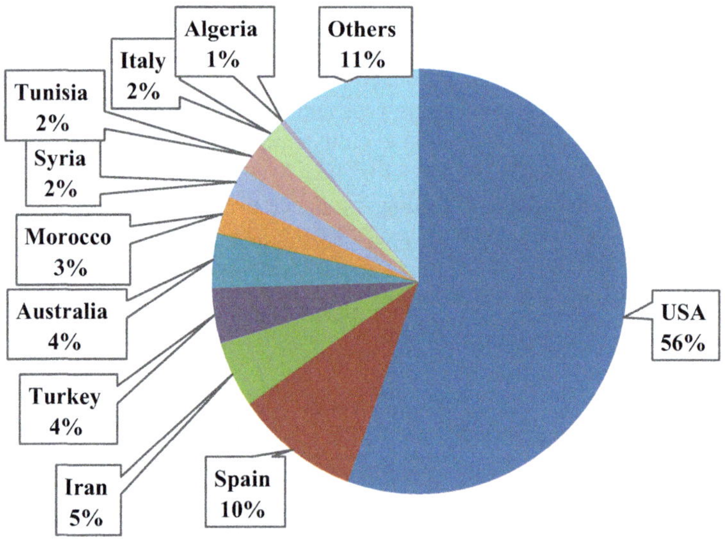

Fig. 1.4 Share of different countries in the almond production during 2020. (Source: FAOSTAT n.d. statistical database)

The productivity of almond in the USA and Spain has been recorded as 4.05 and 0.49 tons/ha, respectively. The other major almond producing countries are Iran (1.77 lakh MT), Turkey (1.50 lakh MT), Australia (1.46 lakh MT) and Morocco (1.02 lakh MT) with a productivity of 2.22, 3.18, 3.86 and 0.54 tons/ha, respectively. The USA and Spain produced 66% of world share in 2020. Australia has doubled its almond production share over the globe during the last 10 years. Both Australia and Spain have surpassed the production benchmark of 1.0 lakh MT in 2020 (Fig. 1.4).

Fig. 1.5 Export of almond in the last 5 years to other countries. (Source: Statistical Year Book 2020–2021)

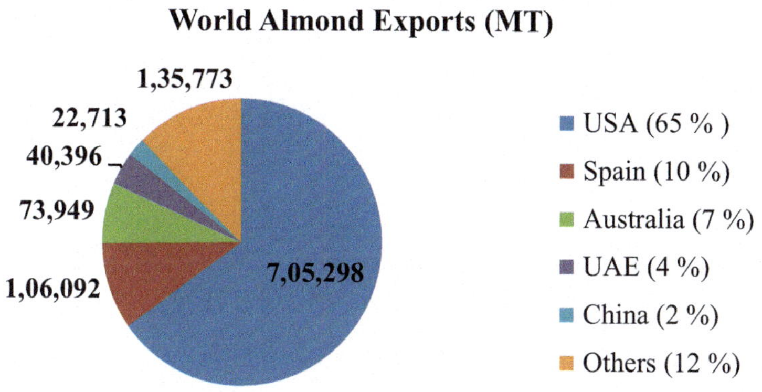

Fig. 1.6 Share of different countries in the export of almond. (Source: Statistical Year Book 2020–2021)

In the year 2019, 10.84 lakh MT of almond was exported by different countries which was 11.96% higher than the previous year and 31.82% higher than which was during the last 5 years. A continuous increase in almond production has been observed in the export of almond in the last decade (Fig. 1.5).

The world export of almond is also led by the USA and continues to dominate in the export with a share of 65%. In 2019, 7.05 lakh MT of almond were shipped mainly to Spain (17%), Germany (8%), The Netherlands (7%) and Japan (7%) (Fig. 1.6). Spain exported 1.06 lakh MT of almond and the main importers were other European countries (85%) headed by Germany, Italy and France. Australia exported a share of 7.0% of in-shell almond to Asian countries and the importing countries are China (71%) followed by India (24%). Other exporting countries are UAE (4%) and China (2%).

Table 1.1 World almond imports (MT) in the last 5 years

Country	2015	2016	2017	2018	2019
Spain	86,150	101,070	111,293	99,562	116,552
India	81,705	86,179	114,790	99,048	115,049
China	55,038	80,625	76,998	85,814	93,771
Germany	85,646	86,826	92,783	84,579	93,765
Italy	42,032	44,598	45,865	48,988	59,281
Netherlands	29,988	28,337	36,497	41,846	49,232
UAE	50,014	42,349	39,301	36,542	46,626
France	38,696	38,083	41,764	43,380	45,175
Japan	38,134	28,059	33,573	37,186	40,322
UK	25,078	25,438	31,371	26,434	30,586
Turkey	15,851	27,468	24,579	18,286	30,210
Canada	26,389	30,292	29,821	30,496	30,131
Korea Rep.	20,010	23,962	25,610	22,927	25,459
Vietnam	38,850	70,179	46,409	59,721	21,131
Belgium	15,728	15,812	16,002	14,344	16,680
Saudi Arabia	9910	10,945	11,792	11,177	13,097
USA	11,424	9161	8111	10,687	12,651
Mexico	11,063	9654	9801	10,447	11,858
Greece	4927	6689	7743	7393	9085
Thailand	4332	4181	2711	4579	8680
Others	136,525	154,538	167,362	174,962	214,879

Source: Statistical Year Book 2020–2021

Main importing countries of almond are Spain (1.16 lakh MT), India (1.15 lakh MT), China (0.93 lakh MT), Germany (0.93 lakh MT), Italy (0.59 lakh MT) and The Netherlands (0.49 lakh MT) (Table 1.1).

3 Walnut

Among temperate nut crops, walnut ranks first because of its nutritive value, popularity, area and production in the world. The two major species of walnuts are English walnut and Black walnut. The former is also known as the Persian walnut since it originated in Persia and the latter traces its origin to North America. The walnut seedlings (*Juglans regia* L.) have been found growing in vast region right from the Carpathian Mountains in eastern Europe, across Turkey, Iraq, Iran, southern USSR and Afghanistan to the Northwestern Himalayas. In 2020, world production of walnut (in shell) was 33.23 lakh MT from an area of 10.21 lakh hectare having a productivity of 3.25 ton/ha. The walnut production in the present year remained 11.32% more than the previous year. In the last 6 years, there is a tremendous expansion in the area of walnut in the world which is 55.69% more in

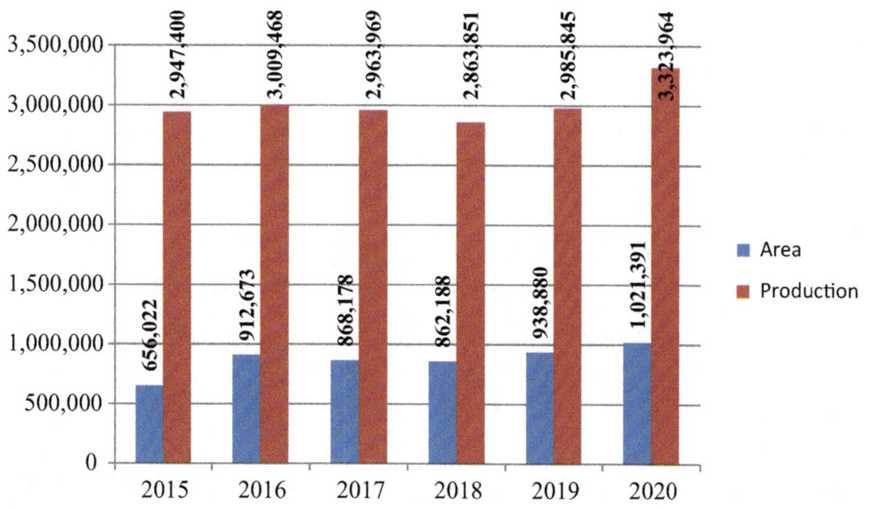

Fig. 1.7 World area and production of walnut during the last 6 years. (Source: FAOSTAT n.d. statistical database)

the year 2020 as compared to 2015, however, production increased only 12.77% during the past 6 years (Fig. 1.7).

Over the past 10 years, production has increased steadily, at a rate of 4.0% per year. But some countries have increased production more than others, and the increase in the USA and Ukraine was around 10%, while it amounted to 20% and 40% in Iran and Chile, respectively.

The important walnut growing countries of the world as per area and production are given in Fig. 1.8. China is the largest producer of walnut in the world contributing 33% share with a production of 11.00 lakh MT from an area of 2.84 lakh hectare. The productivity of walnut in China is 3.86 ton/ha. The USA is the second largest producer with a production of 7.07 lakh MT from an area of 1.53 lakh hectare having a productivity of 4.60 ton/ha. China and the USA together produce more than 54% of the world total walnut production. The other walnut producing countries are Iran (3.56 lakh MT), Turkey (2.86 lakh MT) and Mexico (1.64 lakh MT) with a productivity of 5.95 ton/ha, 2.02 ton/ha and 1.51 ton/ha with a share of 10.73%, 8.6% and 4.9%, respectively (Fig. 1.8).

Ukraine tops in the productivity of walnut with a value of 11.79 in the entire world whereas the share was only 4.7 per cent followed by Iran with a productivity and share of 5.95 and 10.73 per cent. Productivity of walnut in China and USA has been recorded as 3.86 and 4.60 tones/ha, respectively. The Other major walnut producing countries are Iran (3.56 lakh MT), Turkey (2.86 lakh MT), Mexico (1.64 lakh MT), Ukraine (1.58 lakh MT) and Chile (1.13 lakh MT) having productivity of 5.95, 2.02, 1.51, 11.79 and 2.61 tones/ha. China and USA produces more than 54 per cent of world share in 2020. Turkey and Mexico shares 8.6 per cent and 4.9 per cent of the world walnut production over the globe during 2020 (Fig. 1.9).

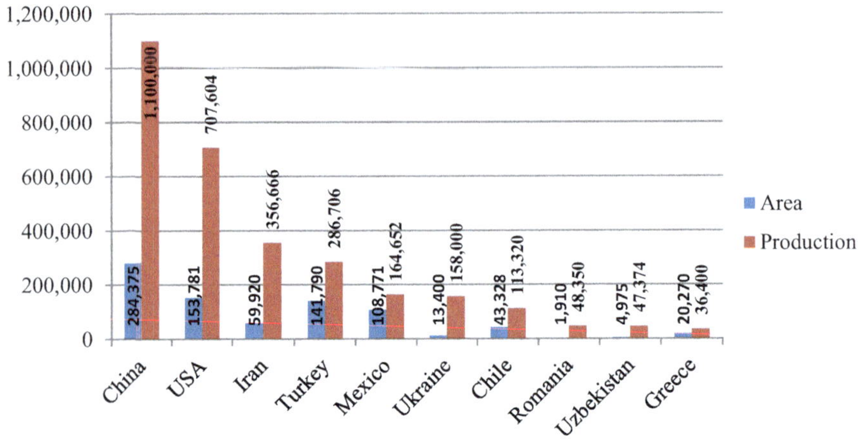

Fig. 1.8 Area and production of walnut in different countries during 2020. (Source: FAOSTAT n.d. statistical database)

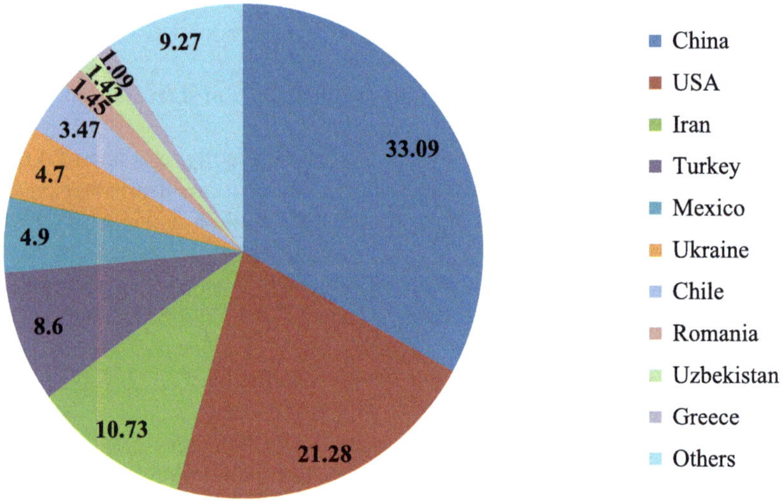

Fig. 1.9 Share of different countries in the walnut production in 2020. (Source: FAOSTAT n.d. statistical database)

Walnuts are the only temperate nut crops of India grown commercially in the hilly states of the country viz. Jammu and Kashmir, Himachal Pradesh, Uttarakhand and Arunachal Pradesh. The total area under the walnut cultivation in India is 108.63 thousand hectares having production of 299.71 thousand MT with a productivity of 1.97 ton/ha. In India, Jammu and Kashmir is the leading walnut producing state with a production of 2.62 lakh MT from 0.84 lakh hectare area. It produces about 90% of the total production in India with an average productivity of 3.09 ton/ha. Other walnut producing states are Himachal Pradesh (2.46 MT), Uttarakhand (21.17 MT).

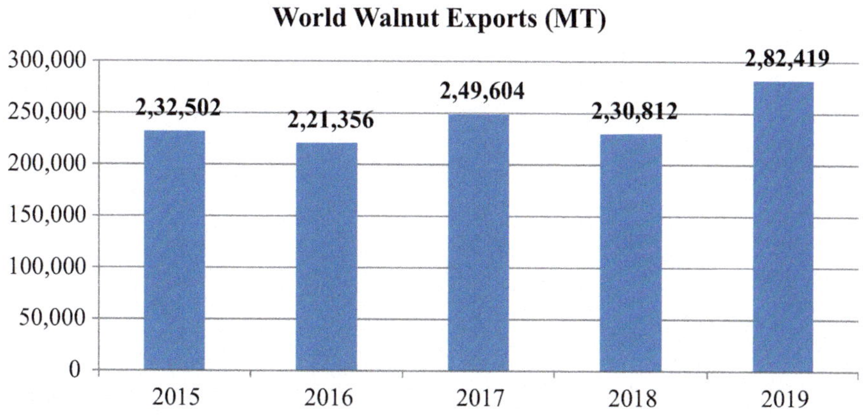

Fig. 1.10 Export of almond in the last 5 years to other countries. (Source: Statistical Year Book 2020–2021)

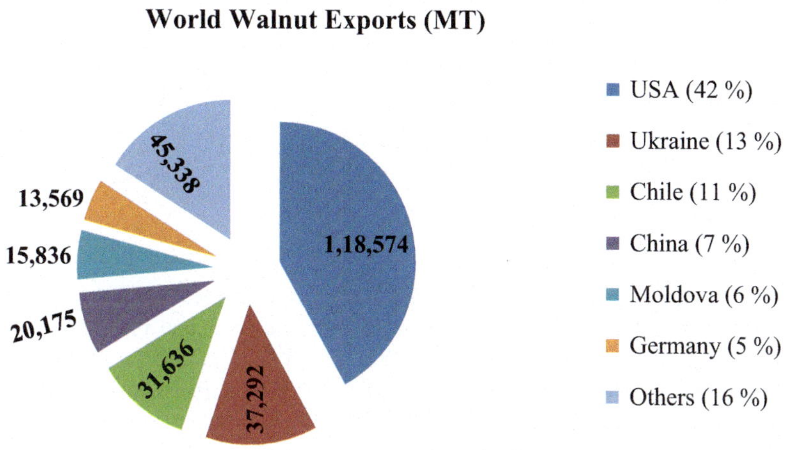

Fig. 1.11 Share of different countries in the export of walnut during 2020. (Source: Statistical Year Book 2020–2021)

Walnuts have always been a highly valued commodity and their consumption has increased considerably in recent years. The walnut market is showing increasing potential, both in terms of household consumption and as a raw material for the food industry, resulting in new opportunities for walnut producers. In 2019, the total shelled walnut shipment was 2.82 lakh MT which was 22.36% more than the previous year and 21.47% more in the last 5 year (Fig. 1.10).

The USA (1.18 lakh MT), Ukraine (0.37 lakh MT) and Chile (0.31 lakh MT) are the major exporter countries in the world with a share of 42%, 13% and 11%. China ranked fourth place in the export of walnut with a share of 7.0% to different countries due to the high consumption of its own production by the domestic market (Fig. 1.11).

Table 1.2 World walnut imports (MT) in the last 5 years

Country	2015	2016	2017	2018	2019
Germany	24,815	25,978	32,667	35,297	41,451
Japan	14,918	18,392	21,172	17,038	19,009
Netherlands	14,026	12,691	17,579	14,535	18,909
Spain	12,204	14,709	14,745	15,983	17,509
Turkey	7827	3892	9195	9093	16,189
France	10,179	9768	13,181	11,215	12,941
Korea Fed.	12,631	16,130	12,685	12,459	12,760
Canada	9386	10,862	10,287	11,160	11,596
Italy	6116	7630	7633	7554	9711
UK	8617	9559	10,552	8892	9508
Israel	4810	5170	5139	5614	6180
UAE	3709	4347	4245	3335	5583
Austria	3654	3621	4791	5033	5122
Australia	5117	4660	5527	4086	4197
Russian Fed.	651	755	2152	2017	4167
Brazil	2659	3187	3287	3583	3855
Greece	2670	2681	3198	3508	3774
Azerbaijan	1611	1872	3182	2580	3467
Moldova	1634	1446	755	329	3336
Saudi Arabia	1299	2135	2137	2196	3063
Others	83,969	61,871	65,497	55,306	70,094

Source: Statistical Year Book 2020–2021

Walnut imports are dominated by the European countries as Germany (41,000 MT) which ranks first in importing walnut followed by Japan (19,000 MT), The Netherlands (18,000 MT) and Spain (17,000 MT). Other major European importers include France, Italy. Canada and Mexico are North America's leading importers, while Turkey and Korea are leading the rest of the world (Table 1.2).

4 Pecan Nut

The world pecan nut production during 2020–2021 was 1.66 lakh MT. The present year production was 7.0% more than the previous year and 39.33% from the previous 6-year average and the highest production was recorded in the last decade (Fig. 1.12).

In 2020, Mexico was the largest producer of pecan nut with a production of 82,000 MT followed by the USA (69,000 MT), South Africa (10,000 MT), Brazil (1700 MT) and Australia (1085 MT) (Fig. 1.13).

Mexico (49.0%) and the USA (41.0%) are the main produces of pecan nut contributing about 90% of the world production and the rest 10% production was

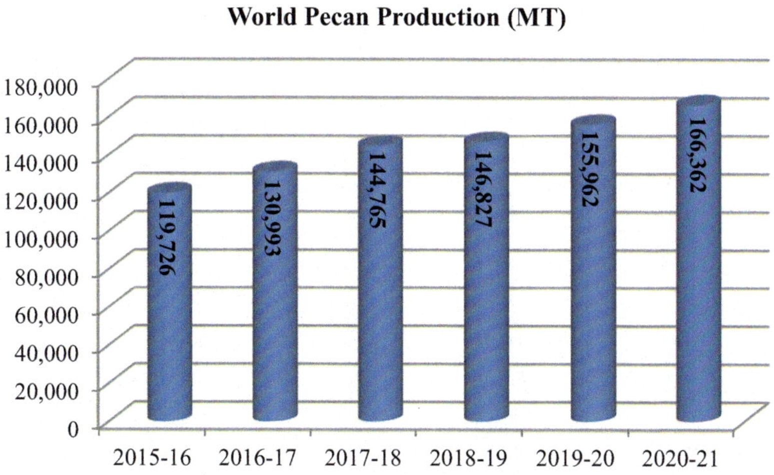

Fig. 1.12 World pecan nut production during the last 6 years. (Source: Statistical Year Book 2020–2021)

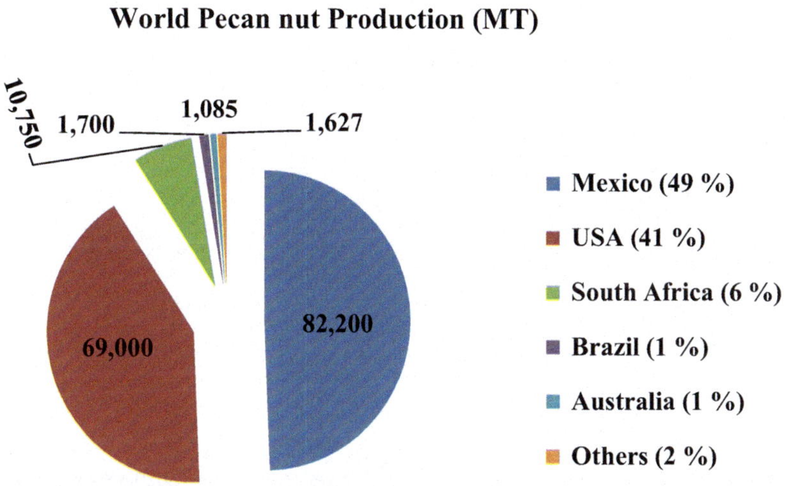

Fig. 1.13 Share of different countries in the pecan nut production in 2020–2021. (Source: Statistical Year Book 2020–2021)

split among South Africa (6.0%), Brazil (1.0%), Australia (1.0%) and other countries (2.0%). Although, South Africa and Brazil produce less quantity of pecan nut, but both countries have seen and consistent increase in the production of pecan nut in recent years and hope this trend continue in the future year also (Fig. 1.13).

In the last 5 years, an increasing trend was observed in the export of pecan nut and in 2019, 82,000 MT of pecan nut was exported which was 8.07% more than the previous year and 48.74% more in the last 5 years (Fig. 1.14). Along with

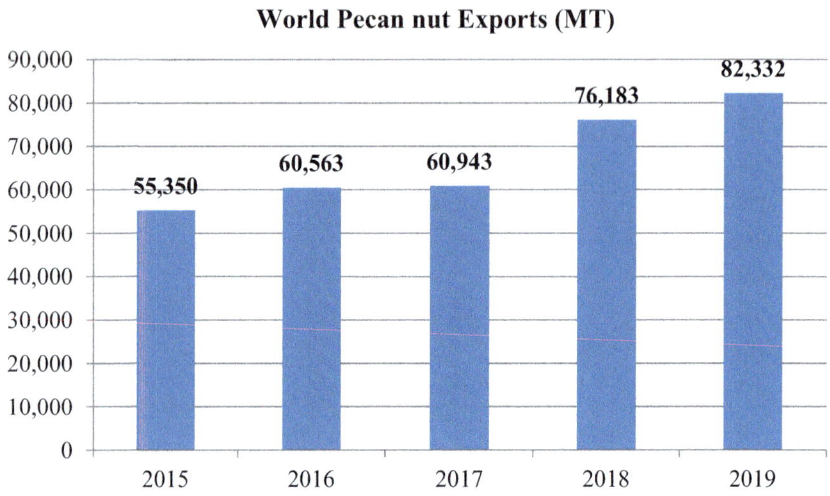

Fig. 1.14 Export of pecan nut in the last 5 years to other countries. (Source: Statistical Year Book 2020–2021)

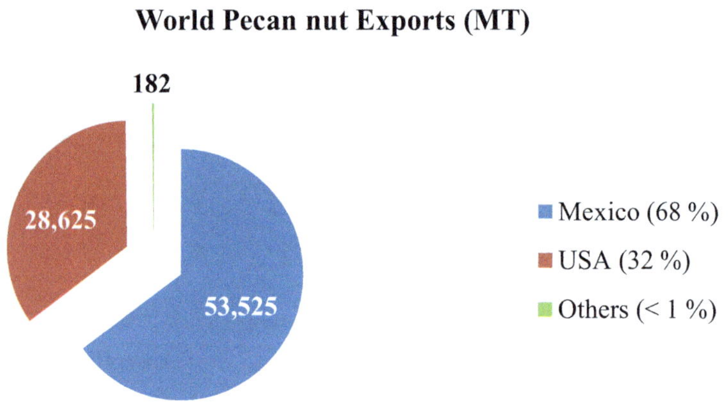

Fig. 1.15 Share of different countries in the export of pecan nut during 2020. (Source: Statistical Year Book 2020–2021)

production, Mexico (53,000 MT) and the USA (28,000 MT) are the main exporter of pecan nut contributing cent percent of the world export. The main destination of the Mexican pecan nut is the USA and the European Union + the United Kingdom (51.0%) and Canada (18.0%) were the main markets for the USA produce (Fig. 1.15).

Among all the importer countries in 2019, the USA led by 53,000 MT of pecan nut followed by The Netherlands (5939 MT), Canada (5223 MT) and Germany (2732 MT) (Table 1.3).

Table 1.3 World pecan nut imports (MT) in the last 5 years

Country	2015	2016	2017	2018	2019
USA	35,252	38,397	37,993	51,767	53,592
Netherlands	3346	4156	4897	4183	5939
Canada	4997	5510	5628	5425	5223
Germany	1186	833	1829	2304	2732
Mexico	1250	779	596	1688	2718
UK	2974	2560	2744	2643	2455
Israel	1316	1531	1681	2077	2383
France	825	999	1027	1195	1099
Belgium	228	373	455	598	933
China	889	2155	690	611	802
Korea Fed.	453	610	650	683	773
Spain	130	191	158	345	737
Saudi Arabia	184	135	299	367	440
UAE	161	147	210	195	340
Australia	162	57	138	160	309
Japan	247	245	211	313	302
Italy	227	306	346	81	248
Lithuania	114	129	150	176	199
Vietnam	176	346	82	70	127
Switzerland	168	231	271	74	87
Others	1067	873	889	1229	896

Source: Statistical Year Book 2020–2021

5 Hazelnut

The world hazelnut production shows fluctuations depending on the climatic conditions from year to year. The global hazelnut production in 2020 was 10.72 lakh MT from an area of 10.15 lakh hectares with a productivity of 1.05 ton/ha which was second largest in the last 10 years and 4.68% less than the previous years (Fig. 1.16).

The main hazelnut producing countries are Turkey, Italy, Spain, the USA and Greece. Although hazelnut is also produced in the Former Soviet Union, Iran, Romania and France, these countries do not have a major input in the world hazelnut trade. Turkey is the major world hazelnut producer and exporter. Turkey ranks first in the area and production of hazelnut in the world which produces 6.65 lakh MT from an area of 7.34 lakh hectare with a productivity of 0.91 ton/ha (Fig. 1.17). Italy is the second largest producer of hazelnut in the world with a production of 1.40 lakh MT from an area of 0.80 lakh hectare with a productivity of 1.75 ton/ha. The other major producing countries are the USA (64,000 MT), Azerbaijan (49,000 MT), Chile (33,000 MT) and Georgia (32,000 MT).

Turkey's crop accounted for 62% of the world total production of hazelnut followed by Italy (13.1%), the USA (6.01%), Azerbaijan (4.61%) and Chile

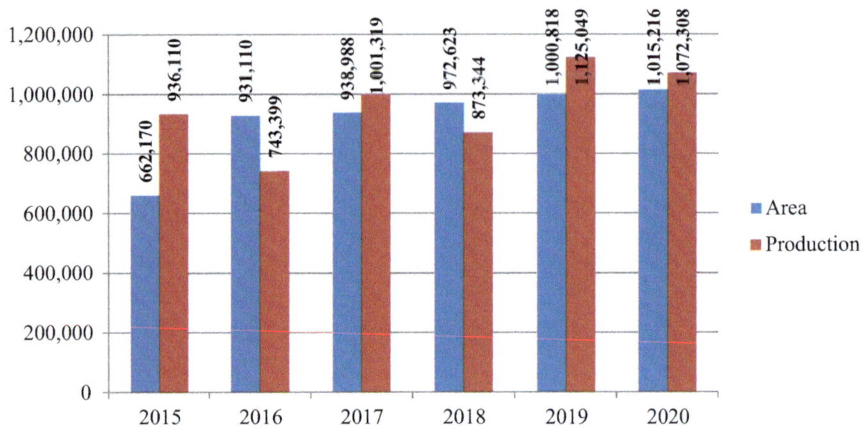

Fig. 1.16 World area and production of hazelnut during the last 6 years. (Source: FAOSTAT n.d. statistical database)

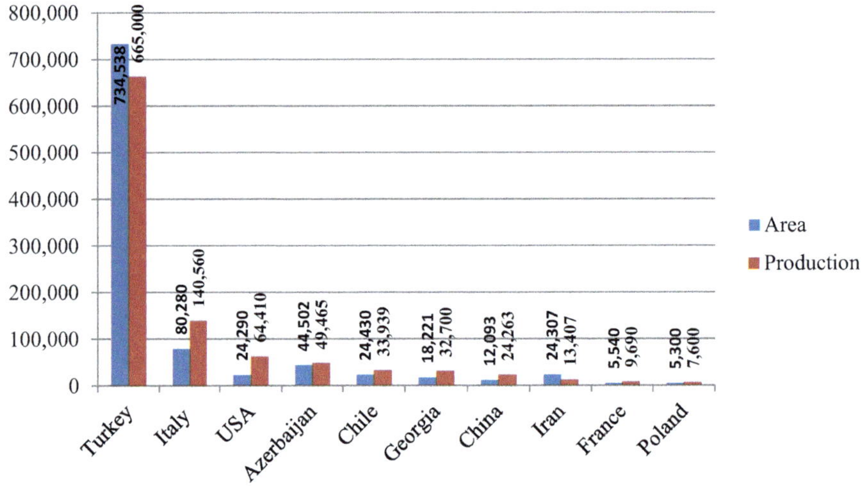

Fig. 1.17 Area and production of hazelnut in different countries during 2020. (Source: FAOSTAT n.d. statistical database)

(3.16%). Among these countries, Italy, the USA and Chile produce a bumper crop reaching record heights in this season (Fig. 1.18).

The total export of hazelnut during the year 2019 was 4.04 lakh MT which was 11.17% more than the previous year. A continous increasing trend of export to different countries was observed during the last 5 years (Fig. 1.19). The hazelnut exports have grown up by an average value of 30.99% from 2015 to 2019.

The most important hazelnut exporter countries are Turkey with an amount of 3.13 lakh MT and a share of 77.2%. The other exporting countries are Italy (25,000

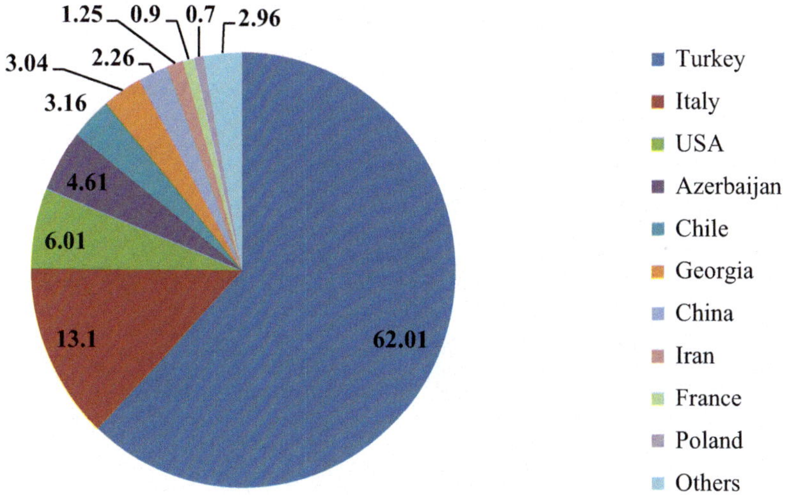

Fig. 1.18 Share of different countries in the hazelnut production during 2020. (Source: FAOSTAT n.d. statistical database)

Fig. 1.19 Export of hazelnut in the last 5 years to other countries. (Source: Statistical Year Book 2020–2021)

MT) with a share of 6.0%, Azerbaijan (22,000 MT) with a share of 5.0%, Georgia (11,000 MT) with a share of 3.0% and Chile (9,000 MT) having a share of 2.0% (Fig. 1.20). The most important hazelnut importing countries in 2019 were Italy (1.03 lakh MT), Germany (0.97 lakh MT), France (0.29 lakh MT), Poland (0.15 lakh MT), Russia (0.15 lakh MT) and Austria (0.12 lakh MT) (Table 1.4).

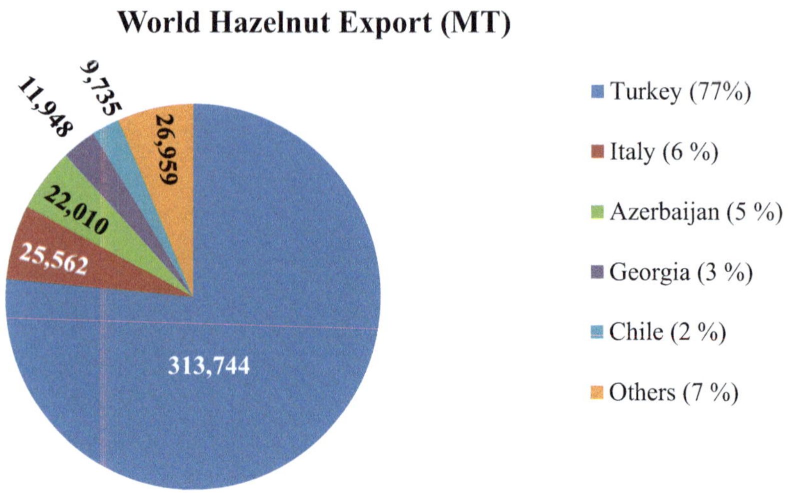

Fig. 1.20 Share of different countries in the export of hazelnut during 2020–2021. (Source: Statistical Year Book 2020–2021)

Table 1.4 World hazelnut imports (MT) in the last 5 years

Country	2015	2016	2017	2018	2019
Italy	61,591	70,000	70,504	68,632	103,985
Germany	83,980	77,087	93,044	91,491	97,617
France	28,617	27,923	29,378	27,803	29,776
Poland	13,214	12,320	14,125	14,805	15,753
Russian Fed.	7923	6229	12,300	15,986	15,557
Austria	10,226	10,277	10,563	10,740	12,121
Netherlands	9525	8639	12,204	11,668	12,036
China	3879	4248	7928	7293	11,313
Switzerland	9962	10,554	11,344	11,125	10,559
Belgium	11,779	11,390	11,402	11,287	10,087
Canada	13,023	11,727	12,139	10,170	9790
Spain	6230	6373	7214	9283	9784
USA	2384	3780	5498	7660	6334
UK	5550	5958	5973	5830	5711
Brazil	3049	2661	3114	4082	4593
Egypt	2939	2829	2391	3778	3859
Australia	2891	3125	3285	3395	3737
Ukraine	1606	1829	2504	3652	3635
Greece	1856	1822	2538	2898	3210
Sweden	2554	2466	2512	2172	2596
Others	87,949	97,838	103,566	109,122	136,892

Source: Statistical Year Book 2020–2021

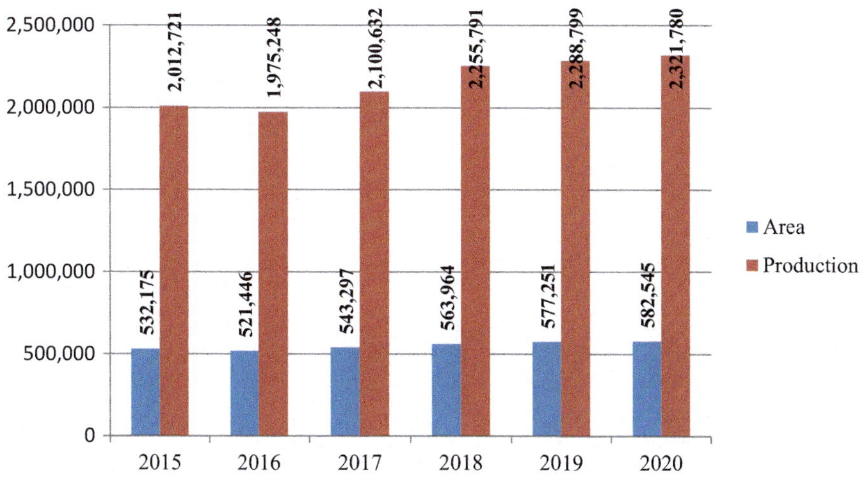

Fig. 1.21 World area and production of chestnut during the last 6 years. (Source: FAOSTAT n.d. statistical database)

6 Chestnut

The current chestnut global distribution is the consequence of natural colonisation together with a long history of human intervention (Freitas et al. 2021). Presently worldwide, the chestnut is grown on an area of 5.82 lakh hectare being China, Southern Europe, Southwestern USA, Korea and Japan are the major distributing countries. Presently, seven species of chestnut is known worldwide developing in subtropical, Mediterranean and temperate forests in the Northern Hemisphere. Chinese chestnut (*Castanea mollissima*), Japanese chestnut (*Castanea crenata*), American chestnut (*Castanea dentata*) and European chestnut (*Castanea sativa*) are widely cultivated species owing to the economic importance of their fruits. In 2020, the world chestnut production was 24.06 lakh MT from an area of 5.82 lakh hectare which was 1.81% more than in the previous year and 20.8% more than 10 years ago (Fig. 1.21).

The major production comes from *Castanea millissima* in Asia followed by *Castanea sativa* in Europe. China is the largest producer in the world with a production of 17.43 lakh MT from an area of 3.05 lakh hectare. Among other producers, Spain ranked second with a production of 1.8 lakh MT from an area of 37,000 hectare. Other major producers are Bolivia (80,000 MT), Turkey (76,000 MT) and Republic of Korea (54,000 MT), etc. (Fig. 1.22). According to production share, China alone produces more than 75% of the world's chestnut production followed by Spain (8.12%), Bolivia (3.48%), Turkey (3.27%), Republic of Korea (2.34%), etc. (Fig. 1.23).

With respect to exporters, China exports maximum (30.2%) chestnut with the main destination to Vietnam. The other major exporters are Italy (23.15%), Spain (9.06%), Turkey (8.87%) and Portugal (8.17%), etc.

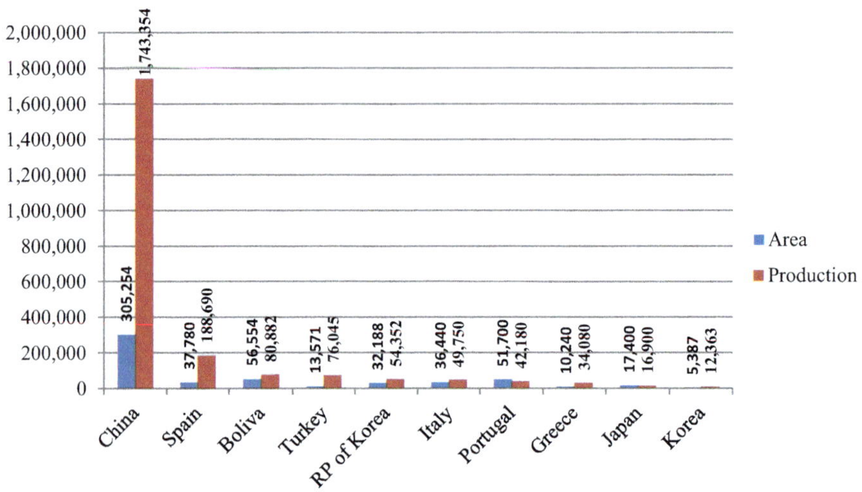

Fig. 1.22 Area and production of chestnut in different countries during 2020. (Source: FAOSTAT n.d. statistical database)

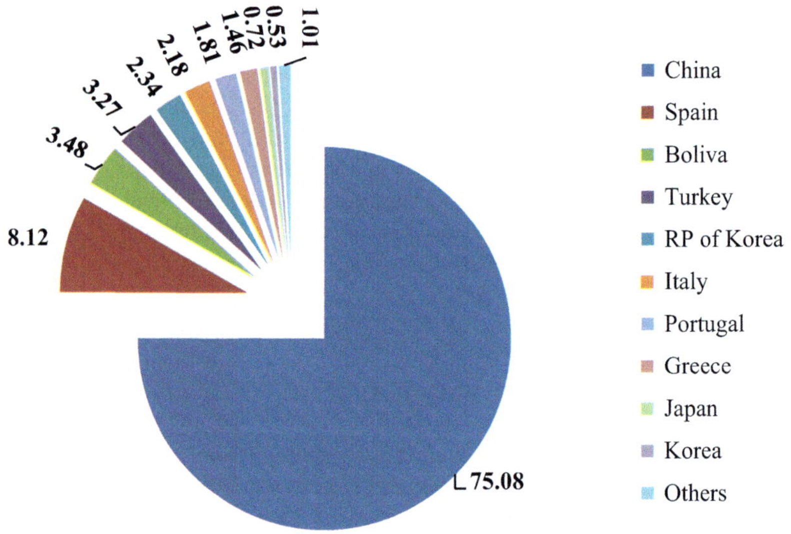

Fig. 1.23 Share of production of chestnut in different countries during 2020. (Source: FAOSTAT n.d. statistical database)

7 Pistachio Nut

The edible pistachio of commerce is the species *Pistacia vera* L. In addition to many named cultivars, significant populations of wild germplasm exist, primarily in Central Asia from Turkey to Afghanistan. Pistachio is grown most intensively in

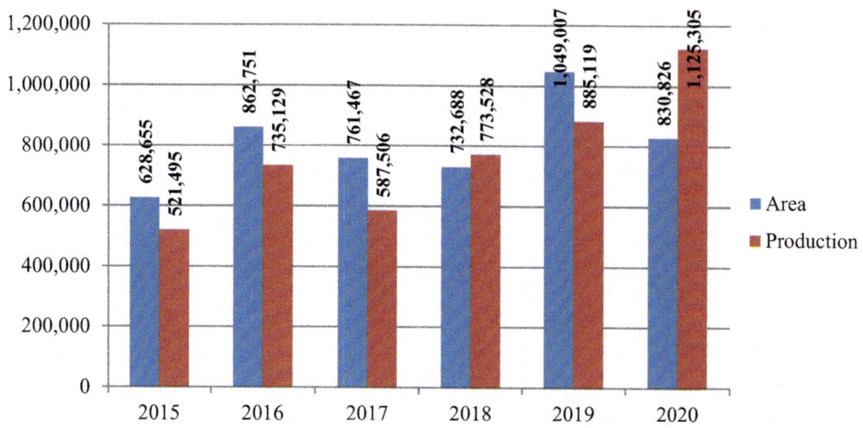

Fig. 1.24 World area and production of pistachio nut during the last 6 years. (Source: FAOSTAT statistical database)

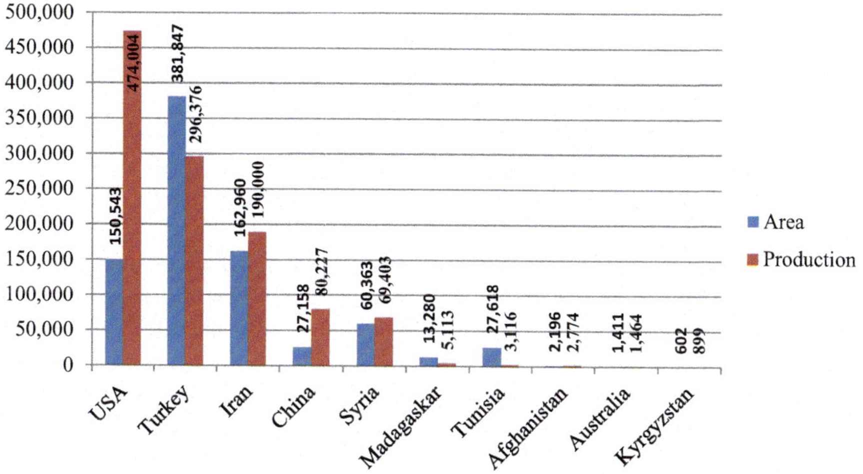

Fig. 1.25 Area and production of pistachio nut in different countries during 2020. (Source: FAOSTAT n.d. statistical database)

USA, Turkey, Iran, China and Syria. The other pistachio producing countries are in the Near East, North Africa and Southern Europe. In 2020, the global production of pistachio nut was 1.1 million MT from an area of 8.3 lakhs hectare (Fig. 1.24).

Based on the nut production data, the USA (4.7 lakh MT) and Turkey (2.96 lakh MT) are the leading producers, together accounting for 68% of the total production from an area of 64% in total area under pistachio cultivation in the world (Fig. 1.25).

There are significant variations in production every other year due to its alternate baring nature. As per the geographic feature or physiology of the plant year 2020 was considered as the 'On year' and as a result, global production totalled over 1.1

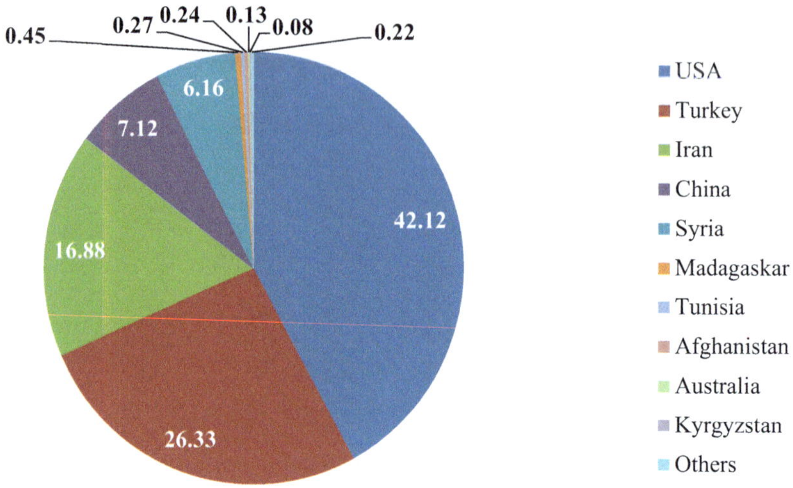

Fig. 1.26 Share of different countries in the pistachio nut production during 2020. (Source: FAOSTAT n.d. statistical database)

million MT (in-shell basis), the highest amount of the last decade, representing a 54% increase from the previous year and 68% increase over the average of the previous 10 year. The USA was the top supplier, accounting for 42.12% of the global share. It was the second 'On year' in a row for Turkey and Iran which showed a particularly high bumper crop with a share of 26.33% and 16.88%, respectively (Fig. 1.26).

The other major pistachio nut producing countries are China (80,000 MT) and Syria (69,000 MT) having an area of 27,000 hectare and 60,000 hectare, respectively under pistachio nut cultivation. China and Syria share 7.12% and 6.16% of pistachio production in the world (Fig. 1.26).

In the last 5 years, increasing and decreasing trends for export were observed, while maximum export was observed in the year 2017, i.e. 4.06 lakh MT. However, in the year 2019, 3.52 lakh MT pistachio nut was exported by different countries (Fig. 1.27).

The USA was the major world pistachio in-shell exporter in 2019, with 2.19 lakh MT sharing 62% of the total export. The top destinations were China (41%) and the European Union + the United Kingdom (34%). Iran was the second top exporter of in-shell pistachios, accounting for the following 20% of the global share with the major export to China (63%), European countries (16%) and the Middle East (12%) in the year 2019. The other exporting countries are China and UAE with a share of 5% and 2%, respectively (Fig. 1.28 and Table 1.5).

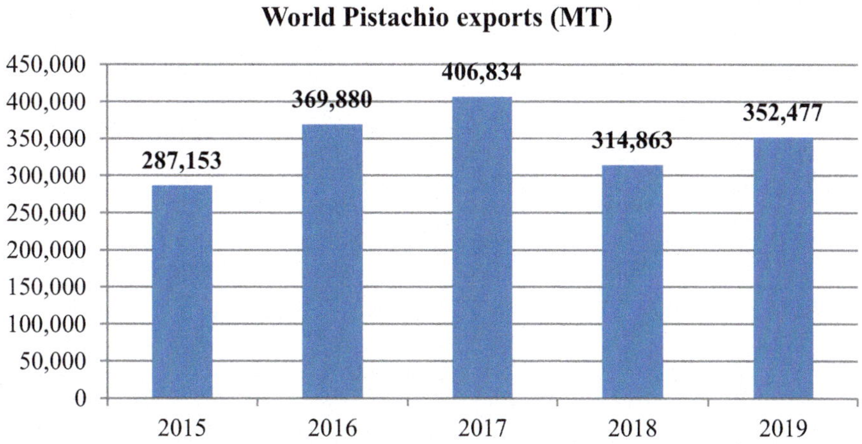

Fig. 1.27 Export of pistachio nut in the last 5 years to other countries. (Source: Statistical Year Book 2020–2021)

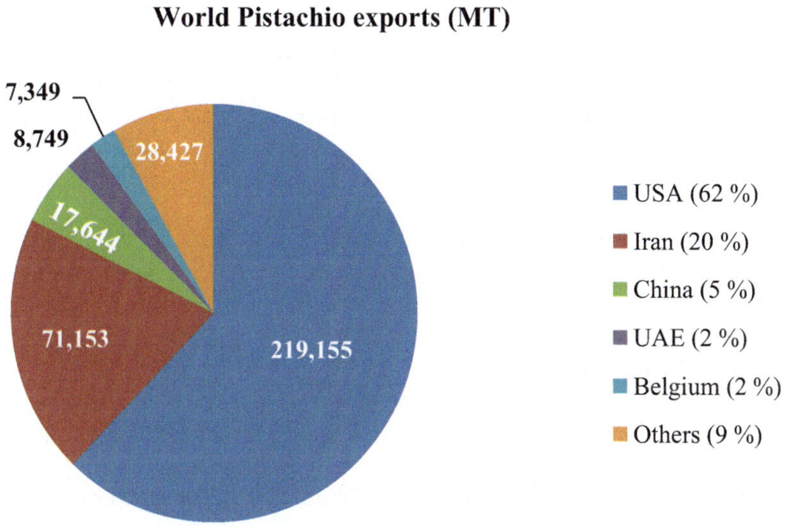

Fig. 1.28 Share of different countries in the export of pistachio nut during 2020. (Source: Statistical Year Book 2020–2021)

8 Conclusion

In the last decades, huge expansion was witnessed in the cultivation of all the temperate nut fruits due to their nutritional value. Developed countries play a major role in area expansion and production targets. Still, there is need of harnessing the potential of nut fruit production in the temperate areas of the world.

Table 1.5 World pistachio nut imports (MT) in the last 5 years

Country	2015	2016	2017	2018	2019
China	67,198	100,082	117,861	91,120	144,406
Germany	28,237	28,743	41,786	37,947	33,648
Belgium	17,826	17,204	16,627	17,377	16,159
Spain	7133	8869	11,753	13,090	15,590
Vietnam	59,935	89,138	69,925	27,669	13,665
Turkey	4929	6126	7936	10,318	9784
Italy	7467	8736	10,209	8709	9380
UAE	13,910	10,909	12,923	9538	9192
India	6241	8106	12,284	8519	9175
France	6928	7151	7334	8342	7027
Saudi Arabia	6752	4004	5261	6094	6913
Netherlands	10,077	9492	10,708	10,341	6678
Russian Fed.	1906	2707	5507	7198	6638
Israel	2376	2685	4107	3291	5029
Thailand	677	851	749	1968	4960
Jordan	1784	2007	3737	3953	4471
UK	3796	3668	4067	5065	3879
Luxembourg	2577	8915	2081	935	3145
Poland	1697	2137	2700	4091	2552
Kyrgyzstan	377	692	960	446	2242
Others	35,329	47,657	58,318	38,853	37,945

Source: Statistical Year Book 2020–2021

9 Future Thrust

- Augmentation of nut germplasm having desired source of resistance, quality, storability, yield, etc.
- The future of nut industry involves the development of an integrated production system based on sound science for an increasing demand of commodities, but to compete in a globalised world.
- Research and development activities on nut must be strengthened.
- Reduction in the juvenile period using dwarfing rootstocks and transgenic techniques.
- To fully realise the potential of the nut industry, it is necessary to enhance competitiveness and profitability, to implement better distribution circuits and control the quality of the products.
- Development of varieties and rootstocks resistant/tolerant to biotic and abiotic stress.
- Development of easy and fast propagation methods, utilising micro-propagation techniques.
- Efforts should be made to communicate with consumers and influence politicians about the value of the nuts as food, timber and other products.

- Technological development, organisation, information, and identification of the new production techniques are fundamental requirements that must be applied along the food or timber chain to conquer the markets and create a modern and efficient nut industry.

References

Food and Agriculture Organization of the United Nations (n.d.) FAOSTAT statistical database. FAO, Rome. https://www.fao.org/faostat/en/#data/QCL

Freitas TR, Santos JA, Silva AP, Fraga H (2021) Influence of climate change on chestnut trees: a review. Plants 10:1463. https://doi.org/10.3390/plants10071463

Statistical Year Book (2020–2021) https://www.nutfruit.org/industry/technical-resources?category=statistical-yearbooks. Accessed 5 Sept 2022

Chapter 2
Nutritional Composition of Temperate Nuts

Nirmal Kumar Meena, Vinod B. R., Menaka M., Ajit Kumar Singh, Gouthami S., Anamika Thakur, and K. Prasad

Abstract Temperate nuts are the dense food for phytonutrients such as unsaturated fatty acids, proteins, carbohydrates, antioxidants and other important vitamins. FDA has considered nuts as 'healthy food' and good for the health. Increased consumption of these nuts could enrich the health promoting compounds in the human body that could be a means to eliminate cardiovascular diseases and diabetes. The epidemiological studies consistently show their inherent nutritional properties and impact on reducing cardiovascular diseases. Temperate tree nuts such as walnut, chestnut, pistachio nut, hazelnut, pecan nut, etc. are prosperous in primary and secondary metabolites and health promoting compounds. Nuts are important and safer for diet and could have wide ranging health benefits. However, certain future studies are required to standardised the nutrient specific doses with respect to age of human. Certain nutrient governing factors need to be addressed as well.

Keywords Hazelnut · Pistachio · Antioxidants · Lipids · Kernel · Phytosterols

1 Introduction

Nuts are the important food of not only temperate region but tropical and subtropical as well; also they constitute a significant part in human diet. Nuts contain less than 50% moisture, and their seed is consumed by people as they are enclosed with a hard nutshell (Verde et al. 2022). The most common temperate nuts such as walnut,

---

N. K. Meena
Division of Food Science and Postharvest Technology, ICAR-Indian Agricultural Research Institute, New Delhi, India

Agriculture University, Kota, Rajasthan, India

V. B. R. (✉) · M. M. · A. K. Singh · G. S. · A. Thakur
Division of Food Science and Postharvest Technology, ICAR-Indian Agricultural Research Institute, New Delhi, India

K. Prasad
Dr. Rajendra Prasad Central Agriculture University, Samastipur, Bihar, India

M. M. Mir et al. (eds.), *Temperate Nuts*,
https://doi.org/10.1007/978-981-19-9497-5_2

Table 2.1 Important temperate nuts and their family

Nuts	Botanical name and family
Walnut	*Juglans* spp. (Juglandaceae)
Hazelnut	*Corylus avellana* (Betulaceae)
Pecan nut	*Carya illinoinensis* (Juglandaceae)
Chestnut	*Castanea mollissima* Blume (Fagaceae)
Pistachio nut	*Pistacia vera* (Anacardiaceae)
Macadamia nut	*Macadamia integrifolia* (Proteaceae)
Pine nut	*Pinus* spp. (Pinaceae)

Table 2.2 Production statics of important temperate tree nuts worldwide (1000 metric tons)

Nuts	Production
Almond	1643.84
Walnuts	996.08
Pistachios	795.3
Hazelnuts	545.15
Pecan nut	129.51
Macadamia	66.44
Pine nuts	40.01

Source: Statista, 2022 (• Nuts: Global Production by Type 2021/22 | Statista) accessed on 24.08.2022

hazelnut, pecan nut, chestnut, chilgoza nut and macadamia are used in snacks, as dessert or as dry fruits. The botanical name and family of important nuts are given in Table 2.1. Nuts are cultivated in the European countries, but they have been distributed all around the world and are now being cultivated in every part of the world including Asia, Australia and Middle East countries. According to the report of FAS, USDA, the European Union is the leading market of tree nuts ($2.75 billion), followed by China ($975.73 million) and India ($885.14 million). The production of tree nuts worldwide is given in Table 2.2.

It is believed that nuts are the integral constituent of human diet owing to their nutraceuticals and numerous health benefits (Eaton and Konner 1997). The important nuts are walnut, hazelnut, pistachio nut, chestnut, chilgoza nut and macadamia nuts. They are consumed as raw or in the form of roasted snacks. Nuts gained importance in recent couple of years due to enormous health benefits, linkage with cardiovascular diseases (CVD), coronary heart diseases, cancer and cholesterol level (Grosso et al. 2015). Nuts are considered to be sources of vitamins, minerals, amino acids, fats and lipids (especially unsaturated fats), dietary fibres, phytosterols, total phenols, antioxidants and many other secondary metabolites (Ros 2010; Segura et al. 2006). Several clinical trials have proven that excess intake of nuts does not lead to obesity and development of cardiovascular diseases. Moreover, nuts are an important source of fatty acids, with total fat content ranging from 46% in pistachio to 76% in macadamia (Ros 2010). They have different varieties rich in vegetable protein. The positive influence of nuts is known to be highly synergistically with all the other bioactive compounds which may favour human physiology. Remarkably, it is reported that nuts are cholesterol free foods but certain phytosterols and non-cholesterol sterols may present in their fatty fraction (Segura et al. 2006). Sometimes, allergic reactions may arise on excess consumption of nuts and due to presence of some allergens. The chemical composition of shell as well as kernel may differ due to genotypes and growing conditions. This chapter gives a brief overlook of nutritional composition of temperate nuts.

2 Carbohydrates

Carbohydrates largely comprise of fiber, sugars and starches. It is well known that carbohydrates serve a major part in human diet as a source of energy. Fruits and vegetables are the main source of carbohydrates but nuts provide higher carbohydrates in the form of monosaccharides and polysaccharides. It is reported that pistachios have the maximum amount of sugar (7.66 g/100 g) among the temperate nuts. Other temperate nuts contain an average sugar content of less than 5 g/100 g. Nearly all (>95%) of the simple sugars in nuts are sucrose, making it the most prevalent kind (Alasalvar et al. 2020a, b). Fructose, glucose, maltose and galactose are some of the other sugars.

Walnut is an important temperate nut and possesses significant rank among all the nuts worldwide. The kernel of walnut is known for its good amount of carbohydrates and sugars. Total carbohydrate content of fresh walnut kernel is reported at 13.7 g/100 g (USDA 2019). Dietary fibres contribute major portion for carbohydrate, i.e. 6.7 g/100 g followed by sucrose (2.43 g/100 g). Fructose (0.09 g/100 g), glucose (0.08 g/100 g) and starch (0.06 g/100 g) also contribute at very low quantity (USDA 2019). Chatrabnous et al. (2018) quantified the total carbohydrate (21 ± 0.2%) in Iranian walnut. A study by Gharibzahedi et al. (2014) revealed that carbohydrate content ranges from 12.8% to 16.7% in different Persian dry walnut cultivars. Tapia et al. (2013) evaluated walnut cultivars of Spain ('Serr', 'Chandler', 'Howard' and 'Hartley') and found a significant difference in carbohydrate content among the cultivars from 13.9% in Serr to 19.4% in Hartley. Similarly, carbohydrate content (in terms of dietary fibre) in 12 New Zeeland cultivars ranges from 3.1 to 5.2 g/100 g DM (Savage 2000).

Pistachio nut contains an appreciable amount of carbohydrates in the form of sugars, fibres and starch. Pistachio nut contains comparatively higher amount of fibre (10% of its mass) in insoluble types and just 3% in soluble types (Terzo et al. 2017). According to USDA (2019), pistachios contain 27.2 g/100 g of carbohydrate which is derived from dietary fibre (10.6 g/100 g), sugars (7.66 g/100 g) and starch (1.67 g/100 g). Sucrose (6.87 g/100 g) is the major sugar abundantly present in pistachios. On the other hand, glucose (0.32 g), fructose (0.24 g) and maltose (0.19 g) are present in minor quantities.

Macadamia nuts are low in total carbohydrates, at around 13.82 g/100 g, with 4.57 g derived from sugars, 1.05 g from starch and 8.6 g from fibre. Glucose and fructose are at very low quantities at 0.07 g (USDA 2019). Mereles et al. (2017) quantified the total carbohydrate, sugars and dietary fibres of three different cultivars of macadamia nuts grown in Paraguay and it ranges from 7.95 to 11.89 g/100 g, 6.08 to 9.62 g/100 g and 5.72 to 7.08 g/100 g, respectively. Another study by Tu et al. (2021) estimated the carbohydrate content and crude fibre in the cultivars Guire-1 (2.26 ± 0.37% and 4.10 ± 1.71%, respectively) and Own choice (2.26 ± 0.37% and 3.60 ± 0.81%, respectively).

Hazelnuts contain relatively low carbohydrates, at around 16.70 g/100 g with 4.34 g derived from sugars and 9.7 g from dietary fibre (USDA 2019). It is reported that carbohydrate content in five American cultivars ranges from 16 to 20 g/100 g (Wang et al. 2018). Sugar content in 24 Italian cultivars were estimated and reported from 3.98 to 5.95 g/100 g (Cristofori et al. 2008). Tian et al. (2012) evaluated the sugar content in different 33 hybrids of hazelnut (*Corylus heterophylla* Fisch. *C. avellana* L.) and reported a range of 12.66 to 19.09 g/100 g FW. They had reported the average total sugar content (12.49 g/100 g) in *C. heterophylla*. A 100-g hazelnut contains around 634 calories (Baysal 1993). Botta et al. (1992) stated that sugar accounts for 2.8–7.9% of DM, with sucrose accounting for 90% of total sugar. Furthermore, they found that around 1% of total sugar is made up of glucose and fructose in hazelnut.

Pecan nut is another important oil crop which belongs to genus Carya and sometimes it is considered as drupe. There are number of health promoting compounds available in pecan nut kernel including sugars, fatty acids, proteins, minerals, flavonoids, saponins and oxalates. It has been reported that the sugar content in pecans is highly influenced by variety. Zhang et al. (2022) reported mean sugar value of 10.7 mg/g in ten different varieties of Southeastern China.

Chestnut is another important temperate nut which contains plenty of sugars. It is a highly rich source of carbohydrate among all the nuts containing around 77.31 g/100 g with 13.3 g/100 g derived from sugars and 11.7 g/100 g derived from fibres (USDA 2019). Sucrose content contributes around one-third of total sugars. Míguez Bernardez et al. (2004) reported that the sucrose content varies from 6.5% to 19.5% in different 15 chestnut varieties of Verín-Monterrei region. Starch is the major constituent of chestnut kernel that accounts for more than 60% of dry weight. Hao et al. (2018) evaluated the total starch content of six Chinese chestnut varieties and found 58.33–63.58 g/100 g DW. The mean starch content of 47 Spanish chestnut cultivars was 57 g/100 g (Pereira-Lorenzo et al. 2006). However, the content of sugars varies with variety, geographical locations and extraction methods.

3 Amino Acid and Protein

Most nuts are known for their inherent content of good quality proteins (GQP). Nut protein contains well balanced chain of essential amino acids. It is believed that nuts have significant amount of plant protein which is easily digestible and functional. The major predominant amino acids are glutamic acid (0.80–6.206 g/100 g), arginine (0.443–4.862 g/100 g) and aspartic acid (0.67–3.146 g/100 g) (USDA 2019). The contribution of these three amino acids to the total amino acids is around

35.56–50.11%. In macadamias, lysine is reported as the limiting amino acid, but in almonds, hazelnuts, macadamias and walnuts, methionine is the limiting amino acid. In addition, protein is the second important composition of hazelnut containing around 20% and more than that according to cultivar and stage. Generally, temperate nuts are a rich source of good quality plant proteins, especially for those who do not eat animal protein.

Walnuts are rich in protein. Persian walnuts have a protein level ranging from 13.0% to 18.1% (Sarikhani et al. 2020). Tapia et al. (2013) revealed the protein content of four cultivars of walnut grown in Spain ranges from 15.1% to 17.4%. Walnut protein possesses high biological value (low lysine/arginine ratio) and digestible, making it an excellent vegetarian diet. Chatrabnous et al. (2018) reported the protein value of 16.5% in Iranian walnut. Wei et al. (2022) reported 15 types of free amino acids (FAA) in walnut kernels during ripening, glutamic acid being the highest (Glu) and methionine (Met) as lowest. The content of free amino acids decreases with the advancement of ripening reason being the tendency of FAA conversion to protein.

Pistachios are a significant source of glutamic acid (4.3 g) and arginine (2.13 g) and are a good source of edible plant protein, accounting for approximately 20.2 g/100 g of total weight (USDA 2019). Tsantili et al. (2010) evaluated eight pistachio nuts varieties from different origins. The crude protein content ranged from 18.99% (Joley) to 21.87% dry weight (Cerasola). Similarly, Seferoglu et al. (2006) reported higher range of protein content (23.2–31.7%) in Uzun pistachio cv. grown in Turkey. Okay (2002) found that the protein content of Uzun pistachio kernels is 19.58% and Martinez et al. (2016) reported 21.6% dry weight of protein content in Kerman pistachio.

Macadamia nut is observed to be lower in protein content (7.91 g/100 g) (USDA 2019), although it contains sufficient levels of all essential amino acids. Among them, glutamic acid (2.27 g) and arginine (1.4 g) are the predominant amino acids, while lysine (0.018 g) is the lowest amino acid present in macadamia nut. Moodley et al. (2007) estimated an average protein content (13.03 g/100 g) in macadamia kernel. Similarly, Mereles et al. (2017) quantified the crude protein in three cultivars of macadamia nuts in Paraguay and result revealed that protein content ranged from 5.19 to 7.56 g/100 g. However, one study revealed that nuts from the 'IAC 9-20X' cultivar had the highest protein content (19.24 g/100 g) (Maro et al. 2012). In a previous report, Tu et al. (2021) estimated the crude protein content Guire-1 ($8.81 \pm 1.25\%$) and Own choice ($8.78 \pm 1.82\%$) macadamia cultivars in China.

Hazelnut kernel is consumed in roasted, blanched and in food products such as dairy and bakery. Turkey contributes the major share in global hazelnut export. A wide range of difference in protein content in hazelnut cultivars have been reported. Several studies found that the protein content of hazelnut kernels varies between 12.6 and 25.9 g/100 g (Peh et al. 2016; Wang et al. 2018). A 100-g hazelnut can satisfy 22% of individual's daily protein needs since it has a total protein ratio of 10–24% (Pala et al. 1996). According to Tian et al. (2012), the total soluble protein content in 33 hazelnut cultivars was reported from 29.22 to 68.57 mg/g. Moreover, the most abundant amino acid, i.e. glutamic acid (2.84–3.71 g/100 g), was reported followed by arginine (1.87–2.21 g/100 g). Koksal et al. (2006) validated the findings and reported similar composition of amino acid profiles in Turkish hazelnut cultivars. Jiang et al. (2021) assessed five hazelnut kernels. Out of them, *C. heterophylla* showed the highest total hydrolysed amino acid (HAA) content from 10.37 to 15.32 g/100 g. Protein content ranges from 14.21% to 20.5% (Balik 2016). Differences in hazelnut cultivar protein and oil ratios were related mostly to genetic and ecological factors, cultural and technical procedures and harvest periods (Balik and Beyhan 2019). Pollinator cultivars have also been demonstrated to influence walnut protein ratios significantly (Golzari et al. 2016).

Similar to macadamia, chestnuts are also poor in protein content (6.39 g/100 g) compared to other temperate nuts. According to the USDA (2019) report, the level of protein differs in American, Chinese and European chestnuts as they have 4.83, 4.20 and 1.98 g/100 g protein level, respectively. Chestnut fruits contain a total amino acid of around 6–9 g/100 g (Borges et al. 2008). They had also reported the amino acid profile of chestnut and revealed that L-aspartic acid, the predominant amino acid, was presented followed by L-glutamic acid.

Pecan nut kernel contains sufficient amount of protein. Zhang et al. (2022) reported average protein content (67.50 mg/g) in different varieties of Southeastern China. Most of pecan cultivars contain 5–9% protein content (Ferrari et al. 2022). Yilmaz et al. (2021) assessed the protein content in different varieties and found them ranging from 7.45% in Mahan to 9.76% in Western.

4 Minerals

Temperate nuts are considered good sources of digestible minerals as they absorb these elements during plant developmental phase. Various metanalysis suggests that the intake of high K is associated with lower risk of cardiac and blood pressure related diseases. Additionally, nut such as walnut contains both macro and micro elements which regulate numerous physiological functions in the human body (Cindrić et al. 2018). The USDA has identified a total of 11 minerals in nuts (USDA 2019). Most of nuts contain high potassium, magnesium and low sodium which contribute toward good health. Nevertheless, certain toxic elements such as arsenic was also reported in nuts.

The mineral content in walnut kernels ranges from 1.7% to 2% (Tapia et al. 2013). A total ash content (1.78 g/100 g) was reported in fresh walnut kernel (USDA 2019) that was mainly bifurcated into potassium (441 mg), phosphorous (346 mg), magnesium (158 mg) and calcium (98 mg). According to Lavedrine et al. (2000), walnut kernel had high levels of potassium (390–700 mg/100 g), phosphorus (310–510 mg/100 g) and magnesium (90–168 mg/100 g), although they had reported lesser amount of sodium and iron (1–15 mg/100 g). Similarly, Tapia et al. (2013), reported high potassium and magnesium and low sodium content. Cultivar wise, Howard and Serr cultivars were rich in sodium, Serr and Hartley cultivars had more magnesium and calcium. On the other hand, Hartley and Chandler cultivars exhibited more manganese. Chatrabnous et al. (2018) reported 3% of ash content in Iranian walnut which comprises of dietary minerals, potassium (451 mg) being the highest and sodium (1.92 mg) the lowest.

Pistachios are another important source of minerals especially for potassium. Pistachios contain high potassium level as well as a high zinc and selenium content which have antioxidant potential (Terzo et al. 2017). Fabani et al. (2013) identified 29 mineral elements from three pistachio cultivars. They have also reported higher K (923–976 mg/100 g), followed by Ca and Mg. Ash content in pistachio nuts is reported from 3.1% to 3.9% in Uzun variety of Turkey (Seferoglu et al. 2006), while Martinez et al. (2016) reported 3.0% ash content in Kerman cultivar.

Macadamia nuts have high levels of essential and non-essential mineral elements. Macadamia is rich in macro minerals such as K (279.10–429.85 mg/100 g), P (154.07–239.85 mg/100 g), Mg (110.42–146.00 mg/100 g) and Ca (45.60–83.23 mg/100 g) (Mereles et al. 2017). Among micro elements, high Fe (1.64–3.11 mg/100 g) and Mn (2.08–4.86 mg/100 g) were predominantly reported (Mereles et al. 2017). Munro and Garg (2008) determined the elements in macadamia and reported low potassium content compared to other nuts. Moodley et al. (2007) reported higher ash content (4.0 g/100 g) in macadamia nut, which was almost three times higher than those of USDA (1.14 g/100 g). Further, Tu et al. (2021) found total ash content in Guire-1 (2.20%) and Own choice (2.40%) in China.

Hazelnut contains approximately 1–3.4% ash and is a good source of minerals. A 100-g hazelnut can provide an individual with their daily Fe, Mg, Cu, Mn, K, P, Zn and Ca requirements (Koksal et al. 2006). The most abundant mineral is potassium, which is followed by phosphorus, calcium and/or magnesium. Several studies have found that the mineral content of hazelnuts is altered by variety, geographic region and harvest year (Bonvehí and Coll 1997). Hazelnut is a great source of copper and selenium, as well as a decent source of iron, magnesium and zinc. The average mineral content of hazelnut varieties as estimated by Koksal et al. (2006) is as follows: K (863 mg/100 g), Mn (186 mg/100 g), Mg (173 mg/100 g), Ca (5.6 mg/100 g), Fe (4.2 mg/100 g), Zn (2.9 mg/100 g), Na (2.6 mg/100 g) and Cu (2.3 mg/100 g). The hazelnut variety 'C akldak' had greater levels of K, Mg, Fe, Mn and Zn compared to the others. The amounts of Ca, P and Cu were greater in 'Palaz' samples. The Na levels in the 'Sivri' and 'Tom bul' kinds were greater than in the other varieties. Hazelnut is desirable for human diets due to the presence of Fe, Zn and Cu, in addition to a high K/Na ratio, and this is especially true for electrolyte balance (Fennema 1985).

Chestnut is a very good source of mineral content. Previous studies suggested that chestnut contains majority of macro and micronutrients, with potassium being the major. Pereira-Lorenzo et al. (2006) investigated the microelement concentration of Spanish chestnuts. On a dry weight basis, the average manganese concentration was 38.7 mg/100 g, followed by 18.1 mg/100 g of iron, 12.3 mg/100 g of zinc and 7.1 mg/100 g of copper. Borges et al. (2008) reported the mineral composition of eight Portugal chestnut cultivars, which was rich in potassium mean value (750 mg/100 g), phosphorus (120 mg/100 g) and magnesium (75 mg/100 g) in dry weight basis.

5 Vitamins

Vitamins are an important constituent of human diet. They regulate various metabolic processes either alone or in cross talk with another element. Temperate nuts contain both water-soluble (betaine, folate, choline, niacin, pantothenic acid, riboflavin, pyridoxine, thiamine and vitamin C) and fat-soluble vitamins (vitamin A, E and K), although nuts are poor in some vitamins such as betaine, choline, vitamin A, C and K (Alasalvar et al. 2020a, b). Some vitamins such as A, C and tocopherol act as an antioxidant and known as antioxidant vitamins.

Struetz et al. (2016) reported a high level of thiamine, riboflavin and pyridoxine in temperate nuts. Pistachio nut contains higher thiamine content than macadamias and hazelnuts. Further, pistachios kernel also contains great amount of pyridoxine (vitamin B6), followed by walnuts and hazelnuts. The content of vitamin A in pistachio present in the form of lutein/zeaxanthin (2760 μg/100 g) and β-carotene (200 μg/100 g). Pistachio nut rich in tocopherol content especially in γ-tocopherol which contributes more than 90% of tocopherol.

Similarly, walnuts kernel is rich in vitamins E, B1, B2 and B6. It was found that walnut kernel oil contains γ-tocopherol (290–435 mg/g oil), which is comparatively higher than the other nuts (Lavedrine et al. 1999). Beyhan et al. (2017) estimated all tocopherol content in 19 walnut genotypes from Turkey. Among them, Gamma (γ) tocopherol was the major isomer of tocopherols presented in most of genotypes. They revealed that the maximum α-Tocopherol (38.04 μg/g) was reported in genotype 18. On the other hand, γ-Tocopherol content was reported from 161.09 to 292.56 μg/g in all genotypes. Zheng et al. (2020) reported an average vitamin E content (42.4 mg/g) in ten Chinese walnut cultivars.

Pistachios are rich in pyridoxin, carotenoids and vitamin E (particularly tocopherol), and have been named one of the top 50 antioxidant-rich foods (Mandalari et al. 2022). Furthermore, pistachio nuts contain vital vitamins such as vitamin A, C, K and B (except vitamin B12) (Khatib and Vaya 2010; Terzo et al. 2017). The pistachio kernel has a high γ-tocopherol concentration, and statistically different groups were detected across the varieties: Kerman had the greatest proportion (399 mg/kg), followed by Mateur (330 mg/kg), while the rest of the varieties had content ranging from 267 to 307 mg/kg. Other isomers detected in trace levels were α- and β-tocopherols, in addition to δ- and γ-tocotrienols (1.2–2.8% of the total quantity) (Ojeda-Amador et al. 2018).

Macadamias are very good source of B vitamins such as niacin (2.47 mg/100 g), thiamine (1.2 mg/100 g), riboflavin (0.162 mg/100 g) and pyridoxine (0.3 mg/100 g). They also have 1.2 mg/100 g of vitamin C and 0.758 mg/100 g of pantothenic acid. Macadamias have the lowest folate level of all nuts, with 11 μg/100 g edible (USDA 2019). Compared to other nuts, macadamia is poor in vitamin content. Mereles et al. (2017) estimated α-tocopherol in three cultivars (HAES 344, Cannon and San Joaquín) of macadamia nut cultivated in Paraguay. The α-tocopherol content ranged from 0.2 to 18.40 mg/100 g fresh weight of kernel, and cultivar San Joaquin exhibited highest α-tocopherol content. In recent study, Abubaker et al. (2017) reported 61.49 mg/100 g of vitamin E content in macadamia kernel.

Hazelnuts are high in both fat-soluble and water-soluble vitamins. Hazelnuts are rich in vitamin E (15.03–25.66 mg/100 g), a lipid-soluble phenolic antioxidant. They are also a good source of biotin (76–80 g/100 g). Only 37–40 g hazelnut per day provides adults with 100% of their AI for biotin. Furthermore, hazelnut offers good source of pyridoxine (0.56–0.63 mg/100 g), pantothenic acid (0.92–1.51 mg/100 g) and thiamine (0.42–0.64 mg/100 g). Hazelnuts have the greatest folate content of any tree nut (113–120 g/100 g edible part) (USDA 2019). Among the 11 different tree nuts, hazelnut has the most vitamin E, biotin and folate content (Alasalvar et al. 2006). The vitamin content of 17 varieties of hazelnut kernels produced in Turkey revealed by Koksal et al. (2006) are as follows: niacin (1.45 mg/100 g), vitamin B1 (0.28 mg/100 g), vitamin B2 (0.05 mg/100 g), vitamin B6 (0.5 mg/100 g), ascorbic acid (2.45 mg/100 g), folic acid (0.043 mg/100 g), retinol (3.25 mg/100 g) and total tocopherol (26.9 mg/100 g).

Chestnut contains high levels of vitamin C, A and folate. European chestnut is rich in vitamin C among the temperate nuts which is 40.2 mg/100 g (USDA 2019). It is also rich in Vitamin A (26 IU/100 g). On a fresh weight basis, Bellini et al. (2004) discovered that chestnut contained 0.6 mg/100 g of δ-tocopherol and 7.9 mg/100 g of γ-tocopherol.

6 Lipids

With a few exceptions, such as chestnuts (4.45 g/100 g), most nuts are high in lipids (varying from 31.41 g/100 g to 79.55 g/100 g); these lipids make up the majority of the energy derived from nuts. The lipid content of mature kernels is influenced by cultivar type, geographical location and growth conditions (Alasalvar et al. 2020a, b). Nuts have a beneficial lipid profile because most of the fatty acids are unsaturated (MUFA and PUFA) rather than saturated (SFA). Except for chestnuts and walnuts, which are high in PUFA, MUFA predominates in most nuts. Oils from hazelnuts, walnuts and macadamia nut have the least proportions of SFA (7.79%), MUFA (15.28%) and PUFA (4.39%), respectively, while oils from macadamia nuts, hazelnuts and walnuts have the highest proportions of SFA (18.18%), MUFA (83.24%) and PUFA (72.96%). In addition, walnut oils are very good sources of omega-3 fatty acids (alpha-linolenic acid; 18:3ω-3) and omega-6 fatty acids (linolenic acid; 18:2ω-6) (13.17% and 59.72%, respectively) (Alasalvar et al. 2020a, b).

Like other nuts, walnuts acquire the majority of their energy from fat. As a result, they are a high-calorie, high-energy snack. They have a total fat content of around 65.2 g/100 g by weight (USDA 2019). Walnut oil contains the fatty acids oleic (18: 1), linoleic (18:2) and linolenic (18:3). Linoleic acid (38.1 g/100 g), an omega-6 fatty acid, is the most abundant. They also include a high amount of alpha-linolenic acid (9.08 g/100 g), a beneficial omega-3 fatty acid (ALA). This accounts for about 8–14% of the overall fat content, and walnuts are the only nuts that contain ALA in considerable proportions (Fatima et al. 2018). Fuentealba et al. (2017) evaluated the lipid content of walnuts cv. Chandler grown in Andes and Coast areas of Chile, varied from 56.0 to 62.8 g/100 g. The major fatty acids found in walnut are unsaturated Oleic acid (C18:1), Linoleic acid (C18:2) and α-Linolenic acid (C18: 3) acids in the range of 69.5–80.8 mg/g, 250.5–320.9 mg/g and 95.3–119.2 mg/g, respectively. Lower content of saturated palmitic acid (C16:0) is also present in the range of 32.2–40.2 mg/g. Wu et al. (2020) investigated 11 walnut cultivars for major fatty acids content. The predominant saturated fatty acids discovered were palmitic acids (C16:0) (4.95–6.21%) and stearic acids (C18:0) (2.86–3.72%), which agrees with prior research (Li et al. 2017; Rabadan et al. 2018). The oils of walnut were abundant in polyunsaturated fatty acids (PUFAs), which are good for human health. The PUFA content in Liao 5 accounted for 73.00% of the fatty acids, but only 58.23% in Liao 7. Rebufa et al. (2022) used data from the literature to calculate the mean, lowest and highest fatty acid content of all walnut varieties from all geographic sources combined. They were palmitic acid (16:0) [range: 2.41–10.41%], stearic acid (18:0) [range: 0.60–4.45%], oleic acid (18:1ω-9) [range: 5.70–31.96%], linoleic acid (18:2 ω-6) [range: 50.15–74.00%] and linolenic acid (18:3 ω-3) [range: 6.08–20.60%].

Fatty acid composition is highly variable in pecan nut and influenced by topography, cultivar and extraction methods. Pecan nut contains Lipid—71.97 g and is the rich source of MUFA—40.80 g, specifically oleic acid (52.0%), good amount of PUFA—21.61 g but it is low in SFA—6.18 g (Alasalvar et al. 2020a, b). Mexican native pecan nuts have 64.55%—oleic acid, 24.40%—linoleic acid, 5.23%—palmitic acid, 2.71%—stearic acid, 2.21%—alfa-linolenic acid, while the Mexican pecan nut variety Western Schley has Oleic acid (53.38%), linoleic acid (34.24%), palmitic acid (6.65%), stearic acid (2.57%), alfa-linolenic acid (1.74%) and Wichita variety pecan nut contains oleic acid (57.28%), linoleic acid (31.50%), palmitic acid (6.56%), stearic acid (2.38%) and alfa-linolenic acid (1.73%) (Rivera-Rangel et al. 2018). Zhang et al. (2022) reported 35.36% oil content in pecan nut kernel of ten important varieties in the form of crude fat. It is well established that kernel of pecan nut is highly dominated by unsaturated fatty acid than those of saturated fatty acids. Furthermore, Zhang et al. (2022) reported that oleic acid (70.02%) was found the dominant monounsaturated fatty acid and linoleic acid (19.58%) as polyunsaturated fatty acid in pecan nut kernel. Among the fatty acids, oleic acid and linoleic acids are the major contributory fatty acids (Reis Ribeiro et al. 2020).

Pistachio nuts are high in fatty acids, which are essential for human nutrition. They have a total fat content of 43.5 g/100 g by weight (USDA 2019). Tsantili et al. (2010) identified 12 fatty acids namely myristic, palmitoleic, palmitic, margaric, oleic, vaccenic, stearic, linoleic, linolenic, arachidic, gondoic and behenic fatty acids (percent in the kernel oil) from eight cultivars of pistachio nut. The main components were monounsaturated oleic acid, polyunsaturated linoleic acid and saturated palmitic acid. The oleic acid contribution to oil (%) was lowest in Kerman (51.6%) and highest in Pontikis (68.33%). Linoleic acid levels in Pontikis range from 11.56% to 27.03%, while palmitic acid levels in Cerasola range from 8.54% to 10.24%. According to Yahyavi et al. (2019), the most prevalent fatty acid in Iranian pistachio nuts was oleic acid (52.5–63.9%), followed by linoleic acid (27.1–37.2%), palmitic acid (4.6–10.3%), palmitoleic acid (0.6–1.2%), stearic acid (0.1–1.3%) and linolenic acid (0.1–1.3%). Noguera-Artiaga et al. (2020) estimated nine fatty acids, four of which were SFAs [myristic acid, palmitic acid, stearic acid and arachidic acid], three of which were MUFAs [palmitoleic acid, oleic acid and eicosenoic acid] and two were PUFAs [linoleic acid and α-linolenic acid].

The macadamia nuts are rich in lipids content [70.1 g/100 g in *M. integrifolia* (Li and Hu 2011) and 69–78 g/100 g in *M. tetraphylla* (Kaijser et al. 2000)], with lower percentage of SFA. The majority of its fatty acids (FA) are monounsaturated, with palmitoleic acids (C16:1) making up 20% and oleic acids (C18:1) making up 60%. Others are lauric acid, myristic acid, palmitic acid, stearic acid, arachidic acid, behenic acid, lignoceric acid, eicosenoic acid, erucic acid, linoleic acid and linolenic acid (Aquino-Bolanos et al. 2016). Monounsaturated fatty acids were more prevalent than saturated fatty acids and polyunsaturated fatty acids overall, in terms of percentage of fatty acids (Liu et al. 2022). The macadamia nut oil's mean MUFA and PUFA values were 82.4 g/100 g and 2.4 g/100 g, respectively (Maguire et al. 2004). The concentration of total lipids in Mexican macadamia nuts ranged from 70.9 to 79.7 g/100 g dry-solids. There are 20 fatty acids, the most common of which are oleic acid, palmitoleic acid and palmitic acid ranging from 40% to 51%, 24% to 36% and 8.4% to 13.1%, respectively. SFA levels in the various types ranged from 17.5% to 25.7% (Aquino-Bolanos et al. 2016).

Total saturated fatty acids (SFA) account for a modest fraction (6.90–8.52%) of total fatty acids in hazelnut oil, while total monounsaturated fatty acids (MUFA) account for the majority (78.90–83.16%). Oleic acid is by far the most prevalent fatty acid in hazelnut (varying from 77.50% to 82.95%), trailed by linoleic acid (varying from 7.55% to 13.69%) and palmitic acid (varying from 4.85% to 5.79%). Over three crop years, Parcerisa et al. (1998) studied the lipid content of four Spanish hazelnut varieties grown in two distinct geographical regions (Reus and Falset) in Spain. Significant changes in hazelnut lipid content were observed as a function of collection period (59.75 1.71 g/100 g) and geographic region (59.75 1.39 g/100 g), but not variety.

Espana et al. (2011) determined fatty acid profile of 11 chestnut varieties from Spain. It was noted that linoleic acid was the predominant fatty acid having a mean value of (43.4%) followed by oleic acid (30.5%) and palmitic acid (14.7%) of total fatty acids. McCarthy and Meredith (1998) reported comparative amount of fats in different country originated chestnut. They envisaged that chestnuts from European origin contained highest fat content (36.0 g/kg DW) followed by American origin (23.4 g/kg DW) and Chinese chestnuts (19.8 g/kg DW).

7 Secondary Metabolites

These are non-nutritive, naturally occurring, biologically active compounds found in plants and plant parts which are extracted from leaves, roots, bark and fruits seeds. Important and predominant phytochemicals include phenolics, antioxidants, alkaloids, terpenoids and some of toxins which involve in the plant defence system and possess high health potential.

7.1 Phenolic Compounds

The primary group of phytochemicals is phenolics which is non-nutritive, and naturally occurring in all plant parts. A diverse range of phenolic compounds have been reported in nuts. Phenolics are mainly consisted of flavonoids, phenolic acids, stilbenes, coumarins and tannins. Flavonoids, phenolic acids and tannins are present in all nuts however, coumarins are negligible in nuts. The ranges of these bioactive compounds are highly varied upon plant parts and extraction methods.

Pistachio has the greatest overall phenolic profiles (1657 mg gallic acid equivalents (GAE)), total flavonoids (15.24 mg/100 g), highest total isoflavones (177 g/100 g), total lignans (199 g/100 g) and total phytoestrogens (383 g/100 g). Furthermore, hazelnuts have the highest total proanthocyanidin content (491 mg/100 g). Most nuts have been shown to contain phytates, with macadamia nuts containing 150 mg/100 g and pistachio nuts containing 290 mg/100 g (Alasalvar et al. 2020a, b). According to Fuentealba et al. (2017), TPC levels in walnuts from the Andes and Coast range from 5.7 to 12.7 mg GAE/g DW. UPLC analysis of phenolic compounds revealed a high concentration of catechin derivatives, along with ellagic and gallic acid derivatives. Several writers have discovered ellagic acid in walnuts, usually in the form of ellagitannins (Colaric et al. 2005; Regueiro et al. 2014). In addition, catechin has been recognised as an important flavan-3-ol present in walnuts (Regueiro et al. 2014). Zheng et al. (2020) quantified the average phenol content of ten cultivars of Chinese walnut is 32.2 mg/g. Jinlong 1 cultivar had the highest polyphenols content of 49.4 mg/g among those 10 cultivars. Mereles et al. (2017) estimated total phenol content (TPC) in three cultivars (HAES 344, Cannon and San Joaquín) of macadamia nut cultivated in Paraguay. The TPC content ranged from 77.90 to 96.31 mg GAE/100 g of kernel. Similarly, the TPC of macadamia nut cultivars namely Guire-1 and Own choice was 60.52 and 68.32 mg GAE/100 g, respectively, as estimated by Tu et al. (2021).

According to Özyurt and Ötles (2018), total phenolics in hazelnut ranged between 1413.32 and 2057.72 GAE mg GAE/g as extracted by different extraction methods. It is reported that 'Foşa' and *C. colurna* L. contained least flavonoid (73.4 mg QE/kg), while 'Kan' exhibited the higher amount of flavonoids (650 mg QE/kg). Previous studies suggest that hazelnut is the only nut that contains dihydrochalcones such as phloretin-2-*O*-glucoside phenols. Chestnut has a total phenolics content of 43.0 mg GAE/100 g (Kalogeropoulos et al. 2013).

It is believed that pecan nut kernel endued several phenolics compounds which has strong pharmaceutical potential. Among the all-temperate nuts, pecan contains highest phenolics. In a study by Tanwar et al. (2020) revealed that the phenolic composition in pecan nut kernel ranges from 47.05 ± 9.85 mg GAE/100 g to 302.67 ± 7.72 mg GAE/100 g. Reis Ribeiro et al. (2020) reported total phenolic range from 19.88 mg GAE/g in 'Desirable' pecans to 45.25 mg GAE/g in 'Imperial' cultivars when compared 11 cultivars in Southern Brazil. Ferrari et al. (2022) evaluated different pecan nut cultivars for their phenolic composition and reported total phenolic compounds ranges from 18 to 41 mg GAE/g, condensed tannin varies from 2 to 12 mg CE/g. They found tocopherol content ranges from 110 to 163 mg/g in different pecan cultivars.

7.2 Antioxidant Activity

Antioxidants are integral part of diet. Antioxidants have capacity to scavenge the free radicles and improve the defence system. Nuts are the rich basket of antioxidants. There are several antioxidants contributory factors present in the nuts. Among these, phenols, flavonoids, ascorbic acid, carotenoids, lycopene and tocopherols are important. Walnuts possess the highest antioxidant activity in DPPH (12,000 μmol Trolox Equivalents (TE)/100 g), ORAC (17,940 μmol TE/100 g), FRAP (23.1 mmol Fe^{2+}/100 g) and LDL + VLDL oxidation inhibition (1.8 μM [IC50]) (Alasalvar et al. 2020a, b). The total polyphenols and antioxidant activities of nine nuts were evaluated by Vinson and Cai (2012) and revealed that walnut exhibited highest antioxidants activities followed by cashew and hazelnuts. Fuentealba et al. (2017) reported that extra light kernel walnut contained 138.1 mmol TE/g DW antioxidant activity, whereas amber walnuts showed 29.8 mmol TE/g DW. The total antioxidant capacity (TAC) in three different cultivars of macadamia nuts grown in Paraguay ranges from 21.0 to 36.81 μM TE/g FW (Mereles et al. 2017). Total DPPH antioxidant in hazelnut genotypes ranged from 1220.8 to 2536.5 mol TE/kg whereas total antioxidant capacity with FRAP method ranged from 2979.6 mol TEk/g to 26996.9 mol TE/kg (Tas and Gokmen 2015).

The antioxidant capacity exhibits a high degree of variation in shell and kernel. Ferrari et al. (2022) reported that shell of pecan exhibited five times higher antioxidant activities ranges from (57.15–578.88 mmol TE/g) than those of kernel 23.15–156.60 mmol TE/g. Jia et al. (2018) studied the antioxidant capacity of different pecan cultivars and reported that *Pawnee* had the highest DPPH of 461.80 μmol TE/g whereas Wichita cultivar showed highest ABTS (201.49 μmol TE/g). They also clued that over-ripeness triggers the activity of antioxidants in most of cultivars. Yilmaz et al. (2021) estimated total antioxidant capacity ranged from 201.36 to 487.89 mg TEAC/g in Choctaw and Burkett cultivars, respectively.

8 Other Secondary Metabolites

Phytosterols (Plant sterols and stanols) are lipid-like substances produced by plants when they are exposed to low temperatures. Walnut oil has the most total phytosterols (307 mg/100 g) followed by pecan oil (283 mg/100 g) (Miraliakbari and Shahidi 2008). Macadamia oil, on the other hand, has the least phytosterol level (128 mg/100 g) of all the nut oils (Kaijser et al. 2000). Phytosterols, primarily β-sitosterol, are found in various patterns and concentrations in nut oils. The most prevalent phytosterol in all nut oils is β-sitosterol, accounting for 53.8–88.4% of total sterols. β-sitosterol (772–2520 mg/kg) is prevalent in walnut oil (Martínez et al. 2011). The next most abundant phytosterol in almond, pecan and walnut oils is stigmasterol, which ranges from 11.67 to 60 mg/100 g (in pecan oil), and followed by campesterol (21 mg/100 g) in pistachio oils. In hazelnut and macadamia oils, Δ^5-avenasterol is the second most prevalent phytosterol after β-sitosterol, with concentrations ranging from 13.27 mg/100 g in macadamia to 13.97 mg/100 g in hazelnut oil (Alasalvar and Pelvan 2011). Abdallah et al. (2015) reported the total sterol levels ranged from 1144 mg/kg (Local) to 1679 mg/kg (Parisienne). β-sitosterol was the most prevalent phytosterol (69.42–89.26%), followed by campesterol (0.33–5.24%) and Δ^5-avenasterol (0.1–7.34%). Similar findings were reported by Fuentealba et al. (2017) in walnuts from both the Andes and the Coast regions have β-sitosterol concentrations ranging from 986 to 1131 mg/g dry weight. In recent study by Tu et al. (2021) estimated the sterol content of two cultivars of macadamia nut namely Guire-1 (107.16 mg/100 g) and Own choice (115.58 mg/100 g) grown in China.

Sphingolipids in nut oils range from 20 to 330 mg/100 g, with hazelnut oil having the lowest quantity and pistachio oil having the highest (Miraliakbari and Shahidi 2008). Fang et al. (2005) analysed the component of sphingolipids in a variety of nuts (almond, cashew, hazelnut, peanut and walnut) and found that cerebroside (d18: 2-C16:0h-Glu) concentrations ranged from 0.021 to 0.068 mg/g nut, with hazelnut having the lowest concentration and almond having the highest.

Flavonoids are phenolic chemicals classified into seven types: anthocyanins, flavones, favan-3-ols, flavonols, favanonols, favanones and isofavones. Flavonoids in nuts are primarily conjugated to sugars or other polyols via O-glycosidic or ester linkages (Alasalvar et al. 2020a, b). Zheng et al. (2020) quantified the average flavonoid content of ten cultivars of Chinese walnut is 16.3 mg/g. Jinlong 1 cultivar had the highest polyphenols content of 27.8 mg/g among those ten cultivars.

Except for macadamias, many favan-3-ols have been identified in nuts. The most favan-3-ols are found in hazelnuts, followed by walnuts. In nuts, catechin is the most prevalent favan-3-ol. Epicatechin is also present in all nuts except pecans. Both hazelnuts and walnuts include procyanidin dimers, procyanidin trimers and procyanidin tetramers (Alasalvar et al. 2020a, b). Flavonols were observed in various levels in almonds, hazelnuts, pistachios and walnuts. The most common flavonol in pistachios is quercetin-3-*O*-rutinoside (Tomaino et al. 2010). Pistachios have also been found to contain rutin, quercetin-3-*O*-glucoside and quercetin-3-*O*-rutinoside. Some flavonols have only been found in walnuts (such as Q-galloyl pentoside 1, Q-galloyl pentoside 2, Q-galloyl pentoside). Furthermore, only hazelnuts contain myricetin-3-*O*-rhamnoside, quercetin pentoside and quercetin-3-rhamnoside (Alasalvar et al. 2020a, b).

Only almonds (cyanidin, delphinidin, procyanidin B_2, and procyanidin B_3), walnuts (cyanidin) and pistachios (cyanidin-3-*O*-galactoside and cyanidin-3-*O*-glucoside) contains anthocyanins (Bellomo and Fallico 2007; Alasalvar et al. 2020a, b). Carotenoids are a category of fat-soluble pigments found in many fruits and vegetables in large quantities having vibrant red, orange and yellow colour. Generally, nuts have a low concentration of carotenoids such α- and β-carotene, β-cryptoxanthin, lutein, lycopene and zeaxanthin. Among nuts, carotenoids are only found in pecan oil (0.014 mg/100 g oil), pistachio oil (6.70 mg/100 g oil) and walnut oil (0.22–0.62 mg/kg oil) (Abdallah et al. 2015; Alasalvar et al. 2020a, b). Stuetz et al. (2017) studied carotenoids content in six nuts. Pistachios had the highest levels of carotenoids, with lutein/zeaxanthin concentrations of around 2757 g/100 g raw nuts and β-carotene concentrations of about 204 g/100 g raw nuts. These values were 16- and 8-fold greater than lutein and β-carotene concentrations in hazelnuts, respectively. Tanwar et al. (2020) reported total average flavonoids content in pecan nut kernel is 44.95 ± 0.23 mg/100 g, trypsin inhibitor activity (1.18 ± 0.03 TIU/mg). They also reported some other secondary metabolites such as saponins (0.49 ± 0.04 g/100 g), alkaloids (0.26 ± 0.03 mg/100 g) and oxalates (8.15 ± 0.58 mg/100 g). Zhang et al. (2022) evaluated ten grafted varieties of pecan nut for their phytochemical evaluation and found that kernel contained mean tannin value 6.07 mg/g. Reis Ribeiro et al. (2020) found some secondary metabolites such as β-sitosterol from 88.74 to 220.42 mg/100 g and squalene ranges from 30.98 to 115.59 mg/100 g, in 11 pecan cultivars.

The aroma of a fruit is created by a complex blend of esters, alcohols, aldehydes and terpenoid compounds (Lanchun et al. 2004). According to Abdallah et al. (2015), the walnut scent is a complex blend of five major compounds: pentanal, hexanal, nonanal and 2-decenal as aldehydes which ranges from 0.07% to 0.12%, 0.26% to 0.80%, 0.34% to 0.89% and 0.25% to 0.68%, respectively, and Hexanol as alcohols which ranges from 0.21% to 1.58%. According to this investigation, the major aromas in walnut oil were C6 aldehydes and alcohols.

9 Conclusion

Nuts are integral part of food diet as they are rich basket of proteins, carbohydrates cholesterol free good fatty acids, phenols, vitamins such as tocopherols, total antioxidants and several other secondary metabolites. Compared to other foods, nuts have strong potential to combat numerous heart and cholesterol related diseases. In addition, nuts have significant amount of macro and micronutrients which are associated with balanced healthy status and serve as deficiency disorders healing agents. The improvement in intake of nuts with diet could serve as safer and more beneficial for malnourished areas.

References

Abdallah IB, Tlili N, Martinez-Force E, Rubio AGP, Perez-Camino MC, Albouchi A, Boukhchina S (2015) Content of carotenoids, tocopherols, sterols, triterpenic and aliphatic alcohols, and volatile compounds in six walnuts (*Juglans regia* L.) varieties. Food Chem 173:972–978. https://doi.org/10.1016/j.foodchem.2014.10.095

Abubaker M, El Hawary SS, Mahrous EA, Abd El-Kader EM (2017) Study of nutritional contents of *Macadamia integrifolia* Maiden & Betche leaves, kernel and pericarp cultivated in Egypt. Int J Pharmacogn Phytochem Res 9(12):1442–1445. https://doi.org/10.25258/phyto.v9i11.11189

Alasalvar C, Pelvan E (2011) Fat-soluble bioactives in nuts. Eur J Lipid Sci Technol 113(8): 943–949. https://doi.org/10.1002/ejlt.201100066

Alasalvar C, Amaral JS, Shahidi F (2006) Functional lipid characteristics of Turkish Tombul hazelnut (*Corylus avellana* L.). J Agric Food Chem 54(26):10177–10183. https://doi.org/10.1021/jf061702w

Alasalvar C, Salas-Salvado J, Ros E, Sabate J (eds) (2020a) Health benefits of nuts and dried fruits. CRC Press

Alasalvar C, Salvado JS, Ros E (2020b) Bioactives and health benefits of nuts and dried fruits. Food Chem 314:126192. https://doi.org/10.1016/j.foodchem.2020.126192

Aquino-Bolanos EN, Mapel-Velazco L, Martín-del-Campo ST, Chavez-Servia JL, Martínez AJ, Verdalet-Guzman I (2016) Fatty acids profile of oil from nine varieties of Macadamia nut. Int J Food Prop 20(6):1262–1269. https://doi.org/10.1080/10942912.2016.1206125

Balik Hİ (2016) Investigations of xenia and metaxenia in kazelnut. Doctoral Dissertation, Ondokuz Mayis University, Samsun, p 258

Balik Hİ, Beyhan N (2019) Pollen compatibility in Turkish hazelnut cultivars. Turk J Food Agric Sci 1(1):12–17

Baysal A (1993) Genel Beslenme (8.Baskı). Hatipoglu Kitabevi, Ankara, p 194

Bellini E, Giordani E, Marinelli C, Perucca B (2004) Marrone del Mugello PGI chestnut nutritional and organoleptic quality. In: III International Chestnut Congress 693, pp 97-102

Bellomo MG, Fallico B (2007) Anthocyanins, chlorophylls and xanthophylls in pistachio nuts (*Pistacia vera*) of different geographic origin. J Food Compos Anal 20(3-4):352–359. https://doi.org/10.1016/j.jfca.2006.04.002

Beyhan O, Ozcan A, Ozcan H, Kafkas NESBE, Kafkas S, Sutyemez M, Ercisli S (2017) Fat, fatty acids and tocopherol content of several walnut genotypes. Notulae Botanicae Horti Agrobotanici Cluj-Napoca 45(2):437–441. https://doi.org/10.15835/nbha45210932

Bonvehí JS, Coll FV (1997) A chemical study of the mineral fraction of Tarragona hazelnuts (*Corylus avellana* L.). Acta Aliment 26:243–253

Borges O, Gonçalves B, de Carvalho JLS, Correia P, Silva AP (2008) Nutritional quality of chestnut (*Castanea sativa* Mill.) cultivars from Portugal. Food Chem 106(3):976–984

Botta R, Gianotti C, Richardson D, Suwanagul A, Sanz CL (1992) Hazelnut variety organic acids, sugars, and total lipid fatty acids. In: III International Congress on Hazelnut 351, pp 693–699. https://doi.org/10.17660/ActaHortic.1994.351.77

Chatrabnous N, Yazdani N, Vahdati K (2018) Determination of nutritional value and oxidative stability of fresh walnut. J Nuts 9(1):11–20

Cindrić IJ, Zeiner M, Hlebec D (2018) Mineral composition of elements in walnuts and walnut oils. Int J Environ Res Public Health 15(12):2674

Colaric M, Veberic R, Solar A, Hudina M, Stampar F (2005) Phenolic acids, syringaldehyde, and juglone in fruits of different cultivars of *Juglans regia* L. J Agric Food Chem 53(16): 6390–6396. https://doi.org/10.1021/jf050721n

Cristofori V, Ferramondo S, Bertazza G, Bignami C (2008) Nut and kernel traits and chemical composition of hazelnut (*Corylus avellana* L.) cultivars. J Sci Food Agric 88(6):1091–1098. https://doi.org/10.1002/jsfa.3203

Eaton SB, Konner MJ (1997) Review paleolithic nutrition revisited: a twelve-year retrospective on its nature and implications. Eur J Clin Nutr 51(4):207–216

Espana MA, Rodríguez Galdon B, Diaz Romero C, Rodriguez E (2011) Fatty acid profile in varieties of chestnut fruits from Tenerife (Spain) Perfil de acidos grasos en variedades de castanas procedentes de Tenerife (Espana). CyTA J Food 9(1):77–81. https://doi.org/10.1080/19476331003686858

Fabani MP, Luna L, Baroni MV, Monferran MV, Ighani M, Tapia A, Wunderlin DA, Feresin GE (2013) Pistachio (*Pistacia vera* var Kerman) from Argentinean cultivars. A natural product with potential to improve human health. J Funct Foods 5(3):1347–1356. https://doi.org/10.1016/j.jff.2013.05.002

Fang F, Ho CT, Sang S, Rosen RT (2005) Determination of sphingolipids in nuts and seeds by a single quadrupole liquid chromatography–mass spectrometry method. J Food Lipids 12(4):327–343

Fatima T, Showkat U, Zameer Hussain S (2018) Nutritional and health benefits of walnuts. J Pharmacogn Phytochem 7(2):1269–1271

Fennema OR (1985) Food chemistry, 2nd edn. Marcel Dekker, Inc., New York. 991 p

Ferrari V, Gil G, Heinzen H, Zoppolo R, Ibáñez F (2022) Influence of cultivar on nutritional composition and nutraceutical potential of pecan growing in Uruguay. Front Nutr 9:868054

Fuentealba C, Hernandez I, Saa S, Toledo L, Burdiles P, Chirinos R, Campos D, Brown P, Pedreschi R (2017) Colour and in vitro quality attributes of walnuts from different growing conditions correlate with key precursors of primary and secondary metabolism. Food Chem 232:664–672. https://doi.org/10.1016/j.foodchem.2017.04.029

Gharibzahedi SMT, Mousavi SM, Hamedi M, Khodaiyan F (2014) Determination and characterization of kernel biochemical composition and functional compounds of Persian walnut oil. J Food Sci Technol 51(1):34–42

Golzari M, Hassani D, Rahemi M, Vahdati K (2016) Xenia and metaxenia in Persian walnut (*Juglans regia* L.). J Nuts 7(2):101–108

Grosso G, Yang J, Marventano S, Micek A, Galvano F, Kales SN (2015) Nut consumption on all-cause, cardiovascular, and cancer mortality risk: a systematic review and meta-analysis of epidemiologic studies. Am J Clin Nutr 101(4):783–793. https://doi.org/10.3945/ajcn.114.099515

Hao H, Li Q, Bao W, Wu Y, Ouyang J (2018) Relationship between physicochemical characteristics and in vitro digestibility of chestnut (*Castanea mollissima*) starch. Food Hydrocoll 84:193–199

Jia X, Luo H, Xu M, Zhai M, Guo Z, Qiao Y, Wang L (2018) Dynamic changes in phenolics and antioxidant capacity during pecan (Carya illinoinensis) kernel ripening and its phenolics profiles. Molecules 23(2):435

Jiang J, Liang L, Ma Q, Zhao T (2021) Kernel nutrient composition and antioxidant ability of Corylus spp. in China. Front Plant Sci 12:690966

Kaijser A, Dutta P, Savage G (2000) Oxidative stability and lipid composition of macadamia nuts grown in New Zealand. Food Chem 71(1):67–70

Kalogeropoulos N, Chiou A, Ioannou MS, Karathanos VT (2013) Nutritional evaluation and health promoting activities of nuts and seeds cultivated in Greece. Int J Food Sci Nutr 64(6):757–767

Khatib S, Vaya J (2010) Fig, carob, pistachio, and health. In: Watson RR, Preedy VR (eds) Bioactive foods in promoting health. Academic, pp 245–263

Koksal AI, Artik N, Simsek A, Gunes N (2006) Nutrient composition of hazelnut (*Corylus avellana* L.) varieties cultivated in Turkey. Food Chem 99(3):509–515

Lanchun N, Jianshe S, Ruihong H (2004) The biosynthesis and affecting factors of aroma in some fruits. Chin Bull Bot 21(5):631–637

Lavedrine F, Zmirou D, Ravel A, Balducci F, Alary J (1999) Blood cholesterol and walnut consumption: a cross-sectional survey in France. Prev Med 28(4):333–339

Lavedrine F, Ravel A, Villet A, Ducros V, Alary J (2000) Mineral composition of two walnut cultivars originating in France and California. Food Chem 68(3):347–351

Li D, Hu X (2011) Fatty acid content of commonly available nuts and seeds. In: Nuts and seeds in health and disease prevention. Academic Press, pp 35–42

Li Q, Yin R, Zhang QR, Wang XP, Hu XJ, Gao ZD, Duan ZM (2017) Chemometrics analysis on the content of fatty acid compositions in different walnut (*Juglans regia* L.) varieties. Eur Food Res Technol 243(12):2235–2242. https://doi.org/10.1007/s00217-017-2925-z

Liu Y, Fan L, Li J (2022) Flavor and compositional analysis of macadamia nuts during long-term storage. J Food Process Preserv 46:e16540

Maguire LS, O'Sullivan SM, Galvin K, O'Connor TP, O'Brien NM (2004) Fatty acid profile, tocopherol, squalene and phytosterol content of walnuts, almonds, peanuts, hazelnuts and the macadamia nut. Int J Food Sci Nutr 55(3):171–178. https://doi.org/10.1080/09637480410001725175

Mandalari G, Barreca D, Gervasi T, Roussell MA, Klein B, Feeney MJ, Carughi A (2022) Pistachio nuts (*Pistacia vera* L.): production, nutrients, bioactives and novel health effects. Plants 11(1): 18. https://doi.org/10.3390/plants11010018

Maro LAC, Pio R, dos Santos Penoni E, de Oliveira MC, Prates FC, de Oliveira Lima LC, das Gracas Cardoso, M. (2012) Chemical characterization and fatty acids profile in macadamia walnut cultivars. Cienc Rural 42(12):2166–2172

Martínez M, Barrionuevo G, Nepote V, Grosso N, Maestri D (2011) Sensory characterisation and oxidative stability of walnut oil. Int J Food Sci Technol 46(6):1276–1281

Martinez ML, Fabani MP, Baroni MV, Huaman RNM, Ighani M, Maestri DM, Feresin GE (2016) Argentinian pistachio oil and flour: a potential novel approach of pistachio nut utilization. J Food Sci Technol 53(5):2260–2269. https://doi.org/10.1007/s13197-016-2184-1

McCarthy MA, Meredith FI (1998) Nutrient data on chestnuts consumed in the United States. Econ Bot 42:29–36

Mereles LG, Ferro EA, Alvarenga NL, Caballero SB, Wiszovaty LN, Piris PA, Michajluk BJ (2017) Chemical composition of *Macadamia integrifolia* (Maiden and Betche) nuts from Paraguay. Int Food Res J 24(6):2599–2608

Míguez Bernardez MM, de la Montana Miguelez J, García Queijeiro J (2004) HPLC determination of sugars in varieties of chestnut fruits from Galicia (Spain). J Food Compos Anal 17(1):63–67. https://doi.org/10.1016/S0889-1575(03)00093-0

Miraliakbari H, Shahidi F (2008) Antioxidant activity and minor components of tree nut oils. Food Chem 111:421–427

Moodley R, Kindness A, Jonnalagadda SB (2007) Elemental composition and chemical characteristics of five edible nuts (almond, Brazil, pecan, macadamia and walnut) consumed in Southern Africa. J Environ Sci Health B 42(5):585–591. https://doi.org/10.1080/03601230701391591

Munro IA, Garg ML (2008) Nutrient composition and health beneficial effects of macadamia nuts. In: Alasalvar C, Shahidi F (eds) Tree nuts. CRC Press, pp 263–272

Noguera-Artiaga L, Sanchez-Bravo P, Hernandez F, Burgos-Hernandez A, Perez-Lopez D, Carbonell-Barrachina AA (2020) Influence of regulated deficit irrigation and rootstock on the functional, nutritional and sensory quality of pistachio nuts. Sci Hortic 261:108994. https://doi.org/10.1016/j.scienta.2019.108994

Ojeda-Amador RM, Fregapane G, Salvador MD (2018) Composition and properties of virgin pistachio oils and their by-products from different cultivars. Food Chem 240:123–130. https://doi.org/10.1016/j.foodchem.2017.07.087

Okay Y (2002) Seasonal changes in fructose, glucose and sucrose content of bark tissues of hazelnut grown in the Black Sea region. Gartenbauwissenschaft 67:225–233

Özyurt VH, Ötles S (2018) Hazelnut testa as a by-product: nutritional composition, antioxidant activity, phenolic compound profile and dietary fiber content. J Faculty Pharm Ankara Univ 42(3):38–57

Pala M, Ackurt F, Loker M, Yildiz M, Omeroglu S (1996) Fındık cesitlerinin bilesimi ve beslenme fizyolojisi bakımından degerlendirilmesi. Turk J Agric For 20:43–48

Parcerisa J, Codony R, Boatella J, Rafecas M (1998) Triacylglycerol and phospholipid composition of hazelnut (*Corylus avellana* L.) lipid fraction during fruit development. J Agric Food Chem 47(4):1410–1415

Peh HY, Tan WD, Liao W, Wong WF (2016) Vitamin E therapy beyond cancer: tocopherol versus tocotrienol. Pharmacol Ther 162:152–169. https://doi.org/10.1016/j.pharmthera.2015.12.003

Pereira-Lorenzo S, Ramos-Cabrer AM, Díaz-Hernández MB, Ciordia-Ara M, Ríos-Mesa D (2006) Chemical composition of chestnut cultivars from Spain. Sci Hortic 107(3):306–314

Rabadan A, Pardo JE, Pardo-Gimenez A, Alvarez-Orti M (2018) Effect of genotype and crop year on the nutritional value of walnut virgin oil and defatted flour. Sci Total Environ 634:1092–1099. https://doi.org/10.1016/j.scitotenv.2018.04.090

Rebufa C, Artaud J, Le Dreau Y (2022) Walnut (*Juglans regia* L.) oil chemical composition depending on variety, locality, extraction process and storage conditions: a comprehensive review. J Food Compos Anal 110:104534. https://doi.org/10.1016/j.jfca.2022.104534

Regueiro J, Sanchez-Gonzalez C, Vallverdu-Queralt A, Simal-Gandara J, Lamuela-Raventos R, Izquierdo-Pulido M (2014) Comprehensive identification of walnut polyphenols by liquid chromatography coupled to linear ion trap–orbitrap mass spectrometry. Food Chem 152:340–348

Ribeiro SR, Klein B, Ribeiro QM, Dos Santos ID, Genro ALG, de Freitas Ferreira D, Hamann JJ, Barin JS, Cichoski AJ, Fronza D, Both V, Wagner R (2020) Chemical composition and oxidative stability of eleven pecan cultivars produced in southern Brazil. Food Res Int 136:109596

Rivera-Rangel LR, Aguilera-Campos KI, García-Triana A, Ayala-Soto JG, Chavez-Flores D, Hernández-Ochoa L (2018) Comparison of oil content and fatty acids profile of Western Schley, Wichita, and native pecan nuts cultured in Chihuahua, Mexico. J Lipids 2018:4781345

Ros E (2010) Health benefits of nut consumption. Nutrients 2(7):652–682. https://doi.org/10.3390/nu2070652

Sarikhani S, Vahdati K, Ligterink W (2020) Biochemical properties of superior Persian walnut genotypes originated from southwest of Iran. Int J Hortic Sci Technol 9(2):13–24. https://doi.org/10.22059/ijhst.2020.309363.392

Savage GP (2000) Chemical composition of walnuts (*Juglans regia* L.) grown in New Zealand. Plant Foods Hum Nutr 56(1):75–82

Seferoglu S, Seferoglu HG, Tekintas FE, Balta F (2006) Biochemical composition influenced by different locations in Uzun pistachio cv.(*Pistacia vera* L.) grown in Turkey. J Food Compos Anal 19(5):461–465

Segura R, Javierre C, Lizarraga MA, Ros E (2006) Other relevant components of nuts: phytosterols, folate and minerals. Br J Nutr 96(S2):S36–S44. https://doi.org/10.1017/BJN20061862

Struetz W, Schlormann W, Glei M (2016) B-vitamins, carotenoids and-/γ-tocopherol in raw and roasted nut. Food Chem 221:222–227. https://doi.org/10.1016/j.foodchem.2016.10.065

Stuetz W, Schlörmann W, Glei M (2017) B-vitamins, carotenoids and α-/γ-tocopherol in raw and roasted nuts. Food Chem 221:222–227

Tanwar B, Modgil R, Goyal A (2020) Nutritional and phytochemical composition of pecan nut [*Carya illinoinensis* (Wangenh.) K. Koch] and its hypocholesterolemic effect in an animal model. Br Food J 123(4):1433–1448

Tapia MI, Sanchez-Morgado JR, Garcia-Parra J, Ramirez R, Hernandez T, Gonzalez-Gómez D (2013) Comparative study of the nutritional and bioactive compounds content of four walnut (*Juglans regia* L.) cultivars. J Food Compos Anal 31(2):232–237. https://doi.org/10.1016/j.jfca.2013.06.004

Tas NG, Gokmen V (2015) Bioactive compounds in different hazelnut varieties and their skins. J Food Compos Anal 43:203–208. https://doi.org/10.1016/j.jfca.2015.07.003

Terzo S, Baldassano S, Caldara GF, Ferrantelli V, Lo Dico G, Mule F, Amato A (2017) Health benefits of pistachios consumption. Nat Prod Res 33(5):715–726. https://doi.org/10.1080/14786419.2017.1408093

Tian WH, Liang LS, Wang GX (2012) Nutrient composition analysis of hazelnut kernel from different varieties. Food Sci 33(8):265–269

Tomaino A, Martorana M, Arcoraci T, Monteleone D, Giovinazzo C, Saija A (2010) Antioxidant activity and phenolic profile of pistachio (*Pistacia vera* L., variety Bronte) seeds and skins. Biochimie 92(9):1115–1122

Tsantili E, Takidelli C, Christopoulos MV, Lambrinea E, Rouskas D, Roussos PA (2010) Physical, compositional and sensory differences in nuts among pistachio (*Pistachia vera* L.) varieties. Sci Hortic 125(4):562–568. https://doi.org/10.1016/j.scienta.2010.04.039

Tu XH, Wu BF, Xie Y, Xu SL, Wu ZY, Lv X, Chen H (2021) A comprehensive study of raw and roasted macadamia nuts: lipid profile, physicochemical, nutritional, and sensory properties. Food Sci Nutr 9(3):1688–1697. https://doi.org/10.1002/fsn3.2143

U.S. Department of Agriculture (USDA) (2019) USDA National Nutrient Database for Standard Reference Legacy Release, 2019. https://ndb.nal.usda.gov/ndb/search/list. Accessed 8 Apr 2022

Verde A, Míguez JM, Leao-Martins JM, Gago-Martínez A, Gallardo M (2022) Melatonin content in walnuts and other commercial nuts. Influence of cultivar, ripening and processing (roasting). J Food Compos Anal 105:104180

Vinson JA, Cai Y (2012) Nuts, especially walnuts, have both antioxidant quantity and efficacy and exhibit significant potential health benefits. Food Funct 3(2):134–140

Wang W, Jung J, McGorrin RJ, Traber MG, Leonard SW, Cherian G, Zhao Y (2018) Investigation of drying conditions on bioactive compounds, lipid oxidation, and enzyme activity of Oregon hazelnuts (*Corylus avellana* L.). LWT 90:526–534. https://doi.org/10.1016/j.lwt.2018.01.002

Wei F, Li Y, Sun D, Chen Q, Fu M, Zhao H, Chen X, Huang Y, Xu H (2022) Odor, tastes, nutritional compounds and antioxidant activity of fresh-eating walnut during ripening. Sci Hortic 293:110–115. https://doi.org/10.1016/j.scienta.2021.110744

Wu S, Ni Z, Wang R, Zhao B, Han Y, Zheng Y, Liu Y (2020) The effects of cultivar and climate zone on phytochemical components of walnut (*Juglans regia* L.). Food Energy Secur 9(2):196. https://doi.org/10.1002/fes3.196

Yahyavi F, Alizadeh-Khaledabad M, Azadmard-Damirchi S (2019) Oil quality of pistachios (*Pistacia vera* L.) grown in East Azarbaijan, Iran. NFS J 18:12–18. https://doi.org/10.1016/j.nfs.2019.11.001

Yilmaz R, Yildirim A, Çelik C, Karakurt Y (2021) Determination of nut characteristics and biochemical components of some pecan nut cultivars. Yuzuncu Yıl Univ J Agric Sci 31(4):906–914

Zhang X, Chang J, Yao X, Wang J, Zhang J, Yang Y et al (2022) Chemical composition in kernels of ten grafted pecan (*Carya illinoensis*) varieties in Southeastern China. Sci 4(2):25

Zheng Y, Wu S, Wang R, Wu Y, Zhang W, Han Y, Liu Y (2020) Analysis and correlationship of chemical components of various walnut (*Juglans regia* L.) cultivars. J Food Meas Charact 14(6):3605–3614. https://doi.org/10.1007/s11694-020-00603-0

Chapter 3
Development and Selection of Rootstocks

Mohammad Maqbool Mir, Mir Uzma Parveze, Umar Iqbal, Munib Ur Rehman, Amit Kumar, Shamim A. Simnani, Nazir Ahmad Ganai, Zaffar Mehdi, Nowsheen Nazir, Aroosa Khalil, Bashir A. Rather, Z. A. Bhat, and M. A. Bhat

Abstract The demand for the nut crops in the world has increased due to their nutritional value and economic returns. The increasing importance of nuts in healthy diet has led to the breeding work being carried out to develop and improve various scions and rootstocks to enhance the productivity and quality of nut crops. There are various challenges associated with the nut crops that make the process of improvement quite tedious and long, reducing their productivity and potential value. The development of suitable rootstocks has been found critical for solving such challenges. In this chapter, the influences of various rootstocks on the vigour of tree, quality and yield of nuts, adaptation to various biotic and abiotic stresses, disease and insect resistance, etc. have been explained. The effect of rootstock on uptake of water and nutrients that are important for growth, development of tree and have direct effect on yield of nut trees, has also been elaborated. Rootstock variability has been found to influence insect and disease sensitivity, scion precocity, fruit quality, tree size, yield at maturity, nutrient and water uptake and sensitivity to soil variables. Initially, open-pollinated seedlings, also known as seedstock, were used, but now a diverse range of clonal rootstocks have been developed through numerous rootstock breeding programmes to address critical issues such as low yield, poor nut quality, high vigour, poor soil, salinity, drought stress, suckering, diseases, graft

M. M. Mir (✉) · M. U. Parveze · U. Iqbal · Munib Ur Rehman · A. Kumar · S. A. Simnani · N. A. Ganai · N. Nazir · A. Khalil
Division of Fruit Science, FOH, Sher-e-Kashmir University of Agricultural Sciences and Technology of Kashmir, Srinagar, Jammu and Kashmir, India

Z. Mehdi
Division of Basic Science & Humanities, FOH, Sher-e-Kashmir University of Agricultural Sciences and Technology of Kashmir, Srinagar, Jammu and Kashmir, India

B. A. Rather
Mountain Research Centre for Field Crops (MRCFC), Sher-e-Kashmir University of Agricultural Sciences and Technology of Kashmir, Khudwani, Jammu and Kashmir, India

Z. A. Bhat · M. A. Bhat
Division of Plant Pathology, FOH, Sher-e-Kashmir University of Agricultural Sciences and Technology of Kashmir, Srinagar, Jammu and Kashmir, India

M. M. Mir et al. (eds.), *Temperate Nuts*,
https://doi.org/10.1007/978-981-19-9497-5_3

incompatibility and climatic change. Rootstock development has potential to open up new avenues for easy propagation and genetic improvement of temperate nut crops.

Keywords Temperate nut · Rootstocks · Walnut · Pecan · Hazelnut · Chestnut · Pistachio

1 Introduction

The kernels or dry fruits that are enveloped in hard or woody shells are called as nuts. These shells are in turn enveloped within outer fibrous husk, which is taken off during harvest. Nut trees, particularly temperate ones, are considered among the important horticultural crops. Due to the great nutritional value of their products and economic returns, both production and consumption are quite huge. Over the last decade, global tree nut production has increased steadily, reaching around 4.6 million metric tonnes in season 2019–2020. Almonds and walnuts were the most abundant crops, accounting for 31% and 21% of global production, respectively, followed by cashews (17%), pistachios (14%) and hazelnuts (12%). The remaining 5% was split between pecans, macadamias, Brazil nuts and pine nuts (INC 2020). There has been rapid increase in the technical knowledge concerned with the production of nut trees and this increase is due to the higher quality and production, different locations of nut fruits in food and consumption industries and increasing significance of nuts in nutritious diet along with protection against numerous diseases (Gervasi et al. 2021).

Walnuts are members of the Juglandaceae family and the genus *Juglans* L., which contains about 60 species, 21 of which are classified as Juglans. The nuts of all species are edible, but those of English or Persian walnut (*Juglans regia* L.) are large and easy to crack. The commercially important and widely produced edible species of walnut in temperate regions of the world is the English or Persian walnut, from Central Asia. These walnuts are cultivated in Asia, North and South America and Europe. However, the rest of the species of walnut are grown either as rootstock (*J. hindsii*) or for the timber purpose (*J. nigra*). The nuts are high in proteins, fats and minerals, and they provide a concentrated source of energy. It grows in temperate regions at altitudes varying from 900 to 3500 m asl. It requires a cool autumn period to promote leaf fall and the physiological process of plant hardening and dormancy induction. During deep dormancy, the plant can tolerate temperatures of –1 °C without suffering serious damage but late spring frosts are harmful.

Pecans are members of the family Juglandaceae and genus *Carya*. It is native to North America. Pecans have numerous benefits, including a rich taste of kernels with no bitterness, high fatty acids, phospholipids, vitamins and proteins content. Such benefits of pecan have led to a great contribution to commercial food industry. Furthermore, it is used in horticulture as a nut crop, oil extraction, wood manufacturing, etc. providing the additional monetary benefits (Xue et al. 2018). Global demand for the nut has increased year after year as the pecan's potential value has become more widely recognised. With the gradual realization of the pecan's potential benefit, global demand for the nut has increased year after year. Pecans are now produced in countries other than its native habitat including Uruguay, Argentina, Australia, Brazil, China and South Africa, and over the next 30 years it is expected that the production will rise more (Fabrizio et al. 2018). Pecan is a monoecious, heterodichogamous, deciduous nut tree that is wind pollinated. The climate in the native pecan area ranges from semi-arid to humid and extremely harsh to mild winters.

The hazelnut is the member of the family Betulaceae and *Corylus* genus. It is native to southern and eastern Europe. Terms 'filbert' and 'hazelnut' are frequently used interchangeably to refer to all plants in the genus *Corylus*. Several species are important commercially for the nut production, and others for ornamental trees and hedgerow. The European filbert (*Corylus avellana*) produces an oil used in soaps and perfumes, food products; the tree is also known for producing soft timber which is red white in color. *C. maxima* and *C. colurna* (Turkish hazel) also have desirable characteristics. The latter has the potential to be a non-suckering rootstock. The hazelnut is a deciduous, monoecious, multi-stemmed bush in its natural state, but should be grown commercially as a single trunk tree. Hazelnuts are high in protein, monounsaturated fat, vitamin E, manganese and a wide range of other vitamins and minerals. Hazelnut grows very well in subtropical climatic conditions with relative humidity as high as 75–80%, average temperatures of 13–14 °C and rainfall ranging from 1500 to 2000 mm, annually (Mirotadze et al. 2009).

The chestnut is the member of the family Fagaceae and the *Castanea* genus. It is native to the northern hemisphere, in the countries such as eastern North America, Asia and Europe. There are seven species in the *Castanea* genus. Because of their large size, *C. sativa*, *C. crenata* and *C. mollissima* are the main cultivated species for nuts. Among these, *C. sativa* is believed to be the most suitable for production of nut; however, *C. sativa* and *C. dentata* are among the most vigorous species, and are thus used for production of timber as well. Chestnuts contain little protein or fat, and their calories are primarily derived from carbohydrates. Trees can be found at altitudes varying from 200 to 1000 amsl. Chestnuts subjected to chill temperatures during the dormant season, produce a better crop. Frosts and snowfalls are favourable to trees rather than detrimental. Chestnut requires evenly distributed rainfall throughout the year up to 800 mm or more.

Pistachio is a member of the family Anacardiaceae and *Pistacia* L. genus. This genus contains 11 species or more; however, pistachio or *Pistacia vera* L. is commercially grown for production of nut. The species of Pistacia belonging to other genera are often used for rootstock purpose or find its use in agroforestry. Pistachios are indigenous to central Asia and middle East. Among the different countries producing pistachio, Iran, the United States, Turkey and Syria are the primary producers, accounting for more than 90% of global production. Pistacia species are dioecious, though there have been a few isolated reports of monoecious individuals. Nuts contain significant amount of potassium and unsaturated fatty acids making them anti-inflammatory and antioxidant. Pistachio is a desert plant that thrives in saline soil. Pistachio trees are fairly hardy, temperatures −10 °C in winter to 48 °C in summer is tolerable. Soil that is well drained and location that is sunny are most favourable.

Almond is the member of the family Rosaceae and the *Prunus* genus. It is native to Iran and the surrounding countries. Almonds grown as nuts may be eaten raw, blanched or roasted. The trees are deciduous, flowering occurs from February to April; however, the nut crop is uncertain in those areas where frost occurs during the flowering. The flowers are self-incompatible therefore, requires pollinator for effective pollination. Almonds are rich in protein and fat, and they also contain trace amounts of iron, phosphorus, calcium and vitamins A, E and B complex.

Rootstocks are currently extensively used in horticulture, silviculture and agriculture. They are mostly used for vegetative propagation of fruit crops (apple, walnut, peach, orange, etc.), vegetables (watermelon, cucumber, eggplant, tomato, etc.) and ornamentals (rose, bonsai, bougainvillaea, etc.) (Lee et al. 2010; Gregory et al. 2013). They effect vegetative growth of scion primarily by affecting the water, hormones and mineral nutrition of grafted tree (Aloni et al. 2010; Goldschmidt 2014; Albacete et al. 2015). Fruit trees performance is determined by the reciprocal interaction of their components (rootstock and scion) with site-specific environmental factors. High stress tolerant rootstocks are extensively used to improve the resistance to biotic (including diseases pests, weeds, etc.) and abiotic stresses (such as low and high temperatures, salinity, drought, heavy metal, flooding, etc.) in grafted plants (Schwarz et al. 2010; Goldschmidt 2014; Nawaz et al. 2016). Rootstock variability also influences insect and disease sensitivity, scion precocity, fruit quality, tree size, yield at maturity, nutrient and water uptake and sensitivity to different variables of soil.

The benefits of selecting rootstocks have been recognised and hence they are used in the nut tree production, but their use for many species is not well known. The research on rootstock is limited to only some of the nut tree species despite of the fact that nut trees are grown all over the world today. The majority of rootstocks were initially open-pollinated seedlings, also known as seedstock. A wide range of clonal rootstocks, in addition to seedstocks, are developed now. Clonal rootstocks have been introduced through many rootstock breeding programmes to resolve critical issues such as low yield, poor nut quality, high vigour, poor soil, salinity, drought stress, suckering, diseases, graft incompatibility and climatic change.

2 Rootstocks of Temperate Nut Crops

2.1 Walnut

In the walnut industry, selecting suitable rootstock and efficient multiplication of seedlings of high quality are the most important steps. They not only ensure high grafting success but are also responsible for enhancing vegetative growth in scion, improving stress resistance of the grafted trees and reducing the time for fruiting (Gregory et al. 2013).

2.1.1 Vigor

Walnut trees have a long juvenility period and are extremely vigorous. Dwarf walnut trees may reduce labour costs increases the yield per hectare due to the greater plant density. Identification of the sources of dwarfing the grafted trees can be of great interest especially in countries rich in genetic diversity such as China, Iran, Central Asia and Turkey despite of the fact that dwarfing has not been the main goal of most of the rootstock breeding programmes. Dwarf walnut trees are said to have a short life span. As a result, breeders at some places are trying to graft slow-growing scions on vigorous rootstocks. The vigour of both juvenile and mature walnut trees is highly heritable (Rezaee and Vahdat 2014). In California orchards, *J. regia* and *J. hindsii* have been largely replaced by open-pollinated 'Paradox' hybrid seedlings (*J. hindsii* × *J. regia*) since 1950 because *J. hindsii* and *J. regia* rooted trees grow slower as compared to Paradox rooted trees. Texas Black, also known as Texas rock walnut, *J. rupestris* (also known as *J. microcarpa*) grows wild in Texas and New Mexico because of the smaller trees, they can be planted at closer spacing (Browne et al. 1977).

Walnut rootstocks (cluster-bearing) reduced length of internodes of scion shoots, scion height and increased lateral shoot formation than the standard ones. In comparison to standard rootstocks, cluster-bearing rootstocks increased precocity as well. Persian walnuts were evaluated in China by Wang et al. (2014), and this evaluation led to the selection of six dwarf walnut rootstocks with potential for breeding such as, Xinwen 724, Xinwen 915, Xinwen 908, Xinwen 917, Xinwen 609 and Xin 916. The evaluation of local genotypes of *Juglans regia* on their own roots for their growth trait, revealed that the climatic and edaphic factors of area influence the growth and development of trees. In addition, dwarf and precocious walnut trees were evaluated in Iran (Vahdati and Mohseniazar 2016). In the preliminary experiments, these genotypes cause dwarfing and precocity in scions, owing to a slow growth rate. They have shorter shoot length, fewer nodes and shorter internodes, as well as low sap flow and hydraulic conductivity, smaller root system, typical characteristics of dwarf rootstocks found in other fruit trees as well. They also possess superior grafting and rooting abilities.

2.1.2 Yield

Rootstock has a significant impact on walnut yield efficiency. Yields of six dwarf walnut rootstocks (Xinwen 908, Xinwen 609, Xinwen 917, Xinwen 724, Xinwen 915 and Xin 916) were evaluated in a study. Out of the six dwarf rootstocks, two to three nuts per inflorescence were produced only by the bearing branches of 'Xinwen 724', while the bearing branches of the other five rootstocks produced single and double nuts. Only 'Xinwen 724' outperformed the control in terms of the average number of nuts on fruiting spurs (Wang et al. 2014).

A study of clonal rootstocks revealed that Chandler on its own roots produced the lowest yield, while Chandler trees grafted on RX1 were found to produce the highest yield. Generally, the yield of Paradox (seedling or clonal) rootstock outperformed the yield of self-rooted 'Chandler' (Grant 2015). In a study, Chandler was propagated on selected seedling rootstock of *J. regia* from diverse sources and Paradox hybrid seedlings; the results indicated that productivity and growth was same among *J. regia* source trees, while Paradox-rooted trees produced more nuts and grew larger than *J. regia* rooted trees (Grant and McGranahan 2006). Another study conducted to compare micropropagated and ungrafted 'Chandler' with the grafted 'Chandler' on Paradox revealed that, ungrafted 'Chandler' have more yield and trunk diameter than grafted trees in the initial years, but after 6 years no significant difference was observed in these factors. The trees on their self-roots showed more dieback and were found more susceptible to nematodes (Hasey et al. 2004).

2.1.3 Nut Quality

Rootstocks may influence nut quality by affecting photosynthesis rate, subsequent assimilation into the crop and water and nutrient uptake, but the influence of rootstock on the nut quality have not been investigated thoroughly. According to Connell et al. (2010), self-rooted 'Chandler' trees produce higher nut quality, lower yield efficiency and fewer catkins as compared to 'Chandler' grafted on *J. regia* cv. Waterloo or Paradox. Wang et al. (2014) compared six dwarf rootstocks of walnut (Xinwen 915, Xinwen 609, Xinwen 908, Xin 916, Xinwen 917 and Xinwen 724) with two controls of high quality (Xinxin 2 and Wen 185). The results showed that all of them possess the nut weight lighter than 'Wen 185' and heavier than 'Xinxin 2' except 'Xinwen 915' that was observed to have the lightest nuts. The kernel yields of Xinwen 915, Xinwen 724 and Xinwen 908 was greater than 60%, while that of the others ranged between 52.92% and 58.71%, which was considered very high.

2.1.4 Nutrient and Water Uptake

The nut tree yield is affected by water and nutrient uptake that is, in turn found to be directly influenced by rootstock. It is governed by intricate interactions between the rootstock and scion, and the long-distance signalling molecules such as hormones, macromolecules and miRNAs regulate nutrient uptake (Nawaz et al. 2016). Rootstock vigour improves water and nutrient uptake. Furthermore, the water and nutrient transfer ability to the scion is governed by the rate at which the vascular bundles develop in a graft union. The influence of rootstock on nutrient density of canopy are affected not only by the physical characteristics of the roots, but also by the environmental conditions and the chemical composition of the soil (Martinez-Ballesta et al. 2010). The root hydraulic conductance of the two walnut rootstocks such as 'RX1' and 'Vlach' was found to be more than 50% as compared to 'VX211'. In a study conducted by Knipfer et al. (2020), it was concluded that it could be one of the reason for tolerance to drought stress by these rootstocks. Under drought conditions, to maintain root biomass, rootstocks 'Vlach' and 'RX1' showed reduction in the root hydraulic conductivity.

Under stress conditions, walnut roots selectively absorb ions. In a study where response of own rooted walnut to salt stress was investigated, revealed that under salt stress conditions, walnut roots absorb and transmit Calcium (Ca) and Pottasium (K). This is more predominant in roots of tolerant varieties than less tolerant ones. In addition, salt-tolerant walnut roots not only absorb more Ca and K, but they also transfer it more to the leaves (Lotfi et al. 2009).

2.1.5 Resistance to Abiotic Stress

Drought tolerance in walnut genotypes is caused by a number of physiological processes such as maintenance of net assimilation and photosynthetic rate, cavitation resistance via stomatal regulation, proline and TSS accumulation, increased antioxidative enzyme activity and improved water use efficiency (Lotfi et al. 2019). Considering the association of water use efficiency (WUE) with drought tolerance, researchers studied the Persian walnut population that is considered diverse in WUE, to identify antioxidant responses, drought stress-responsive genes involved in ABA signalling, osmotic adjustment, leaf development, environmental signal transduction and stomatal regulation. The study concluded that the genes involved in ABA signalling and stomatal regulation were found close to identified markers, confirming their role in WUE and drought adaptation (Arab et al. 2020).

Seedlings of three Juglans rootstocks (*J. mandshurica* Maxim, *J. regia* L. cv. Jizhaomian and *J. nigra* L.) were studied, and the results indicated that *J. mandshurica* and *J. regia* had better drought stress adaptability than *J. nigra* and were associated with higher WUE (Liu et al. 2019). Common clonal walnut rootstock hybrids of *J. hindsii* × *J. regia* ('Vlach' and 'VX211') and *J. microcarpa* × *J. regia* ('RX1'), have a good drought stress response and can maintain root biomass under stress conditions. Drought tolerance in RX1 and VX211 was associated with increased leaf turgor and leaf water use efficiency, as well as decreased hydraulic conductivity in the root system (Knipfer et al. 2020).

A study on the susceptibility of salinity stress in Juglans species discovered that *J. Hindsii* and its hybrid (Paradox) are more tolerant to salinity than the Persian walnut (Caprile and Grattan 2006). Drought and salt tolerance in walnut have been successfully induced through genetic transformation. Under salinity and osmotic stress conditions, the Persian walnut that was modified genetically with a gene namely flavodoxin (fld), showed improved growth (Sheikh Beig Goharrizi et al. 2016). Waterlogging in walnut, especially with poorly drained soils and spring rains may cause root asphyxiation as well as *Phytophthora* damage later on. Other than the Chinese wingnut (*Pterocarya stenoptera*), Juglans species are extremely sensitive to waterlogging. During waterlogging, ABA transfer to the leaves thus increases the content of ABA in them and plays a crucial role in reducing growth (Shaybany and Martin 1977).

2.1.6 Resistance to Biotic Stress

Crown and root rots are among the most severe diseases of walnut all over the world and it is caused by Phytophthora species, with *P. citricola* and *P. cinnamomi* as the most virulent. Although no Juglans species or hybrids with good tolerance or resistance to *P. cinnamomi* have been identified, Paradox hybrid rootstocks are markedly more resistant to other species of Phytophthora than English walnut seedling rootstocks or Northern California black walnut rootstock. Commercially available paradox hybrids are diverse, and resistance to *P. citricola* may vary significantly because of this diversity. The results, obtained under disease conducive conditions, indicate that moderately tolerant clones JX-2 and VX-211 (maternal parent M = *J. hindsi*), AZ-2, AZ-3, and NZ-1 (M = [*J. major* × *hindsii*] × *nigra*) and RX-1 (M = *J. microcarpa*) have great promise as improved Paradox rootstocks in *P. citricola* affected orchards (Browne et al. 2006). Trees on Paradox rootstock grows more vigorously than trees on *J. regia* and *J. hindsii* and are found more tolerant to lesion nematode (*Pratylenchus vulnus*) and more resistant to Phytophthora crown and root rot disease, the two soil borne problems commonly found in California (Browne et al. 1977). Since around 1900, the Northern California Black Walnut, *J. hindsii*, has been the standard rootstock for California walnuts. It is believed to be resistant to oak root fungus and tolerable to saline conditions. In the nursery, it grows quickly and uniformly, and it is less susceptible to crown gall than other rootstocks (Browne et al. 1977).

In 1960, it was discovered that blackline disease in walnut was caused by the Cherry leaf roll virus (CLRV), the most prevalent disease in many districts of California producing walnuts. The disease was found to be the cause of the death and decline of walnuts grafted on black walnut and Paradox rootstocks, but no such effect was observed on *J. regia* rooted trees, which renewed the attention of using *J. regia* seedlings as rootstocks in California. A hypersensitive reaction is caused by CLRV infection at the graft union, girdling and eventually killing Paradox and black walnut rooted trees. This reaction does not take place in *J. regia* as they are CLRV tolerant (Mircetich et al. 1980). Another particularly serious problem associated with Paradox rootstock that is the most popular and widely used rootstock in California is Crown gall. The causal organism of crown gall is *Agrobacterium tumefaciens*. This disease reduces productivity, kills young orchard trees and can sometimes destroy nursery production. However, *J. major*, *J. macrocarpa* and *Pterocarya* spp. seedlings have revealed some promising results (Leslie and McGranahan 2014).

Paradox has been found very susceptible to Armillaria root disease. Some clonally propagated Paradox rootstocks (AX1, Px1, RR4 11A, RX1, Vlach, VX211) were examined for their resistance to *Armillaria mellea*. It has been observed that there is not even a single walnut rootstock possibly to solve all root-disease related problems. In greenhouse evaluations, AX1, the most resistant rootstock to Armillaria, showed a high degree of susceptibility to *P. citricola* and *P. cinnamomi* (Browne et al. 2012) while VX211, most susceptible Paradox rootstock to Armillaria (clone VX211), is highly tolerant to both *Pratylenchus vulnus* (lesion nematode) and *Meloidogyne incognita* (root-knot nematode). Thus, 'VX211', which grows very quickly even under nematode infestation, has been approved for commercial usage. It is found to be tolerant to lesion nematode and not resistant (Buzo et al. 2009).

2.2 Pecan

2.2.1 Vigor

Currently, the breeding of pecan to produce dwarf grafts helps in the management of land resources, while for timber production, the breeding of fast-growing grafted trees is in high demand. It is well known that using rootstocks with varying effects on the performance of scion is an important approach for cultivating of grafts. The effect of rootstock on scion vigour of pecan (*Carya illinoinensis*) has shown that common pecan rootstocks have a diverse impact on scion growth, and they differ by geographic region. Open-pollinated seedstocks were commonly used for vegetative propagation of commercially important pecan cultivars with varying growth behaviour prior to the introduction of clonal rootstocks.

According to study, rootstock may influence tree vigour by affecting mineral nutrition, hormonal balance and/or water relations (Fallahi et al. 2001). Although some researchers have studied gene expression patterns in scions with different growth vigour on various rootstocks (Prassinos et al. 2009) but Grauke and Pratt (2019) revealed that rootstock had a significant impact on scion growth. It was found that the bud growth of Candy on Elliot and Curtis rootstocks was greater than the bud growth on Apache, Sioux, Burkett and Riverside rootstocks from the analysis of bud growth of pecan cultivars (Candy, Stuart and Cape Fear) on open-pollinated seedstocks (Riverside, Curtis, Apache, Burkett, Elliott, Sioux and Moore).

MicroRNAs (miRNAs) are known to be involved in many plant development processes as well as stress responses (Sunkar et al. 2012). Several miRNAs have previously been shown to function as long-distance transport signals in grafted plants. Liu et al. (2020) investigated miRNAs that are responsive to grafting and are involved in grafted pecan growth regulation. As a result of this investigation some miRNAs were identified that regulate grafted pecan by controlling auxin transport, cell activity and acquisition of inorganic phosphate (Pi).

2.2.2 Yield

Water hickory (*C. aquatica*) has been used as one of the rootstocks for pecan scions, but results have been mixed. Pecan scions on water hickory rootstock grow slower, have chlorotic foliage more often, and yield less than pecan scions on pecan rootstocks (Grauke and O'Barr 1996).

2.2.3 Nut Quality

For rootstock production, the commercial pecan nursery industry is dependent on open-pollinated seed. Nut size, nut fill, nut shape, seed availability, stand uniformity, root characteristics and seedling vigor are the factors influencing seedstock selection. Local availability is also an important consideration when selecting seedstock. Nuts that are well-filled leads to better germination. Small nuts are quite often preferred, particularly when purchasing seed, because having more nuts per pound increases potential production. Because of enhanced performance in some mechanical planters, round nuts are generally preferred over long nuts. The USDA released 'Apache', a controlled cross (Burkett × Schley), in 1962. It generally contains 60% kernel and about 50 nuts per lb. The nuts of 'Burkett' are typically large with 43 nuts per lb, round in shape and may approximately contain up to 55% kernel. 'Curtis' is a seedling selection with small nuts about 89 nuts per lb, but they are usually filled properly with approximately 57% kernel. 'Elliott' is a seedling selection from Florida since pecans are not indigenous to this place, so the parents of 'Elliott' were introduced in Florida. Nuts are of medium size with 68 nuts per lb, round in shape and contain approximately 54% kernel. 'VC1-68' is a seedling selection and it is used as a rootstock in the west, particularly California. The nuts are large with about 41 nuts per lb but with a thick shell, due to this they have low kernel (43%) percentage (Grauke 2010).

2.2.4 Nutrient and Water Uptake

Miyamoto et al. (1985) studied the effects of sodium (Na), chlorine (Cl) and salinity on the ion uptake and growth of pecan seedlings that were used as rootstock. Seedlings used for the evaluation were Riverside, Apache and Burkett. Concentrations of Na in soil solutions as well as the content of Na in leaves, were found to be inversely related to seedling growth, while the Cl contents in leaf did not appear to correlate with Cl concentrations in soil solutions or with seedling growth. In another study, Reid (1997) examined the leaves of two pecan scions, Pawnee and Posey on ten rootstocks such as Colby, Chickasaw, Mohawk, Dooley, Giles, Greenriver, Peruque, Posey, Shoshoni and Major. He came to the conclusion that rootstock had an effect on K and zinc (Zn) concentration. The trees grown on seedling rootstock 'Chickasaw' had the most Zn concentration, while those grown on 'Major' seedlings had the lowest, in addition to this, the trees on 'Posey' seedlings accumulated the most concentration of K, while scions on 'Greenriver' seedlings accumulated the lowest.

2.2.5 Resistance to Abiotic Stress

One of the important abiotic factors that affects the growth of pecan trees at different stages of development is the temperature, particularly low or freezing conditions and it is influenced by rootstock. In Georgia, the most common pecan rootstocks are 'Elliott' and 'Curtis' seedlings. They provide better germination and develop rapidly large stem callipers. Among the different rootstocks tested, 'Giles', 'Apache' and 'Peruque' were least affected by freezing temperatures (−2 and −5 °C, respectively). Among the scion cultivars 'Kanza' was least damaged than 'Mount', 'Mohawk' or 'Creek' (Smith et al. 2001). 'Kanza' showed no sign of injury while other cultivars showed severe signs of injury at the time of winter and autumn freeze in Oklahoma, and thus it is regarded as the most cold hardy (Smith 2002).

For both fall and midwinter freeze damage, Pawnee has been found resistant but as it is one of the first cultivars in which bud break early in the spring, it is extremely vulnerable to damage by spring frost (Smith et al. 2001). It was observed that scion cultivar influences the cold hardiness of the rootstock above-ground parts, also 'Pawnee' and 'Kanza' scions reduced the rootstock's resistance to cold during the early autumn freeze because they enter into the dormancy later in the autumn (Thomas and Reid 2006).

Besides autumn freezes, very cold winters may cause considerable damage to pecan trees. Browning and splitting of the rootstock, inner bark and phloem, browning and death of the cambium, as well as delayed bud break and splitting are typical symptoms (Wood and Reilly 2001). Cultivars such as 'Wichita' and 'Choctaw' grafted on 'Apache' seedling rootstock showed the damage of one-third than the same cultivars grafted onto 'Riverside' seedling rootstock. Apache' rootstock resulted in cold-hardy, fast-growing trees. Pecan rootstocks of northern origin, such as 'Giles,' 'Peruque,' or 'Colby,' are thought to be more resistant to late-spring frost than those of southern origin, such as 'Riverside' or 'Moore' (Carroll and Smith 2017). The 'Stuart' seedlings start growing late in the spring, providing certain protection from spring frost. Early spring growth of 'Elliott' seedlings that are known for excellent quality of nuts makes them more vulnerable to spring frost damage as compared to 'Moore' rootstock (Grauke and Pratt 2019). Pecan rootstocks were tested under drought conditions and the drought tolerance order from high to low. It was as follows: 'Posey', 'Peruque', 'Riverside', '87MX5-1.7', 'VC1-68', 'Elliott', '87MX1-2.2', 'San Felipe', 'Moore', 'Major', 'Giles' and 'Frutoso' (Cao et al. 2019). Due to the salt tolerance, 'Riverside' is commonly used as a rootstock in the western pecan growing region (Miyamoto et al. 1985).

2.2.6 Resistance to Biotic Stress

Xylella fastidiosa, a xylem-limited bacterial pathogen with a broad host range, can infect pecan trees. Infection causes leaf scorch disease, which can result in pronounced defoliation, lower nut yield and decrease growth of tree. *X. fastidiosa* is rapidly transmitted into newly emerging grafted trees through infected rootstocks. Sanderlin (2015) tested seven rootstocks, out of those, 'Riverside', 'Curtis' and 'Elliott' rootstocks were found less susceptible as compared to highly susceptible 'Cape Fear' rootstock and one standard rootstock VC1-68. Rootstocks from 'Stuart', 'Apache', 'Moore' and 'VC1-68' seed, on the other hand, had a susceptibility level equivalent to 'Cape Fear', and thus may not be the rootstock of choice in those geographic areas where *X. fastidiosa* is widespread. *Carya cathayensis* (Chinese hickory) trunk canker, caused by *Botryosphaeria dothidea*, is a devastating disease in China, and presently there are no resistant varieties or effective pesticides available (Yao et al. 2019).

In pecan, *Meloidogyne partityla* is primary nematode. Open-pollinated rootstocks of pecan (*C. illinoinensis*), water hickory (*C. aquatica*), and their hybrids were tested for resistance to nematode (*Meloidogyne partityla*). Grauke and Starr (2014) studied difference in gall formation severity in different seedstocks. Among all seedstocks, 'Elliott' was noted to be quite susceptible to gall formation and thus most severely affected. Rest of the seedstocks had moderate to severe nematode damage except 'Riverside' seedling, which was found to be gall-free.

2.3 Hazelnut

2.3.1 Vigor

In Oregon, hazelnut rootstock breeding began in 1968. Open-pollinated seedlings of *C. colurna* with intermediate traits between *C. avellana* and *C. colurna* were selected and propagated in nursery rows. Over the course of 20 years, from 20,000 seedlings about 150 potential rootstocks were chosen. Among these, two non-suckering clonal rootstocks ('Dundee' and 'Newberg') were released, these were found to impart vigor to scions (Lagerstedt 1993). Both rootstocks are considered interspecific hybrids because their nut and husk features vary from their maternal parent. Salimi and Hoseinova (2012) collected seeds from 14 local genotypes of *Corylus avellana* L. with relative tolerance to environmental stresses and sowed them with seedlings of controls. It was observed that Negret seedlings had low growth vigour and thus is not considered as ideal rootstock for grafting of hazelnut. However, the genotypes such as Pashmineh, Mahalli Karaj, Shirvani and Nakhon Rood, on the other hand, are suitable to be used as seedling rootstocks especially in those hazelnut growing areas with climatic conditions like that of Iran, and this is attributed to their high graft success, high growth vigor, leaf scorch tolerance and good germination percentage. Rovira et al. (2014) grafted the

'Negret-N9' selection on four clonal rootstocks (two open pollinated *C. colurna* seeds 'Newberg' and 'Dundee' and the low suckering cultivars 'IRTA-MB69' and 'Tonda Bianca') and compared it with the self-rooted 'Negret-N9'. Results revealed that all grafted cultivars were more vigorous than the self-rooted cultivars. These findings are consistent with other studies (Lagerstedt 1993) in which own-rooted 'Negret' and 'Ennis' trees were less vigorous than cultivars grafted on other rootstocks.

2.3.2 Yield

The seedlings of Turkish hazelnut often take 2 years to reach a size suitable for grafting and are difficult to propagate. Furthermore, on *C. colurna* rootstocks, more variability is found in size and yield of hazelnut trees than the self-rooted *C. avellana* trees. A trial in Barcelona using scion cultivar revealed that the nut yield decreased with age (20–25 years) (Cerovic et al. 2007). Further, Rovira et al. (2014) found that the cumulative yield efficiency of the less vigorous trees ('Negret N9' own rooted) was significantly higher. In another study, 25 hazelnut cultivars were evaluated for compatibility with the Turkish hazelnut rootstock in Germany. The results showed that all the evaluated cultivars were compatible with the Turkish hazelnut rootstock and the yielding capacity of some cultivars grafted on this rootstock increased (Ninic-Todorovic et al. 2009).

2.3.3 Nut Quality

Own-rooted 'Negret N9' produced significantly higher quality nuts and kernels (larger weight and size) than grafted 'Negret N9'. However, kernel oil stability was more in grafted trees than own-rooted trees, grafted trees also favoured kernel skin colour (Rovira et al. 2014). In Eastern Serbia, four hazelnut cultivars were planted self-rooted and grafted on *Corylus colurna* rootstock. Because of the vigorous root system of *Corylus colurna*, the kernel and fruit size of cultivars grafted on *C. colurna* was higher than those cultivars grown self-rooted (Miletic et al. 2009).

2.3.4 Nutrient and Water Uptake

In a study of rootstocks of hazelnut, it was observed that 'Newberg' and 'Dundee' have more resistance to iron chlorosis, thus trees retain leaves for extended period of time during the season. It is an important factor to consider because the trees can then only take soil nutrients for a prolonged period of time (Rovira et al. 2014).

2.3.5 Resistance to Abiotic Stress

C. colurna cultivars have been shown to be more cold hardy and drought tolerant than *C. avellana* cultivars in the United States. The *C. colurna* was found to be deeply rooted, non-suckering and graft compatible with *Corylus* species and all cultivars of *C. avellana*, implying that it could be used as a rootstock. Selected *C. colurna* seeds are currently used for raising rootstock of hazelnut, due to their strong root system they are resistant to frost and drought, thus making them more suitable for dry regions, trees are more productive and vigorous than own-rooted trees and have wider adaptability to varying soil conditions (Miletic et al. 2009).

2.3.6 Resistance to Biotic Stress

C. colurna can be used widely as a rootstock in hazelnut due to the benefits of grafting on this rootstock over the traditional methods of breeding (Miletic et al. 2009). The advantages of using this rootstock are the lack of suckering (reducing the maintenance costs), improved quality of fruit and kernel, improved resistance to pests and diseases. The seeds for rootstock (*C. colurna*) production are harvested from parent plants that have not been protected or fertilized as no insect or pathogen attack has been observed (Ninic-Todorovic et al. 2012).

2.3.7 Suckering

Suckering is a major problem of hazelnut, necessitating spray of four to five herbicide per year in commercial fruit orchards, as well as removal of suckers by hand in the winter. Thus, suckers increase the orchard management costs associated with their removal and most importantly they are responsible for diverting nutrients and water away from the main trunk. However, the use of non-suckering rootstocks can be helpful to improve the situation and currently, there are three types of hazelnut rootstocks used such as *C. avellana* seedlings, *C. colurna* seedlings and two clonal selections from open pollinated *C. colurna*, i.e. 'Newberg' and 'Dundee' (Lagerstedt 1993). Sucker emission varies greatly with the rootstock, as well as in self-rooted 'Negret N9'. In a study conducted by Rovira et al. (2014), the rootstocks with the lowest suckers emission were 'MB-69', 'Newberg' and 'Dundee', respectively. The use of such rootstocks may improve the management of orchard by reducing the cost and time on removal of suckers and enabling mechanised harvesting.

2.4 *Pistachio*

2.4.1 Vigor

Growers and breeders of pistachios are searching for vigorous rootstocks as the graft union produced by highly vigorous rootstocks are more uniform and they also decrease the bark damage caused by uneven graft unions. According to Kallsen and Parfitt (2011), UCB1 (University of California Berkeley 1—*P. atlantica* × *P. integerrima*) is a better rootstock than *Pistacia integerrima* for 'Kerman' because it is found to produce smoother trunk. Caruso et al. (2005) studied eight in vitro propagated clonal (*P. integerrima* clone 1, 3, 5 and 8 and *P. atlantica* clone 2, 4, 5 and 6) and one seedling (*P. terebinthus*) pistachio rootstocks, it was found that rootstock had a significant impact on growth rate of scion and nut yield. However, *P. integerrima* and *P. atlantica* clones were found extremely to moderately vigorous rootstocks.

'Bianca', a pistachio cultivar grew significantly better on *P. integerrima* seedling rootstock than *P. atlantica* or *P. terebinthus* clonal rootstocks. The vigour of scion was lowest on *P. terebinthus* rootstocks. Furthermore, 'Bianca' scions budded on eight clonal rootstocks propagated under in vitro conditions for 4 years revealed that the trunk cross-sectional area was three times greater on *P. terebinthus* rootstocks than on *P. integerrima* rootstocks (Barone et al. 1997). Ak and Turker (2006) reported that cultivars 'Siirt' and 'Kirmizi' grafted on *P. atlantica, P. khinjuk* and *P. vera* showed variation in budbreak, vegetative growth and flowering time. It was observed that flowering in *P. vera* was earlier while *P. khinjuk* and *P. atlantica* were having greater stem diameter.

2.4.2 Yield

The impact of four different rootstocks was investigated on the marketable yield of pistachio trees (*P. vera* cv. 'Kerman'), it was observed that trees grown on seedling rootstock UCB1 produced 45.3%, 19.1% and 15.1% greater marketable yield than the trees grown on *P. atlantica, P. integerrima* and PGII rootstocks, respectively (Ferguson et al. 2002). An examination of the yield components of pistachio such as nuts per cluster, nut size and clusters per tree revealed that UCB1 seedling rootstock produced higher yield due to larger size of the trees and that resulted in greater number of clusters per tree but not necessarily a higher cluster density or more number of nuts per cluster. This observation suggested the trees on various rootstocks may yield equally when they are pruned to maintain the same canopy size, or that trees cultivated at high densities on less vigorous but more efficient rootstocks will be more productive than trees planted on UCB1 (Ferguson et al. 2016). In a study, Rahemi and Tavallali (2007) investigated effect of 'Sarakhs' (wild *P. vera*), 'Beneh' (*P. mutica*) and 'Badami' (*P. vera*) seedling rootstocks on yield, nut quality and growth of three Iranian cultivars ('Kalleh-Ghouchi', 'Ahmad-Aghaei' and

'Ohadi'). The 'Badami' rootstocks produced the maximum yield along with the best nut quality, while the vigour of 'Sarakhs' seedlings was found to be the lowest.

2.4.3 Nut Quality

The quality factors for pistachio include nut weight, nut size, frequency of blank nuts and percentage of split nuts. Many trials in California on pistachio revealed that rootstocks have little influence on characteristics of nuts, implying that scion cultivar breeding primarily improve the quality of nuts (Ferguson et al. 2016). Another study revealed that rootstock had a significant influence on nut quality of 'Kerman' cultivar, including nut size, oil content, splits, flavour, aftertaste and colour. As compared to *P. integerrima* or *P. terebinthus, P. atlantica* rootstock led to increase in the sensory attributes, kernel mineral content and consumer satisfaction (Carbonell-Barrachina et al. 2015). Commercially, nut split is very important trait in pistachio, and it is primarily influenced by scion cultivar's genetic factor as well as rootstock and cultural practices. According to Tajabadipour et al. (2006), early cracking and splitting of pistachio hull are one of the main aspects leading to contamination of aflatoxins in the orchard, while cultivar and rootstock are the two major factors influencing early cracking and splitting of pistachio hull. Pistachios with smooth and soft hulls or dry and shrivelled hulls mostly splitted earlier on *P. atlantica* and Baneh rootstocks than on Ahli. The scions grafted on Ahli rootstocks produced the lowest cracking percentage on the pistachio hull, but there were no significant variations when compared to other rootstocks. Turker and Ak (2010) studied the impact of pistachio rootstocks (*P. atlantica*, *P. khinjuk* and *P. vera*) budded on the cultivars 'Siirt' and 'Ohadi'. Cultivars on rootstock *P. atlantica* produced highest number of filled nuts and splits, with only few blank nuts.

2.4.4 Nutrient and Water Uptake

The ability of pistachio rootstocks to absorb nutrients varies. This is evident from a study concluding that *P. integerrima* rootstock is less efficient in the uptake of copper (Cu) and zinc (Zn), while more efficient in the uptake of boron (B), chloride (Cl) and sodium (Na) compared to *P. terebinthus* or *P. atlantica*. However, in a saline conditions, high absorption and translocation of Cl and Na ion takes place by the leaves and it can prove to be harmful to the scion of *P. integerrima*. As compared to *P. terebinthus*, Zn and Cu is more efficiently taken up by PGII rootstock (*P. integerrima* × *P. atlantica*), *P. atlantica* is intermediate while *P. integerrima* and UCB1 being the least efficient. Boron uptake efficiency is found similar in UCB1, *P. atlantica* and *P. terebinthus*. However, compared to *P. integerrima* rootstocks, PGII is slightly less efficient in the uptake of boron, while slightly more efficient than UCB1. The nutrient level was often found highest in the leaves of trees grafted on *P. terebinthus* rootstock (Brown et al. 1994). Barone et al. (1997) conducted a study in which 'Bianca' was used as budded scion for different in-vitro propagated clonal rootstocks. In this study, it was revealed that *P. atlantica* and *P. integerrima* clones were most efficient in uptake of magnesium (Mg), least in K uptake, while *P. terebinthus* was most efficient in uptake of K and less efficient in Mg uptake.

Tavallali and Rahemi (2007) observed that the uptake of phosphorus (P), potassium (K) and zinc (Zn) is highest in the cultivar grafted on 'Baneh' rootstock than cultivars grafted on 'Sarakhs' and 'Badami' rootstocks. The leaves of pistachio cultivar on Badami rootstock have higher Ca levels, while the leaves of Sarakhs cultivar have a higher content of copper (Cu). The cultivars on 'Sarakhs' rootstock produced the kernel with more P, Mg, Fe, K, Zn and Cu than cultivars grafted on other rootstocks. *P. atlantica* seedling rootstock grafted trees rarely develop deficiencies of B, Ca and Zn. Grafting of 14 pistachio cultivars on *P. khinjuk* seedling rootstock was studied by Surucu et al. (2020), the results showed that the 'Haciserifi' scion cultivar accumulated the highest N, P and K, while highest Fe and Zn was accumulated by 'Vahidi'. Ca, Mg and Cu by 'Mumtaz' and the highest Mn accumulated by 'Sel-15'. These findings suggest that identifying the suitable scion and rootstock for a given set of environmental conditions is a significant consideration that can influence orchard growth and production.

2.4.5 Resistance to Abiotic Stress

The pistachio tree is salinity and drought tolerant. The orchards of pistachio in California are irrigated, while it is grown in non-irrigated or less irrigated conditions in many parts of Iran. Drought in unirrigated areas is thus one of the major stresses affecting pistachio yield and cultivation. In a study, Gijon et al. (2010) used three rootstocks (*P. atlantica*, hybrid *P. atlantica* × *P. vera* and *P. terebinthus*) for cultivar Kerman to find the drought resistance of this cultivar on these rootstocks. *P. terebinthus* was found as the most drought resistant while *P. atlantica* have high sensitivity to drought stress conditions. In another study, Moriana et al. (2018) also used three rootstocks (*P. terebinthus*, *P. atlantica* and UCB1) for cultivar Kerman to investigate the effect of water stress on grafted 'Kerman'. Dehydration was observed in all three rootstocks, resulting in reduced number of leaves and vegetative growth, while there is increased root weight. UCB1 had highest drought tolerance and in addition, drought tolerance of *P. atlantica* and *P. terebinthus* was found to be more than *P. integerrima* rootstock.

Pistacia vera, *Pistacia terebinthus*, *Pistacia atlantica* and *Pistacia khinjuk*, are the four most widely used rootstocks in Turkey. They are not commonly used as seedlings, except for *P. vera*, due to their low seed germination percentage. However, *P. vera* seedlings are more homogeneous and robust than other rootstocks. The cultivars budded on this rootstock showed slow growth during first year than accelerated in next years. They can withstand high lime content, salinity and soil drought but show sensitivity to high soil moisture. The trees on *P. terebinthus* rootstock grow typically as bushes. They possess extensive and robust root system. As a result, they can be grown in stony and rocky soils. They are hardier than

P. atlantica and as hardy as *P. vera* and thus it is preferable to *P. atlantica* in cold climates. *P. terebinthus* also grows well in soils rich in Ca along the Mediterranean coast. *P. atlantica* has been found to be an excellent rootstock for *P. vera* and it is best suited for irrigated plantations but can also be grown in heavier soils (Kaska 1995).

Soil salinity is a major impediment to production of pistachio in many growing areas of Iran and throughout world. The balance of ions is affected by salinity as plants experience hyperosmotic stress, which causes competition between sodium (Na) and potassium (K) ions. Important metabolic enzymes are inhibited when K levels fall (Kurum et al. 2013). Due to the high vigour and resistance of UCB1 to wide range of abiotic and biotic stresses it is a popular rootstock. According to Ferguson et al. (2002), UCB1 was found as the most salt-tolerant, followed by 'Badami', 'Ghazvini', 'Kaleh-Ghouchi' and 'Akbari'. Jamshidi Goharrizi et al. (2020a, b) investigated the impact of drought, salt and salt + drought stress on seedling rootstocks of pistachio ('UCB1', 'Ghazvini', 'Badami' and 'Kale-Ghouchi'). It was found that the order of the rootstocks with the tolerance to these stresses were UCB1, 'Badami', 'Ghazvini' and 'Kaleh-Ghouchi'. Cold stress is another issue, particularly in some growing regions of Iran. *Pistachio terebinthus* is the most cold hardy of the common pistachio rootstocks followed by *Pistachio atlantica* and UCB1, while as, *Pistachio integerrima* is the least (Ferguson et al. 2016).

2.4.6 Resistance to Biotic Stress

Several fungal and bacterial diseases have been observed to infest the pistachio tree. The three most serious soilborne fungal diseases worldwide are Phytophthora root and crown rot (*Phytophthora* spp.), Verticillium wilt (*Verticillium dahlia*) and Armillaria root rot (*Armillaria mellea* Vahl.). For commercial purpose, large edible nuts are produced only by *P. vera* (Sheikhi et al. 2019). Initial studies revealed that *P. vera* seedling trees are susceptible to soil borne pathogens such as *Verticillium dahlia*, *Phytophthora* spp. and nematodes; however, the use of resistant or tolerant rootstocks may provide the best protection. Few *P. integerrima* seedlings were found to have tolerance to Verticillium wilt, although the trees were infested but with no mortality and only few symptoms were observed. Because of this tolerance, Pioneer Gold 1 (PG1), the *P. integerrima* seedling rootstock was quickly commercialised. Also, UCB1, which was created from *P. integerrima* (tolerant to Verticillium), has moderate resistance to the disease and just like PG1, showed few symptoms with no mortality. On the other hand, *P. terebinthus* and *P. atlantica* rootstocks are susceptible to Verticillium wilt (Ferguson et al. 2016). The resistance of four rootstocks (PGII, UCB1, *P. integerrima* and *P. atlantica*) to *Verticillium dahlia* was studied by Epstein et al. (2004), among these four rootstocks, UCB1 showed the few symptoms, while both *P. integerrima* and UCB1 have shown the highest tree vigor.

Root and crown rot in pistachio is caused by *Phytophthora* spp. UCB1 has become the dominant rootstock in California due to its resistance to *Phytophthora* and *Verticillium*. Armillaria root rot is another serious fungal soilborne disease that can occasionally affect pistachio; however, the use of resistant rootstocks can provide the effective defence (Ferguson et al. 2016). UCB1 and *P. terebinthus* are tolerant to the pathogen causing this disease, while *P. integerrima* and *P. atlantica* are susceptible (Holtz and Teviotdale 2016). In Kern County, California, Nouri et al. (2020) studied *Macrophomina phaseolina* as a new pathogen causing rootstock crown rot as well as foliage wilting. This pathogen is now posing a serious threat to California's pistachio crop.

2.5 Chestnut

2.5.1 Vigor

Dwarfing is a desirable trait for increasing the profitability of tree nut production. The huge tree canopy makes pest control, training, pruning quite expensive and difficult. As a result, using dwarfing rootstocks may be a solution to decrease cost of production and thereby increasing profitability. The dwarfing rootstocks have an impact on soil and climatic adaptability, susceptibility to disease, anchorage and precocity in addition to tree size. In the United States, Anagnostakis et al. (1998) tried to breed chestnut rootstocks that induce dwarfness and proposed that hybrids with *Castanea seguinii* might be a dwarfing source. In Missouri, standard vigorous rootstocks to be used for the purpose of commercial cultivation of Chinese chestnut trees are the seedlings of 'AU-Cropper'. To avoid incompatibility and improve union formation, Jaynes (1979) suggested that chestnut cultivar should be grafted onto the same cultivar's seedling rootstock. 'Little Giant', 'King Arthur' and 'Hope'(complex hybrids of *C. mollissima* × *C. seguinii* Dode) have recently been found to be a potential source of genetic dwarfing thus leading to increase in yield and density of trees in orchards (Warmund 2011). Dwarf chestnut rootstocks are still being studied for graft compatibility, vegetative growth and productivity.

2.5.2 Yield

Dwarfing culture and close planting can improve *C. mollissima* yield and quality, which is important for growers, but no suitable cultivars were available. As a result, rapid selection for new dwarf varieties and close planting became the new breeding goal in Turkey. Wild Turkish rootstock was used to graft some complex hybrid chestnut seedlings (*C. sativa* Mill). A-25 and A-41, among complex hybrids, may be useful as dwarfing rootstocks thereby increasing yield (Pereira-Lorenzo et al. 2016).

2.5.3 Nut Quality

Hybridisation was used to introduce the trait of easy peeling in nuts of Japanese cultivar. As a result, hybrid clones were released in the United States, Europe (Spain, Portugal and France) and Japan, with some commercialised for nut production, timber production and rootstocks. Asian species exhibited phenotypic traits that were not accepted by growers and consumers, such as poor nut quality and timber stature. As a result, one of the first breeding programmes in forest/fruit species in France, Portugal and Spain was the hybridisation of Japanese chestnuts with European chestnuts (Pereira-Lorenzo et al. 2016). Casval is a seedling rootstock that can be used as a cultivar or in intensive orchards in chestnut areas. The fruit is medium in size (10 g). The success rate for budding or grafting is greater than 83%. It is compatible with Romanian and French cultivars, and trees grafted on this rootstock bear nuts earlier. It is also suitable for processing due to its white coloured nuts and pleasant taste (Botu et al. 2014).

2.5.4 Nutrient and Water Uptake

C. sativa seedlings were induced with growth limiting conditions in a study conducted by Camison et al. (2020), to determine differences in starch, soluble sugars and total non-structural carbohydrates (NSC) in stem, leaves and roots. The findings revealed that degradation of chlorophyll was only observed in plants that had been waterlogged. The highest N content was found in the leaves of drought-stressed plants, while the roots of controlled plants have the highest C/N ratio. Under drought conditions, soluble sugars were obtained from the rapid depletion of starch and then it remained constant, but there was no change observed in total NSC, implying that plants were able to rectify water stress induced xylem embolisms. On the other hand, waterlogging resulted in a net gain of NSC in plant roots and stem with time, indicating that the plants were not able to use them.

2.5.5 Resistance to Abiotic Stress

To maintain nut production, Japanese chestnut trees were introduced in south-western France due to their resistance to ink disease (*Phytophthora* spp.). They have also been used as rootstocks for grafting vulnerable cultivars. However, it was observed that they are unsuitable for the environmental conditions because of their susceptibility to early frost damage due to earlier blooming (Pereira-Lorenzo et al. 2016). Chinese chestnut seedlings, on the other hand, are used as rootstock because they have resistance to chestnut blight and cold injury. Another abiotic stress affecting chestnut orchards is increased drought stress resulting from climate change (Conedera et al. 2010). The problem has further aggravated by the use of inter-specific hybrid rootstock clones (*C. sativa* × *C. crenata*) that have resistance to

Phytophthora cinnamomi but show low drought tolerance in comparison to native rootstocks of *C. sativa*. Rootstocks tolerant to drought can be used to reduce the impact of climatic change on the cultivation of chestnut. In *C. sativa*, drought tolerance was investigated using rootstocks and scions with tolerance to drought obtained from xeric (X) and humid (H) populations from Spain. To improve drought tolerance in chestnut, the findings of the investigation has suggested the use of xeric rootstocks and scions (Camison et al. 2021). In 1997 and 1998, Soylu and Serdar (2000) tested a total of 23 genotypes, out of these five genotypes (SE 18-2, 554-1, SE 21-9, SA 5-1 and SE 23-9) were chosen as seedling rootstock. These rootstocks can also be used in studies of breeding for resistance to salt and *Phytophthora* species.

2.5.6 Resistance to Biotic Stress

The cultivation of chestnut is affected by root rot (*Phytophthora* spp.) and chestnut blight (*Cryphonectria parasitica*) diseases. Root rot was the first pandemic disease in chestnut. There are two species responsible for root rot (*P. cambivora* and *P. cinnamomic*), these species are widely spread in Asia Minor and Europe. The use of resistant rootstock has been found effective against the disease. The two species of chestnut (*C. crenata* and *C. mollissima*) have been found resistant to root rot. Thus, in the early 19th century, these were introduced to Europe but it was observed that they were sensitive to frost and their nut quality was also poor as a result they have been used as rootstocks but they showed graft incompatibility (Pereira-Lorenzo et al. 2016). It was observed that *Castanea seguinii* seems to have some resistance to chestnut blight disease. In a study, some superior genotypes were planted in an orchard infected with *Phytophthora* spp. In 1946, Marsol, Marigoule, Ferosacre, Precoce. 'Marigoule' and 'Maraval' were obtained as natural hybrids of *C. crenata* × *C. sativa* (Pereira-Lorenzo et al. 2016). Among them, 'Maraval' and 'Marsol' were found as resistant rootstocks. However, 'Marigoule' has been used in forest areas because it grows fast. Due to its tolerance to the chestnut blight (*C. parasitica*) and resistance to root rot (*Phytophthora* spp.) 'Marigoule' is now used as rootstock in many countries but as a scion cultivar, 'Marigoule' seedlings are not resistance or tolerant to such diseases. The observations made for 10 years have demonstrated that 'Marigoule' seedling survival (affected by *Phytophthora* spp.) is only 10% more as compared to European chestnut seedlings.

It is recommended to use five French rootstocks with resistance ranging from very high to low (Breisch 1992) and four Spanish hybrid clones with resistance ranging from very high to medium to Ink disease (*Phytophthora cinnamomi*) as mentioned in Table 3.1 (Pereira-Lorenzo and Fernandez-Lopez 1997). Among the Spanish rootstocks, 'CHR-151' ('HS') is widely used because in vitro propagation is easier in this rootstock (Miranda and Fernandez 1992). To obtain root rot resistant rootstocks, hybridisation was done in several countries, including Italy, Portugal, USA and Australia. Most of the hybrids developed as a result of hybridisation showed graft

Table 3.1 Characteristics of rootstocks for temperate nut fruits

Rootstock	Characteristics	References
Walnut		
The Royal Hybrid (*J. nigra* × *J. hindsii*)	Extremely vigorous	Browne et al. (1977)
'RX1' (clonal rootstock)	Resistant to *Phytophthora*	Browne et al. (2015)
Paradox	Higher uptake of N, P, Ca, Mg and Mn than *J. hindsii*	Reil et al. (1992)
J. macrocarpa	More drought resistant than *J. ailantifolia* or *J. hindsii*	Knipfer et al. (2018)
'RX1'	Resistant to *P. cinnamomic*	Browne et al. (2012)
Seedlings of *J. cathayensis* and *J. macrocarpa*	Nematode resistant	Buzo et al. (2009)
Pecan		
Riverside	Adaptable in saline areas	Miyamoto et al. (1985)
'Curtis'	Resistant to cold	Wells (2014)
'Curtis' and 'Elliott'	Resistant to Pecan scab disease	Grauke (2010)
C. illinoensis, American pecan	Resistant to trunk canker in Chinese hickory hybrids	Yao et al. (2019)
Hazelnut		
C. colurna	No suckering	Miletic et al. (2009)
'MB-69', *C. avellana* clonal rootstock	No suckering	Tous et al. (1994)
Self-rooted 'Negret N9'	High quality nuts and kernels	Rovira et al. (2014)
Pistachio		
Pistacia integerrima	Vigorous rootstock Least cold tolerant	Ferguson et al. (2016)
P. terebinthus	Efficient absorption of Cu, Zn and other micronutrients	Brown et al. (1994)
P. khinjuk	Hardy rootstock More drought resistant than *P. terebinthus* Suitable for much higher elevation and arid regions	Kaska (1995)
P. atlantica	Salt tolerant rootstock followed by 'UCB1' and *P. integerrima*	Ferguson et al. (2016)
P. atlantica and UCB1	More tolerant to *Phytophthora* spp. than *P. integerrima*	Ferguson et al. (2016)
UCB1	Highly susceptible to *M. phaseolina*	Nouri et al. (2020)
Chestnut		
'Maridonne' and 'Marlhac'	Combat root rot	Pereira-Lorenzo et al. (2016)
'Menzies' (*C. sativa* × *C. crenata*)	Resistance to root rot Used as a seedling rootstock in Australia	Avanzato (2009)
		Breisch (1992)

(continued)

Table 3.1 (continued)

Rootstock	Characteristics	References
Ferosacre CA90 Maraval CA74 Marigoule CA15 Marlhac CA118 Marsol CA07	French rootstocks resistant to ink disease (ranging from very high to low)	
CHR-162 (7521) CHR-151 (HS) CHR-168 (110) CHR-161 (100)	Spanish hybrid clones resistant to ink disease (ranging from very high to medium)	Pereira-Lorenzo and Fernandez-Lopez (1997)
Chilgoza		
Chir pine	2 year old chir pine rootstock and scion wood from medium sized tree produces better grafting success	Khan (2006)

incompatibility therefore, very few of them were used commonly. Chestnut production in Asia is mostly from *C. mollissima* and *C. crenata*, they are found to have natural resistance to root rot and chestnut blight but possess sensitivity to Asian chestnut gall wasp (*D. kuriphilus*) (Pereira-Lorenzo et al. 2012).

2.6 Almond

2.6.1 Vigor

The evaluation of different rootstocks such as Almond, Peach, GF 305 and Peach × Almond GF 677 on two varieties Ferragnes and Tuono revealed that almond as best rootstock in terms of qualitative characteristics; however, larger scion diameter was obtained on Peach rootstock (Barbera et al. 1993). Parvaneh et al. (2011) investigated three seed rootstocks such as bitter almond, peach seedling and sweet almond seedlings on three different Iranian almond cultivars to study the effect of rootstock on vegetative growth (plant height, trunk diameter and extension width). Results revealed that vegetative growth with rootstock of bitter almond was less than the other rootstocks, while peach rootstock showed more vegetative growth. Khadivi-Khub and Anjam (2016) evaluated rootstock-scion combinations of Rabiee grown on *P. scoparia* (a wild almond species) and Estahban (*P. dulcis*) rootstocks. Result of the study revealed that the vegetative characters of scion (Rabiee) including trunk diameter, tree height, internode length and annual year's growth was significantly affected by rootstock combinations. The lowest tree height was observed for trees grafted onto *P. scoparia* (110.00 cm), while the highest was observed in trees grafted on Estahban rootstock (516.00 cm). Thus, *P. scoparia* may reduce height of Rabiee scion and can be used as a dwarfing rootstock for almond.

2.6.2 Yield

Godini and Palasciano (1998) in a 13 year study observed that, among the three rootstocks such as GF 677, sweet almond seedling Don Carlo and GF 305, the lowest annual kernel yield was produced by GF 305 as compared to Don Carlo and GF 677. However, no difference in cumulative kernel yield was observed in all the years on rootstock GDF 677 and Don Carlo. Under non-irrigated conditions, *P. scorparia* rootstock have been found to produce significantly more nut yield than *P. dulcis* cv. 'Estahban', owing to greater tolerance exhibited by *P. scorparia* to drought stress (Khadivi-Khub and Anjam 2016).

2.6.3 Nut Quality

Khadivi-Khub and Anjam (2016) investigated rootstock-scion combinations of Rabiee grown on *P. scoparia* (a wild almond species) and Estahban (*P. dulcis*) rootstocks. The study revealed significant differences in nut width, length and kernel thickness on different rootstocks. Trees grafted on *P. scoparia* had higher nut length (35.66 mm), width (26.59 mm) and weight (4.97 g). In another study, Romero et al. (2017) grafted the almond cultivar Marinada on ten different rootstocks (INRA-GF-677, Garnem, Cadaman, Ishtara, IRTA-1, IRTA-2, Puebla del Soto, ROOTPAC-70-R, ROOTPAC-40 and ROOTPAC-20). It was concluded that vigorous rootstocks (Garnem, INRA-GF-677 and Cadaman) improved fruit quality but increased the risk of outer shell breakage. Dwarfing rootstocks (ROOTPACK-20, Ishtara and ROOTPAC-40), on the other hand, produced low quality kernels.

2.6.4 Nutrient and Water Uptake

Parvaneh et al. (2011) conducted a study on three seed rootstocks (bitter almond, peach seedling and sweet almond seedling) on almond cultivars to evaluate the effect of rootstock on the vegetative growth and absorption of nitrogen, phosphorus and potassium. Results indicate that rootstock had a significant impact on nutrient absorption – greatest amount of nitrogen absorption was obtained in peach rootstock, while the highest amount of potassium absorption was obtained in sweet almond rootstock.

Three prunus rootstocks were evaluated for iron chlorosis such as Adesoto (*Prunus insititia*), GF 677 (*Prunus amygdalus* × *Prunus persica*) and Barrier (*P. persica* × *Prunus davidiana*). The rootstock Adesoto showed higher total organic and amino acid concentrations, while the lower concentration of total amino acid and phosphoenolpyruvate carboxylase activity values were reported in Barrier rootstock. Thus, these results indicate the tolerance level of rootstocks to iron chlorosis under iron deficiency (Jimenez et al. 2011). Felinem, Monegro and Garnem are the three almond × peach hybrid rootstocks that can adapt to calcareous soils (Felipe 2009).

Almond and peach scions grafted on Marianna plum rootstock accumulated less Cl and showed less growth inhibition than those grafted on Lovell peach rootstock (Bernstein et al. 1956). In a comparison of rootstock effects on B tolerance, Hansen (1955) observed that French prune on Marianna plum or peach roots showed more stem injury than those on almond or Myrobalan plum rootstocks. As a result, almond rootstocks are better suited for areas with the problem of excessive boron.

2.6.5 Resistance to Abiotic Stress

Seedlings of Marcona are generally used as rootstocks to prevent frost injury (Mahhou and Dennis 1992). In wet and poorly drained soils, almond rootstocks are susceptible to nematode and fungal diseases, as well as root asphyxia. As a result, other rootstock species particularly plum, peach and their interspecific hybrids have been used. The two myrobalan (*Prunus cerasifera* Erhr) rootstocks are tolerant to waterlogging such as P.2175 and P.2980, while two almond × peach [*Prunus amygdalus* Batsch × *Prunus persica* (L.) Batsch] interspecific hybrids (Garnem and Felinem) are susceptible ones (Amador et al. 2012). Plum is also used as almond rootstock due to its more tolerance to pathogen and more resistance to waterlogging conditions than peach and almond rootstocks. Therefore, these rootstocks are more suitable for high bulk density, poorly aerated and fine textured soils (Rubio-Cabetas et al. 2017). Garnem, an almond × peach hybrid rootstock is well adapted to drought conditions as the abscisic acid (ABA) levels during drought conditions increased (Bielsa et al. 2019).

2.6.6 Resistance to Biotic Stress

In Italy, European plums such as Penta and Tetra are used as rootstocks because they have been found more resistant to root asphyxia and root parasites (nematodes, *Phytophthora*, *Armillaria*, *Agrobacterium*, etc.) than almond and peach seedlings (Rubio-Cabetas et al. 2017). In Mediterranean conditions, one of the major problems is of root-knot nematode, however, three almond × peach hybrid rootstocks (Felinem, Garnem and Monegro) are found to have resistance to root-knot nematodes (Felipe 2009). *P. scoparia* rootstock can be used as a source of resistance to fungus (*Aspergillus flavus*) and insects for cultivated almond (Gharaghani and Eshghi 2014). Furthermore, the rootstock and scion interaction is also one of the important factor for determining the production and productivity of temperate nuts. The compatibility of rootstocks and scions is shown in Table 3.2. Moreover, various studies on rootstocks and their characteristics has been conducted in different institutes of the world and are indexed in Table 3.3.

Table 3.2 Rootstock-scion compatibility

Rootstock	Scion	Compatibility	References
UCB1	*P. vera* (cv. 'Kerman')	Incompatible	Ferguson et al. (2016)
California black walnut and their hybrids	Persian walnut	Long-delayed incompatibility	Grant and McGranahan (2006)
Turkish filbert (*C. colurna*)	cv. Contorta	Incompatible	Cerovic et al. (2007)
Water Hickory	Pecan scion	Least successful grafting	Sitton and Dodge (1938)
C. illinoensis	*C. cathayensis*	High compatibility	Yao et al. (2019)
Corylus colurna L.	cv. Tonda Gentile Romana, Rimski, Istarski dugi, Cosford *Corylus avellana* cv. Atropurpurea	High compatibility	Cerovic et al. (2007), Ninic-Todorovic et al. (2012)
Chinese wingnut (*Pterocarya stenoptera*)	Chandler	Incompatible	Browne et al. (2011)

3 Conclusion

In the recent years, much progress has been made to develop suitable rootstocks for newly emerging challenges as a result of climatic change. These changes have increased the importance of development and selection of various rootstocks. Selection of rootstocks is an important aspect for the establishment of fruit orchard. Trends in the production of fruit trees include reduced plant size to accommodate more trees per unit area, reduced use of chemical fertilisers and pesticides, high yield and quality production, reduced losses due to various environmental stresses, etc. Rootstock breeding must focus on these trends. The breeding of rootstocks has many objectives. However, their application require more understanding on the nature of interaction between environment, rootstock and scion.

4 Future Strategies

- Considering that rootstocks are used for production of superior commercial cultivars, studies of the impact of rootstock on the growth of tree, quality and yield of fruits/nuts, compatibility of rootstock and scion, the response of plant to various environmental stresses, etc. are required to be done. Rootstock approaches will open new ways for easy propagation and genetic improvement of temperate nut crops in future.

Table 3.3 Institutes associated with rootstock research of nut crops

Institute	Rootstock	Characteristics	Reference
Walnut			
University of California, Davis	AX-1, GZ-1 and PX-1	Highly susceptible clones to *Phytophthora citricola*	Browne et al. (2006)
	AZ-2, AZ-3, NZ, JX-2, VX-211 and RX-1	Moderately tolerant clones to *Phytophthora citricola*	
Kearney Agricultural Center, Parlier, California	VX211	Grow roots faster, less nematodes attack	Buzo et al. (2009)
University of California, Davis	RX1 (*Juglans microcarpa* × *Juglans regia*)	Resistance to *Phytophthora*	Leslie and McGranahan (2014)
	VX211, very vigorous (*Juglans hindsii* × *J. regia*)	Nematode tolerance	
Hazelnut			
Instituto de Investigaciones Agropecuarias (INIA) in Temuco (Araucanía region, southern Chile)	BA-5 (clone of Chilean Barcelona)	Significant influence on plant growth and nut production	Ellena et al. (2014)
United States Department of Agriculture in Corvallis, Oregon	Two selections released as clonal rootstocks: 'Newberg' (USOR 7-71) and 'Dundee' (USOR 15-71)	Impart vigor to the scion cultivar, non-suckering	Lagerstedt (1993)
	C. colurna seedlings	Non-suckering, deep root system	Rovira (2021)
Chestnut			
Centro de Investigaciones Forestales de Lourizan, Xunta de Galicia, Spain	Hybrid clone CHR151	Resistant to ink disease, amenable to micropropagation	Pereira-Lorenzo and Fernandez-Lopez (1997))
Uludag University, Turkey	554-1, SE 18-2, SE 21-9, SE 23-9 and SA 5-1	Selected as seedling rootstock	Soylu and Serdar (2000)
Pistachio nut			
University of California, Davis	*P. atlantica*	Superior in enhancing leaf concentrations of the elements, boron (B), copper (Cu), zinc (Zn) and phosphorus (P) from a range of soil types	Brown et al. (1994)
University of California, Berkeley	UCB-1		Ferguson et al. (2002)

(continued)

Table 3.3 (continued)

Institute	Rootstock	Characteristics	Reference
		Moderate resistance to Verticillium wilt, frost tolerance	
University of Tehran, Iran	*P. khinjuk*	Resistant to harsh conditions, such as dry weather and salinity	Behboodi (2003)
Pistachio Research Institute, Rafsanjan, Iran	Badami seedling rootstock	Greatest cumulative production and yield efficiency, lowest blankness of nuts and highest percentage of splitting of nuts	Rahemi and Tavallali (2007)
University of Harran Turkey	*P. atlantica*	High shell splitting rate and total filled nuts rate (%)	Turker and Ak (2010)
Pecan			
LSU Pecan Research Extension Station, Shreveport	*C. lecontei*	More graft success	Grauke and O'Barr (1996)
USDA	Apache seedstock (Burkett × Schley)	Trees usually yield well, and nuts are filled	Grauke (2010)
USDA-ARS, Brownwood, Texas	Wichita seedstock (Halbert × Mahan)	Nuts tend to be well filled, germinate well and seedlings grow vigorously	Grauke (2010)

- It can be achieved through the integration of breeding with new technologies such as use of the Molecular markers to identify useful traits in the rootstocks, reducing the time and cost involved in long term field trials.
- Improvement in the procedure for screening of rootstocks is to be assured which in turn helps in the identification of useful characteristics of these rootstocks.
- Development of transgenic rootstocks should be undertaken to transfer transgene product into the scion. Transgenic rootstock development will facilitate the transformation of only few rootstocks as compared to the need of transforming large number of scion cultivars.
- Emphasis should be given on the influence of wild species of temperate nuts as rootstocks to screen them for biotic and abiotic stresses.
- Clonal propagation of rootstocks for easy multiplication hold key for exploiting the beneficial effects of rootstocks for commercialization of nuts to boost horticulture-based economy of the region.

References

Ak BE, Turker S (2006) Effects of different rootstocks on phenological stages and vegetative growth of some pistachio cultivars. Acta Hortic 726:209–214

Albacete A, Martinez-Andujar C, Martinez-Perez A, Thompson AJ, Dodd IC, Perez-Alfocea F (2015) Unravelling rootstock×scion interactions to improve food security. J Exp Bot 66(8): 2211–2226

Aloni B, Cohen R, Karni L, Aktas H, Edelstein M (2010) Hormonal signaling in rootstock-scion interactions. Sci Hortic 127(2):119–126

Amador ML, Sancho S, Bielsa B, Gomez-Aparisi J, Rubio-Cabetas MJ (2012) Physiological and biochemical parameters controlling waterlogging stress tolerance in Prunus before and after drainage. Physiol Plant 144(4):357–368

Anagnostakis SL, Gordon P, Hebard FV (1998) Identification of American chestnut trees. Annu Rep Northern Nut Grower Assoc 89:1–4

Arab MM, Marrano A, Abdollahi-Arpanahi R, Leslie CA, Cheng H, Neale DB, Vahdati K (2020) Combining phenotype, genotype, and environment to uncover genetic components underlying water use efficiency in Persian walnut. J Exp Bot 71(3):1107–1127

Avanzato D (2009) Following chestnut footprints (*Castanea spp.*) cultivation and culture, folklore and history, tradition and uses. International Society for Horticultural Science (ISHS), Leuven, Belgium

Barbera G, Di Marco L, La Mantia T, Schirra M (1993) Effect of rootstock on productive and qualitative response of two almond varieties. Int Congr Almond 373:129–134

Barone E, Sottile F, Palazzolo E, Caruso T (1997) Effect of rootstock on trunk growth and foliar mineral content in cv. Bianca pistachio (*Pistacia vera* L.) trees. Acta Hortic 470:394–401

Behboodi B (2003) Ecological distribution study of wild pistachios for selection of rootstock. Options Mediterr Ser A 63:61–67

Bernstein L, Brown JW, Hayward HE (1956) The influence of rootstock on growth and salt accumulation in stone-fruit trees and almonds. Proc Am Soc Hortic Sci 68:86–95

Bielsa B, Sanz MÁ, Rubio-Cabetas MJ (2019) Uncovering early response to drought by proteomic, physiological and biochemical changes in the almond × peach rootstock 'Garnem'. Funct Plant Biol 46(11):994–1008

Botu M, Achim G, Baciu A, Alecu A, Manthos I (2014) 'Romval' and 'Casval'-one cultivar and one rootstock for sweet chestnut. Acta Hortic 1043:139–144

Breisch H (1992) Compatibility tests between the main French varieties of chestnut trees and ink resistant hybrid rootstocks. In: World Chestnut Industry conference, Morgantown, West Virginia, pp 41–53

Brown PH, Zhang Q, Ferguson L (1994) Influence of rootstock on nutrient acquisition by pistachio. J Plant Nutr 17(7):1137–1148

Browne L, Brown LC, Ramos D (1977) Walnut rootstocks compared. Calif Agric 31(7):14–16

Browne GT, McLaughlin ST, Hacket WP, McGranahan GH, Leslie CA (2006) Evaluation of resistance to Phytophthora citricola among diverse clones of paradox hybrid rootstocks. Acta Hortic 705:395–400

Browne GT, Grant JA, Schmidt LS, Leslie CA, McGranahan GH (2011) Resistance to Phytophthora and graft compatibility with Persian walnut among selections of Chinese wingnut. HortScience 46:371–376

Browne G, Bhat R, Schmidt L, Leslie C, Michailides T, Hackett W, Robinson R, Kluepfel D, Aradhya M, Beede B, Buchner R, Connell J, Fichtner E, Grant J, Hasey J (2012) Etiology and management of crown and root rots of walnut. Walnut Research Reports. California Walnut Board, University of California, Davis

Browne GT, Leslie CA, Grant JA, Bhat RG, Schmidt LS, Hackett WP, Kluepfel DA, Robinson R, McGranahan GH (2015) Resistance to species of phytophthora identified among clones of juglans microcarpa × J. Regia. HortScience 50(8):1136–1142

Buzo T, Mckenna J, Kaku S, Anwar SA, Mckenry MV (2009) VX211, A vigorous new walnut hybrid clone with nematode tolerance and a useful resistance mechanism. J Nematol 41(3): 211–216

Camison A, Angela Martin M, Dorado FJ, Moreno G, Solla A (2020) Changes in carbohydrates induced by drought and waterlogging in *Castanea sativa*. Trees 34(2):579–591

Camison A, Martin MA, Flors V, Sanchez-Bel P, Pinto G, Vivas M, Rolo V, Solla A (2021) Exploring the use of scions and rootstocks from xeric areas to improve drought tolerance in Castanea sativa Miller. Environ Exp Bot 187:1–10

Cao F, Wei Y, Wang X, Li Y, Peng F (2019) A study of the evaluation of the pecan drought resistance of grafted "Pawnee" trees from different seedling rootstocks. HortScience 54(12): 2139–2145

Caprile J, Grattan S (2006) Evaluation of the salt tolerance of walnut rootstocks. HortScience 41(4): 1040

Carbonell-Barrachina AA, Memmi H, Noguera-Artiaga L, Gijon-Lopez MDC, Ciapa R, Perez-Lopez D (2015) Quality attributes of pistachio nuts as affected by rootstock and deficit irrigation. J Sci Food Agric 95(14):2866–2873

Carroll B, Smith MW (2017) Pecan varieties for Oklahoma. Division of Agricultural Sciences and Natural Resources, Oklahoma State University

Caruso T, Barone E, Marra FP, Sottile F, La Mantia M, De Pasquale C (2005) Effect of rootstock on growth, yield and fruit characteristics in cv. Bianca pistachio (*Pistacia vera* L.) trees. In: XIII GREMPA meeting on pistachios and almond

Cerovic S, Ninic-Todorovic J, Golosin B, Ognjanov V, Bijelic S (2007) Production technology of young hazelnut trees grafted on turkish filbert (*Corylus colurna*). Acta Hortic 732:355–357

Conedera M, Barthold F, Torriani D, Pezzatti GB (2010) Drought sensitivity of *Castanea sativa*: case study of summer 2003 in the Southern Alps. Acta Hortic 866:297–302

Connell JH, Olson WH, Limberg J, Metcalf SG (2010) Effects of various roots on Chandler walnut catkin and pistillate bloom, tree growth, yield, and nut quality. Acta Hortic 861:237–244

Ellena M, Sandoval P, González A (2014) Effect of type of propagation on earliness of flowering and fruiting on 'Tonda di Giffoni' and 'Daviana' cultivars. Acta Hortic 1052:221–224

Epstein L, Beede R, Kaur S, Ferguson L (2004) Rootstock effects on pistachio trees grown in *Verticillium dahliae* infested soil. Phytopathology 94(4):388–395

Fabrizio GC, Van der Watt E, Gesine MC (2018) Propagation of pecan (*Carya illinoensis*): a review. Afr J Biotechnol 17(18):586–605

Fallahi E, Chun IJ, Neilsen GH, Colt WM (2001) Effects of three rootstocks on photosynthesis, leaf mineral nutrition, and vegetative growth of "BC-2 Fuji" apple trees. J Plant Nutr 24(6):827–834

Felipe AJ (2009) 'Felinem', 'Garnem', and 'Monegro' almond × peach hybrid rootstocks. HortScience 44(1):196–197

Ferguson L, Beede RH, Reyes H, Metheney P (2002) California pistachio rootstock evaluations. Acta Hortic 591:63–66

Ferguson L, Beede RH, Reyes H, Sanden BL, Grattan SR, Epstein L (2016) Pistachio rootstocks. In: Ferguson L, Haviland D (eds) Pistachio production manual, vol 3545. University of California Division of Agriculture and Natural Resources Publications, California, pp 65–74

Gervasi T, Barreca D, Lagana G, Mandalari G (2021) Health benefits related to tree nut consumption and their bioactive compounds. Int J Mol Sci 22(11):5960

Gharaghani A, Eshghi S (2014) *Prunus scoparia*, a potentially multi-purpose wild almond species in Iran. In: II International symposium on wild relatives of subtropical and temperate fruit and nut crops, 1074, pp 67–72

Gijon MDC, Gimenez C, Perez-Lopez D, Guerrero J, Couceiro JF, Moriana A (2010) Rootstock influences the response of pistachio (*Pistacia vera* L. cv. Kerman) to water stress and rehydration. Sci Hortic 125(4):666–671

Godini A, Palasciano M (1998) Growth and yield of four self-fruitful and four self-unfruitful almonds on three rootstocks: a thirteen-year study. In: II International symposium on pistachios and almonds, 470, pp 200–207

Goldschmidt EE (2014) Plant grafting: new mechanisms, evolutionary implications. Front Plant Sci 5:727

Grant J (2015) Performance of clonal paradox and blackline tolerant walnut rootstock in San Joaquin Country. Walnut Research Report. California Walnut Board, University of California, Davis, pp 1–3

Grant JA, McGranahan GH (2006) Orchard performance of "Chandler" walnut on selected *Juglans regia* seedling rootstocks. Acta Hortic 705:365–370

Grauke LJ (2010) Pecan seed stock selection - regional implications. Proc SE Pecan Grow Assoc 103:42–51

Grauke LJ, O'Barr RD (1996) Initial survival of pecan grafts on seedling rootstocks of pecan, water hickory, and their interspecific hybrid. HortTechnology 6(1):45–48

Grauke LJ, Pratt JW (2019) Pecan bud growth and freeze damage are influenced by rootstock. J Am Soc Hortic Sci 117(3):404–406

Grauke LJ, Starr JL (2014) Phenotypic screening of pecan seedling rootstocks in search of nematode resistance. Trees 28(5):1333–1341

Gregory PJ, Atkinson CJ, Bengough AG, Else MA, Fernandez-Fernandez F, Harrison RJ, Schmidt S (2013) Contributions of roots and rootstocks to sustainable, intensified crop production. J Exp Bot 64(5):1209–1222

Hansen CJ (1955) Influence of rootstock on injury from excess boron in Nonpareil almond and Elberta peach. Proc Am Soc Hortic Sci 65:128–132

Hasey JK, Westerdahl BB, Lampinen B (2004) Long-term performance of own-rooted Chandler walnut compared to Chandler walnut on paradox rootstock. Acta Hortic 636:83–87

Holtz BA, Teviotdale BL (2016) Soil-borne diseases. In: Ferguson L, Haviland D (eds) Pistachio production manual. University of California Agriculture and Natural Resources, Davis, pp 257–264

INC (2020) International nut and dried fruit council. Statistical Yearbook 2019–2020

Jamshidi Goharrizi K, Amirmahani F, Salehi F (2020a) Assessment of changes in physiological and biochemical traits in four pistachio rootstocks under drought, salinity and drought + salinity stresses. Physiol Plant 168(4):973–989

Jamshidi Goharrizi K, Baghizadeh A, Kalantar M, Fatehi F (2020b) Combined effects of salinity and drought on physiological and biochemical characteristics of pistachio rootstocks. Sci Hortic 261:108970

Jaynes RA (1979) Chestnuts. In: Jaynes RA (ed) Nut tree culture in North America. Northern Nut Growers Association, Hamden

Jimenez S, Ollat N, Deborde C, Maucourt M, Rellan-Alvarez R, Moreno MA, Gogorcena Y (2011) Metabolic response in roots of Prunus rootstocks submitted to iron chlorosis. J Plant Physiol 168(5):415–423

Kallsen CE, Parfitt DE (2011) Comparisons of scion/rootstock growth rates among U.S. pistachio cultivars. HortScience 46(2):197–200

Kaska N (1995) Pistachio nut production growing in Turkey. Acta Hortic 419:161–164

Khadivi-Khub A, Anjam K (2016) Prunus scoparia, a suitable rootstock for almond (*Prunus dulcis*) under drought condition based on vegetative and fruit characteristics. Sci Hortic 210:220–226

Khan M (2006) Inter-species grafting of chalghoza pine scion-wood on chir pine root-stock. Pak J For 56:1

Knipfer T, Barrios-Masias FH, Cuneo IF, Bouda M, Albuquerque CP, Brodersen CR, Kluepfel DA, McElrone AJ (2018) Variations in xylem embolism susceptibility under drought between intact saplings of three walnut species. Tree Physiol 38(8):1180–1192

Knipfer T, Reyes C, Momayyezi M, Brown PJ, Kluepfel D, McElrone AJ (2020) A comparative study on physiological responses to drought in walnut genotypes (RX1, Vlach, VX211) commercially available as rootstocks. Trees 34(3):665–678

Kurum R, Ulukapi K, Aydinşakir K, Onus AN (2013) The influence of salinity on seedling growth of some pumpkin varieties used as rootstock. Notulae Botanicae Horti Agrobotanici Cluj-Napoca 41(1):219–225

Lagerstedt HB (1993) Newberg and Dundee, two new filbert rootstocks. Proc Nut Grow Soc Oregon 78:94–101

Lee JM, Kubota C, Tsao SJ, Bie Z, Echevarria PH, Morra L, Oda M (2010) Current status of vegetable grafting: diffusion, grafting techniques, automation. Sci Hortic 127(2):93–105

Leslie CA, McGranahan GH (2014) The California walnut improvement program: scion breeding and rootstock development. Acta Hortic 1050:81–88

Liu B, Liang J, Tang G, Wang X, Liu F, Zhao D (2019) Drought stress affects on growth, water use efficiency, gas exchange and chlorophyll fluorescence of Juglans rootstocks. Sci Hortic 250: 230–235

Liu Z, Li F, Peng F, Tan P, Zhu K, Feng G, Mo Z, Li Y (2020) Identification of grafting-responsive MicroRNAs associated with growth regulation in pecan [*Carya illinoinensis* (Wangenh.) K. Koch]. Forests 11(2):1–16

Lotfi N, Vahdati K, Kholdebarin B, Ashrafi EN (2009) Germination, mineral composition, and ion uptake in walnut under salinity conditions. HortScience 44(5):1352–1357

Lotfi N, Soleimani A, Vahdati K, Cakmakci R (2019) Comprehensive biochemical insights into the seed germination of walnut under drought stress. Sci Hortic 250:329–343

Mahhou A, Dennis FG (1992) The almond in Morocco. HortTechnology 2(4):488–492

Martinez-Ballesta MC, Alcaraz-Lopez C, Muries B, Mota-Cadenas C, Carvajal M (2010) Physiological aspects of rootstock-scion interactions. Sci Hortic 127(2):112–118

Miletic R, Mitrovic M, Rakicevic M (2009) Contrasting fruit properties of hazelnut cultivars grown on different rootstocks. Acta Hortic 845:283–286

Miranda ME, Fernandez J (1992) Micropropagation as a nursery technique for chestnut hybrid clones. In: International chestnut conference. West Virginia University Press, Morgantown, USA, pp 101–103

Mircetich SM, Sanborn RR, Ramos DE (1980) Natural spread, graft-transmission, and possible etiology of walnut blackline disease. Phytopathology 70:962–968

Mirotadze N, Gogitidze V, Mikadze N, Goginava L, Mirotadze M (2009) Agro-ecological zones of hazelnut in Georgia. Acta Hortic 845:291–294

Miyamoto S, Gobran GR, Piela K (1985) Salt effects on seedling growth and ion uptake of three pecan rootstock cultivars. Agron J 77(3):383–388

Moriana A, Memmi H, Centeno A, Martin-Palomo MJ, Corell M, Torrecillas A, Perez-Lopez D (2018) Influence of rootstock on pistachio (*Pistacia vera* L. cv. Kerman) water relations. Agric Water Manag 202:263–270

Nawaz MA, Imtiaz M, Kong Q, Cheng F, Ahmed W, Huang Y, Bie Z (2016) Grafting: a technique to modify ion accumulation in horticultural crops. Front Plant Sci 7:1457

Ninic-Todorovic J, Cerovic S, Ognjanov V, Golosin B, Bijelic S, Jacimovic G, Kurjakov A (2009) Rootstocks of *Corylus colurna* L. for nursery production. Acta Hortic 845:273–278

Ninic-Todorovic J, Ognjanov V, Keserovic Z, Cerovic S, Bijelic S, Cukanovic J, Kurjakov A, Cabilovski R (2012) Turkish hazel (*Corylus colurna* L.) offspring variability as a foundation for grafting rootstock production. Bulgarian J Agr Sci 18(6):883–888

Nouri MT, Lawrence DP, Kallsen CE, Trouillas FP (2020) Macrophomina crown and root rot of pistachio in California. Plants 9(2):1–17

Parvaneh T, Afshari H, Ebadi A (2011) A study of the influence of different rootstocks on the vegetative growth of almond cultivar. Afr J Biotechnol 10(74):16793–16807

Pereira-Lorenzo S, Fernandez-Lopez J (1997) Propagation of chestnut cultivars by grafting: methods, rootstocks and plant quality. J Hortic Sci 72:731–739

Pereira-Lorenzo S, Ballester A, Corredoira E, Vieitez AM, Agnanostakis S, Costa R, Bounous G, Botta R, Beccaro LG, Kubisiak TL, Conedera M, Krebs P, Yamamoto T, Sawamura Y, Takada N, Gomes-Laranjo J, Ramos-Cabrer AM (2012) Chestnut. In: Badenes ML, Byrne DH (eds) Fruit breeding. Springer, Boston, pp 729–769

Pereira-Lorenzo S, Costa R, Anagnostakis S, Serdar U, Yamamoto T, Saito T, Ramos-Cabrer AM, Ling Q, Barreneche T, Robin C, Rotta R, Contessa C, Conedera M, Martin A, Gomes-Laranjo J,

Villani F, Carlson JE (2016) Interspecific hybridization of chestnut. In: Mason AS (ed) Polyploidy and hybridization for crop improvement. CRC Press, Boca Raton, pp 377–407

Prassinos C, Ko JH, Lang G, Iezzoni AF, Han KH (2009) Rootstock-induced dwarfing in cherries is caused by differential cessation of terminal meristem growth and is triggered by rootstock-specific gene regulation. Tree Physiol 29(7):927–936

Rahemi M, Tavallali V (2007) Effects of rootstock on Iranian pistachio scion cultivars. Fruits 62(5): 317–323

Reid W (1997) Rootstock influences yield, nut quality, and leaf analysis of pecan trees. HortScience 32:474–475

Reil W, Sibbett S, Ramos D (1992) Walnut cultivar nutritional evaluation. Annual walnut research report. California Walnut Board, University of California, Davis, pp 1–13

Rezaee R, Vahdat K (2014) Morphological variation, heritability and phenotypic correlation of traits related to the vigor in Persian walnut (Juglans regia L.). J Crop Prod Process 4(12): 259–271

Romero A, Batlle I, Miarnau X (2017) Almond physical traits affected by rootstocks in 'Marinada' cultivar. VII Int Symp Almonds Pistachios 1219:31–36

Rovira M (2021) Advances in hazelnut (*Corylus avellana* L.) rootstocks worldwide. Horticulturae 7(9):267

Rovira M, Cristofori V, Silvestri C, Celli T, Hermoso JF, Tous J, Romero A (2014) Last results in the evaluation of "Negret" hazelnut cultivar grafted on non-suckering rootstocks in Spain. Acta Hortic 1052:145–150

Rubio-Cabetas MJ, Felipe AJ, Reighard GL (2017) Rootstock development. In: Socias i Company, Gradziel TM (eds) Almonds. Botany, production and uses, pp 209–227

Salimi S, Hoseinova S (2012) Selecting hazelnut. Crop Breed J 2(2):139–144

Sanderlin RS (2015) Susceptibility of some common pecan rootstocks to infection by Xylella fastidiosa. HortScience 50(8):1183–1186

Schwarz D, Rouphael Y, Colla G, Venema JH (2010) Grafting as a tool to improve tolerance of vegetables to abiotic stresses: thermal stress, water stress and organic pollutants. Sci Hortic 127(2):162–171

Shaybany B, Martin GC (1977) Abscisic acid identification and its quantitation in leaves of Juglans seedlings during waterlogging. J Am Soc Hortic Sci 12(3):300–302

Sheikh Beig Goharrizi MA, Dejahang A, Tohidfar M, Izadi Darbandi A, Carillo N, Hajirezaei MR, Vahdati K (2016) Agrobacterium mediated transformation of somatic embryos of Persian walnut using flavodoxin (fld) gene for osmotic stress tolerance. J Agric Sci Technol 18(2): 423–435

Sheikhi A, Mirdehghan SH, Ferguson L (2019) Extending storage potential of de-hulled fresh pistachios in passive-modified atmosphere. J Sci Food Agric 99(7):3426–3433

Sitton BG, Dodge FN (1938) Growth and fruiting of three varieties of pecan on different seedling rootstock. J Am Soc Hortic Sci 38:121–125

Smith MW (2002) Damage by early autumn freeze varies with pecan cultivar. HortScience 37(2): 398–401

Smith MW, Cheary BS, Carroll BL (2001) Rootstock and scion affect cold injury of young pecan trees. J Am Pomol Soc 55(2):124–128

Soylu A, Serdar U (2000) Rootstock selection on chestnut (*Castanea sativa* Mill.) in the middle of black sea region in Turkey. Acta Hortic 538:483–487

Sunkar R, Li YF, Jagadeeswaran G (2012) Functions of microRNAs in plant stress responses. Trends Plant Sci 17(4):196–203

Surucu A, Acar I, Demirkiran AR, Farooq S, Gokmen V (2020) Variations in nutrient uptake, yield and nut quality of different pistachio cultivars grafted on *Pistacia khinjuk* rootstock. Sci Hortic 260:108913

Tajabadipour A, Panahi B, Zadehparizi R (2006) The effects of rootstock and scion on early splitting and cracked hull of pistachio. Acta Hortic 726:193–198

Tavallali V, Rahemi M (2007) Effects of rootstock on nutrient acquisition by leaf, kernel and quality of pistachio (*Pistacia vera* L.). Am Eurasian J Agric Environ Sci 2(3):240–246

Thomas AL, Reid WR (2006) Hardiness of black walnut and pecan cultivars in response to an early hard freeze. J Am Pomol Soc 60:90–94

Tous J, Girona J, Tasias J (1994) Cultural practices and costs in hazelnut production. Acta Hortic 351:395–418

Turker S, Ak BE (2010) Effects of different rootstocks on physical traits of Siirt and Ohadi pistachio cultivars. Options Mediterranean's 94:245–250

Vahdati K, Mohseniazar M (2016) Early bearing genotypes of walnut: a suitable material for breeding and high density orchards. Acta Hortic 1139:101–105

Wang GA, Zhang Q, Huang MM, Yakup A (2014) The breeding of six Xinjiang dwarf walnut cultivars. Acta Hortic 1050:151–160

Warmund MR (2011) Chinese chestnut (*Castanea mollissima*) as a niche crop in the central region of the United States. HortScience 46(3):345–347

Wells L (2014) Budding and grafting of pecan. Univ Georgia Cooperative Ext Bull 1376:1–8

Wood BW, Reilly CC (2001) A typical symptoms of cold damage to pecan. HortScience 36(2): 298–301

Xue TT, Liu J, Shen YB, Liu GQ (2018) The microstructure of the suture in the pecan endocarp. J Hortic Sci Biotechnol 93(4):441–448

Yao J, Mei L, Jiang H, Hu G, Wang Y (2019) Evaluation of *Carya cathayensis* resistance to Botryosphaeria trunk canker using grafting on pecan. Sci Hortic 248:184–188

Chapter 4
Cultivars and Genetic Improvement

Kourosh Vahdati, Abdollatif Sheikhi, Mohammad Mehdi Arab, Saadat Sarikhani, Asaad Habibi, and Hojjat Ataee

Abstract In recent years, temperate nut tree breeding programmes have been accelerated as a result of the increasing demand for their production. High yield with desirable kernel characteristics and disease-resistance are the main breeding objectives of temperate nut trees cultivars. In addition, leading challenges especially global warming and climate change dictate new objectives such as low-chilling requirements and tolerance to late-spring and early-autumn frosts, etc. Various conventional and molecular strategies are used to achieve these breeding objectives. The utilisation of genetic resources and targeted hybridisation are the main conventional breeding strategies of temperate nut trees which along with molecular breeding can help to accelerate the breeding programmes. The advancement of genome sequencing technologies, high-quality and quantity of molecular data accelerated breeding programmes and lowered their price for temperate tree nut crops. Therefore, future researches in breeding of tree nut trees should be directed towards construction of fully annotated reference genomes, discovery of genes controlling traits of interest and finding molecular markers linked to these genes to be used for marker assisted selection (MAS) or genomic selection (GS), through constructing saturated linkage maps, QTLs mapping, genome-wide association mapping, along with omics techniques. In this chapter, some important breeding objectives and strategies of temperate nut trees including walnuts, pistachios, almonds, pecans, hazelnuts and chestnut have been discussed with emphasis on recent findings.

K. Vahdati (✉) · M. M. Arab · S. Sarikhani · A. Habibi
Department of Horticulture, College of Aburaihan, University of Tehran, Tehran, Iran

A. Sheikhi
Department of Horticulture, Faculty of Agriculture, Vali-e-Asr University of Rafsanjan, Rafsanjan, Kerman, Iran

H. Ataee
Department of Horticultural Sciences, Gorgan University of Agricultural Sciences and Natural Resources, Gorgan, Iran

M. M. Mir et al. (eds.), *Temperate Nuts*,
https://doi.org/10.1007/978-981-19-9497-5_4

Keywords Almond · Chestnut · Conventional breeding · Hazelnut · Molecular breeding · Pecan · Pistachio · Walnut

1 Introduction

As a source of unsaturated fatty acid, proteins, minerals, vitamins and phytochemicals, nut trees are considered nutrient-rich food for human health (Habibi et al. 2017). Therefore, the demand for their production is significantly increasing due to the high economic value and ease of storage and handling (Habibi et al. 2021; Vahdati et al. 2021). The development of nut trees in different climate conditions and increasing the negative consequences of climate change have led to run different nut tree breeding programmes worldwide to improve the commercial and compatible rootstocks and cultivars. As a result of these breeding programmes, different nut tree cultivars and rootstocks have been released, as listed in previous reports (Al-Khayri et al. 2019; Vahdati et al. 2021). Some of the most important and newest cultivars and rootstocks of nut trees with their characteristics are presented in Table 4.1. Advanced strategies for improving cultivars and rootstocks of nut trees have been reviewed in previous studies (Mehlenbacher 2003; Al-Khayri et al. 2019; Vahdati et al. 2021). However, new results have been achieved due to the high progress of breeding programmes. In this chapter, some important breeding objectives and strategies of nut trees including walnuts, pistachios, almonds, pecans, hazelnuts and chestnut are mentioned with emphasis on recent findings. This chapter consists of four sections. The first section demonstrates the important breeding objectives of temperate nut trees rootstocks and cultivars. The second and third sections deal with traditional and molecular breeding strategies, respectively. At the end, conclusions and prospects for nut trees breeding are presented.

2 Breeding Objectives

The general objectives of temperate nut trees breeding are to release productive scion with high kernel quality and disease-resistance, along with rootstocks tolerance to biotic and abiotic stresses. In other words, high yield and kernel quality are the most important objectives of temperate nut trees (Al-Khayri et al. 2019). Of course, leading challenges, such as the negative consequences of climate change, dictate new objectives to maintain and develop the cultivation area and production, but high yield and crop quality are the main objectives for all crops (Habibie et al. 2017, 2019; Vahdati et al. 2019).

In addition to high yield and quality, late-leafing date, early harvest, lateral bearing, precocious production, ease of kernel removal, thin shell thickness and disease resistance (blight, anthracnose) are the main walnut cultivars breeding objectives (Bernard et al. 2018; Vahdati et al. 2019). The main breeding objectives and traits of pistachio include high yield and nut quality, reduction in the percentage

Table 4.1 Some of the main and new cultivars of nut trees around the world

Cultivars	Main characteristics
Almond	
Nonpareil	Self-incompatible, soft-shell, the attractive kernel of medium size, uniform shape, light-coloured skin
Carmel	High-quality nut, good shell seal, soft-shell, medium-sized nuts, light colour and sweet flavour
Avalon	Self-incompatible, medium kernel, early blooming, paper-shell, harvest approx. 8 days after 'Nonpareil'
Sweetheart	Self-compatible, mid-blooming, large kernel, semi-softshell, resistant to postharvest worm damage
Winters	Self-incompatible, early blooming, large kernel, paper-shell, good bloom overlaps with early 'Nonpareil' bloom
Guara	Self-compatible, late flowering, early ripening, high yield
Felisia	Self-compatible, very late-blooming, medium-hard shell, small kernel, very low alternate, early-medium ripening
Lauranne	Self-compatible, medium-hard shell, medium vigor, late-blooming, early-medium ripening, some double kernels
Mandaline	Self-compatible, late-blooming, medium ripening, hard shell, medium to upright growth
Shefa	Self-incompatible, vigorous, early-blooming, highly adapted to Israel conditions, softshell, large kernel, early ripening
Penta	Self-compatible, extra-late flowering, medium vigor, hard shell, early ripening
Tardona	Self-compatible, ultra-late flowering (the latest flowering almond cultivar ever released), hard shell, small kernel, medium vigor
Vairo	Self-compatible, late flowering, vigorous, tolerant to 'fusicoccum' (Phomopsis amygdali) and 'red leaf blotch' (Polystigma ochraceum)
Makako	Self-compatible, extra-late flowering, vigorous, large kernel
Marta	Self-compatible, late flowering, vigorous, hard shell, medium ripening
Marinada	Self-compatible, late flowering, medium growth, medium ripening, highly productive, very precocious
Soleta	Self-compatible, late flowering, large kernel, average-late ripening, high productivity
Marcona	Self-incompatible, small, precocious habit, large trees, exceptional quality due to its higher oil content and smooth texture
Isabellona	Self-compatible, late flowering, early harvest, vigorous, average nut size, high productivity
Walnut	
Fernor	High yield (less than 'Chandler'), lateral bearing, late leafing, extra light kernel colour
Ferbel	High yield, lateral bearing, large nut, thin shell, good kernel quality
Ferouette	High yield, lateral bearing, extra light kernel colour, large nut
Feradam	High yield, lateral bearing, extra light kernel colour, medium leafing
Fertignac	High yield, lateral bearing, thin shell, extra light kernel colour
Chandler	High yield, lateral bearing, medium leafing, extra light kernel colour, thin shell
Howard	High yield, lateral bearing, medium leafing, large nut, thin shell, light kernel colour

(continued)

Table 4.1 (continued)

Cultivars	Main characteristics
Tulare	High yield, lateral bearing, large nut, shell ticker than Chandler, light amber kernel colour
Ivanhoe	High yield, lateral bearing, very precocious, very early harvest, thin shell, extra light kernel colour
Solano	High yield, lateral bearing, extra light kernel colour
Durham	Good yield, early harvest, lateral bearing, large nut, light kernel colour
UC Wolfskill	High yield, early harvest, light kernel colour, thin shell
Persia	High yield, lateral bearing, late leafing, extra light kernel colour, medium to early harvest
Caspian	High yield, lateral bearing, late leafing, extra light kernel colour, medium to early harvest
Chaldoran	High yield, lateral bearing, precocious, amber kernel, early harvest
Alvand	High yield, lateral bearing, early to medium leafing, light kernel colour, low vigor, early harvest, relatively sensitive to anthracnose, bark canker and blight
Maras 18	Moderate yield, lateral bearing, light kernel colour, high kernel percentage, very early harvest
Sütyemez 1	Moderate yield, lateral bearing, extra-large nut, light kernel colour, very early harvest
Kaman 1	High yield, lateral bearing, thin shell, light kernel colour
Diriliş	High yield, lateral bearing, late leafing, light kernel colour, high kernel percentage, thin shell, early harvest
15 Temmuz	High yield, lateral bearing, very late leafing, light kernel colour, high kernel percentage thin shell, early harvest
Pistachio	
Male	
Peters	Large quantity of pollens, 2 weeks pollen shed, irregular flowering during low-chill years
Randy	Early flowering, long bloom period, suitable for early flowering cultivars such as Golden Hills and Lost Hills
Famoso	Large quantity of pollens, synchronised bloom period with Kerman scion cultivar
Tejon	Large quantity of pollens, early flowering, synchronised bloom period with Gumdrop scion cultivar
Female	
Kerman	High nut quality and productivity, high alternate bearing tendency, high blank percentage, high unsplit percentage
Golden Hills	Early flowering and fruit ripening, high yield, low unsplit and blank nuts percentage, balanced shell-hinge strength
Lost Hills	Early flowering and fruit ripening, large nut size, high yield, low unsplit nuts percentage, low shell-hinge strength
Gumdrop	Earlier flowering and ripening compared to Golden Hills and Lost Hills
Akbari	Large nut size, high yield and fruit quality, late flowering and fruit ripening, susceptible to pistachio psylla
Ahmad-Aghaei	Large nut size, high yield and fruit quality, a high alternate bearing tendency

(continued)

Table 4.1 (continued)

Cultivars	Main characteristics
Kalle-Ghoochi	Large nut size, high yield, early flowering, susceptible to spring-frost
Ohadi (Fandoghi)	Early ripening, low alternate bearing tendency, tolerant to pistachio psylla (*A. pistaciae*)
Hazelnut	
McDonald	Resistance to eastern filbert blight (EFB), high yield, high kernel percentage, early nut maturity, excellent kernel quality
Tonda Pacifica	Smaller trees, high yield, smaller nuts and kernels, high kernel percentage, early nut maturity, excellent kernel quality
Yamhill	Resistance to EFB, high yield, early nut maturity, high kernel percentage, good kernel quality
Sacajawea	Resistance to EFB, excellent kernel quality and acceptable nut yields, slightly smaller trees, earlier nut maturity
Dorris	High resistant to EFB, resistant to bud mites, high kernel quality, large kernel size, high yield
Felix	High resistant to EFB, low yield
Webster	High resistance to EFB and bud mite, non-suckering
Eta	High resistance to EFB, moderately vigorous, small nut size and late maturity
Pecan	
Wichita	Vigorous, often with a late flush of growth, very susceptible to scab
Mahan	Large nut size, thin shell, high kernel percentage
Pawnee	Large nut size, early harvest, resistance to aphids, high kernel and nut quality, susceptible to scab
Avalon	Scab resistant, late leafing, early to mid-season harvesting, susceptibility to black pecan aphid
Byrd	Large size nut, very early harvest date
Huffman	Resistant to insect, highly resistant to scab disease
Morrill	Large nut size, good kernel quality, moderate resistance to scab, resistance to powdery mildew and black pecan aphid
Tom	High yield, early maturity, excellent kernel colour, resistance to scab
Barton	High yield, late leafing, protandrous, precocious, early to mid-season harvesting, resistant to scab
GraTex	High yield, protogynous, good resistant to scab, mid-season harvesting, average nut size, vigorous
Peruque	Kernels golden, protandrous, early to mid-season flowering, very early ripening, good resistant to scab, golden kernel colour
Comanche 4M	Protogynous, very susceptible to scab, golden kernel colour
GraKing	Large nut size, mid-season harvesting, protogynous, medium vigor
Choctaw	Late harvesting, vigorous, protogynous, large nut size
Apache	Medium vigor, high yield, protogynous, golden kernel colour, precocious, late harvesting
Mohawk	Vigorous, protogynous, mid-season harvesting, susceptible to scab
Stuart	Vigorous, mid-season harvesting, brown kernel colour, resistant to scab, Susceptible to downy spot, black pecan aphids and yellow aphids

(continued)

Table 4.1 (continued)

Cultivars	Main characteristics
Navaho	High productivity, vigorous, precocious, good kernel quality, golden to light brown kernel colour, susceptible to scab and vein spot
Western (schley)	Golden to light brown kernel colour, very susceptible to scab and downy spot
10J	Late harvesting, medium vigor, protogynous, large nut size
Ukulinga	High yield, medium nut size, late season harvesting, very precocious
Chestnut	
Huaqiao 2	Resistance to chestnut blight, large nut, high yield
Jenny	Vigorous, large nuts, very good flavour
Emalyn's Purple	Large nut, good flavour, mid-late season harvesting
Nanjing Special	Superb flavour, mildly sweet, complexity of flavour, erect grower tree, dull brown nut colour
Hong Kong	Very good flavour, sweetness, mid-season harvesting
YGF	Medium to large nuts with excellent flavour
Ness	Large nuts with excellent flavour, low vigor tree
Payne	Large nut, good flavour, mid-late season harvesting
Szego	High nut quality, sweet and flavourful nut, resistant to *Phytophthora* root rot

of blank or unfilled nuts, increasing the green colour of the kernel at maturity, late flowering and early harvesting, precocity and low-chilling requirements (Sheikhi et al. 2019a).

The specific almond breeding objectives are self-compatibility, late-blooming, frost tolerance, disease resistance and tree architecture (Martínez-García et al. 2019).

The main pecan breeding objectives are high yield and kernel quality, precocious and consistent production, resistance to pecan scab and other leaf diseases (Wells and Conner 2015). The specific objectives of chestnut cultivars are resistant to blight and *Phytophthora* root rot and gall wasp (Hill Craddock and Taylor Perkins 2019). Hazelnut breeders focused on high yield and nut and kernel quality, high kernel percentage precocity, early maturity, round nut shape in their breeding projects (Botta et al. 2019).

Most of the rootstock breeding programmes of temperate nuts focused on tolerance to drought and salinity stresses, compatibility with scion, higher nutrient uptake efficiency and positive effect on scion performance (Vahdati et al. 2021). However, blackline disease tolerance (caused by Cherry leaf roll virus (CLRV)) and dwarfness are other objectives of walnut rootstock breeding (Vahdati et al. 2019). Dwarfness is also a desirable characteristic of chestnuts. In contrast, pistachio breeders looking for high-vigorous, disease-resistant (*Phytophthora*, *Armillaria* and *Verticillium*) and cold tolerant rootstocks. In addition to biotic and abiotic resistance, non-suckering is a specific issue in hazelnut rootstock breeding (Vahdati et al. 2021).

As above mentioned, global warming and climate change dictate new breeding objectives, and in addition to all the above objectives, temperate nut trees breeders pay special attention to some objectives that alleviate the adverse effects of climate change, such as low-chilling requirement and tolerance to late-spring and early-autumn frost, drought, salinity and heat stresses (Al-Khayri et al. 2019).

3 Conventional Breeding

3.1 Germplasm Utilisation

Persian walnut (*Juglans regia* L.), with a long history of cultivation, had a huge genetic diversity in its origin and diversity centers due to high heterozygosity and sexual propagation over thousands of years (Vahdati and Khorami 2021; Habibi et al. 2022). The utilisation of genetic resources along with molecular breeding can help accelerate the breeding programmes (Vahdati et al. 2019). Evaluating genetic diversity and identifying superior genotypes is one of the fundamental approaches in walnut breeding programmes (Kouhi et al. 2020). Persian plateau (including Iran and some central Asian countries) is the primary origin center of walnut. In addition, there is a worth diversity in walnut populations in East Asia and Europe (Vahdati et al. 2019). This variation has provided ideal conditions for walnut growers to release commercial cultivars and rootstocks. So that some commercial walnut cultivars in the world, such as cultivars released in France, Iran, China and Turkey are the result of exploiting this genetic diversity (Vahdati et al. 2019). In addition, the selected superior genotypes originated from Iran, Turkey and China are the ancestors of many commercial walnut cultivars from the UC Davis breeding programme (Tulecke and McGranahan 1994).

Walnut populations in different countries such as Iran (Sarikhani et al. 2012; Khorami et al. 2018; Sarikhani et al. 2021), Turkey (Bozhuyuk et al. 2020; Yildiz et al. 2021; Orhan et al. 2020; Sütyemez et al. 2021), China (Yuan et al. 2018; Wambulwa et al. 2021), Romania (Cosmulescu and Ionescu 2021), France (Bernard et al. 2018, 2020a), etc. have been evaluated to achieve cultivars with desirable nut characteristics. Also, this genetic diversity has been evaluated to introduce cultivars and rootstocks adapted to climate change conditions such as low-chill requirements (Hajinia et al. 2021), tolerance to drought, cold, salinity and heat stresses (Arab et al. 2020; Liu et al. 2020; Ebrahimi et al. 2020).

Genus *Carya* is a member of the Juglandaceae family. This genus has 20 species that 13 *Carya* species, including pecan, are native to the USA, and about 7 species are cultivated for their nuts. Due to a wide genetic diversity, these species are scattered in different climatic conditions (Sparks 2005). This genetic diversity has enabled these species to grow in all regions, including temperate and subtropical regions. This adaptability has led the pecan cultivated in South Africa, Brazil, Argentina, Chile, Peru, Uruguay, India, Egypt, China and Australia. Louis D. Romberg began collecting pecan cultivars in 1930. He then grafted the cultivars to trees and in 1980 created the first pecan germplasm genetic repository in the United States. Breeding programmes at the University of Georgia (UGA) and USDA on nut quality and productivity pioneer genetic improvement in pecan trees. The important breeding goal of pecan is to produce cultivars with high yield and resistance to biotic stresses (Hill Craddock and Taylor Perkins 2019).

The genus *Pistacia* L. belongs to the *Anacardiaceae* family along with mango (*Mangifera indica* L.), cashew (*Anacardium occidentale* L.) and pepper tree (*Schinus* spp.) (Hormaza and Wünsch 2007; Sheikhi et al. 2019b). The genus *Pistacia* consists of 11 or more species; *Pistacia vera* L. is the only commercially important species producing large edible nuts. Wild and undomesticated *P. vera* forests extend from the Kopet Dagh mountain range of southern Turkmenistan and northern Afghanistan to the Khorasan district and Sarakhs region in northeastern Iran, where can be considered as the origin and center of diversity for this species (Hormaza and Wünsch 2007). These wild relatives of cultivated pistachio have excellent potential for genetic improvement to produce superior scion and male cultivars and rootstocks. Pistachio is still relatively unexamined genetically relative to other fruit trees, given its nutraceutical and commercial importance. Relying on a few cultivars and rootstocks makes the pistachio industry vulnerable to new diseases and pests.

Additionally, most pistachio cultivars have undesirable characteristics such as a high percentage of unsplit and blank nuts or extreme alternate bearing tendency. Therefore, improving these undesirable traits should be an essential component of future breeding programmes (Vahdati et al. 2021). Genetic diversity evaluation is the first step for utilising the germplasm potential of any crop improvement programme. Several studies have been conducted concerning the genetic diversity of *P. vera* cultivars using morphological markers, isozymes and DNA markers, including restriction fragment length polymorphisms (RFLPs), random amplified polymorphic DNA markers (RAPDs), amplified fragment length polymorphisms (AFLPs), simple sequence repeats (SSRs), single nucleotide polymorphisms (SNPs), etc. (Sheikhi et al. 2019a). It appears that the *P. vera* var. Sarakhs, which is wildly growing in forests in the Sarakhs border region of Iran and Turkmenistan, plays a vital role in the evolutionary history of the commercial pistachio cultivars and maybe an essential source of new alleles for pistachio improving programmes (Arabnezhad et al. 2011; Talebi et al. 2016). More information on intra-specific and inter-specific genetic relations of pistachios is available in Parfitt et al. (2012) and Hormaza and Wünsch (2007).

Almond [*P. dulcis* (Mill.) D.A.Webb; syn. *P. amygdalus* Batsch] is a species of the genus *Prunus* and Rosaceae family. This nut crop's edible and tasty kernel is the main reason for its commercial production worldwide (Gradziel and Martínez-Gómez 2013). Today, almond nuts are produced in more than 50 countries (http://faostat.fao.org), of which 95% of them are cultivated in California, Australia and the Mediterranean region (Martínez-García et al. 2019). The arid mountainous regions of Central Asia are believed to be the original center of the cultivated almond (Grasselly 1976). Ladizinsky (1999) proposed *A. fenziliana* Fiitsch as the wild ancestor of the cultivated almond. More information on the different theories about almonds' origin center can be found in Martínez-García et al. (2019) and Socias I Company et al. (2012). Despite a high level of genetic variability in natural almond populations, due to its self-incompatibility, early studies reported a limited variability in cultivated almonds; using a few parental genotypes selected for their desirable traits could be the main reason for the limited genetic base (Martínez-García et al. 2019; Kester and Gradziel 1996).

Local cultivars and wild relatives of almond crop represent valuable germplasm in terms of tree performance, yield, biotic/abiotic stress resistance and fruit quality traits for addressing future almond cultivation challenges and so should be preserved, evaluated and incorporated into breeding programmes (Martínez-García et al. 2019). To date, the major almond germplasm collections has been established worldwide; at Nikistki Botanical Garden in Yalta, Crimea (Rikhter 1972), in France (Grasselly and Crossa-Raynaud 1980), in Spain (Espiau et al. 2002) and in California (Kester and Gradziel 1996). In California almond breeding programme, a greater emphasis was on the introgression of interspecific traits from wild relatives into advanced breeding lines. For example, self-compatibility in Californian almond cultivars has been derived from *P. mira* Koehne, *P. persica* Batsch, *P. davidiana* Carr. (Franch) and Yugoslavian accessions of *P. webbii* (Gradziel 2008). Wild species, such as *P. spartioides* (Spach) Schneid and *P. webbii*, have been directly used as rootstocks for almond cultivation in dry land in Iran and Turkey, respectively. Recently, under non-irrigation conditions, almond × peach hybrids are showing promising performance. Under high input, well-irrigated conditions, the shallower peach and plum rootstocks have shown more efficiency for almonds (Martínez-García et al. 2019). Self-compatibility has been the main objective in many almond breeding programmes. The main source of self-compatibility alleles for European cultivars have been 'Tuono' (Socias i Company 2002). Late-blooming is another important breeding objective to prevent spring frost. The main genomic source for this trait has been cultivars like 'Tardy Nonpareil' mutant (Socias i Company et al. 1999).

Hazelnut is one of the most important tree nuts products in the world, its geographical distribution extends from Europe and North Africa to the Caucasus and Asia Minor. Among the cultivars, European hazelnut is one of the most important nut products in terms of area under cultivation and global production (Ferreira et al. 2010). Due to the very high natural distribution range and cultivation in different climates, rich genetic diversity has been reported in hazelnut germplasm (Boccacci et al. 2013). Both phenotypic and molecular markers have been used to evaluate hazelnut germplasm. The most important advantages of molecular markers for identifying cultivars include high differentiation power at a relatively low cost, clarifying homonymy and synonymy cases, studying the origin of cultivars, determining the lineage and evaluating genetic relationships between cultivars, which is an excellent way to accurately identify hazelnut cultivars (Boccacci et al. 2013). Genetic diversity studies in hazelnuts have shown a large genetic diversity in *Corylus avellana*, independently in six different areas: the British Islands, central Europe, Spain, Italy, the Black Sea region and Iran (Boccacci et al. 2015; Mehlenbacher 2018).

There are several chestnut species in Asia, Europe and North America, which indicates the compatibility of this plant species in different environments (Martín et al. 2017). Wild and domesticated chestnut genetic resources spread from the

Caucasus to Portugal, southern Britain, the Canary Islands and the Azores archipelago. Preservation of intraspecific and interspecific diversity of chestnut is one of the main factors in cultivating this crop in different areas. In the distant past, chestnut trees were used to supply timber and firewood. Historical evidence shows that these species' first artificial selection and domestication occurred thousands of years ago in East Asia and Western Europe. Cultivation of this crop began in the early Middle Ages in Western Europe, especially Italy. Chestnut diversity can also contribute to the selection of varieties more tolerant or resistant to biotic and abiotic stresses. Genetic and eco-physiological studies on *Castanya sativa* have shown that these species adapt to different climatic conditions (Hill Craddock and Taylor Perkins 2019).

3.2 Hybridisation

After germplasm utilisation, hybridisation considers as a main walnut breeding strategy. This is the main strategy in UC Davis and France walnut breeding programme, which led to the release of different cultivars and rootstocks (McGranahan and Leslie 2005). In recent years, walnut hybridisation programmes have been implemented in other countries, especially Iran (Fallah et al. 2022) and Turkey (Özcan et al. 2020). Improving yield, nut and kernel characteristics, lateral bearing, early harvesting, late leafing and disease resistance have been the most important objectives for walnut hybridisation programmes (Tulecke and McGranahan 1994; Bernard et al. 2018; Vahdati et al. 2019). Grafting, micropropagation and somatic embryogenesis techniques have been used for propagation of hybrid walnuts (Rezaee et al. 2008; Vahdati et al. 2006, 2009).

Viable hybrids can be made via intra- and interspecific cross-pollination between different cultivars and species of genus *Pistacia*. There is no reproduction barrier, except the difference in blooming time, to obtain hybrid seedlings between *Pistacia* species such as *P. vera*, *P. eurycarpa* Yalt., *P. atlantica* Desf, *P. terebinthus*, etc. (Hormaza and Herrero 1998; Kafkas and Kaska 1997). Procedures for hybridisation and controlled pollinations in pistachios are explained in detail by Sheikhi et al. (2019a). The pistachio industry's main breeding objectives and traits include nut quality, yield, lower chilling requirements, abiotic and biotic stress resistance, lower alternate bearing tendency, and vigor and rapid growth. Most pistachio cultivars are naturally occurring hybrids and selections made by local growers in primary pistachio-growing regions. In the past, there was a major attempt at germplasm collection and evaluations in Iran, where cultivars 'Akbari', 'Ahmad-Aghaei', 'Kalle-Ghoochi' and 'Ohadi' ('Fandoghi') are highly commercialised (Sheikhi et al. 2019a).

Some superior pistachio cultivars have been released through breeding programmes in California (Chao et al. 1997; Parfitt et al. 1994), Turkey (Kafkas and Kaska 1997), Spain (Vargas et al. 2001) and Australia (Maggs 1990). Sirora was released in Australia from a selection of Red Aleppo progenies at CSIRO Merbein

Laboratories (Maggs 1990). Kerman, Damghan, Lassen, Joley, Golden Hills, Lost Hills and Gumdrop are the scion cultivars, and Peters, Randy, Famoso and Tejon are the male cultivars introduced from California pistachio breeding programmes on different occasions (Sheikhi et al. 2019a).

The most effective pistachio hybridisation effort to produce a superior rootstock tolerant to biotic and abiotic stresses has been conducted by Lee J. Ashworth introduced a new rootstock named University of California Berkeley 1, or UCB1. UCB1 is a hybrid of a specific *P. atlantica* female and a specific *P. integerrima* male. The male parent used in this hybridisation is a genomic source of *Verticillium* wilt resistance commercialised as Pioneer Gold I (PGI). Hybridisation of a female *P. integerrima* and male *P. atlantica* named Pioneer Gold II (PGII) was not commercially successful due to lack of *Verticillium* resistance; however, the same cross named Platinum was produced by a commercial nursery (Ferguson et al. 2016).

The traditional almond breeding techniques involve controlled crosses between parents selected for traits of interest, followed by hybrid seed handling and field evaluations (Kester and Gradziel 1996). To obtain a new almond cultivar, two cultivars can be crossed with complementary characteristics to assemble the interesting phenotypic traits of both parents or obtain progeny exhibiting a more extreme phenotype than either genitor. It is relatively easy to perform interspecific crosses in *Prunus* species because of their interspecific compatibility, which can be very useful for the development of new almond cultivars/rootstocks (Martínez-García et al. 2019). Almond has a short juvenile period relative to the other tree nut trees. Most seedlings enter the reproductive stage during the third year after seed germination. This period could be shortened by germinating the immature embryos and top-budding the buds from the seedlings onto rootstocks (Kester and Gradziel 1996). The average time to introduce a new almond variety is usually 12 years. The most important objectives in almond genetic improvement programmes are as follows: self-compatibility, late-blooming, yield, resistance to pests and diseases, kernel quality and ripening time. Desirable traits for new rootstocks include ease of micropropagation and propagation by hardwood cuttings, high vigor, tolerance to heavy soils and water-saturated soils, resistance to soil-borne diseases (Martínez-García et al. 2019).

In hazelnuts, the strategies used to improve breeding programmes are based on *C. avellana* intraspecific crosses (Botta et al. 2019). The purpose of hazelnut breeding programmes is to increase and improve the superior quality of nuts, yield and other characteristics of its commercial production compared to uncultivated species in this genus (Botta et al. 2019). In 1983, hybridisation-based breeding was started to achieve important qualitative and quantitative traits of the hazelnut tree. The experiment results by Okay and Özenc (2000) reported that about 15,000 hybrid seedlings were selected, and four high-yielding were proposed as a new cultivar. In a breeding programme, eight species of *Corylus* L. (hazelnut) were intercrossed in all possible combinations to reveal genetic relationships. The results showed that the following crosses were successful; *C. californica* × *C. avellana*, *C. chinensis* × *C. avellana*, *C. americana* × *C. heterophylla*, *C. cornuta* × *C. heterophylla*, *C. californica* × *C. colurna* and

C. americana × *C. sieboldiana*, but the reciprocals were not (Erdogan and Mehlenbacher 2000).

The first hybridisation of chestnut was done in the United States in 1894, then in Spain and Japan in 1926 and 1917, respectively. Hybridisation of chestnut trees was carried out in Europe in the 19th century due to the combat ink disease caused by the fungus Phytophthora (Hill Craddock and Taylor Perkins 2019). These crosses have been done to increase the quality of crops and forests and create resistance varieties to diseases. Asian chestnut species are resistant to diseases, unlike European species, but they were not suitable for cultivation in European orchards due to their climatic conditions. In 1909, however, the Costana Sernata cultivar was introduced from one of the French Pyrenees valleys, which was well adapted. Then commercial cultivars, 'Ederra', 'Marki', 'Ipharra', were selected and registered in the catalog of French fruit species. Hybrid clones introduced as cultivars have no special place than natural cultivars (Hill Craddock and Taylor Perkins 2019). Some hybrid cultivars are used to produce nuts (France, USA and Japan), timber (Spain) and rootstocks for resistance to *Phytophthora* spp. (Europe). Nuts of hybrids are often big; their seedlings grow more rapidly and maintain the wood quality. European chestnuts and Asian chestnut hybrids are sensitive to spring frosts. Chestnut hybrids will revive wood-producing chestnuts in the future (Jacobs et al. 2013).

4 Molecular Breeding

4.1 Molecular Markers

Like other perennial woody plants, developing new commercial cultivars of nut trees is a long-time effort due to the long juvenile phase, which incurs a considerable cost, energy and time to generate, maintain and evaluate breeding populations. Thus, developing rapid-cycle breeding and efficient genotype selection pipelines has become crucial to conduct pistachio improvement programmes at a reasonable cost and time (Vahdati et al. 2019; Sheikhi et al. 2019a). Using advanced new technologies in molecular plant breeding through the application of next-generation sequencing (NGS) techniques, followed by quantitative trait loci (QTLs) mapping of commercially important traits, marker-assisted selection (MAS), genomic selection (GS) and genetic transformation, have the potential to shorten the length of the breeding cycle in fruit crops while increasing the efficiency and accuracy of the selections (Iwata et al. 2016).

In the last few decades, various molecular markers systems consecutively applied for evaluation of population structure and genetic diversity of walnut and assessment uniformity of its cultivars (Vahdati et al. 2019). Genetic markers have been extensively applied in walnut breeding (genetic mapping), genetic diversity analysis, evolutionary studies and fingerprinting walnut cultivars as follow: Isozymes (Busov et al. 2002; Ninot and Aleta 2003; Vyas et al. 2003), restriction fragment length polymorphism (RFLP) (Fjellstrom and Parfitt 1994), randomly amplified polymorphic DNA (RAPDs) (Francesca et al. 2010; Erturk and Dalkilic 2011; Shah et al. 2019), amplified fragment length polymorphism (AFLPs) (Ma et al. 2010; Xu et al. 2012; Ali et al. 2016), inter simple sequence repeats (ISSRs) (Pollegioni et al. 2003; Christopoulos et al. 2010; Ji et al. 2014; Shah et al. 2019), simple sequence repeat (SSRs) (Gunn et al. 2010; Zhang et al. 2013; Ebrahimi et al. 2017; Pollegioni et al. 2017; Roor et al. 2017; Vischi et al. 2017; Bernard et al. 2018; Guney et al. 2021), EST-SSR (Qu et al. 2011; Zhang et al. 2013; Yuan et al. 2018), start codon targeted (SCoT) (Tabasi et al. 2020) and sequence-related Amplified Polymorphism (SRAP) (Yildiz et al. 2021). Recent advances in sequencing techniques and dramatically decreased costs facilitate the rapid discovery of single nucleotide polymorphisms (SNP) markers (Marrano et al. 2019a). SNP markers are eventually distributed across the genome, and their use is growing in population genomic studies in walnut. Initially, the SNP markers were developed based on internal transcribed spacers (ITS) sequences and applied for 18 *J. regia* L. cultivars (Ciarmiello et al. 2011). In another study, a 6K Infinium SNP genotyping array was developed based on the 'Chandler' cultivar and applied in walnut genetic mapping studies (You et al. 2012). Arab et al. (2019) genotyped a diverse panel of Iranian walnut populations using the Axiom *J. regia* 700K SNP array. Genetic structure analysis divided this panel into four main groups, reflecting their geographic distributions. Bernard et al. (2020b) compared the utility of SSR and SNP markers for inferring the population genetic structure in INRAE walnut (*Juglans regia* L.) germplasm collection. Their results showed that the first level of structure was observed equally using 13 SSRs or 364,275 SNPs. They also highlighted that a few multi-allelic SSR markers could capture most of the SNP allelic diversity and best suit the core collections establishment (Bernard et al. 2020b). Also, restriction-associated DNA (RAD) sequencing approaches have been applied for SNP discovery in 41 walnut cultivars from China (Wang et al. 2020). The population structure analysis showed that the 6357 SNP markers divided the 41 walnut cultivars into two main clusters. Recently, phylogenetic analysis of 815 walnut genotypes (*J. regia* and *J. sigillata*) from China, Iran and Pakistan using whole-genome resequencing data (at an average sequencing depth of approximately 19.0×) showed that these two species could be divided into different clades, and they also found one intermediate group between these two species (Ji et al. 2021).

Different molecular markers have been extensively employed in assessing genetic diversity and genetic mapping studies in pecan, including AFLP, RAPD, ISSR, SSR and SNP. Several studies have shown that SSRs are more effective than other markers (AFLP, RAPD and ISSR) in their high allelic variation, good reproducibility and high informative content (Hill Craddock and Taylor Perkins 2019).

The first linkage map of pistachio was developed by Turkeli and Kafkas (2013) using ISSR, SRAP and AFLP markers. A major use of molecular markers in pistachio crops is too early detection of seedling gender by MAS. *Pistacia* species are all dioecious; however, some other sex types have been reported (Crane 1974; Kafkas et al. 2000). Dioecy, where 50% of the progeny are male and 50% female, is a breeding disadvantage as nut characters of male parents cannot be evaluated directly. In addition, keeping the male progeny in evaluation blocks when breeding

Table 4.2 Sex-linked DNA markers in pistachio species

Species	Gender	Marker name	Primer sequence	Marker type	Reference
P. vera	Female	OPO08945	CCTCCAGTGT	RAPD	Hormaza et al. (1994)
P. eurycarpa	Male	BC156	GCCTGGTTGC	RAPD	Kafkas and Perl-Treves (2001)
P. eurycarpa	Female	BC360	CTCTCCAGGC	RAPD	Kafkas and Perl-Treves (2001)
P. atlantica	Female	OPAK09	AGGTCG GCGT	RAPD	Kafkas and Perl-Treves (2001)
Pistacia spp.	Female	BC1200	GCCTGATTGC	RAPD	Esfandiyari et al. (2012)

for scion cultivars is a considerable waste of time, labour, land and breeding resources (Kafkas et al. 2015; Turkeli and Kafkas 2013). To date, many researchers have developed sex-linked DNA markers in pistachio species (Table 4.2). Khodaeiaminjan et al. (2017) constructed a linkage map of the sex chromosome in pistachio using sex-linked single nucleotide polymorphisms (SNPs) and expressed sequence tag-derived simple sequence repeats (EST-SSRs). The resulting consensus map showed a total length of 65.19 cM with the sex locus in the center of the chromosome at 31.86 cM. Kafkas et al. (2015) developed eight SNPs markers using restriction site-associated DNA sequencing (RAD-seq), which could distinguish sex with 100% accuracy in pistachio. They also reported a ZZ/ZW sex-determination system for the first time in pistachio, in which females are the heterogametic sex.

The advances in the development of molecular markers linked to traits of interest (such as self-incompatibility, late-blooming, disease resistance and kernel flavour) can increase efficiency of almond genetic improvement programmes, because they provide fast, accurate and environment-independent evaluations at the seedling stage. A detailed review of molecular markers research in almonds has recently been provided by Martínez-García et al. (2019) and Socias I Company et al. (2012). Three main linkage maps have been reported for almonds from the linkage analysis conducted on three different populations (Viruel et al. 1995; Foolad et al. 1995; Arús et al. 1994a, b). The availability of high-density linkage maps has helped to map position of important genes such as self-incompatibility (Ballester et al. 1998), kernel flavour (Sánchez-Pérez et al. 2010) blooming time (Ballester et al. 2001), shell hardness (Arús et al. 1999), flower colour (Jáuregui 1998) and genes involved in rootknot nematode resistance (Dirlewanger et al. 2004). Ricciardi et al. (2018) developed CAPS markers linked to the sweet kernel locus, controlling amygdalin synthesis in almond. Forcada et al. (2012) developed genetic map for the chemical components of the almond kernel and reported 12 QTLs controlling their metabolism. PCR-based markers have been successfully used to identify different self-incompatibility alleles (S-alleles) of almond genotypes (Gradziel et al. 2001; Kodad et al. 2010).

In hazelnut breeding programmes, molecular markers have been used to identify the cultivar, analyse genetic relationships in germplasms, and identify self-incompatible individuals (Botta et al. 2019). Amplified fragment length polymorphism (AFLPs), random amplified polymorphic DNA (RAPDs) and especially simple sequence repeats (SSRs) have been extensively used for germplasm characterisation in hazelnut breeding programmes (Botta et al. 2019). A study was conducted to evaluate the levels of polymorphisms identified by ISSR markers in hazelnuts in northern Spain (Ferreira et al. 2010). The results of this study clustered the population into two main groups including cultivars and local germplasm from other regions (Ferreira et al. 2010). In another study, 17 microsatellite markers were used to evaluate the genetic diversity and population structure of 323 unique hazelnut accessions, including a panel of known germplasms (accessions representing a wide diversity of Corylus cultivars, breeding selections and interspecific hybrids), in comparison to EFB-resistant germplasm of unknown origin. The results of this study showed that there is strong evidence that EFB resistance is widespread throughout the genus *Corylus* (Muehlbauer et al. 2014).

A small number of SSR markers are proving useful information in *Castanea* genetic studies. SSR and SNP markers are derived into four different sources: bacterial artificial chromosomes (BACs), genomic DNA libraries enriched for repeated sequences, whole-genome sequencing and expressed sequence tags (ESTs). Currently, 83 SSR markers have been obtained from *Castanea*, including 46 SSR primer pairs developed from *C. sativa*,15 SSR primer pairs developed from *C. crenata*, 22 SSR primer pairs from *C. mollissima* (Hill Craddock and Taylor Perkins 2019). These SSR markers are used to examine genetic diversity analysis and cultivar identification. More than 400 SSR markers were developed in Japan for *C. crenata*. Furthermore, Nishio et al. (2011) developed a total of 366 SSR markers derived from *C. crenata*, comprising 220 SSRs from enriched genomic libraries and 146 EST–SSRs from large-scale EST sequencing analysis.

4.2 Genome Sequencing and Association Mapping

In fruit and nut trees, many genomic resources have been generated in a short period due to the recent advances in next-generation sequencing (NGS) and third-generation sequencing technologies. These high-throughput technologies have reduced the cost and duration of genome-scale sequencings, thus will play an important role in preparing reference genomes and identifying genes controlling traits of importance to devise genetic improvement strategies in fruit tree crops (Savadi et al. 2021). Reference genomes are highly valuable to plant breeders as they enable the assembly of new genomes or transcriptomes more precisely, quickly, and cheaply, in addition to a detailed understanding of the structural organisation of genomes such as regulatory elements, repetitive regions and gene structure and number (Marrano et al. 2020).

Release the walnut reference genomes (Martínez-García et al. 2016; Zhu et al. 2019; Marrano et al. 2020) has facilitated the application of genomic tools in walnut genetic improvement programmes. A new era of walnut genetic studies began with developing the new Applied Biosystems™ Axiom™[1] *J. regia* 700K SNP genotyping array (Marrano et al. 2019a). The availability of a high-quality reference genome with fully assembled chromosomes allows whole-genome resequencing, genotyping by sequencing (GBS) and genetic mapping of economically important traits in walnut (Arab et al. 2019; Marrano et al. 2019a; Büküсü et al. 2020). Recently, QTL-mapping and genome-wide association study (GWA) as a classical and advanced genetic mapping approaches has been employed to dissect the genetic architectures of complex traits in Persian walnut.

Several studies of QTL and association mappings have been conducted in Persian walnut using the new Axiom *J. regia* 700K SNP genotyping array. For instance, Marrano et al. (2019b) performed both QTL mapping and GWAS to dissect the genetic control of phenology, yield and pellicle colour traits in walnut. They identified a genomic region at the beginning of Chr1 associated with both leafing and harvest date as well as found a large genomic region on Chr11 linked to lateral fruit-bearing in walnut (Marrano et al. 2019b). In another study, Sideli et al. (2020a, b) carried out both QTL mapping and GWAS to unravel the genetic architecture of shell suture strength and pellicle pigment in walnut. They found a major QTL for suture strength on LG05 and two minor QTLs on LG01 and LG11 and confirmed the identified QTLs with GWAS on corresponding chromosomes (Sideli et al. 2020a). This SNP genotyping array was also applied to genotype the INRAE[2] walnut germplasm collection and Iranian panel to dissect the genetic control of nut-related traits (Arab et al. 2019; Bernard et al. 2021), identifying several marker-trait associations on different chromosomes. In addition, Bernard et al. (2020c) have performed association and linkage mapping to decipher the genetic control of phenological traits and lateral bearing in walnut. They found a peak SNPs at the beginning of chromosome 1 associated with bud break and female flowering dates and developed a Kompetitive Allele-Specific PCR (KASP) marker for this trait (Bernard et al. 2020c). Recently, 188 walnut accessions from Turkey were genotyped using DArT-seq, and GWAS analysis of phenological traits identified a major QTL with pleiotropic effects linked to leaf budburst and flowering dates in walnut (Büküсü et al. 2020). However, few studies used genomic approaches to dissect the genetic basis of abiotic stress tolerance in walnut due to the difficulties in evaluating a large population under various stress conditions (Famula et al. 2019; Arab et al. 2020). In the beginning, Famula et al. (2019) genotyped 60 scion genotypes from the UC Davis walnut breeding programme through the Axiom *J. regia* 700K SNP array, then conducted a GWAS for WUE-related traits and identified four SNP markers linked to the carbon isotope discrimination ($\Delta\ ^{13}C$). In addition, GWAS was also carried out in walnut for drought-related traits such as water use efficiency as estimated by carbon isotope discrimination (Arab et al. 2020). Although several linkage and association mapping studies have been conducted in

[1]The Axiom Genotyping Solation is the platform of choice for large-scale genotyping studies.

[2]National Research Institute for Agriculture, Food and the Environment (France).

walnuts, and some major effect QTLs have been identified, extensive studies with different populations are needed to help accelerate breeding programmes.

Recently, genotyping by sequencing (GBS) of pecan cultivars with known flowering patterns discovered 87,446 SNPs and genome-wide association study identified 17 SNPs which were significantly associated with the flowering (Bentley et al. 2019).

Although reference genome sequences of many fruit trees have become available (Savadi et al. 2021), the chromosome-scale reference genome of pistachio has not been reported yet, and there is very little information on genome sequencing. Motalebipour et al. (2016) reported a genome size of 600 Mb with a high heterozygosity rate for *P. vera* using whole-genome shotgun sequencing. Zeng et al. (2019) assembled a draft genome of pistachio and re-sequenced 107 whole genomes of different wild and domesticated *Pistacia* species. According to their study, pistachio was domesticated 8000 years ago, and it seems that key genes for domestication are those involved in tree and seed size, which experienced artificial selection.

Thus far, genomic resources for almonds have been limited despite their economic importance. Recently, D'Amico-Willman et al. (2022) reported whole-genome sequence and methylome profiling of the almond cultivar 'Nonpareil'. Using Illumina, PacBio and optical mapping technologies, they generated a 615.89X coverage genome sequence. Genome-wide association study (GWAS) in almond crop using genotyping-by-sequencing (GBS) data (Pavan et al. 2021) and Illumina Infinium® 18 K Peach SNP array (Di Guardo et al. 2021) resulted in the identification of genomic associations with fruit quality traits of interest. Prudencio et al. (2018) used epi-genotyping by sequencing (epi-GBS) for DNA methylation analysis of dormancy release in almond flower buds. Their results provided information about genomic regions linked to early and late flowering methylation markers and the chill accumulation process. Genotyping-by-sequencing (GBS) has also been applied to discover and map single nucleotide polymorphisms in the almond cultivars 'Nonpareil' and 'Lauranne'. The sequence-based linkage maps provide the basis for mapping loci for agriculturally relevant traits and selecting markers for almond breeding applications (Goonetilleke et al. 2018). In the future, the availability of high-coverage sequence data of important almond cultivars will be useful to breeders and geneticists in identifying quantitative trait loci (QTLs) or developing marker sets for marker-assisted selection (MAS) or genomic selection (Velasco et al. 2016).

The early fire blight-resistant hazelnut accession 'Jefferson' genome sequenced by Rowley et al. (2018).Therefore, these genomic resources will help identify molecular markers related to traits of interest for hazelnut breeding (Rowley et al. 2018). In one study, an existing linkage map developed from a population of F_1 progeny bred from the cross 'Tonda Gentile delle Langhe' × 'Merveille de Bollwiller', consisting of 11 linkage groups were used to identify quantitative trait sites (QTLs) for phenology-related traits, such as male and female flowering time, bisexuality and the period required for the nuts to reach maturity. This study identified 42 QTL-harbouring regions and 22 candidate genes underlying the studied traits, most of which were involved in the phenology process (Valentini et al. 2021).

4.3 Transcriptomics and Proteomics

Sequencing and assembling the entire genome of a crop species is a costly and complex process because of the high level of heterozygosity and polyploidy. However, sequencing only coding genes, i.e. transcriptome profiling, is easier and more useful. The main focus of transcriptome analysis is investigating the genes expressed in a specific biochemical pathway at a particular plant tissue, developmental stage, or environmental condition (Savadi et al. 2021). First, transcriptome analyses in different tissues in walnut focused on generating ESTs to develop molecular markers such as EST-SSRs resulting from cDNA libraries (Zhang et al. 2013; Dang et al. 2016). Recently several transcriptome analyses have been conducted in walnut under different stages of tree development and abiotic and biotic stresses conditions. In a study, the expression of flowering genes (*FT*, *SOC1*, *CAL*, *LFY* and *TFL1*) during the growing season was investigated, and the results showed FT gene expression activated downstream floral meristem identity genes (Hassankhah et al. 2020). In another study, comparative transcriptome analysis of leaf and peel colour change in red and green *J. regia* identified the genes involved in anthocyanin accumulation (Li et al. 2018). Gene expression analysis of heat stress transcription factors (HSFs) genes in walnut under heat, salinity and drought stresses showed that they play a crucial role in walnut response to abiotic stresses by regulating the expression of stress-responsive genes (Liu et al. 2020). Cloning and expression analysis of the transcription factor gene JrCBF in *J. regia* L. demonstrated that this gene might play a key role in cold resistance mechanisms (Xu et al. 2014). Feng et al. (2021) carried out a comparative transcriptome analysis of F26 (anthracnose-resistant) and F423 (anthracnose-susceptible) fruit bracts to identify differentially expressed long non-coding RNAs (lncRNAs). Enrichment analysis results exhibited that the target genes of upregulated lncRNAs were involved in immune-related processes (Feng et al. 2021). Recently, high-throughput transcriptome analysis through the RNA-Seq method has been incorporated in walnut transcriptomics studies. RNA-Seq profile obtained from twenty different tissues revealed that a large family of resistance genes (NBSLRR) in Persian walnut (*J. regia*) is probably involved in plant-microbe interactions (Chakraborty et al. 2016). Transcriptome analysis of *J. regia* L. embryos at different stages of development identified most of the genes involved in oil biosynthesis (Huang et al. 2020). The result provides new insights into the metabolic engineering of walnuts to increase oil contents. Transcriptome profiling of flower bud development through Illumina- and SMRT-based RNA-seq showed that investigating key genes related to the circadian clock could help to understand the mechanisms underlying flower bud development in walnut (Ma et al. 2021). Transcriptome analysis of *J. regia* L. cv. Chandler leaf revealed 921, 1035 differentially expressed genes (DEGs) between well-watered and water deficit conditions in nine and 18-day after beginning of experiment, respectively (Sadat-Hosseini et al. 2020).

One of the first studies on the whole transcriptome survey of *P. vera* was conducted by Jazi et al. (2017). They created a robust pooled transcriptome assembly from leaf, stem and root of *P. vera* under control and salinity conditions. They observed the contrasting expression pattern of NCED3 and SOS1 genes between salt-sensitive and salt-tolerant cultivars. Martinelli et al. (2017) conducted an RNA-Seq analysis to investigate alternate bearing mechanisms in *P. vera*. They reported that genes associated with bud abscission might directly or indirectly involve in alternate bearing. Recently, changes in the transcriptome of inflorescence buds (Benny et al. 2020) and fruits (Benny et al. 2022) of pistachio cv. Bianca has been investigated using the RNA-seq technique to explore molecular mechanisms causing premature inflorescence bud abscission in the pistachio trees. According to the results of their studies, changes in genes expressed in pistachio 'ON' vs. 'OFF' inflorescence buds and fruits triggers a cascade of events in nutrient and stress signalling pathways involving trehalose-6-phosphate and rapamycin (TOR) signalling, SnRK1 complex, polyamines, hormones and ROS which end, through programmed cell death and autophagy phenomena, resulting in the abscission of inflorescence buds. These findings have important implications for breeding pistachio scion cultivars with less intensity of alternate bearing behaviour.

As the central dogma states that genes are eventually expressed as proteins, proteomic analysis of a crop of interest provides a complementary tool for transcriptomics and genomics to discover the important genes for crop breeding advancements (Chen and Harmon 2006; Das et al. 2015). Goharrizi et al. (2019) investigated the proteomic profile of UCB-1 pistachio rootstock leaves under 100 mM salinity stress. They showed that the highest number of variations of protein expression in response to salinity stress was assigned to proteins involved in photosynthesis, including 50S ribosomal protein L13, ribulose bisphosphate carboxylase/oxygenase activase 1 and phosphoribulokinase, ribulose-1,5-bisphosphate carboxylase/oxygenase (RuBisCO) large subunit, and ribulose bisphosphate carboxylase small chain. Polcalcin Phl p 7-like, ribonucleoside-diphosphate reductase small chain and Golgin subfamily A member 5 are three proteins that were detected for the first time in response to salinity stress, and likely they are the key proteins in salinity tolerance of UCB-1 rootstock. Proteomic profile changes of leaf tissue of UCB-1 rootstock under drought stress have been reported by Pakzad et al. (2019). This fundamental information on the salinity and drought tolerance of UCB-1 rootstock provides technical support and a theoretical basis for future genetic improvement of abiotic stress-tolerant pistachio rootstocks.

Among the new omics approaches, proteomics was used to study pollen-pistil interactions, to disclose self-(in) compatibility mechanisms in almonds (Martínez-García et al. 2015; Gómez et al. 2015). Transcriptomic analysis has also been used to study different aspects of almond tree biology, including discovering candidate genes for components of gametophytic self-incompatibility (Gómez et al. 2019), flowering time (Prudencio et al. 2021), freezing stress (Mousavi et al. 2014), fruit drop (Guo et al. 2021) and bud failure (D'Amico-Willman et al. 2021). Non-infectious bud-failure (BF) in almonds is an age-related disorder affecting vegetative bud development and nut yield. This characteristic is hypothesised to be associated with loss of juvenility mediated by genome-wide DNA-(de) methylation (Fresnedo-Ramírez et al. 2017; Gradziel and Shackel

2021). D'Amico-Willman et al. (2021) used integrated methylome and transcriptome analysis to study BF in almonds. Their results support the hypothesis that BF is associated with hypomethylation in almonds.

Transcriptome sequencing analysis was performed to dissect the main molecular mechanisms underlying bud burst in hazelnut. Gene expression profiles showed that signal transduction, phenylpropanoid and phytohormones biosynthesis pathways and transcription factors (MADSbox, MYB and bHLH) play an important role during bud burst (Kavas et al. 2019). One of the common problems with hazelnuts is reduced yield due to the high ratio of blank fruit. In a study, RNA-seq analysis results showed that the *PYL*, *PP2C* and *ABF* genes involved in the abscisic acid pathway and the *CTR1*, *EIN3*, *ERF1* and *ERF2* genes involved in the ethylene signal pathway in aborted eggs were upregulated, and these changes may affect the abscisic and ethylene signal pathways to abortive ovule aging and dormancy or senescence of ovules (Cheng et al. 2015). MicroRNAs (miRNAs) are short, single-stranded, non-coding RNAs with a length of 18-25 nucleotides generally involved in developmental processes and response processes to biotic and abiotic stresses (Willmann and Poethig 2007). There are few studies on the miRNA in hazelnuts, but a recent study has shown that some of the known conserved miRNAs are also expressed in two different types of hazel species *Corylus avellena* L. and *Corylus colurna* L. (leaf, bud, male and female flower), including miR159, miR160, miR171, miR396, miR2919 and miR8123 (Yirmibeş et al. 2021). Transcriptomic-proteomic integrative analysis of ovary abortion during ovary development in hazelnut showed that hormonal pathways include ethylene, jasmonic acid and salicylic acid signal transduction pathways may involve in the regulation of the abortive ovary formation in hazelnut by up-regulating ethylene-responsive transcription factor ethylene 1, transcription factor MYC2, transcription factor (TGA), jasmonate ZIM domain-containing protein and pathogenesis protein 1(Liu et al. 2018).

In recent years, most of the QTL mapping studies in chestnut focused on resistances to blight and root rot, adaptive traits and nut quality-related traits (Hill Craddock and Taylor Perkins 2019). Recently, genomic resources have been used in genetic mapping studies and molecular breeding in chestnut (Westbrook et al. 2020). In 2015, Staton et al. used the first interspecific genetic linkage map of *C. sativa* × *C. crenata* to identify the quantitative trait locus (QTL) for resistance to *Phytophthora cinnamomic*. This study genotyped chestnut populations through SSR and SNP markers and identified two QTLs for ink disease resistance. In several studies, QTLs related to agronomic traits have been identified. In another study, Ji et al. (2018) constructed a high-density linkage map by genotyping by sequencing (GBS) of an F_1 mapping population of 259 progenies and identified QTLs for five nut traits (Hill Craddock and Taylor Perkins 2019).

4.4 Gene Transformation

In the 1980s, the production of somatic embryos in walnuts was reported by Tulecke and McGranahan (1985), which facilitated the use of genetic engineering techniques to accelerate walnut breeding. For the first time in woody plant species, walnut somatic embryos successfully transformed using *Agrobacterium* and express foreign genes (McGranahan et al. 1988). Dandekar et al. (1994) introduced the *cry1A*(c) gene of *Bacillus thuringiensis* into walnut to produce codling moth (*Cydia pomonella* L., CM) resistance plants, but their transgenic work was not successful because of low levels of expression of wild type *cry1A*(c) gene in transgenic walnut somatic embryos. To overcome this problem, walnut somatic embryos were transformed with the full-length synthetic version of the *cry*IA(c) gene (Dandekar et al. 1998). Escobar et al. (2000) used GFP as a scorable marker for efficient and quick visual selection of transgenic embryos. The *rolABC* genes were inserted into somatic embryos of a 'Paradox' hybrid using the *A. tumefaciens*. The rooting of the transgenic plant was investigated (Vahdati et al. 2002). The results showed that these genes induced a shorter internode length and a more fibrous root system but did not affect rooting potential. Zhang et al. (2015) used a red fluorescent protein (DsRED) as a reporter for selecting transformed walnut somatic embryos. The results demonstrated that this approach is more reliable and efficient than green fluorescent protein (GFP) and β-glucuronidase (GUS) for selecting transgenic walnut embryos. Tobacco lines transformed using *JrPPO1* gene showed significantly higher PPO activity and lower disease severity to the bacterial blight (Khodadadi et al. 2020). Sheikh Beig Goharrizi et al. (2016) inserted *fld* gene to walnut somatic embryos via agrobacterium to improve osmotic stress tolerance of transgenic plants. In addition, *Agrobacterium*-mediated transformation of walnut somatic embryos using BADH gene conferred salt and drought tolerance in transgenic plants (Rezaei Qusheh Bolagh et al. 2021).

To the best of our knowledge, there has been no report of genetically engineered pistachio cultivars. On the other hand, somatic embryogenesis from kernel (Onay et al. 1995; Ghadirzadeh-Khorzoghi et al. 2019), flower (Onay et al. 2004) and leaf explants (Onay 2000) of pistachio have been reported, which could be considered as the first step for genetic engineering of pistachio cultivars. This provides an opportunity to use new genome-editing methods such as transcription activator-like effector nucleases (TALENs), zinc finger nucleases (ZFNs) and clustered regulatory interspaced short palindromic repeat (CRISPR)/Cas-based methods. Finally, considering the long juvenility period, as one of the most important barriers of producing improved pistachio cultivars, can be addressed by developing a FastTrack breeding system in which progenies of a perennial woody plant transformed by an early-flowering gene, i.e. the Poplar FT (*PtFT*) gene, rapidly inter flowering phase a few months after seed germination (Sheikhi et al. 2019a).

The development of genetically engineered almond cultivars depends upon the availability of effective regeneration protocols coupled with the methods that permit efficient DNA delivery, selection of transformed cells and recovery of transgenic plants. Other relatives of almond, including *P. domestica* has already been successfully transformed (Petri et al. 2018), and will lay a solid foundation for future gene manipulations in almond fruit trees. Although new genetic transformation procedures such as CRISPR/Cas9-mediated approaches have been used in horticultural crops (Xin et al. 2022), Agrobacterium-mediated transformation is the only method implemented in almonds (Costa et al. 2007; Sabbadini et al. 2019). However, there is a long road ahead to genetic engineered almond cultivars. One of the main obstacles is that the regeneration of plantlets from regular cultivar cells is very difficult mostly due to the recalcitrance of calli cells from almond explants to initiate the organogenesis (Costa et al. 2007; Santos et al. 2009). Recently, Zong et al. (2019) reported successful adventitious shoot regeneration and agrobacterium-mediated transformation of almond × peach hybrid rootstock 'Hansen 536'.

Chang et al. (2018) studied in vitro cultivation and molecular biology in trees, especially chestnuts. Earlier, the American Chestnut Foundation, which had a cross-breeding programme, worked with scientists at the University of New York, College of Plant Science to increase the resistance to blight in chestnuts using genetic engineering. Somatic embryo cultures of *C. dentata* are an important breakthrough in this area. The main gene conferring blight tolerance is a gene encoding oxalate oxidase produced in various plants, such as peanut, strawberry, beet, cereal grains and apricot (Chang et al. 2018). In vitro study of oxalate oxidase gene expression in transformed plants showed high *C. parasitica* infection tolerance (Hill Craddock and Taylor Perkins 2019). Santos et al. (2017) used plantlets generated by micropropagation to study the genetics of resistance to ink disease. Researchers in Europe are also looking for using genetic engineering to increase resistance to both blight and ink disease in *C. sativa*.

5 Conclusions and Future Strategies

In recent decades, demand for temperate nut trees production has been strangely increased due to high economic and nutritional value. In line with increasing demand, temperate nut tree breeding programmes have been accelerated, especially due to molecular techniques, which led to the release of various cultivars and rootstocks. Like other fruit trees, high productive scions with desirable kernel characteristics and disease-resistance and rootstocks tolerance to biotic and abiotic stresses are the main breeding objectives of temperate nut trees. In addition, global warming and climate change dictate new breeding objectives such as low-chilling requirements and tolerance to late-spring and early-autumn frosts, drought, salinity and heat stresses. Temperate nut tree breeders have been applying different conventional and molecular strategies to achieve the aforementioned objectives. With the advancement of genome sequencing technologies, high-quality and quantity of molecular data will be produced in an accelerated rate and lowered prices for tree nut crops viz. walnut, pecan, pistachio, almond, hazelnut and chestnut. Therefore, future researches in breeding of tree nut crops should be directed towards construction of fully annotated reference genomes, discovery of genes controlling traits of interest and finding molecular markers linked to these genes to be used for marker assisted selection (MAS) or genomic selection (GS), through constructing saturated linkage maps, QTLs mapping, genome-wide association mapping, along with omics techniques.

Acknowledgement We would like to thank Abdul Wahab at Shanghai Center for Plant Stress Biology and Center of Excellence in Molecular Plant Sciences, Chinese Academy of Sciences, Shanghai, China for editing the book chapter. We also appreciate Iran National Science Foundation (INSF), Iran's National Elites Foundation (NEF), Center of International Scientific Studies & Collaboration (CISSC), Ministry of Science Research and Technology of Iran, Center of Excellence of Walnut Improvement and Technology of Iran, and the University of Tehran for their supports.

References

Ali AM, Zubair SJ, Abbas AM, Jubrael JM (2016) Genetic diversity among Walnuts (*Juglans regia* L.) population in Kurdistan region–Iraq using AFLP-PCR. ZANCO. J Pure Appl Sci 28:50–55

Al-Khayri JM, Jain SM, Johnson DV (2019) Advances in plant breeding strategies: nut and beverage crops. Springer, Switzerland

Arab MM, Marrano A, Abdollahi-Arpanahi R, Leslie CA, Askari H, Neale DB, Vahdati K (2019) Genome-wide patterns of population structure and association mapping of nut-related traits in Persian walnut populations from Iran using the Axiom *J. regia* 700K SNP array. Sci Rep 9:1–14

Arab MM, Marrano A, Abdollahi-Arpanahi R, Leslie CA, Cheng H, Neale DB, Vahdati K (2020) Combining phenotype, genotype, and environment to uncover genetic components underlying water use efficiency in Persian walnut. J Exp Bot 71:1107–1127

Arabnezhad H, Bahar M, Pour AT (2011) Evaluation of genetic relationships among Iranian pistachios using microsatellite markers developed from Pistacia khinjuk stocks. Sci Hortic 128:249–254

Arús P, Olarte C, Romero M, Vargas FJ (1994a) Linkage analysis of ten isozyme genes in F1 segregating progenies of almond. J Am Soc Hortic Sci 119:339–344

Arús P, Messeguer R, Viruel MA, Tobutt K, Dirlewanger E, Santi F, Quarta R, Ritter E (1994b) The European Prunus mapping project. Euphytica 77:97–100

Arús P, Ballester J, Jáuregui B, Joobeur T, Truco MJ, Vicente MC (1999) The European Prunus mapping project: update of marker development in almond. Acta Hortic 484:331–336

Ballester J, Bošković R, Batlle I, Arús P, Vargas F, De Vicente MC (1998) Location of the self-incompatibility gene on the almond linkage map. Plant Breed 117:69–72

Ballester J, Socias i Company R, Arus P, De Vicente MC (2001) Genetic mapping of a major gene delaying blooming time in almond. Plant Breed 120:268–270

Benny J, Marra FP, Giovino A, Balan B, Caruso T, Martinelli F, Marchese A (2020) Transcriptome analysis of *Pistacia vera L* inflorescence buds in bearing and non-bearing shoots reveals the molecular mechanism causing premature flower bud abscission. Genes 11:851

Benny J, Giovino A, Marra FP, Balan B, Martinelli F, Caruso T, Marchese A (2022) Transcriptomic analysis of the *Pistacia vera* (L.) fruits enable the identification of genes and hormone-related gene linked to inflorescence bud abscission. Genes 13:60

Bentley N, Grauke LJ, Klein P (2019) Genotyping by sequencing (GBS) and SNP marker analysis of diverse accessions of pecan (*Carya illinoinensis*). Tree Genet Genomes 15:1–17

Bernard A, Barreneche T, Lheureux F, Dirlewanger E (2018) Analysis of genetic diversity and structure in a worldwide walnut (*Juglans regia* L.) germplasm using SSR markers. PLoS One 13:e0208021

Bernard A, Barreneche T, Donkpegan A, Lheureux F, Dirlewanger E (2020a) Comparison of structure analyses and core collections for the management of walnut genetic resources. Tree Genet Genomes 16:1–14
Bernard A, Barreneche T, Lheureux F, Dirlewanger E (2020b) SSR genetic diversity assessment of the INRAE's walnut (*Juglans* spp.) germplasm collection. Acta Hortic 1297:377–384
Bernard A, Marrano A, Donkpegan A, Brown PJ, Leslie CA, Neale DB, Dirlewanger E (2020c) Association and linkage mapping to unravel genetic architecture of phenological traits and lateral bearing in Persian walnut (*Juglans regia* L.). BMC Genomics 21:1–25
Bernard A, Crabier J, Donkpegan AS, Marrano A, Lheureux F, Dirlewanger E (2021) Genome-wide association study reveals candidate genes involved in fruit trait variation in Persian walnut (*Juglans regia* L.). Front Plant Sci 11:607213
Boccacci P, Aramini M, Valentini N, Bacchetta L, Rovira M, Drogoudi P, Silva AP, Solar A, Calizzano F, Erdoğan V, Cristofori V, Ciarmiello LF, Contessa C, Ferreira JJ, Marra FP, Botta R (2013) Molecular and morphological diversity of on farm hazelnut (*Corylus avellana* L.) landraces from southern Europe and their role in the origin and diffusion of cultivated germplasm. Tree Genet Genomes 9:1465–1480
Boccacci P, Beltramo C, Prando S, Lembo A, Sartor C, Mehlenbacher SA, Botta R, Torello Marinoni D (2015) In silico mining, characterization and cross-species transferability of EST-SSR markers for European hazelnut (*Corylus avellana* L.). Mol Breed 35:21
Botta R, Molnar TJ, Erdogan V, Valentini N, Marinoni DT, Mehlenbacher SA (2019) Hazelnut (*Corylus* spp.) breeding. In: Al-Khayri JM, Jain SM, Johnson DV (eds) Advances in plant breeding strategies: nut and beverage crops. Springer, Cham, pp 157–219
Bozhuyuk MR, Ercisli S, Orhan E, Koc A (2020) Determination of the genetic diversity of walnut (*Juglans regia* L.) cultivar candidates from Northeastern Turkey using SSR markers. Mitteilungen Klosterneuburg 70:269–277
Büküçü ŞB, Sütyemez M, Kefayati S, Paizila A, Jighly A, Kafkas S (2020) Major QTL with pleiotropic effects controlling time of leaf budburst and flowering-related traits in walnut (*Juglans regia* L.). Sci Rep 10:1–10
Busov VB, Rink G, Woeste K (2002) Allozyme variation and mating system of black walnut (*Juglans nigra* L.) in the central hardwood region of the United States. For Genet 9:315–322
Chakraborty S, Britton M, Martínez-García PJ, Dandekar AM (2016) Deep RNA-Seq profile reveals biodiversity, plant–microbe interactions and a large family of NBS-LRR resistance genes in walnut (*Juglans regia* L.) tissues. AMB Express 6:1–13
Chang S, Mahon EL, MacKay HA, Rottmann WH, Strauss SH, Pijut PM, Powell WA, Coffey V, Lu H, Mansfield SD, Jones TJ (2018) Genetic engineering of trees: progress and new horizons. In Vitro Cell Dev Biol 54:341–376
Chao CT, Parfitt DE, Ferguson L, Kallsen C, Maranto J (1997) Breeding and genetics of pistachio: the California Program. Acta Hortic 470:152–161
Chen S, Harmon AC (2006) Advances in plant proteomics. Proteomics 6:5504–5516
Cheng Y, Liu J, Zhang H, Wang J, Zhao Y, Geng W (2015) Transcriptome analysis and gene expression profiling of abortive and developing ovules during fruit development in hazelnut. PLoS One 10:e0122072
Christopoulos MV, Rouskas D, Tsantili E, Bebeli PJ (2010) Germplasm diversity and genetic relationships among walnut (*Juglans regia* L.) cultivars and Greek local selections revealed by Inter-Simple Sequence Repeat (ISSR) markers. Sci Hortic 125:584–592
Ciarmiello LF, Piccirillo P, Pontecorvo G, De Luca A, Kafantaris I, Woodrow P (2011) A PCR based SNPs marker for specific characterization of English walnut (*Juglans regia* L.) cultivars. Mol Biol Rep 38:1237–1249
Cosmulescu S, Ionescu M (2021) Phenological and pomological properties of promising walnut (*Juglans regia* L.) genotype with cluster fruiting from selected native population in Oltenia, Romania. Genet Resour Crop Evol 68:2289–2297
Costa MC, Miguel C, Oliveira MM (2007) Improved conditions for Agrobacterium-mediated transformation of almond. Acta Hortic 738:575–581

Crane J (1974) Hermaphroditism in pistacia. Calif Agric 28:3–4

D'Amico-Willman KM, Niederhuth C, Willman MR, Gradziel TM, Ouma WZ, Meulia T, Ramirez JF (2021) Integrated analysis of the methylome and transcriptome of twin almonds (*Prunus dulcis* [Mill.] DA Webb) reveals genomic features associated with non-infectious bud failure. BioRxiv. https://doi.org/10.1101/2021.02.08.430330

D'Amico-Willman KM, Ouma WZ, Meulia T, Sideli GM, Gradziel TM, Fresnedo-Ramírez J (2022) Whole-genome sequence and methylome profiling of the almond (*Prunus dulcis* [Mill.] DA Webb) cultivar 'Nonpareil'. G3 12(5):jkac065

Dandekar AM, McGranahan GH, Vail PV, Uratsu SL, Leslie C, Tebbets JS (1994) Low levels of expression of wild type Bacillus thuringiensis var. Kurstaki cryIA (c) sequences in transgenic walnut somatic embryos. Plant Sci 96:151–162

Dandekar AM, McGranahan GH, Vail PV, Uratsu SL, Leslie CA, Tebbets JS (1998) High levels of expression of full-length cryIA (c) gene from Bacillus thuringiensis in transgenic somatic walnut embryos. Plant Sci 131:181–193

Dang M, Zhang T, Hu Y, Zhou H, Woeste KE, Zhao P (2016) De novo assembly and characterization of bud, leaf and flowers transcriptome from *Juglans regia* L. for the identification and characterization of new EST-SSRs. Forests 7:247

Das A, Paudel B, Rohila JS (2015) Potentials of proteomics in crop breeding. In: Al-Khayri JM, Jain SM, Johnson DV (eds) Advances in plant breeding strategies: breeding, biotechnology and molecular tools. Springer, Cham, pp 513–537

Di Guardo M, Farneti B, Khomenko I, Modica G, Mosca A, Distefano G, Gentile A (2021) Genetic characterization of an almond germplasm collection and volatilome profiling of raw and roasted kernels. Hortic Res 8:1–19

Dirlewanger E, Cosson P, Howad W, Capdeville G, Bosselut N, Claverie M, Voisin R, Poizat C, Lafargue B, Baron O, Laigret F, Kleinhentz M, Arús P, Esmenjaud D (2004) Microsatellite genetic linkage maps of myrobalan plum and an almond-peach hybrid - location of root-knot nematode resistance genes. Theor Appl Genet 109:827–832

Ebrahimi A, Zarei A, McKenna JR, Bujdoso G, Woeste KE (2017) Genetic diversity of Persian walnut (*Juglans regia* L) in the cold-temperate zone of the United States and Europe. Sci Hortic 220:36–41

Ebrahimi A, Lawson SS, McKenna JR, Jacobs DF (2020) Morpho-physiological and genomic evaluation of *Juglans* species reveals regional maladaptation to cold stress. Front Plant Sci 11: 229

Erdogan V, Mehlenbacher SA (2000) Interspecific hybridization in hazelnut (Corylus). J Am Soc Hortic Sci 125:489–497

Erturk U, Dalkilic Z (2011) Determination of genetic relationship among some walnut (*Juglans regia* L.) genotypes and their early-bearing progenies using RAPD markers. Rom Biotechnol Lett 16:5944–5952

Escobar MA, Park JI, Polito VS, Leslie CA, Uratsu SL, McGranahan GH, Dandekar AM (2000) Using GFP as a scorable marker in walnut somatic embryo transformation. Ann Bot 85:831–835

Esfandiyari B, Davarynejad GH, Shahriari F, Kiani M, Mathe A (2012) Data to the sex determination in Pistacia species using molecular markers. Euphytica 185:227–231

Espiau MT, Ansón JM, Socias i Company R (2002) The almond germplasm bank of Zaragoza. Acta Hortic 591:275–278

Fallah M, Vahdati K, Hasani D, Rasouli M, Sarikhani S (2022) Breeding of Persian walnut: aiming to introduce late-leafing and early-harvesting varieties by targeted hybridization. Sci Hortic 295: 110885

Famula RA, Richards JH, Famula TR, Neale DB (2019) Association genetics of carbon isotope discrimination and leaf morphology in a breeding population of *Juglans regia* L. Tree Genet Genomes 15:1–13

Feng S, Fang H, Liu X, Dong Y, Wang Q, Yang KQ (2021) Genome-wide identification and characterization of long non-coding RNAs conferring resistance to Colletotrichum gloeosporioides in walnut (*Juglans regia*). BMC Genomics 22:1–17

Ferguson L, Beede RH, Reyes H, Sanden BL, Grattan SR, Epstein L (2016) Pistachio rootstocks. In: Ferguson L, Haviland D (eds) Pistachio production manual, 5th edn. Fruit and Nut Research and Information Center, University of California, Davis, pp 65–73

Ferreira JJ, Garcia-González C, Tous J, Rovira M (2010) Genetic diversity revealed by morphological traits and ISSR markers in hazelnut germplasm from northern Spain. Plant Breed 129: 435–441

Fjellstrom RG, Parfitt DE (1994) Walnut (*Juglans* spp.) genetic diversity determined by restriction fragment length polymorphisms. Genome 37:690–700

Foolad MR, Arulsekar S, Becerra V, Bliss FA (1995) A genetic map of Prunus based on an interspecific cross between peach and almond. Theor Appl Genet 91:262–269

Forcada ICF, Martí IÀF, Company IRS (2012) Mapping quantitative trait loci for kernel composition in almond. BMC Genet 13:1–9

Francesca PI, Pamfil D, Raica P, Petricele IV, Sisea C, Vas E, Botu M (2010) Assessment of the genetic variability among some Juglans cultivars from the Romanian National Collection at SCDP Vâlcea using RAPD markers. Rom Biotechnol Lett 15:41–49

Fresnedo-Ramírez J, Chan HM, Parfitt DE, Crisosto CH, Gradziel TM (2017) Genome-wide DNA-(de) methylation is associated with noninfectious bud-failure exhibition in almond (*Prunus dulcis* [Mill.] DA Webb). Sci Rep 7:1–12

Ghadirzadeh-Khorzoghi E, Jahanbakhshian-Davaran Z, Seyedi SM (2019) Direct somatic embryogenesis of drought resistance pistachio (*Pistacia vera* L.) and expression analysis of somatic embryogenesis-related genes. S Afr J Bot 121:558–567

Goharrizi KJ, Baghizadeh A, Kalantar M, Fatehi F (2019) Assessment of changes in some biochemical traits and proteomic profile of UCB-1 pistachio rootstock leaf under salinity stress. J Plant Growth Regul 39:608–630

Gómez EM, Federico D, Pedro JMG, Encarnación O (2015) iTRAQ-based quantitative proteomic analysis of pistils and anthers from self-incompatible and self-compatible almonds with the S f haplotype. Mol Breed 35:1–14

Gómez EM, Buti M, Sargent DJ, Dicenta F, Ortega E (2019) Transcriptomic analysis of pollen-pistil interactions in almond (*Prunus dulcis*) identifies candidate genes for components of gametophytic self-incompatibility. Tree Genet Genomes 15:1–13

Goonetilleke SN, March TJ, Wirthensohn MG, Arús P, Walker AR, Mather DE (2018) Genotyping by sequencing in almond: SNP discovery, linkage mapping, and marker design. G3 8:161–172

Gradziel TM (2008) Almond (*Prunus dulcis*). In: Bhaskaran S, Prabhudesai VR (eds) Breeding of plantation crops. Springer, Boston, pp 1–33

Gradziel TM, Martínez-Gómez P (2013) Almond breeding. In: Janick J (ed) Plant breeding reviews, vol 37, pp 207–258

Gradziel TM, Shackel KA (2021) Propagation of an epigenetic age-related disorder in almond is governed by vegetative bud ontogeny rather than chimera-type cell lineage. Horticulturae 7:190

Gradziel TM, Martínez-Gómez P, Dicenta F, Kester DE (2001) The utilization of related 4 species for almond variety improvement. J Am Pomol Soc 55:100–108

Grasselly C (1976) Les espèces sauvages d'amandiers. Options Méditerranéennes; Revue des Problèmes. Agricole Méditerranéennes 32:28–43

Grasselly C, Crossa-Raynaud P (1980) L'amandier. G.P. Maisonneuve et Larose, Paris, p XII, 446

Guney M, Kafkas S, Keles H, Zarifikhosroshahi M, Gundesli MA, Ercisli S, Bujdoso G (2021) Genetic diversity among some walnut (*Juglans regia* L.) genotypes by SSR markers. Sustainability 13:6830

Gunn BF, Aradhya M, Salick JM, Miller AJ, Yongping Y, Lin L, Xian H (2010) Genetic variation in walnuts (*Juglans regia* and *J. sigillata*; *Juglandaceae*):species distinctions, human impacts, and the conservation of agrobiodiversity in Yunnan, China. Am J Bot 97:660–671

Guo C, Wei Y, Yang B, Ayup M, Li N, Liu J, Wang H (2021) Developmental transcriptome profiling uncovered carbon signaling genes associated with almond fruit drop. Sci Rep 11:1–12

Habibi A, Yazdani N, Koushesh Saba M, Vahdati K (2017) Ascorbic acid preserved phenolic compounds of cold stressed fresh walnut kernels. In: I International conference and X National Horticultural Science Congress of Iran, 1315, pp 665–668

Habibi A, Yazdani N, Chatrabnous N, Koushesh Saba M, Vahdati K (2021) Inhibition of browning via aqueous gel solution of Aloe vera: a new method for preserving fresh fruits as a case study on fresh kernels of Persian walnut. J Food Sci Technol 31:1–10

Habibi A, Hamedpour-Darabi M, Vahdati K (2022) Local cultural values of Persian walnut in Iran. The cultural value of trees. Taylor & Francis, London, pp 162–173

Habibie A, Yazdani N, Saba MK, Vahdati K (2017) Limitation of access to oxygen for inhibition of browning in fresh walnut kernels. In: VIII International symposium on walnut, cashew and pecan, 29, pp 137–140

Habibie A, Yazdani N, Saba MK, Vahdati K (2019) Ascorbic acid incorporated with walnut green husk extract for preserving the postharvest quality of cold storage fresh walnut kernels. Sci Hortic 245:193–199

Hajinia Z, Sarikhani S, Vahdati K (2021) Exploring low-chill genotypes of Persian walnut (*Juglans regia* L.) in west of Iran. Genet Resour Crop Evol 68:2325–2336

Hassankhah A, Rahemi M, Ramshini H, Sarikhani S, Vahdati K (2020) Flowering in Persian walnut: patterns of gene expression during flower development. BMC Plant Biol 20:1–10

Hill Craddock J, Taylor Perkins M (2019) Chestnut (*Castanea spp. Miller*) breeding. In: Al-Khayri JM, Jain SM, Johnson DV (eds) Advances in plant breeding strategies: nut and beverage crops. Springer, Cham, pp 105–156

Hormaza J, Herrero M (1998) Pollen effects on fruit and seed characteristics in pistachio (*Pistacia Vera* L.). Ann Appl Biol 132:357–364

Hormaza J, Wünsch A (2007) Pistachio. In: Kole C (ed) Fruits and nuts. Genome mapping and molecular breeding in plants, vol 4. Springer, Berlin, pp 243–251

Hormaza J, Dollo L, Polito V (1994) Identification of a RAPD marker linked to sex determination in Pistacia Vera using bulked segregant analysis. Theor Appl Genet 89:9–13

Huang R, Zhou Y, Zhang J, Ji F, Jin F, Fan W, Pei D (2020) Transcriptome analysis of walnut (*Juglans regia* L.) embryos reveals key developmental stages and genes involved in lipid biosynthesis and polyunsaturated fatty acid metabolism. J Agric Food Chem 69:377–396

Iwata H, Minamikawa MF, Kajiya-Kanegae H, Ishimori M, Hayashi T (2016) Genomics-assisted breeding in fruit trees. Breed Sci 66:100–115

Jacobs DF, Dalgleish HJ, Nelson CD (2013) A conceptual framework for restoration of threatened plants: the effective model of American chestnut (*Castanea dentata*) reintroduction. New Phytol 197:378–393

Jáuregui B (1998) Localización de marcadores moleculares ligados a caracteres agronómicos en un cruzamiento interespecífi co almendro × melocotonero. PhD Thesis, Univ. Barcelona, Spain

Jazi MM, Seyedi SM, Ebrahimie E, Ebrahimi M, De Moro G, Botanga C (2017) A genome-wide transcriptome map of pistachio (*Pistacia vera* L.) provides novel insights into salinity-related genes and marker discovery. BMC Genomics 18:1–21

Ji A, Wang Y, Wu G, Wu W, Yang H, Wang Q (2014) Genetic diversity and population structure of North China mountain walnut revealed by ISSR. Am J Plant Sci 5:3194

Ji F, Wei W, Liu Y, Wang G, Zhang Q, Xing Y, Zhang S, Liu Z, Cao Q, Qin L (2018) Construction of a SNP-based high-density genetic map using genotyping by sequencing (GBS) and QTL analysis of nut traits in Chinese chestnut (*Castanea mollissima* Blume). Front Plant Sci 9:816

Ji F, Ma Q, Zhang W, Liu J, Feng Y, Zhao P, Song X, Chen J, Zhang J, Wei X, Zhou Y (2021) A genome variation map provides insights into the genetics of walnut adaptation and agronomic traits. Genome Biol 22:1–22

Kafkas S, Kaska N (1997) Pistachio rootstock breeding by crossing different wild species grown in Turkey. Acta Hortic 470:219–225

Kafkas S, Perl-Treves R (2001) Morphological and molecular phylogeny of Pistacia species in Turkey. Theor Appl Genet 102:908–915

Kafkas S, Perl-Treves R, Kaska N (2000) Unusual Pistacia atlantica Desf. (*Anacardiaceae*) monoecious sex type in the Yunt Mountains of the Manisa Province of Turkey. Isr J Plant Sci 48:277–280

Kafkas S, Khodaeiaminjan M, Güney M, Kafkas E (2015) Identification of sex-linked SNP markers using RAD sequencing suggests ZW/ZZ sex determination in *Pistachio vera* L. BMC Genomics 16:98

Kavas M, Kızıldoğan AK, Balık Hİ (2019) Gene expression analysis of bud burst process in European hazelnut (*Corylus avellana* L.) using RNA-Seq. Physiol Mol Biol Plants 25:13–29

Kester DE, Gradziel TM (1996) Almonds. In: Janick J, Moore JN (eds) Fruit breeding. Nuts, vol III. Wiley, New York, pp 1–97

Khodadadi F, Tohidfar M, Vahdati K, Dandekar AM, Leslie CA (2020) Functional analysis of walnut polyphenol oxidase gene (*JrPPO1*) in transgenic tobacco plants and PPO induction in response to walnut bacterial blight. Plant Pathol 69:756–764

Khodaeiaminjan M, Kafkas E, Güney M (2017) Development and linkage mapping of novel sex-linked markers for marker-assisted cultivar breeding in pistachio (*Pistacia vera* L.). Mol Breed 37:98

Khorami SS, Arzani K, Karimzadeh G, Shojaeiyan A, Ligterink W (2018) Genome size: a novel predictor of nut weight and nut size of walnut trees. Am Soc Horticult Sci 53(3):275–282

Kodad O, Alonso JM, Fernández i Martí À, Oliveira MM, Socias i Company R (2010) Molecular and physiological identify cation of new S-alleles associated with self-(in)compatibility in local Spanish almond cultivars. Sci Hortic 123:308–311

Kouhi M, Rezaei A, Hassani D, Sarikhani S, Vahdati K (2020) Phenotypic evaluation and identification of superior Persian walnut (*Juglans regia* L.) genotypes in Mazandaran province, Iran. J Nuts 11(4):315–326

Ladizinsky G (1999) On the origin of almond. Genet Resour Crop Evol 46:143–147

Li Y, Luo X, Wu C, Cao S, Zhou Y, Jie B, Wu G (2018) Comparative transcriptome analysis of genes involved in anthocyanin biosynthesis in red and green walnut (*Juglans regia* L.). Molecules 23:25

Liu J, Xing J, Fang J, Ai P, Cheng Y (2018) New insight into ovary abortion during ovary development of hazelnut through a combined proteomic and transcriptomic analysis. Sci Hortic 234:36–48

Liu X, Meng P, Yang G, Zhang M, Peng S, Zhai MZ (2020) Genome-wide identification and transcript profiles of walnut heat stress transcription factor involved in abiotic stress. BMC Genomics 21:1–13

Ma Q, Qi J, Pei D (2010) FISH-AFLP analysis of genetic diversity of early-fruiting walnut cultivars. For Res 23:631–636

Ma K, Luo X, Han L, Zhao Y, Mamat A, Li N, Wang J (2021) Transcriptome profiling based on Illumina-and SMRT-based RNA-seq reveals circadian regulation of key pathways in flower bud development in walnut. PLoS One 16:e0260017

Maggs D (1990) The Australian pistachio 'Sirora'. Fruit Varieties J 44:178–179

Marrano A, Martínez García PJ, Bianco L, Sideli GM, Di Pierro EA, Leslie CA, Neale DB (2019a) A new genomic tool for walnut (*Juglans regia* L.): development and validation of the high-density Axiom™ J. regia 700K SNP genotyping array. Plant Biotechnol J 17:1027–1036

Marrano A, Sideli GM, Leslie CA, Cheng H, Neale DB (2019b) Deciphering of the genetic control of phenology, yield, and pellicle color in Persian walnut (*Juglans regia* L.). Front Plant Sci 10: 1140

Marrano A, Britton M, Zaini PA, Zimin AV, Workman RE, Puiu D, Neale DB (2020) High-quality chromosome-scale assembly of the walnut (*Juglans regia* L.) reference genome. Gigascience 9: 50

Martín MA, Mattioni C, Cherubini M, Villani F, Martín LM (2017) A comparative study of European chestnut varieties in relation to adaptive markers. Agrofor Syst 91:97–109

Martinelli F, Marchese A, Balan B, Giovino A, Caruso T, Fretto S, Marra FP (2017) RNA-Seq analysis to investigate alternate bearing mechanism in *Pistacia Vera* L. Int Symp Flower Fruit Set Alternate Bearing 1229:71–78

Martínez-García PJ, Gãmez EM, Casado-Vela J, Elortza F, Dicenta F, Ortega E (2015) Differential protein expression in compatible and incompatible pollen-pistil interactions in almond [Prunus dulcis (Miller) DA Webb] by 2D-DIGE and HPLC-MS/MS. J Hortic Sci Biotechnol 90:71–77

Martínez-García PJ, Crepeau MW, Puiu D, Gonzalez-Ibeas D, Whalen J, Stevens KA, Neale DB (2016) The walnut (*Juglans regia* L.) genome sequence reveals diversity in genes coding for the biosynthesis of non-structural polyphenols. Plant J 87:507–532

Martínez-García PJ, Prudencio AS, Sánchez-Pérez R, De Los Cobos FP, Kodad O, Gouta H, Martínez-Gómez P (2019) Almond [*Prunus dulcis* (Miller) DA Webb] breeding. In: Al-Khayri JM, Jain SM, Johnson DV (eds) Advances in plant breeding strategies: nut and beverage crops, vol 4, pp 3–29

McGranahan GH, Leslie CA (2005) Advances in genetic improvement of walnut at the University of California, Davis. Acta Hortic 705:117–122

McGranahan GH, Lesli CA, Uratsu SL, Martin LA, Dandekar AM (1988) Agrobacterium-mediated transformation of walnut somatic embryos and regeneration of transgenic plants. Biotechnology 6:800–804

Mehlenbacher SA (2003) Progress and prospects in nut breeding. Acta Hortic 622:57–79

Mehlenbacher SA (2018) Advances in genetic improvement of hazelnut. Acta Hortic 1226(1):12

Motalebipour ZE, Kafkas S, Khodaeiaminjan M, Çoban N, Gözel H (2016) Genome survey of pistachio (*Pistacia Vera* L.) by next generation sequencing: development of novel SSR markers and genetic diversity in *Pistacia* species. BMC Genomics 17:998

Mousavi S, Alisoltani A, Shiran B, Fallahi H, Ebrahimie E, Imani A, Houshmand S (2014) De novo transcriptome assembly and comparative analysis of differentially expressed genes in *Prunus dulcis* Mill. in response to freezing stress. PLoS One 9:e104541

Muehlbauer MF, Honig JA, Capik JM, Vaiciunas JN, Molnar TJ (2014) Characterization of eastern filbert blight-resistant hazelnut germplasm using microsatellite markers. J Am Soc Hortic Sci 139:399–432

Ninot A, Aleta N (2003) Identification and genetic relationship of Persian walnut genotypes using isozyme markers. J Am Pomol Soc 57:106

Nishio S, Yamamoto T, Terakami S, Sawamura Y, Takada N, Saito T (2011) Genetic diversity of Japanese chestnut cultivars assessed by SSR markers. Breed Sci 61:109–120

Okay AN, Özenc N (2000) Hazelnut improvement through hybridization. Acta Hortic 556:235–240

Onay A (2000) Histology of somatic embryogenesis in cultured leaf explants of pistachio (*Pistacia Vera* L). Turk J Bot 24:91–96

Onay A, Jeffree CE, Yeoman MM (1995) Somatic embryogenesis in cultured immature kernels of pistachio, *Pistacia Vera* L. Plant Cell Rep 15:192–195

Onay A, Pirinç V, Yıldırım H, Basaran D (2004) In vitro micrografting of mature pistachio (*Pistacia vera* var. Siirt). Plant Cell Tissue Organ Cult 77:215–219

Orhan E, Eyduran S, Poljuha D, Akın M, Weber T, Ercişli S (2020) Genetic diversity detection of seed-propagated walnut (*Juglans regia* L.) germ plasm from Eastern Anatolia using SSR markers. Folia Hortic 32(1):37–46

Özcan A, Sütyemez M, Nar A, Yıldırım E, Süslüoglu Z (2020) Similarities of leafing and leaf fall date of 'Pedro' walnut variety with its progenies in breeding programs. Physiol Mol Biol Plants 26(10):1945–1959

Pakzad R, Fatehi F, Kalantar M, Maleki M (2019) Evaluating the antioxidant enzymes activities, lipid peroxidation and proteomic profile changing in UCB-1 pistachio rootstock leaf under drought stress. Sci Hortic 256:108617

Parfitt DE, Badenes ML, Ferguson L, Maranto J, Almehdi A, Guo Yang L (1994) Pistachio breeding and genetics program at the University of California. In: Proceedings of the I international symposium on pistachio, pp 279–280

Parfitt D, Kafkas S, Batlle I, Vargas FJ (2012) Pistachio. In: Badenes M, Byrne D (eds) Fruit breeding, Handbook of plant breeding, vol 8. Springer, Boston, pp 803–826

Pavan S, Delvento C, Mazzeo R, Ricciardi F, Losciale P, Gaeta L, Lotti C (2021) Almond diversity and homozygosity define structure, kinship, inbreeding, and linkage disequilibrium in cultivated germplasm, and reveal genomic associations with nut and seed weight. Hortic Res 8:1–12

Petri C, Alburquerque N, Faize M, Scorza R, Dardick C (2018) Current achievements and future directions in genetic engineering of European plum (*Prunus domestica* L.). Transgenic Res 27: 225–240

Pollegioni P, Bartoli S, Cannata F, Malvolti ME (2003) Genetic differentiation of four Italian walnut (*Juglans regia* L.) varieties by inter simple sequence repeat (ISSR). J Genet Breed 57: 231–240

Pollegioni P, Woeste K, Chiocchini F, Del Lungo S, Ciolfi M, Olimpieri I, Malvolti ME (2017) Rethinking the history of common walnut (*Juglans regia* L.) in Europe: its origins and human interactions. PLoS One 12:e0172541

Prudencio ÁS, Werner O, Martínez-García PJ, Dicenta F, Ros RM, Martínez-Gómez P (2018) DNA methylation analysis of dormancy release in almond (*Prunus dulcis*) flower buds using epi-genotyping by sequencing. Int J Mol Sci 19:3542

Prudencio ÁS, Hoeberichts FA, Dicenta F, Martínez-Gómez P, Sánchez-Pérez R (2021) Identification of early and late flowering time candidate genes in endodormant and ecodormant almond flower buds. Tree Physiol 41:589–605

Qu J, Hao Y, Zhu Y, Wu C, Wang W, Leng P (2011) Studies on germplasm of Juglans by EST-SSR markers. Acta Hortic 38:441–448

Rezaee R, Vahdati K, Grigoorian W, Valizadeh M (2008) Walnut grafting success and bleeding rate as affected by different grafting methods and seedling vigor. J Hortic Sci Biotechnol 83:94–99

Rezaei Qusheh Bolagh F, Solouki A, Tohidfar M, Zare Mehrjerdi M, Izadi-Darbandi A, Vahdati K (2021) Agrobacterium-mediated transformation of Persian walnut using BADH gene for salt and drought tolerance. J Hortic Sci Biotechnol 96:162–171

Ricciardi F, Del Cueto J, Bardaro N, Mazzeo R, Ricciardi L, Dicenta F, Lotti C (2018) Synteny-based development of CAPS markers linked to the sweet kernel LOCUS, controlling amygdalin accumulation in almond (*Prunus dulcis* (Mill.) DA Webb). Genes 9:385

Rikhter AA (1972) Biological basis for the creation of almond cultivars and commercial orchards (in Russian). Ed. AN SSSR, Glavny Bot. Sad, Moscow, Russia

Roor W, Konrad H, Mamadjanov D, Geburek T (2017) Population differentiation in common walnut (*Juglans regia* L.) across major parts of its native range—insights from molecular and morphometric data. J Hered 108:391–404

Rowley ER, VanBuren R, Bryant DW, Priest HD, Mehlenbacher SA, Mockler TC (2018) A draft genome and high-density genetic map of European hazelnut (*Corylus avellana* L.). BioRXiv 2018:469015

Sabbadini S, Ricci A, Limera C, Baldoni D, Capriotti L, Mezzetti B (2019) Factors affecting the regeneration, via organogenesis, and the selection of transgenic calli in the peach rootstock Hansen 536 (*Prunus persica* × *Prunus amygdalus*) to express an RNAi construct against PPV virus. Plants 8:178

Sadat-Hosseini M, Bakhtiarizadeh MR, Boroomand N, Tohidfar M, Vahdati K (2020) Combining independent de novo assemblies to optimize leaf transcriptome of Persian walnut. PLoS One 15: e0232005

Sánchez-Pérez R, Howad W, Garcia-Mas J, Arús P, Martínez-Gómez P, Dicenta F (2010) Molecular markers for kernel bitterness in almond. Tree Genet Genomes 6:237–245

Santos AM, Oliver MJ, Sanchez AM, Oliveira M (2009) An integrated strategy to identify key genes in almond adventitious shoot regeneration. J Exp Bot 60:4159–4173

Santos C, Duarte S, Tedesco S, Fevereiro P, Costa RL (2017) Expression profiling of Castanea genes during resistant and susceptible interactions with the oomycete pathogen *Phytophthora cinnamomi* reveal possible mechanisms of immunity. Front Plant Sci 8:515

Sarikhani KS, Arzani K, Roozban MR (2012) Identification and selection of twelve walnut superior and promising genotypes in Fars province, Iran. Seed Plant Prod J 28-1:277–296

Sarikhani S, Vahdati K, Ligterink W (2021) Biochemical properties of superior Persian walnut genotypes originated from southwest of Iran. Int J Hortic Sci Technol 8:13–24

Savadi S, Mangalassery S, Sandesh MS (2021) Advances in genomics and genome editing for breeding next generation of fruit and nut crops. Genomics 113:3718–3734

Shah UN, Mir JI, Ahmed N, Fazili KM (2019) Genetic diversity analysis of walnut (*Juglans regia* L.) from Kashmir Valley using RAPD and ISSR markers. Agrotechnology 8:185

Sheikh Beig Goharrizi MA, Dejahang A, Tohidfar M, Izadi Darbandi A, Carrillo NJ, Hajirezaei MR, Vahdati K (2016) Agrobacterium mediated transformation of somatic embryos of Persian walnut using fld gene for osmotic stress tolerance. J Agric Sci Technol 18:423–435

Sheikhi A, Arab MM, Brown PJ, Ferguson L, Akbari M (2019a) Pistachio (*Pistacia* spp.) breeding. In: Al-Khayri JM, Jain SM, Johnson DV (eds) Advances in plant breeding strategies: nut and beverage crops. Springer, Cham, pp 353–400

Sheikhi A, Mirdehghan SH, Karimi HR, Ferguson L (2019b) Effects of passive- and active-modified atmosphere packaging on physio-chemical and quality attributes of fresh in-hull pistachios (*Pistacia vera* L. cv. Badami). Foods 8:564

Sideli GM, Marrano A, Montanari S, Leslie CA, Allen BJ, Cheng H, Neale DB (2020a) Quantitative phenotyping of shell suture strength in walnut (*Juglans regia* L.) enhances precision for detection of QTL and genome-wide association mapping. PLoS One 15:e0231144

Sideli GM, McAtee P, Marrano A, Allen BJ, Brown PJ, Butterfield TS, Neale DB (2020b) Genetic analysis of walnut (*Juglans regia* L.) pellicle pigment variation through a novel, high-throughput phenotyping platform. G3 10:4411–4424

Socias i Company R (2002) Latest advances in almond self-compatibility. Acta Hortic 591:205–212

Socias i Company R, Felipe AJ, Gómez Aparisi J (1999) A major gene for flowering time in almond. Plant Breed 118:443–448

R. Socias i Company Alonso JM, Kodad O, Gradziel TM (2012) Almond. In Alonso JM, Kodad O, Gradziel TM, R. Socias i Company Fruit breeding. Springer, Boston 697-728

Sparks D (2005) Adaptability of pecan as a species. HortScience 40:1175–1189

Staton M, Zhebentyayeva T, Olukolu B, Fang GC, Nelson D, Carlson JE, Abbott AG (2015) Substantial genome synteny preservation among woody angiosperm species: comparative genomics of Chinese chestnut (*Castanea mollissima*) and plant reference genomes. BMC Genomics 16:1–3

Sütyemez M, Bükücü ŞB, Keleş Ö et al (2021) Phenological differences, genetic diversity, and population structure of genotypes obtained from seeds of Kaman-1 walnut cultivar. J Food Qual 2021:5579875

Tabasi M, Sheidai M, Hassani D, Koohdar F (2020) DNA fingerprinting and genetic diversity analysis with SCoT markers of Persian walnut populations (*Juglans regia* L.) in Iran. Genet Resour Crop Evol 67:1437–1447

Talebi M, Akbari M, Zamani M et al (2016) Molecular polymorphism in *Pistacia vera* L. using non-coding regions of chloroplast DNA. J Genet Eng Biotechnol 14:31–37

Tulecke W, McGranahan G (1985) Somatic embryogenesis and plant regeneration from cotyledons of walnut, *Juglans regia* L. Plant Sci 40:57–63

Tulecke W, McGranahan G (1994) The walnut germplams collection of the University of California. University of California, Davis, CA

Turkeli Y, Kafkas S (2013) First genetic linkage map in pistachio constructed using an interspecific cross between *Pistacia Vera* L. and monoecious *Pistacia atlantica* Desf. Sci Hortic 151:30–37

Vahdati K, Khorami SS (2021) The past, present and future of walnut genetic improvement and propagation. Acta Hortic 1318:251–258

Vahdati K, McKenna JR, Dandekar AM, Leslie CA, Uratsu SL, Hackett WP, McGranahan GH (2002) Rooting and other characteristics of a transgenic walnut hybrid (*Juglans hindsii* × *J. regia*) rootstock expressing rolABC. J Am Soc Hortic Sci 127:724–728

Vahdati K, Jariteh M, Niknam V, Mirmasoumi M, Ebrahimzadeh H (2006) Somatic embryogenesis and embryo maturation in Persian walnut. Acta Hortic 705:199–205
Vahdati K, Razaee R, Mirmasoomi M (2009) Micropropagation of some dwarf and early mature walnut genotypes. Biotechnology 8(1):171–175
Vahdati K, Arab MM, Sarikhani S, Sadat-Hosseini M, Leslie CA, Brown PJ (2019) Advances in Persian walnut (*Juglans regia* L.) breeding strategies. In: Al-Khayri JM, Jain SM, Johnson DV (eds) Advances in plant breeding strategies: nut and beverage crops. Springer, Cham, pp 401–472
Vahdati K, Sarikhani S, Arab MM, Leslie CA, Dandekar AM, Aleta N, Mehlenbacher SA (2021) Advances in rootstock breeding of nut trees: objectives and strategies. Plants 10(11):2234
Valentini N, Portis E, Botta R, Acquadro A, Pavese V, Cavalet EG, Torello DM (2021) Mapping the genetic regions responsible for key phenology-related traits in the European hazelnut. Front Plant Sci 12:749394–749394
Vargas F, Romero M, Vargas I (2001) Flowering precocity in pistachio progenies. Acta Hortic 591: 297–303
Velasco D, Hough J, Aradhya M, Ross-Ibarra J (2016) Evolutionary genomics of peach and almond domestication. G3 6:3985–3993
Viruel MA, Messeguer R, Vicente MC, Garcia-Mas J, Puigdomènech P, Vargas F, Arús P (1995) A linkage map with RFLP and isozyme markers for almond. Theor Appl Genet 91:964–971
Vischi M, Chiabà C, Raranciuc S, Poggetti L, Messina R, Ermacora P, Testolin R (2017) Genetic diversity of walnut (*Juglans regia* L.) in the Eastern Italian Alps. Forests 8:81
Vyas D, Sharma SK, Sharma DR (2003) Genetic structure of walnut genotype using leaf isozymes as variability measure. Sci Hortic 97:141–152
Wambulwa MC, Fan PZ, Milne R, Wu ZY, Luo YH, Wang YH, Wang H, Gao LM, Xiahou ZY, Jin YC, Ye LJ (2021) Genetic analysis of walnut cultivars from southwest China: implications for germplasm improvement. Plant Divers. https://doi.org/10.1016/j.pld.2021.08.005
Wang JY, Yan SY, Hui WK, Gong W (2020) SNP discovery for genetic diversity and population structure analysis coupled with restriction-associated DNA (RAD) sequencing in walnut cultivars of Sichuan Province, China. Biotechnol Biotechnol Equip 34:652–664
Wells L, Conner P (2015) Pecan varieties for Georgia Orchards. University of Georgia Cooperative Extension Service, p Circular 898
Westbrook JW, Zhang Q, Mandal MK et al (2020) Genomic selection for blight-tolerance in American chestnut (*Castanea dentata*) backcross populations. Evol Appl 13(1):31–47
Willmann MR, Poethig RS (2007) Conservation and evolution of miRNA regulatory programs in plant development. Curr Opin Plant Biol 10:503–511
Xin T, Tian H, Ma Y, Wang S, Yang L, Li X, Yang X (2022) Targeted creating new mutants with compact plant architecture using CRISPR/Cas9 genome editing by an optimized genetic transformation procedure in cucurbit plants. Hortic Res 9:uhab086
Xu Z, Hu T, Zhang F (2012) Genetic diversity of walnut revealed by AFLP and RAPD markers. J Agric Sci 4:271–276
Xu L, Chen X, Zhang LS, Liu QZ (2014) Molecular cloning and expression analysis of the transcription factor gene JrCBF from *Juglans regia* L. Acta Hortic 1050:41–47
Yildiz E, Pinar H, Uzun A, Yaman M, Sumbul A, Ercisli S (2021) Identification of genetic diversity among *Juglans regia* L. genotypes using molecular, morphological, and fatty acid data. Genet Resour Crop Evol 68:1425–1437
Yirmibeş B, Aydin HN, Ulu ZÖ, Ulu S, Özgentürk NÖ (2021) Analysis of some conserved miRNAs in hazelnut (*Corylus avellena* L. and *Corylus colurna* L.) by real-time PCR. Sigma J Eng Nat Sci 39:170–176
You FM, Deal KR, Wang J, Britton MT, Fass JN, Lin D, Dvorak J (2012) Genome-wide SNP discovery in walnut with an AGSNP pipeline updated for SNP discovery in allogamous organisms. BMC Genomics 13:1–16

Yuan XY, Sun YW, Bai XR, Dang M, Feng XJ, Zulfiqar S, Zhao P (2018) Population structure, genetic diversity, and gene introgression of two closely related walnuts (*Juglans regia* and *J. sigillata*) in southwestern China revealed by EST-SSR markers. Forests 9:646

Zeng L, Tu XL, Dai H, Han FM, Lu BS, Wang MS, Nanaei HA, Tajabadipour A, Mansouri M, Li XL, Ji LL (2019) Whole genomes and transcriptomes reveal adaptation and domestication of pistachio. Genome Biol 20:1–13

Zhang JB, Li RQ, Xiang XG, Manchester SR, Lin L, Wang W, Wen J, Chen ZD (2013) Integrated fossil and molecular data reveal the biogeographic diversification of the Eastern Asian-Eastern North American disjunct hickory genus (*Carya* Nutt.). PLoS One 8:e70449

Zhang Q, Walawage SL, Tricoli DM, Dandekar AM, Leslie CA (2015) A red fluorescent protein (DsRED) from *Discosoma* sp. as a reporter for gene expression in walnut somatic embryos. Plant Cell Rep 34:861–869

Zhu T, Wang L, You FM, Rodriguez JC, Deal KR, Chen L, Luo MC (2019) Sequencing a *Juglans regia*× *J. microcarpa* hybrid yields high-quality genome assemblies of parental species. Hortic Res 6:55

Zong X, Denler BJ, Danial GH, Chang Y, Song GQ (2019) Adventitious shoot regeneration and Agrobacterium tumefaciens-mediated transient transformation of almond × peach hybrid rootstock 'Hansen 536'. HortScience 54:936–940

Chapter 5
Improved Propagation Techniques in Temperate Nuts

Nowsheen Nazir, Iftisam Yaseen, Tabish Jehan Been, Aroosa Khalil, Umar Iqbal, Mohammad Maqbool Mir, Munib Ur Rehman, Shafat A. Banday, A. R. Malik, and Shahzad Bhat

Abstract In simple terms, plant propagation can be defined as reproduction or multiplication of plants. Various procedures of plant propagation need to be developed in order to commercialize the crops. Each method has its own advantages and disadvantages. The response of every plant is different to various methods of propagation. A number of propagation techniques have been developed with the aim of developing uniform crops, early bearers with increased production, resistance against pests and diseases, and introducing certain characters in new generation. It is because of these objectives that the plant propagation has become challenging and interesting. Plant propagation is both a science and an art through which the nurseries can be commercialized and income can be increased. It also helps in maintaining in the rootstock of the fruit trees. There are two methods of plant propagation, i.e. sexual and asexual. Whereas, on the one hand, sexual method includes propagation through seeds, asexual methods involve using vegetative methods like cutting, layering, grafting and budding which needs technical skilled. An advanced method of vegetative propagation in which the tissues are grown in controlled conditions is known as tissue culture. This technique is used to produce a large number of saplings that are virus free and true to type. Among the temperate fruit species, nuts hold a special position and in the past few years, their contribution has increased in the international market which has made it imperative to propagate them various means. Important nut crops like walnut (budding, bench grafting, micropropagation, hot callusing) pecan nut (hardwood and softwood cuttings, layering, etc.), chestnut (budding and grafting, direct seeding), pistachio (budding, grafting, micropropagation), hazelnut (layering, root cutting, etc.), almond

---

N. Nazir (✉) · I. Yaseen · T. J. Been · A. Khalil · U. Iqbal · M. M. Mir · Munib Ur Rehman · A. R. Malik · S. Bhat
Division of Fruit Science, FOH, Sher-e-Kashmir University of Agricultural Sciences and Technology of Kashmir, Srinagar, Jammu and Kashmir, India

S. A. Banday
Krishi Vigyan Kendra, Sher-e-Kashmir University of Agricultural Sciences and Technology of Kashmir, Ganderbal, Jammu and Kashmir, India

M. M. Mir et al. (eds.), *Temperate Nuts*,
https://doi.org/10.1007/978-981-19-9497-5_5

(budding), chilgoza (vegetative propagation and seed propagation) are mostly propagated by the above-mentioned vegetative methods. Therefore, vegetative propagation can be considered to be the best method in continuing preferred traits over generations.

Keywords Propagation · Asexual · Micropropagation · Nuts

1 Introduction

Plant propagation is one of the most basic activities of mankind. Ancient man may have started civilization when he learnt to sow and grow plants that satisfied his nutritional requirements. Cultivated plants are mostly the result of three different methods. Some plants were selected straight from their wild forms, but as a result of selection, they evolved into the types that are different from wild species. Other plants, such as pear, prunes and strawberry evolved as hybrids with chromosomal modifications. Another class of plants naturally occur as unique monstrosities. They may be beneficial to man despite lack of adaptation to their native habitat (Srivastava 1966). Fruit plants do not come true when grown from seed, hence they are usually propagated by vegetative methods, such as layering or cutting techniques on own roots, or by grafting and budding methods on roots of some other plants (Sharma 2002).

Plant propagation refers to the reproduction or multiplication of plants. Plant propagation methods and procedures have evolved as a result of the commercialization of the crops. Plant propagation is defined as the controlled multiplication or perpetuation of plants with the goals of increasing the number of plants and preserving basic traits of plant. There are two main types of propagation: sexual propagation and asexual or vegetative propagation. Sexual propagation means multiplication through seeds, whereas asexual or vegetative propagation relies on the use of plant vegetative parts to raise new plants. Plants propagated by vegetative propagation skip immature seedling stage and attain mature stage sooner. For commercial production of plant, this method saves money and time. The fundamental benefit of vegetative method of propagation is the resultant plants carrying genetic information of one parent only, making them clones of parent plant. This means plant with beneficial characteristics can reproduce continuously these beneficial traits when growing conditions remain same. This is critical for commercial producers who reproduce high-quality plants and assure the uniformity of a plant for sale.

Each plant reacts to diverse types of propagation in a unique way. Various propagation techniques have been devised with the goal of achieving crop uniformity, improved production, early bearing, pest and disease resistance and the introduction of specific traits into subsequent generations. Plant propagation has become more intriguing and demanding as a result of these goals.

Plant propagation entails the skillful integration of art and science. Through commercialized nurseries, fundamental knowledge and expertise might be a greater source of revenue. It aids in the conservation of plant stocks and the preservation of endangered (extinct) species. Growth and development of tissues under controlled conditions is recent and advanced technique of vegetative method of propagation and is called as tissue culture. It is technical and specialized method of propagation. This technique allows for the rapid and mass production of virus-free and true-to-type plants.

2 Walnut Propagation Methods

2.1 Budding

Budding is a popular method for propagating Persian walnuts. Patch, I-shield, and chip budding are the most common budding methods. Patch budding is an old and popular technique for propagating plants in an outdoor nursery (Kuniyuki and Forde 1985) (Fig. 5.1). In different countries, the success of grafting process varies (Nedev et al. 1976; Özkan et al. 2001). It is a more complex single bud grafting technique (Nedev et al. 1967). Patch budding needs a unique double-bladed knife with parallel blades for removing bark symmetrically from rootstock as well as scion (Mauricio-de-Almeida 2020) sometimes, two knives are used, smaller to remove bark from rootstock and larger to remove bark from scion, resulting in tighter fit between rootstock and scion. Several knives possess a brass tip on top of handle for lifting the bud (White 2009).

According to Solar et al. (2001), the patch budding method for walnut propagation has a success rate of only 16% in Slovenia. In Turkey, the success of patch budding is 88.3% (Özkan et al. 2001). Chandel et al. (2006) found that the best time

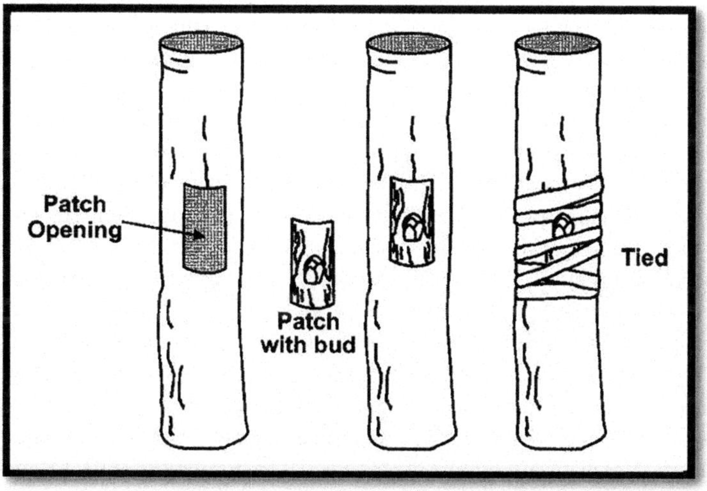

Fig. 5.1 Patch budding in Walnut (*Juglans regia*)

for grafting is from the mid to end June in Himalayan regions with success rate of 50%. Late spring and winter frosts decrease percentage of survival rate but they are not the sole limiting factors. The temperature of the air after grafting is also crucial (Mauricio-de-Almeida 2020). According to Lagerstedt and Roberts (1972), grafting in the open may fail due to lower temperatures after grafting, which make callus formation difficult or impossible. Gandev and Dzhuvinov (2006) reported that growing walnut in the open field conditions of South Bulgaria reduces the percentage of the success rate. That is why the large temperature range between day and night in European countries makes grafted tree survival difficult in the open field conditions. Scion growth and callus formation as well as rate of success were highest in patch budding (91.00%), followed by the shield budding (31.10%) and chip budding (9.10%) (Ebrahimi et al. 2007). Different methods of budding have an impact on bud-take percentage both in greenhouse and open field (Rezaee et al. 2014). Patch budding proved to be the most efficient method for propagating walnuts. But, the success of the patch budding method is highly dependent on the inoculation season. June budding was responsible for 80% of effective inoculation. Moreover, cutting off leaves 20 days before taking the buds for budding resulted in even better results, with an inoculation rate of 87% (Hodaj et al. 2014). Ring budding and flute budding are two types of modified patch budding methods (Garner and Bradley 2013). When the contact between the top and lower regions of the rootstock needs to be partially disrupted, flute budding is performed (Garner and Bradley 2013).

2.2 Micropropagation

Micropropagation is the most efficient way to propagate all plants on a large scale. It is used to produce healthy, disease-free plants and for multiplication of genotypes with ideal characteristics (Vahdatia and Aalifar 2016). Nodal segments, immature cotyledons, mature embryos, ovules and meristems can all be used to micropropagation of *J. regia* (Vahdati et al. 2004; Avıĺes et al. 2009). The number of axillary shoots emerging from microshoots was maximum in dwarf and semi-dwarf cultivars as compared to highly vigorous cultivars, and the rooting ability and cultivar vigour had a positive relationship. The major hurdles in its way are the stabilization of the tips that are formed from the in vitro culture, decreasing multiplication rate, difficulty in the proliferation of roots, apart from very low success achieved during the plant adaptation. However, in the past decade, success has been achieved in rooting and acclimatization of various cultivars of walnut (Nacheva 2012). Still this method cannot be much relied upon and therefore it has not found industrial acceptance for the production of planting material of the given species.

Physiological and morphological abnormalities are common in plantlets cultivated in traditional tissue culture techniques. The formation of leaf wax and the survival of plants in vitro culture can both benefit from forced ventilation (Zobayed et al. 2001). Similarly, Hassankhah et al. (2014) found that natural ventilation improved the majority of growth parameters in Persian walnut (Vahdatia and Aalifar

2016). Sucrose is a common carbon source in plant culture media, usually at a proportion of 2–5% (Hazarika et al. 2004). Hassankhah et al. (2014) found that using 15 g/L sucrose resulted in healthy plantlets. Vahdati et al. (2004) also studied that increasing the sucrose concentration of root induction medium increased rooting. The mineral of medium has been shown to influence morphogenic processes. Explants grown on mDKW media showed no symptoms of deficiency. Explants cultured on mDKW also developed the longest stems and the more auxiliary buds (Ashrafi et al. 2010). Higher concentrations of Cu and myoinositol affect the growth rate and percentage of rooting of explants (Najafian-Ashrafi et al. 2009). According to Vahdati et al. (2008), the optimum germination treatment contained 2 mg/L abscisic acid and resulted in a 41% transition of embryos into plantlets. He also outlined the role of sucrose in increasing the number of secondary embryos.

2.3 Bench Grafting

Bench grafting is a traditional method for propagating or topworking Persian walnuts. Whip, saddle, omega and side grafting are the most common bench grafting techniques, and callus quality, grafting survival and graft take are all significantly affected by these techniques (Vahdatia and Aalifar 2016). In four varieties of walnut ("Hartley", "Serr", "Pedro" and "Z53"), the omega method had the best success of grafting out of all combinations, with the highest callus quality (2.5 out of 4), grafting survival (84.33%), graft take (67.77%) and scion growth (12.9 cm), followed by the side stub and the whip grafting method (Dehghan et al. 2009).

In comparison to side and whip grafting, omega grafting yielded maximum callus rating (2.6 out of 4.0), the highest count of callused plants (82%) and the maximum (81%) rate of graft survival (Dehgan et al. 2010). Three weeks after grafting, the percentage of graft take for both whip and cleft grafting methods was 80% (Rezaee et al. 2008). Several scions failed to grow and died within a year. This failure could be caused by strong root pressure on walnut trees in the spring or winter frost damage caused by poor or too fast scion growth. The modified bark grafting conducted in mid-April, on the other hand, had the highest rate of grafting success (100%) and percentage of survival (96.3%) (Fig. 5.2).

2.4 Hot Callusing

Heating of the graft union with the hot callusing pipe has been effectively utilized in recent years (Achim and Botu 2001; Avanzato et al. 2006; Gandev 2007). A temperature range of 24–28 °C is maintained at the grafting site with the help of a heating cable. The grafted plants are horizontally placed with the graft union on the heating cable in this technology. Using a hot callusing pipe and heating the graft union, a success rate of 86.00% for tongue and whip grafting and 89.50% for cleft grafting was obtained (Achim and Botu 2001). Gandev (2007) noted 74.20%

Fig. 5.2 Side grafting in Walnut (*Juglans regia*)

propagation success in cultivar Izvor-10 using hot callus and cleft grafting, and Avanzato et al. (2006) obtained 7–100% plant emergence success. The percentage of plant emergence success depends on the phytosanitary condition of mother plants as well as genotypes. The variation between genotypes is described on the basis of chemical contents, according to Pinghai and Rongting (1993). Grafting survival rate depends on the quality of a scion, C/N ratio, concentration of starch and soluble sugars (Pinghai and Rongting 1993). Anaerobic conditions are created because of excessive bleeding at graft junction thus lowering the survival rate of graft. This study generated new idea of tying plastic string at graft junction (Pinghai and Rongting 1993).

Hot callusing technique is utilized to produce roughly 20,000 plants annually in Turkey, with a success rate of 82% (Erdogan 2006). Callus proliferation develops most readily in late winter and immediately before spring or during "budbreak" stage in spring. This is attributed to auxin gradient that is decreasing in summer and early autumn and increasing in the late winter and early spring (Hartmann et al. 1997). Avanzato (1999) introduced a new method for heating the grafting area without removing the rootstocks from the soil, which included the use of an electric line. As the rootstocks are not removed from the soil, stress is avoided, and also growth of graftage following grafting operation is favoured. Nacheva and Gandev (2009) used radioisotopes to study the transport and spread of 14C-photo assimilates into walnut plants propagated using the hot callus method and no negative impact was observed on transport of photo assimilates.

2.5 *Bark Grafting*

Bark grafting is a type of scion grafting employed in walnut. In this method, success rate of 80% has been obtained in South Africa (Rotondo Walnuts 2004). The major

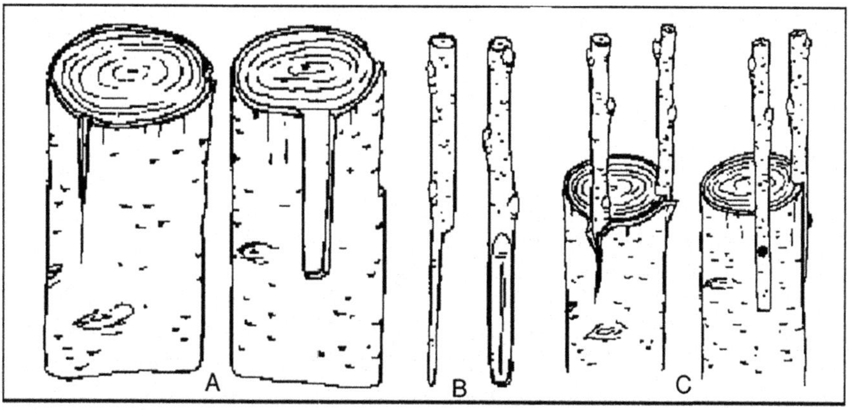

Fig. 5.3 Bark grafting in Walnut (*Juglans regia*)

problem this method faces is the long time period it takes to grow the rootstock in to the required thickness of 30–100 mm which takes around 3–4 years (Hartmann et al. 2002). This is the reason why this method is not made use of for the large-scale production. It can however be practiced when one wants to grow the walnuts in controlled atmosphere. Out of all the various factors needed to be controlled for the proper growth of the tree, temperature is the most important as it directly affects the callus formation especially in the temperate fruit species (Hartmann et al. 2002). A number of studies carried out on deciding the best temperature for callus formation in walnut is somewhat around 26–27 °C, and a temperature below 20 °C results in unsatisfactory callus formation (Hartmann et al. 2002). Callus formation starting time is also decided by the temperature. Whereas, the temperatures of 27 °C enables callus formation after fifth day, it might start on the fourth day if a temperature of 32 °C is kept but in this case, the callus formation is less. Therefore, the most widely used temperature that has been employed mostly is 27 ± 2 (Avanzato et al. 2006; Vahdati and Zareie 2006). Optimal time for spring budding is when both the stock and scion are dormant. Also, the air humidity of around 80% should be maintained (Solar et al. 2001) (Fig. 5.3).

2.6 Scion Budding

The best time for doing scion budding is during the winter season when both the scion and the rootstock are in a state of dormancy (Hartmann et al. 2002). However, grafting can be done at the start during or towards the end of the dormancy. A study conducted by Ebadi et al. (2002) showed that when the grafting was conducted in the month of December, the survival percentage was comparatively higher as compared to when the grafting was conducted in January. Vahdati and Zareie (2006) concluded that it is better to carry out the grafting in the month of March, i.e. towards the end of the dormancy.

2.7 Chip Budding

Chip budding is yet another method of cold multiplication in the open field. Chandel et al. (2006) studied that the best time for grafting in the high Himalayan region starts from the first week of June. With respect to the patch budding, the best time is from mid-June till end of the month. *Juglans regia* is the most common rootstock used and the buds are collected in the dormant state for grafting where the survival rate of 89% is obtained in case of chip budding and 50% success is obtained for patch budding. However, a study conducted by Polat and Ördek (2006) showed that a success of 13% was obtained for chip budding while 43% was obtained for patch budding in the Turkish climatic conditions. The age of the rootstock and your removal after the completion of the grafting have an effect on the survival rate. If the rootstock of *Juglans regia* is taken and planted in the month of March or early spring, forced and then grafting is also performed in the same year, 78% chances of survival are seen. On the other hand, when the rootstocks are used from last year and they are grafted in the current year, their survival percentage lowers down to 40%. Both the above methods show a decrease survival rate of the rootstock when they're being cut immediately after grafting. Therefore, it is recommended to cut the rootstocks 15 days after grafting is being performed. If the plants to be grafted are kept under controlled conditions, then chip budding can be performed both when the trees in a dormant state during vegetation as well as when the trees are in the dormant state (Bayazit et al. 2005). A study conducted by Porebski et al. (2002) showed that the survival percentage was only 26.90% when the chip budding was performed in the winter months. They suggested that the survival percentage could be maximized with the winter grafting, if the rootstocks are in full vegetation.

2.8 Hypocotyl Grafting

Hypocotyl grafting is one of the other methods for the propagation of walnut (Avanzato 2001). It has gained quite a lot of importance from the past few years. In this method grafting is performed during the vegetative state of the plant. The growing tip of the rootstock is removed and replaced with the tip of the scion cultivar to be grafted onto it. The plants are being provided with for temperature of 26 °C for 4 weeks after which they are kept in the shade for about 2–3 weeks. After this they are taken into the open field. Gandev and Dzhuvinov (2006) showed that this method of grafting obtained the survival percentage of 83.00%.

2.9 Epicotyl Grafting

Epicotyl grafting is another method which can be used for the propagation of walnut. In this method in place of the grafting the plant for a period of 3–4 weeks, it is being substituted by a temperature of 27 °C and a high air humidity (Arnaudov 2011). When performing epicotyl grafting, the bud sticks are normally collected during dormancy and they can be stored until they are to be used. When performing grafting, the cambium of both the scion and the stock are fixed into each other. Plastic foil is kept around the graft which should not be much tight so as to allow the expulsion of extra moisture at the grafting point.

3 Pecan nut Propagation Methods

3.1 Hardwood Cuttings

According to McEachern (1973), the most essential aspect in inducing roots is the time of collection. In addition, the main time to harvest pecan hardwood cuttings is during the dormant season or after they have received 200–400 h of field chilling (7 ° C).

Cuttings directly placed in mist beds after a basal dip for 5 s with 1% IBA was the best combination of IBA/cold treatment that generated the best rooting percentage (65%). The 2 or 4 weeks of cold treatment at 4 °C followed by the IBA treatment resulted in more than 60% rooting. When the cuttings were given a 2-week cold treatment before the IBA treatment, the proportion of cuttings that survived was much higher (48%). Although “Shawnee” cultivar survived the best, where no significant difference in rooting percentage among the cultivars was found (Allan et al. 1980).

According to Spencer (1980), the difficulties of establishing cuttings once they are rooted can be solved by taking hardwood cuttings in the summer before the buds become dormant and exposing them to basal heat for quick root development. Huang et al. (2006) used 3-year-old pecan seedling hardwood cuttings. The best rooting was obtained using 200 ppm NAA (naphthalene acetic acid) and 100–500 ppm IBA among the concentrations tested (Fig. 5.4).

Li et al. (2013) investigated the effects of parent tree age and cutting thickness on the rooting capability of the pecan. One year old wood with diameters of 0.5 cm, 0.5–0.8 cm, and 0.8 cm was collected from parent trees that were 2, 5, 16, and 25 years old, respectively. IBA and NAA concentrations or combinations were applied to cuttings. The callus and rooting percentages of cuttings from 2 year old parent trees were 87.70% and 14.40%, respectively, which were significantly higher than other tree ages. The cuttings with a diameter of 0.8 cm produced the best results, with callus and rooting percentages of 33.1% and 9.7%, respectively. After treating the cuttings with a thickness of 0.8 cm, the treatment of IBA 1.0 g/L + NAA 0.25 g/L

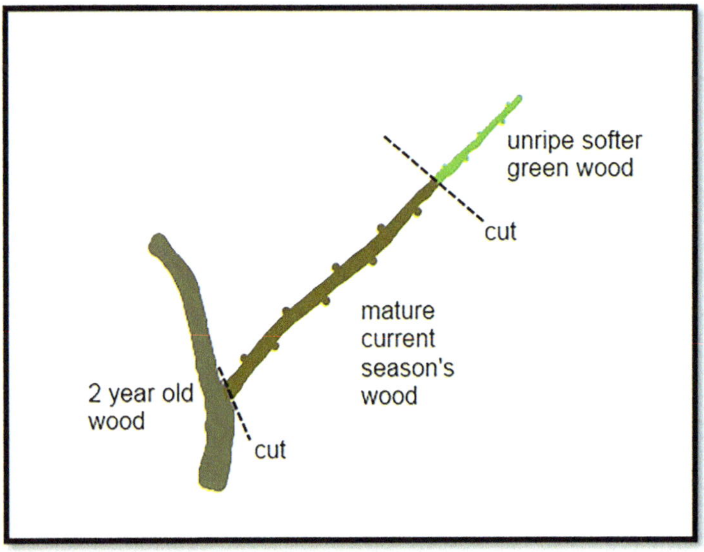

Fig. 5.4 Hardwood cutting in pecan nut (*Carya illinoensis*)

produced the best results, with callus and rooting percentages of 43.0% and 14.6%, respectively. The callus and rooting percentages of the cuttings were positively correlated to the thickness of the cutting at the same ages as the parent trees.

Zhang et al. (2015) investigated the influence of auxins (0.03%, 0.06%, or 0.09% IAA or IBA and 0.06%, 0.09%, or 0.12% NAA) as well as varied media concentrations and air temperatures on the rooting ability of hardwood cuttings taken from 1 year old pecan. As a control, the basal ends of each cutting were soaked in distilled water. The rooting substrate was made of peat, perlite, coarse sand, and silver sand. Cuttings were then planted into the root substrate at the following media/ambient temperatures: 2 peat, 4 perlites, 1 coarse sand, 1 silver sand (1) Both the media and the ambient temperature were set at 13.2 °C. (2) Temperatures of the medium (25 ° C + 2 °C) and ambient (13 °C + 2 °C). (3) Both the media and the ambient temperature were set to 25 °C + 2 °C. The best results were obtained with the 25 ° C media and 13 °C ambient temperature, both treated with 0.09% NAA (82%) or 0.06% IBA (80%) rooting.

3.2 *Softwood Cuttings*

Gossard (1944) was one of the first to successfully root softwood pecan cuttings. Despite obtaining roots under continual mist, no plant survived in the end. To stimulate the roots of softwood cuttings, Shreve (1974) employed six 1-year-old pecan seedlings. To force growth from lateral visible buds, all visible buds were removed from three cuttings and all terminal buds were removed from the stems of the other three cuttings from the six seedlings. The cuttings were placed in 1:1 peat: perlite pots. Twelve cuttings (two from each seedling) were set (six from visible buds and six from adventitious shoots) and kept moist and sprayed with a 4-4-8 Bordeaux mixture. After 15 days, the six cuttings from adventitious shoots were rooted, and after 35 days, shoot growth began. Cuttings from the visible buds took 30–70 days to grow roots, but none of them produced shoots.

3.3 Air Layering

Another method for obtaining clonal pecan rootstocks is air layering. It's done with a sharp knife, with two parallel incisions around the stem and through the bark and cambium layer about 2 cm apart. One lengthy cut is represented by the two parallel cuts. The outer bark ring is removed, exposing the interior woody tissue. To prevent a bridge of callus tissue from developing, the newly bared ring is scraped to remove the cambial tissue. The rooting hormone is administered at this time, and the moss is wrapped in plastic or aluminium foil and kept in place with twist ties or electrician's tape (Beckford n.d.). This process is completed during rainy seasons when humidity levels are at their highest (Anonymous 2007). Litchi, guava, macadamia, mango, and pecan (Pokorny and Sparks 1967; Abou-Taleb et al. 1992) are all propagated using this method.

3.4 Mound Layering

Mound layering is an old method of propagation that was developed to mass propagate apple clonal rootstock but was later used for other fruit tree species such as quince, currants, gooseberries, and pecans (Carlson and Tukey 1955; Brase and Way 1959; Duarte and Medina 1971; Medina 1981; Garner 1988). The shoots are cut back to ground level with this method, and soil or rooting medium is mounded around them to induce rooting at the base. If stool shoots have enough roots by the end of the growing season, they are separated from the parent plant. Rooted shoots are cut at their bases and delivered to customers as "rooted liners" to be transplanted into the nursery (Hartmann et al. 2002).

3.5 Budding and Grafting

Tissue culture and cuttings have made significant advances in pecan propagation over the years. The commercial pecan industry is still reliant on grafting and budding for propagation. Grafting and budding have a success rate of greater than 75%

(Nesbitt et al. 2002). The disadvantage of these techniques is that seedlings require two to three seasons of growth before reaching an acceptable stem diameter to be grafted (Zhang et al. 2015). Ajamgard et al. (2016) investigated three different grafting methods and grafting times in a study conducted in Iran. The best time for cleft grafting with polyethylene bags was from mid-February to late March. The "Wichita" scions had the highest percentage of graft success (92%) and the lowest in "10J" scions (less than 10%). The side-stub grafting procedure was ineffective, with grafting success rates of less than 20% for various cultivars.

3.5.1 Patch Budding

Patch budding is a method of propagating pecan nursery stock that requires taking a section of bark from the scion with a dormant bud and inserting it into a gap cut into the rootstock's bark. When the bud begins to grow, the shoot above it is clipped to allow the grafted bud to take its place as the primary shoot. Patch budding is most commonly done in the late summer, but it can also be done in the spring. The diameter of the rootstock and scion wood should be the same, ranging from 1.5 to 2.5 cm. Scion wood made out of current season's shoots should be healthy (Wells 2014).

3.5.2 Whip Grafting

Seedling trees and nursery stock with a diameter of up to 2.5 cm are commonly used for whip grafting. It is possible to do it in the late winter, when the buds are still dormant. The diameter of the rootstock and scion wood should be the same. The scions should be strong and at least 18–25 cm in length (Wells 2014). An oblique cut is made in both the rootstock and the scion with a sharp knife so that the two cuts are face to face and properly overlapped. After that, parafilm is wrapped around the cut areas and the scion to seal the cuts and preserve the scion. In 3–4 weeks, successful grafts begin to grow (Wells 2014).

3.5.3 Bark Grafting

The bark graft is a good approach to propagate pecan cultivars onto rootstocks with a diameter of 5–10 cm (Reid 2010). To execute the grafting, a place on the rootstock is chosen that is above the first whorl of branches. Bark grafting should be done in the spring, about 2–3 weeks after the first signs of growth appear (Wells 2014). To implant the scion, the top of the rootstock is cut with a saw, and the outer layer of the bark is removed with a sharp knife. At least three buds and a diameter of 1 cm should be present on the scion. A shallow cut is made on the back of the scion, which is whittled down to less than half its initial thickness. The scion is carved down to less than half its original thickness, and a shallow angled cut is made on the back of the scion (Reid 2010).

After the second cut, the scion is in the shape of a wedge, and after the third cut, it is in the shape of a triangle. After making a chisel point at the end of the bud stick, it is ready for grafting. The rootstock's bark is then lifted away from the wood with a sharp knife, and the scion is inserted between the bark and the rootstock's wood. The scion is tapped until the shoulder of the deep cut makes contact with the wood of the rootstock. On smaller trees, a staple gun can be used to secure the graft union, while brad nails can be used to secure the bark graft on larger trees (Reid 2010). Following grafting, the graft union is wrapped in aluminium foil and a polyethylene bag.

3.6 Micro Propagation

Organogenesis is a plant-specific process that involves the creation of meristematic adventitious centres called meristemoids to produce unipolar structures (shoots or roots) from non-meristematic cell aggregates or plant tissues. In a media containing 50 mM NAA, Obeidy and Smith (1993) stimulated organogenesis from cotyledon segments to generate adventitious roots. Without producing cotyledon abscission, a regeneration medium containing 20 and 5 M IBA induced profuse auxiliary shoot development from the embryonic axis. After a pretreatment in dark conditions on a medium with 20 mM IBA, 30% of the microshoots rooted on auxin-free medium. Plantlets with roots were successfully transplanted to soil. Payghamzadeh and Kazemitabar (2010) used immature pecan embryos as explants in organogenesis experiments. Immature fruits were grown on a modified DKW basal medium after disinfection. On modified basal media supplemented with 1 mg/L BAP, 0.05 mg/L IBA, and 2 mg/L GA_3 and dark culture conditions, a high frequency of plantlets was obtained for immature embryo cultivation. A callus is an undifferentiated mass of tissue with thin-walled parenchyma cells that emerge from proliferating cells in the parent tissue (Dodds and Robert 1985). Due to the usage of phytohormones (auxins and cytokinins), callus development in tissue culture can be generated using a small percentage of plant tissue (Skoog and Armstrong 1970; Letham 1974; Akiyoshi et al. 1983). In both monocotyledonous and dicotyledonous plants, callus cultivation is an important approach for establishing clonal populations, plant regeneration, and genetic modification (Reinert and Bajaj 1976). To stimulate embryogenesis, Rodriguez and Wetzstein (1994) used immature pecan seeds, as well as the entity of callus formed in proportion to the kind and concentration of auxin. In cultures induced on 2,4-D, callus development was higher than in cultures induced on NAA. Higher levels of both auxins resulted in more callus formation than lower levels. For organogenesis research, Payghamzadeh and Kazemitabar (2010) used immature pecan seeds as explants. They studied that varying concentrations of BAP and IBA commonly promoted callus development, which was likewise suppressed by adding GA_3 to the culture medium.

One of the most significant invasions in the tissue culture industry has been somatic embryogenesis. It has a variety of uses, including mass clonal propagation, genetic transformation, and embryo development research. It is feasible to establish clonal pecan rootstocks with genes of economic importance such as dwarfing for size control, increased nutrient uptake, alternative bearing control, saline tolerance, worm resistance, and growth uniformity through somatic embryogenesis (Wetzstein et al. 1996). Merkle et al. (1987) developed somatic embryos using young nuts as explants, which was the first study of somatic embryogenesis. Wetzstein et al. (1989) obtained an embryogenic frequency of 85 using immature pecan seeds cv. Stuart collected 15 weeks after pollination, which is considered the optimal stage for embryogenic induction. To improve the rooting of somatic embryos, Wetzstein et al. (1990) used a cold treatment followed by a desiccation therapy. Immature pecan seeds were obtained from eight cultivars at various phases of maturity by Yates and Reilly (1990). The embryogenic response differed between cultivars and was genotype-dependent, according to the study.

4 Chestnut Propagation Methods

Castanea genus of the family Fagaceae comprises approximately 13 species throughout the world. *Castanea dentate* (North American); *C. sativa* P. Mill (Europe); *C. mollissima* Blume (China) and *C. crenata* Siebold (Japan) are the four major species of this genus (Littsle 1979) and *C. dentate* produce excellent flavoured nuts (Vossen 2000). *Castanea sativa* and *Castanea dentate* are two species that produce delicious nuts (Vossen 2000). *Castanea sativa* is the only species found in Kashmir (Pandit et al. 2009). Chestnut fruit is healthy and nutritious. Chestnut is in high demand both locally and internationally, and it attracts a higher price than other nut crops. The crop can adapt to a wider range of geographical and climatic conditions. Chestnut is cultivated in India from the Himalayas to Assam and Meghalaya, at altitudes of 2000–3000 m above sea level (Amardeep 2008). Quality planting material appears to be most important requirement for commercial and profitable production. Rooting of cuttings, layering, budding, grafting and micropropagation processes are some of the most common propagation methods used to obtain high-quality planting material.

4.1 Budding and Grafting

Budding and grafting are most productive and feasible methods (Serdar and Demirsoy 2010). Many factors influence the budding and grafting methods used in chestnut, including the timing of budding and grafting, growth and age, rootstock vigour, technical know-how and quantity and quality of scion wood (Serdar and Soylu 1999; Duman and Serdar 2006).

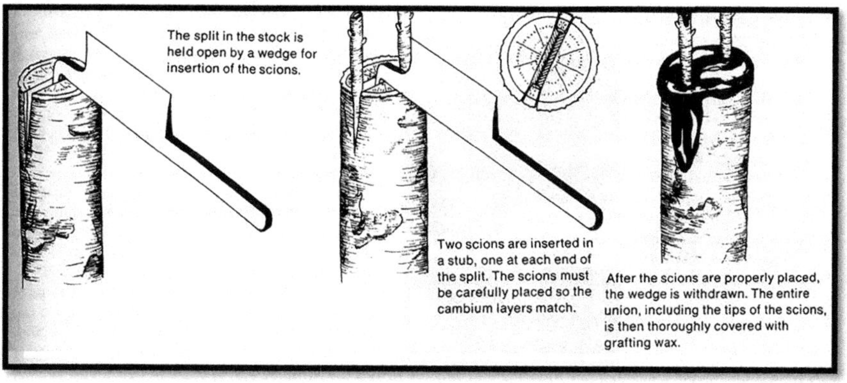

Fig. 5.5 Cleft grafting in Chestnut (*Castanea sativa*)

A viable cloning technology maintains and propagates trees exploited in breeding programmes and preserves recessive genes present in local chestnut populations thus ensuring that genetic diversity is preserved.

In *Castanea* genus, grafting is most widely employed vegetative method of propagation. But, grafting has several drawbacks, including incompatibility between rootstock and scion, union is prone to blight infections, weak trunk at grafting union and sudden death of developed tree (Key 1978; Huang et al. 1994; Oraguzie et al. 1998). With success rate of 47.50% and sprouting rate of 60.83%, **cleft grafting** was superior than tongue grafting (Sharbat et al. 2016) (Fig. 5.5).

4.2 Softwood Cuttings

Propagation with softwood cuttings is an ideal method in European chestnut (Jinks 1995; Osterc et al. 2005). Cuttings obtained from young mother trees of European Chestnut or stump shoots of the older trees root easily than the cuttings obtained from crown shoots of the mature trees because the shoots collected near root-shoot junction are mostly juvenile than the shoots obtained far away (Osterc 2001; Hartmann et al. 2002). Physiological maturity (Vieitez and Vieitez 1976; Osterc 2001), stem sclerenchyma ring anatomy (Vieitez et al. 1980), endogenous inhibitors (Vieitez et al. 1987) and environmental conditions all influence adventitious rooting (Hartmann et al. 2002). The propagation conditions can be manipulated to maximize the percentage of rooted cuttings more effectively than endogenous parameters. Maintaining a low-evaporation environment, minimizing loss of water from cuttings, ensuring optimal temperatures for regeneration of roots and providing adequate levels of light for photosynthesis during formation of root and establishment of plant are all requirements of propagation systems (Hartmann et al. 2002) (Fig. 5.6).

Fig. 5.6 Softwood cuttings in Chestnut (*Castanea sativa*)

4.3 Direct Seeding

In comparison to planting of seedlings, direct seeding is a popular method owing to its low cost and easy installation. However, seed predation increases the mortality rates of the direct seeded trees, thus lowering the overall efficiency (McCarthy et al. 2010; Fields-Johnson et al. 2012). A meta-analysis of study found that planted seedlings are susceptible to biotic and abiotic stressors, with an estimated mortality rate of 48% (Palma and Laurance 2015). Planting lesser-quality seedlings with reduced chance of surviving raises the effective cost due to higher labour expenses and lower survival rates. Planting higher quality seedlings with a higher survival rate may be more effective method, reducing the amount of labour necessary to attain the same end planting density (Zaczek et al. 1995; Ward et al. 2000). Seedling quality refers to set of phenotypic characteristics associated with physiology and morphology of seedling that determine survival and field performance (Grossnickle and MacDonald 2018a).

Morphological features such as height, diameter of root collar, shoot:root ratio, architecture of root, as well as their cumulative effect on quality of seedling affect the performance of seedling (Rose et al. 1990; Dey et al. 2010; Grossnickle and MacDonald 2018a, b). Nursery growers can assess seedlings for overall quality and anticipated field performance through parameters like height or root collar diameter (Rose et al. 1990; Dey et al. 2008). *Cryphonectria parasitica*, an imported parasite, has destroyed American chestnut (*Castanea dentata*) all across its natural range (Russell 1987; Paillet 2002). For decades, breeding procedures have now been carried out to develop blight-resistant chestnuts, and hybrid chestnut seedlings have lately been examined to determine viability for replanting efforts all across the native range of American chestnut (Clark et al. 2009, 2012; Pinchot et al. 2015; Skousen et al. 2018). The life of chestnut seedling and its earliness in growth is only possible if the seedling material is of good quality. If the quality of the seedling is very good, it will bridge the gap between successful nursery production and the revival programs taken for the revitalization of the species. The method of multiplication and the type of the medium being used over two good choices in order to change the quality of the seedling so that you are able to improve them to a great extent. Growing the seeds in beds saves a lot of money if one wants to obtain planting material in big quantities in comparison to those seedlings which are grown in containers (Wilson et al. 2007). However, it has been experienced that the transplantation shock to which the seedlings raised in beds are exposed might be greater and their survival might be less as compared to the container grown seedlings because of size of the root system when being taken out and increase in the size of the shoot as compared to the root (Wilson et al. 2007; Struve 2009). In fact, it has been shown that the performance of the bed grown and container grown methods shows a great variation when the seedlings are being grown under stress, particularly with dry soil which results in greater death rate among the bed grown seedlings (Landhausser et al. 2012). Also, the damage that is being done to the seedlings during the time of their removal and handling may result in their exposure to disease and even death when stored (Grossnickle and El-Kassaby 2016). Therefore, it is extremely important to decrease these which can be done by taking out the seedlings in a way that the root gets damaged very minutely (Grossnickle and El-Kassaby 2016). On the other hand, it has been seen that when the bedroom seedlings are allowed to grow in proper aerobic conditions, they are able to perform better as compared to the container grown seedlings (Grossnickle and El-Kassaby 2016). Van-Sambeek et al. (2016) studied that the less transplantation shock seen in case of the container grown seedlings is because of their strong root system which allows them to grow weather greater pace and so low death rate in the field conditions. Also, it might be possible that the container seedlings which have a height of more than 1.5 m might escape the browse height which in turn results in lower death rate (Clark et al. 2012). Even though, there are a lot of advantages in propagating the seedlings in containers, one has to deal with its own problems. If the size of the container is not in accordance with a particular species, the seedlings formed as a result might suffer from deformities in the root system which include circling, matted or J shaped roots which have to be pruned before planting. This whole process therefore decreases the benefits that are being provided by transferring the whole root systems (Arnold and Struve 1993). Not only this, the seedlings which are grown in containers might not be economically viable because of the additional money that has to be spent on purchasing the materials needed in comparison to the bed grown seedlings, making the container grown seedlings a more expensive dealing.

4.4 Root Propagation

Another method which is used for the propagation of chestnut is the root propagation method which is a complex procedure in which highly rated media is formed in order to produce the seed links data container grown but have a high volume of roots and a large diameter. The procedure followed in this propagation method includes growing the seedlings in containers without any bottom in order to make the root pruning easy, decreasing the taproots, in order to allow the formation of large numbers of lateral roots which improves the overall seedling quality (Lovelace 2002). Grossman et al. (2003) found that RPM (root production method) seedlings were able to survive more than 95% even after 2 years in comparison to the bed grown seedlings were in the survival rate decreased from 95% to 77.4% in the same area. RPM seedlings are subjected to a lot of labour and are difficult to produce, even though the cost required for every seedling cannot be considered less when we look at the quality of the seedling. Also the seedling cost has been shown to get reduced when these are being planted in the field (Spetich et al. 2002).

4.5 Multicontainer Method

Yet another method used for the propagation of chestnut is the multicontainer method or the MCM method. It is the method which is modification of the Lovelace's (2002) RPM method, in which the pots of a standard size are being used in the nursery.

An ideal propagation method is one which is able to produce a large number of roots with less labour involved, less material required and less inputs involved. Highly porous media, bottomless container, bottomless raised beds, also known as air-prune beds, have the capacity to produce plantlets with highly fibrous root systems with the convenience of separation of container grown plantlets and the transport abilities of bed grown plantlets. The containers which have the air-prune route openings do not allow the root tips to develop and cause their mortality when they come into contact with the outer atmosphere. This in turn results in the formation of the stronger roots in the inside of the root ball because of the removal of apical dominance in the roots (Arnold and Struve 1993; Amoroso et al. 2010). Air-prune beds should alleviate root deformities difficulties associated with container grown seedlings by expanding container size from a single pot to a raised bed. Hybrid chestnut seedlings reflect the massive quantity of resources that can be dedicated to save a single species, underscoring the need for more research into how to best employ plant material in the restoration of rare species in any ecosystem. In this regard, The United States National Forest system has come to this conclusion that the most successful method to protect this species from the biotic and abiotic stresses would be to improve the quality of its seedling (Clark et al. 2014; Collins et al. 2017). However, more study needs to be done in order to standardize the nursery protocols which can result in the production of good quality seedlings of chestnut that could be further used in the revival projects. This research could then be further applied to improve the propagation of those species that are difficult to find.

5 Pistachio Propagation Methods

Anacardiaceae family includes over 70 genera and 600 species, the majority of which are subtropical, tropical and temperate trees and shrubs (Wannan 2006). The *Pistacia* genus comprises 11 species of evergreen or deciduous trees and shrubs that are classified as xerophytic trees. Hormaza and Wunsch (2011) reported that only *Pistacia vera* is grown commercially, and other species are used as pistachio rootstocks. *Pistacia* is distinguished by its dioecious reproductive system, which means that the male and female flowers are borne on separate tress. Pistachios are propagated by grafting or budding methods with selected scions on seedling stocks of *Pistacia terebinthus* and *Pistacia atlantica*. These rootstocks are resistant to biotic and abiotic stresses and have good vigour. Using these old methods, however, has hindered the expansion of pistachio plants. The industry could benefit greatly from the development of technologies for quick in vitro micropropagation. Horticulture, agriculture and plant breeding have all benefited from tissue culture technologies. Its importance stems from its application in bulk multiplication of elite genotypes, secondary metabolite production, viral eradication and plant in vitro cloning. This technique provides a viable alternative to vegetative methods and has the potential to obtain significant multiplication rates.

5.1 *Micropropagation*

Explants from part of the plant can be utilized to regenerate plantlets cultivated in vitro. Selection of right explant is critical before tissue culture. Inappropriate explants increase the risk of contamination. The explant must be obtained at proper physiological development stage. Effect of contamination frequency and percent responsiveness of explants was studied by Benmahioul (2009). Contamination percentages of 38.7 and 84.7 were observed in mercuric chloride and sodium hypochlorite, respectively. Tissue browning is common phenomenon in in vitro growth and development of explants. Because tissues oxidize quickly after plating, management of pistachio micropropagation with nodal segments or buds is difficult. Subsequently subculturing minimized this problem (Benmahioul 2009).

Rooting is a critical phase in micropropagation success. This process entails not only the rooting of shoots, but also the nurturing of plants to improve their survival and adaptation capability. Many herbaceous species may easily induce adventitious root formation, whereas most woody species cannot (Bajaj 1991). Micropropagated *P. vera* has been successfully rooted and acclimatized using both in vitro and ex vitro methods. Ninety-two percent rooting was observed in 1 year old softwood cuttings of *P. chinensis* when treated with IBA @ 5000 mg/L, as compared to higher concentrations (Pair and Khatamian 1980). Eighty-eight percent rooting was obtained in softwood cuttings of *P. vera* in a mist system with auxin concentration of 35,000 mg/L after 6 weeks (Al-Barazi and Schwaba 1982). Different rooting responses have been observed in micropropagation of pistachio in semi-solid media (Tilkat et al. 2008; Tilkat and Onay 2009). 3.72 roots with rooting percent of 27.5 were observed in shoot culture with IBA @ 2 mg/L and rooting percent of 20 and 7.5 was recorded with 1 mg/L and 4 mg/L, respectively (Akdemir et al. 2014). Rooting percentage of 75 and 82.5 was obtained with media containing BA and mT (Benmahioul et al. 2012; Vibha et al. 2014).

The media composition, specifically plant growth regulators (PGR), has been demonstrated to have an impact on tissue differentiation and growth. In fact, the application of PGR in tissue culture of pistachio is critical (Benmahioul et al. 2015). Meta-topolin (mT) was found to be effective for axillary shoot proliferation in vitro. PGR depleted media with 0.2% activated charcoal and 2 mg/L of mT enhanced plantlet elongation (Benmahioul et al. 2012).

In vitro micro-environmental factors, like light, have been demonstrated to alter culture response (Lee et al. 2007; Aggarwal et al. 2012). Plants grown in moderate light have longer internodes and leaves compared to the plants growing in intense light (Barber and Anderson 1992). According to Benmahioul (2009), light intensity of 90 mol/m^2/s affected the growth of *Pistacia vera* seedlings, with 40 mol/m^2/s providing the best results.

5.2 Layering

Vegetative propagation is a great way for propagating *Pistacia vera* L. It is considered to be the best method because asexual method enables the sex determination of the newly produced plants as their genetic behaviour remains preserved as in the rootstock plant. It is quite cumbersome to propagate the pistachio by vegetative means, because most of these trees are heterozygous in nature and the cuttings are even difficult to root even though marketable varieties can be propagated by budding and grafting, but resin creates a problem in union formation of stock and scion. *Pistacia vera* can also be propagated by the tissue culture technique, however, it is a very costly affair and requires many instrumentation facilities.

Therefore, another method which can be used for the propagation of pistachio is layering. This method is preferable for the difficult to root species, in which the adventitious root formation occurs while the roots are still joined to the mother plant (Beyl and Trigiano 2016). A study conducted by Dunn and Cole (1995) suggested the use of mound layering for the multiplication of the species. Air layering is a type of layering method that has been successfully used for the propagation of a number of temperate fruit species including peaches, plums, cherries, etc. (Reddy et al. 2014). The basic methodology for propagation by layering includes the assembly of carbohydrates, proteins, plant hormones like auxin and ethylene at the point of girdling, which in turn leads to high respiration and call us formation, which ultimately leads to rooting and plant propagation by layering (Hartmann et al. 2011). As a result of large number of experiments carried out by the scientists from time to time, it has been seen that applications of plant growth hormones either increase the physiological states of the roots for adventitious root formation or either the auxin activity has been seen to enhance rooting (Davies and Hartmann 1988; Castro and Silveira 2003). A study carried out by Pacholczak et al. (2005) showed that plant spray with 50–500 mg/dm^3 of IBA or NAA was found to be effective in the propagation of ornamental shrubs. Similarly, a study carried out by Ford et al. (2002) found that application of gibberellin (GA_3) improved rooting as well as the shoot growth of the cuttings in cherry.

5.3 Budding and Grafting

In October, with a maximum ambient temperature ranging from 15 to 25 °C and a minimum temperature ranging from 5 to 10 °C, substantial success was achieved employing the chip bud approach in the field. Rootstocks with pencil-thin stems that are actively growing and have some mature leaves are ideal. Under glass house circumstances with temperatures ranging from 15 to 30 °C, success has also been achieved using the chip bud method on actively growing rootstocks throughout the year, provided the budwood is in good condition. Buds are wrapped in such a way that their tips remain exposed, allowing for unrestricted growth.

The upper third of the rootstock above the bud is clipped off during budding, but at least ten basal leaves are preserved to ensure the root system remains healthy. Some operators prefer field budding with a T-bud technique in February and early March, but a chip bud may also be used. The selection of plump buds is critical for success with these procedures. After budding, a semi-cincture with a single knife cut above the implanted bud was employed to offset the pistachio's strong apical dominance and resulted in enhanced bud take. After around 3 weeks, shoots emerge from the buds and should be attached to the rootstock if exposed to wind. After 9–12 months, the residual rootstock above the point of bud insertion is removed with a loping cut about 15° from the horizontal. Pistachio roots are especially sensitive and easily damaged by dry air. As a result, in order to avoid transplanting losses with this species, it is critical to keep root exposure to air or desiccation to an absolute minimum (Fig. 5.7).

6 Hazelnut Propagation Methods

Hazelnuts have traditionally been propagated by layering and suckers on their own roots (Tous et al. 2009), but the rate of propagation is poor and trees possess long juvenile phase. Thus, different propagation methods, like grafting, layering, micropropagation, and cutting are now commonly used (Olsen and Smith 2013). Although these propagation methods have good results, but have not been commercially used (Tous et al. 2009; Ellena et al. 2014).

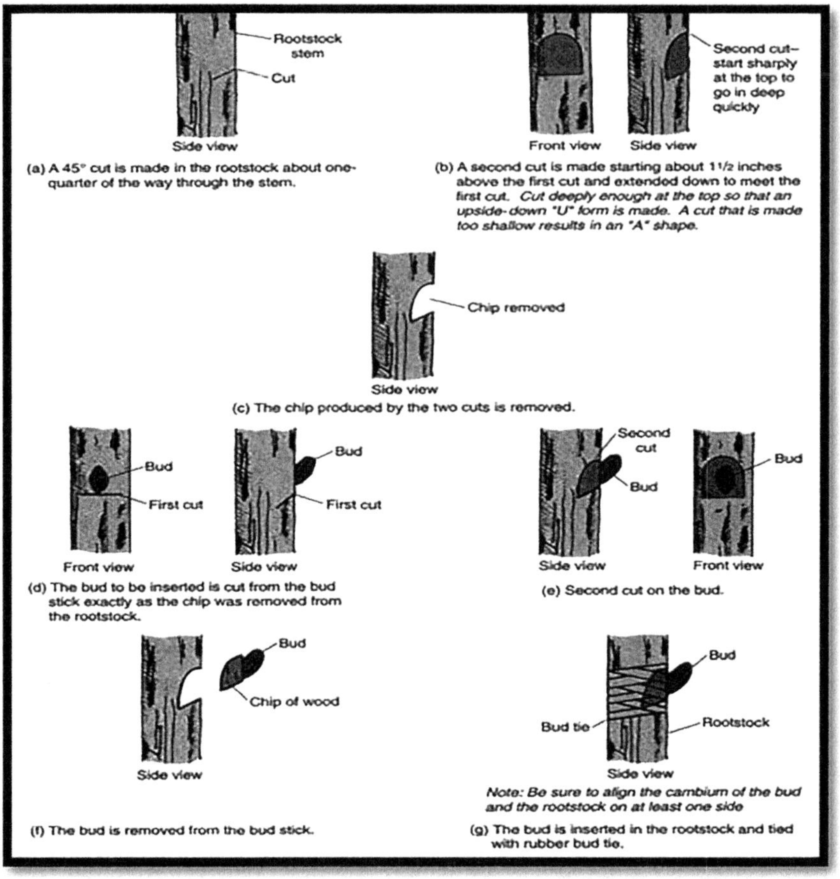

Fig. 5.7 Chip budding in Pistachio nut (*Pistacia vera*)

Grafting method of propagating hazelnuts allows rapid multiplication of varieties, enables using of rootstocks, thus prevents suckering, produces fruits earlier, and may improve disease resistance or winter hardiness of cultivars that are grafted on resistant and hardy rootstocks (Hartman et al. 1990; Janick and Paull 2008). But grafting is problematic due to the slow formation of callus in hazelnut. High temperatures at graft union have improved success of grafting, reviving interest of using vegetative rootstocks in hazelnuts (Tous et al. 2009; Janick and Moore 1996). Seedlings can also be used as grafting rootstock for hazelnut cultivars (Salimi and Hoseinova 2012). Morphological variations are present in seedling and vegetative rootstocks. Vegetative rootstocks have less primary roots, shallow root system and no tap roots (Hartman et al. 1990). *C. colurna* seedling rootstocks yield deep-rooted trees which do not topple in windstorms like *C. avellana* having shallow root system (Janick and Moore 1996).

Drought resistance is enhanced in deep-rooted and non-irrigated orchards (Rahemi and Yadollahi 2006). Accessions of native hazelnuts (*C. americana*) are cold hardy and exhibit a high degree of resistance to Filbert Blight (Coyne et al. 1998). When grown as rootstock, these accessions transfer these useful traits to scion (Hartman et al. 1990). The success rate of chestnut has been improved by grafting on the hypocotyl seedlings (Serdar et al. 2005; Duman and Serdar 2006; Serdar 2009; Georgi et al. 2013; Galic et al. 2014). Hypocotyl grafting also has the advantages of grafting hazelnuts throughout the year, reducing juvenility of trees by utilizing mature scion, having a limitless supply of the scion wood, and quick rate of multiplication. Furthermore, compared to sexual propagation, vegetative propagation enables plants to be grown more quickly (Hubert 1977).

6.1 Layering

One of the most effective methods of propagating hazelnuts is layering (Baron and Stebbins 1969; Howes 1948). Mound layering and simple layering are mostly used for propagating hazelnuts. Simple layering involves bending hazelnut stems into a 20–25 cm deep trench and covering them with soil when they are still dormant. Two to three buds persist above soil surface on stem tips (Roots of Peace 2007). Twenty-six thousand to 29,600 layers per hectare are obtained (Achim et al. 2001). In traditional orchards, layering beds are formed by pollarding trees at the ground level in the winter and enabling new shoots to sprout in the next spring (Hubert 1977). In the beginning of summer, emerging shoots are coppiced at base and rooting hormones are applied above girdle immediately. To maintain moisture in rooting area, 100–200 L of sawdust are stacked up. After dormancy, freshly rooted plants are transplanted in following spring (Braun 2010). Trees are propagated through layering while they are attached to their mother plants. Till their root systems are developed, the young plants absorb nutrients and water from mother plants. This technique is efficient and results in a high rate of survival (Hubert 1977) (Fig. 5.8).

In case of **layering**, one requires a lot of labour apart from large land for the multiplication, even though the yield obtained is comparatively less. The grower is required to maintain block for the safekeeping of the mother plants, while there might be no market availability for these trees. It is also observed that the plants generated as a result of simple layering produce a lot of suckers. This even increases the input cost for your control.

6.2 Root Cuttings

Root cuttings is a comparatively easier and cheap method for propagation. However, it is affected by the surrounding environment, the effect of the chemical apart from

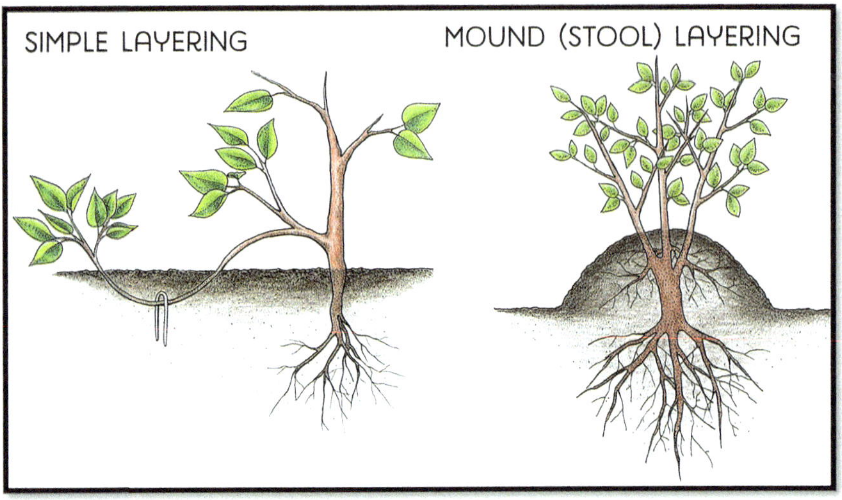

Fig. 5.8 Layering in Hazel nut (*Corylus avellana*)

the age of the cuttings (Rhodes 1968). In order to make the effect of propagation more consistent, both the types of cuttings are taken normally for the propagation of hazelnut, i.e. hardwood and softwood cuttings. While the softwood cuttings are being taken in the month of June, the hardwood cuttings are taken towards the end of the winter season. After taking the softwood cuttings, they are being treated with 1000–3000 ppm of IBA followed by a treatment with 1000 ppm of IBA in the middle of July.

6.3 Grafting

Grafting is propagation method that is followed in hazelnut mostly for the nursery growing (Roots of peace 2007) but is nearly used as compared to the above-mentioned methods (Fideghelli and De-Salvador 2009). One of the advantages of this method is the elimination of the cost of the suckers. The most common rootstocks used for the propagation of hazelnut are *Corylus colurna* and *Corylus chinensis* which are being used because they improve the drought in frost tolerance apart from improving the improving the size and nut of the kernel (Ninic-Todorovic et al. 2008; Miletic et al. 2008). For successful grafting, it is important that design should be without any disease and should be collected in the dormant and stored till it is to be used for the purpose.

6.4 Micropropagation

It is an improved method for the propagation of hazelnut which allows a person to obtain a large population of genetically similar plants. It is an improvement over the traditional methods of propagation. For carrying out the process, one has to go through a number of steps which include the formation of aseptic cultures creating a large number of shoots, root proliferation from the shoots, hardening and then the movement of the plants formed to the field conditions. The growth media used is prepared in the lab itself which acts as a supplier of carbohydrates, various nutrients and plant growth hormones (Payghamzadeh and Kazemitabar 2011). Hazelnuts can be multiplied by micropropagation because this method allows to select number of characters which is not possible by the traditional methods (Nas and Read 2001). The success of this method is determined by the state of the explants, the shoot which is taken in the spring from the dormant plants are considered to be good for large-scale multiplication. Apart from this, the success is being determined by the substances used in the artificial medium and the proportion of various hormones used (Bacchetta et al. 2008). In a study it was observed that when the shoots of hazelnut were being treated with 1000 ppm IBA solution for period of 10 s and the medium used had peat pellets in it, 97% success was obtained within a 3 month time period (Nas and Read 2004).

7 Almond Propagation Methods

The origin of majority of almond orchards are of seedling origin that grow to enormous sizes, making orchard management difficult and producing inadequate and inferior quality almonds. Since almonds are heterozygous because of self-incompatibility, they are propagated through seed and thus do not provide uniform and identical planting material. Budding is a popular method of vegetative propagation that improves quality of the planting material and shortens juvenile phase. This aids in the production of true-to-type plants, resulting in high-quality fruits. Almond is mostly propagated through T-budding on the seedling rootstocks in either fall, June or spring (Hartman et al. 1990; Riel and Sutter 1996). The most crucial time for budding in the propagation of fruit nurseries is early fall and late summer.

7.1 Budding

A common method for propagation of almond is **T-budding** which is performed in the month of July and August in Turkey (Dokuzoguz and Gulcan 1979). The success percentage that was obtained when the budding was performed in the month of June was around 66.2. The reason for the higher bud take can be accounted to the suitable

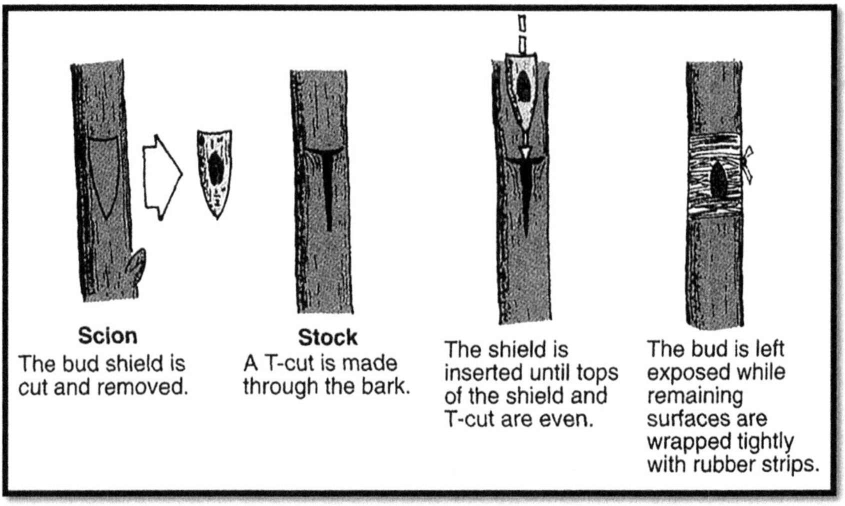

Fig. 5.9 T-budding in Almond (*Prunus amygdalus*)

temperature and the easy soft flow (Mir et al. 2016). Maryam et al. (2015) also reported that conducive temperature and good percentage of humidity are responsible for a proper union between rootstock and the scion. An experiment conducted by Euro (2012) showed a success percentage of 85 when the budding was performed in the month of June, and a success percentage of 80% was recorded when budding was performed in the month of July (Ahmad et al. 2012). The success obtained in budding is being affected by the method used. About 65.6% of success was obtained in case of T-budding. In comparison with chip budding, higher success was obtained in case of T-budding, which might be accounted to the greater contact of the cambium observed in case of T-budding (Abou-Rayga et al. 2009; Mir et al. 2016). Another factor affecting the bud success is the part of the scion used for budding. It was observed that when the middle portion of the scion was used, it resulted in maximum success percentage of 64.8%. This might be a counter to the fact that the amount of cell division taking place in the younger buds is more as compared to the buds present in the older part of the scion (Mir et al. 2016) (Fig. 5.9).

We can say that the method of budding, time of budding, the type of variety used and the scion part used decide the percentage of success to be opened in case of budding in almond plants. In the method of budding, T-budding was found to be more successful as compared to chip. In case of T-budding, the better union present between the stock and scion is produced because of the active division which is seen in the cambium, as a result of which thin-walled cells are produced on both the sides of the cambium (Hartmann et al. 2007). With respect to the timings of budding, the plants which were budded in the month of June showed greater success percentage while as the minimum success (40.8%) percentage was found in those plants in which the burning was performed in the month of September (Mir et al. 2016). Similarly, Kasmi et al. (2013) and Maryam et al. (2015) reported that about 80% of success was obtained when the budding was performed in June. The study carried out by Mir et al. (2016) showed that the performance of Non Pariel was better as compared to Shalimar as far as the characteristic properties that was studied, which leads to the conclusion that the differences among the varieties or the differences in the genomes of the varieties play a key role in deciding the success percentage in a propagation method (Mir et al. 2016). Another study by Ullah et al. (2000) revealed that the higher success percentage of budding was found in the almond cultivar "Tuono" while minimum was found in "Genco".

8 Chilgoza Propagation Methods

The propagation of chilgoza can be carried out by two methods which include the seed and the vegetative methods. These methods can be explained as follows:

8.1 Seed Propagation

The seeds of the chilgoza removed from the cone after which they are taken for storage in the month of October. However, the conditions for storage should be completely dry, after which the seeds are sown in the month of April. The reason for storing the seed at very low temperatures is because the viability of the seed is lost very easily. The proper time for sowing is towards the end of autumn, i.e. in the month of October and November. It is important to sow the seeds in the polythene bags in the first year without sowing them in the beds. It is because of the reason that the roots of the tree grow very vigorously in the first year. It is important to change the polythene bags after every month in order to avoid the roots getting fixed underground soil. After this, the seed is stratified for around 20 days at a very low temperature of 2 °C. In order to make the mixture for the polythene bags, the soil sand and compost should be taken in the ratio of 1:2:1. It is very important to cover the polythene bags with a proper net made of wire so as to avoid eating of cotyledons by the animals, rats, flying birds immediately after they germinate. In order to transplant the seedlings, one has to wait for around 3–4 years until they reach a height of 25–30 cm.

8.2 Vegetative Propagation

Two root stocks which are used for the propagation of chilgoza scion are "Chud" and "Kail". But it has been observed that over the years, the rootstock becomes very thick resulting in slow growth of chilgoza. Other methods which have also been found successful for the propagation of chilgoza include earlier ring and tissue culture.

9 Conclusion

Clonal propagation is one of the best ways to increase the production per unit area and to create uniformity in the woody species. During the past decade, vegetative methods have proven to be the successful methods given to the industry of horticulture. These methods including the technique of tissue culture have been successfully used in a lot of other crops as well. It is quite obvious that vegetative methods show a large number of advantages including fast multiplication rate, plants free from diseases, germplasm conservation, etc. The best means of taking maximum advantages from these vegetative methods of propagation are to optimize the conditions for growth of the plant which in turn improves the quality of the plant material obtained. In the method of micropropagation, it has been seen that meta-topolin is suitable for effective in vitro growth of axillary shoots. Similarly, ex vitro rooting is also helpful for the development of roots which is considered even more advantageous for the commercial production of cuttings. For the long-term preservation, cryopreservation can be employed which involves 2 h of dehydration followed by dropping in liquid nitrogen. This method has been developed for embryonic axes.

10 Future Strategies

- Unlike apple orchards, growers don't have regular nut orchards in India. Therefore, we need to increase area under nut plantation.
- Establishing regular orchards of nuts like apple, by providing farmers good quality, disease-free planting material.
- Establish nurseries and mother orchards of lateral bearing and high yielding varieties.
- The nut should come forward united in form of nut grower associations to formulate strategies with government to get incentives under some schemes or missions.
- To bring more area under nut cultivation in cluster mode, the state should adopt end to end approach involving multiplication of nuts.
- Providing technical assistance to growers on various nut crops propagation methods.
- Systematic evaluation of genetically diverse germplasm of nut crops needs to be taken up immediately for selecting superior genotypes for building gene banks.

References

Abou-Rayga MS, Kasim NE, Shaheen MA, Yehia TA, Ali EL (2009) Morphological and anatomical evaluation of different budding and grafting methods and times of Neplus ultra almond cultivar. J Appl Sci Res 5:253–263

Abou-Taleb S, Wood BW, Fouad MM, Yates IE (1992) Propagation of pecan by layering. Hortic Sci 27(6):691

Achim G, Botu I (2001) Results in walnut propagation by using different methods. Acta Hortic 544: 503–509

Achim G, Godeanu I, Baciu A (2001) Research on clonal propagation of hazelnut in Valcea-Romania. Acta Hortic 556:281–286

Aggarwal D, Kumar A, Sharma J, Reddy MS (2012) Factors affecting micropropagation and acclimatization of an elite clone of *Eucalyptus tereticornis* Sm. In Vitro Cell Dev Biol Plant 48:521–529

Ahmad I, Cheng Z, Liu T, Nan WC, Ejaz M, Khan AK, Wasila H (2012) Effect of different time of budding on the bud take success of peach on peach rootstock. Adv Environ Biol 6:1848–1852

Ajamgard F, Rahemi M, Vahdati K (2016) Development of improved techniques for grafting of pecan. Sci Hortic 204:65–69

Akdemir H, Suzerer V, Onay A, Tilkat E, Ersali Y, Ozden-Ciftci Y (2014) Micropropagation of the pistachio and its rootstocks by temporary immersion system. Plant Cell Tissue Organ Cult 117: 65–76

Akiyoshi DE, Morris RO, Hinz R, Mischke BS, Kosuge T, Garfinkel DJ, Gordon MP, Nester EW (1983) Cytokinin/auxin balance in crown gall tumors is regulated by specific loci in the T-DNA. Proc Natl Acad Sci U S A 80:407–411

Al-Barazi Z, Schwaba WW (1982) Rooting softwood cutting of adult *Pistacia vera*. J Hortic Sci 57(2):247–252

Allan P, Brutsch O, Smith IE, Wolstenholme BN (1980) Improved techniques for pecan propagation by hardwood cuttings. Pecan Q 14(2):7–14

Amardeep K (2008) Survey and selection of promising chestnuts in district Srinagar. M.Sc. Thesis submitted to Sher-e-Kashmir University of Agriculture Science and Technology of Kashmir, Shalimar Srinagar, p 58

Amoroso G, Frangi P, Piatti R, Ferrini F, Fini A, Faoro M (2010) Effect of container design on plant growth and root deformation of little leaf linden and field elm. Hortic Sci 45(12):1824–1829

Anonymous (2007) Vegetative propagation techniques perennial crop support series, Jalalabad, Afghanistan. https://www.sas.upenn.edu/~dailey/VegetativePropagationTechniques.pdf

Arnaudov V (2011) Propagation method of epicotyl grafting in walnut (*Juglans regia* L.) under production condition. Bulgarian J Agric Sci 17:173176

Arnold M, Struve D (1993) Root distribution and mineral uptake of coarse-rooted trees grown in cupric hydroxide-treated containers. Hortic Sci 28:988–992

Ashrafi NE, Vahdati K, Ebrahimzadeh H, Mirmasoumi M (2010) Analysis of invitro explants mineral contents to modify medium mineral composition for enhancing growth of Persian walnut (*Juglans regia* L.). J Food Agric Environ 8:325–329

Avanzato D (1999) Un sistema mobile di riscaldamento locallizzato del punto d'innesto applicato in situ su semenzali di noce. Riv Fruttic Ortofloric 61(11):74–76

Avanzato D (2001) Effect of different hygro-thermic environments on growth of potted grafted seedlings. Acta Hortic 544:546–464

Avanzato D, Ducci D, Gorian F, Gui L, Major A, Malvolti E, Mezzalira G, Pollegioni P, Proietti R (2006) Propagation ability of selected walnut hybrids (*Juglans regia* L. x *J. nigra* L.). Acta Hortic 705:359–364

Avıĺes F, Rios D, Gonzalez R, Sanchez-Olate M (2009) Effect of culture medium in callogenesis from adult walnut (*Juglans regia* L.) leaves. Chilean J Agric Res 69(3):460–467

Bacchetta L, Aramini M, Bernardini C, Rugini E (2008) *In vitro* propagation of traditional Italian hazelnut cultivars as a tool for the valorization and conservation of local genetic resources. Hortic Sci 43(2):562–566
Bajaj YPS (1991) Automated micropropagation for en masse production of plants. In: Bajaj YPS (ed) Biotechnology in agriculture and forestry, High tech and micropropagation 1, vol 17. Springer, Berlin, pp 3–16
Barber J, Anderson B (1992) Too much a good thing: light can be bad for photosynthesis. Trends Biochem Sci 17:61–66
Baron LC, Stebbins RL (1969) Growing filberts in Oregon. Oregon State University Extension Service: Bull no. 39
Bayazit S, Imrak B, Küden A (2005) Determination of grafting times and methods of walnut under Adana ecological conditions. Bançe Ceviz 34(1):231–234
Beckford, R. Plant propagation methodologies. Seeding, grafting, budding, cuttings and layering. https://www.slideshare.net/debbieanhall/plantpropagationmethodologies
Benmahioul B (2009) Amélioration de la micropropagation in vitro du pistachier (*Pistacia vera* L.) en vue de l'extension des vergers en Algérie. Ph.D. Thesis, University of Sciences and Technology of Oran, Oran, p 129
Benmahioul B, Dorion N, Kaid-Harche M, Daguin F (2012) Micropropagation and *ex vitro* rooting of pistachio (*Pistacia vera* L.). Plant Cell Tissue Org Cult 108:353–358
Benmahioul B, Kaid-Harche M, Daguin F (2015) Cryopreservation of *Pistacia vera* embryonic axes. J For Sci 61:182–187
Beyl CA, Trigiano RN (2016) Plant propagation concepts and laboratory exercises. CRC Press
Brase KD, Way RD (1959) Rootstocks and methods used for dwarfing fruit trees. Agricultural Experiment Station, New York, p 783
Braun L (2010) How to propagate hybrid hazelnuts by mound layering. Department of Agronomy and Plant Genetics, University of Minnesota
Carlson RF, Tukey HB (1955) Cultural practices in propagating dwarfing rootstocks in Michigan. Michigan Agric Exp Stat Q Bull 37:492–497
Castro LASD, Silveira CAP (2003) Vegetative propagation of peach by air layering technique. Rev Bras Frutic 25(2):368–370
Chandel JS, Gautam DR, Sharma NC (2006) Chip budding: an excellent method of propagation of walnut (*Juglans regia* L.). Acta Hortic 705:335–339
Clark SL, Schweitzer CJ, Schlarbaum SE, Dimov LD, Hebard FV (2009) Nursery quality and first-year response of American chestnut (*Castanea dentata*) seedlings planted in the southeastern United States. Tree Plant 53:13
Clark SL, Schlarbaum SE, Saxton AM, Hebard FV (2012) Nursery performance of American and Chinese chestnuts and backcross generations in commercial tree nurseries. Forestry 85(5): 589–600
Clark SL, Schlarbaum SE, Pinchot CC, Anagnostakis SL, Saunders MR, Thomas-Van Gundy M, Williamson TS (2014) Reintroduction of American chestnut in the national forest system. J For 112(5):502–512
Collins K, Fitzsimmons S, Saielli T, Westbrook J (2017) American chestnut restoration. Acta Hortic 1174:215–218
Coyne CJ, Mehlenbacher SA, Smith DC (1998) Sources of resistance to Eastern Filbert Blight in hazelnut. J Am Soc Hortic Sci 123:253–257
Davies FT, Hartmann HT (1988) The physiological basis of adventitious root formation. Acta Hortic 227:113–120
Dehgan B, Vahdati K, Rezaee R, Hassani D (2010) Walnut grafting success as affected by different grafting methods, cultivars and forcing treatments. Acta Hortic 861:345–352
Dehghan B, Vahdati K, Rezaee R, Hassani D (2009) Persian walnut (*Juglans regia* L.) grafting as influenced by different bench grafting methods and scion cultivars. J Appl Hortic 11:56–58

Dey DC, Jacobs D, McNabb K, Miller G, Baldwin V, Foster G (2008) Artificial regeneration of major oak (Quercus) species in the eastern United States: a review of the literature. For Sci 54(1):77

Dey DC, Gardiner ES, Kabrick JM, Stanturf JA, Jacobs DF (2010) Innovations in afforestation of agricultural bottomlands to restore native forests in the eastern USA. Scand J For Res 25(8): 31–42

Dodds JH, Robert LW (1985) Experiment in plant tissue culture, 2nd edn. Cambridge University Press, Cambridge

Dokuzoguz M, Gulcan R (1979) Almond growing and the problems. Tubitak Publication No. 432, TOAG No. 90, Ankara (in Turkish)

Duarte O, Medina C (1971) Propagation of citrus by improved mound layering. Hortic Sci 6:567

Duman E, Serdar U (2006) Research on shortening the nursery period in grafted chestnut. Hortic Sci 33(1):16–22

Dunn DE, Cole JC (1995) Propagation of Pistacia chinensis by mound layering. J Environ Hortic 13(2):109–112

Ebadi A, Solgi M, Zamani Z (2002) Effects of date of grafting and kind of callusing bed on grafting success of side and saddle grafting in Persian walnut (*Juglans regia* L.). Seed Plant 18(3): 294–305

Ebrahimi A, Vahdati K, Fallahi E (2007) Improved success of Persian walnut grafting under environmentally controlled conditions. Int J Fruit Sci 6(4):3–12

Ellena M, Sandoval P, Gonzalez A, Grau P (2014) Development and situation of the hazelnut in Chile. FAO-CIHEAM-NUCIS Newsl 16:3–7

Erdogan V (2006) Use of hot callusing cable in walnut propagation. Acta Hortic 705:313–317

Fideghelli C, De Salvador FR (2009) World hazelnut situation and perspectives. Acta Hortic 845: 39–52

Fields-Johnson C, Burger J, Evans D, Zipper C (2012) American chestnut establishment techniques on reclaimed Appalachian surface mined lands. Ecol Restor 30:99–101

Ford YY, Taylor JM, Blake PS, Marks TR (2002) Gibberellin A3 stimulates adventitious rooting of cuttings from Cherry (*Prunus avium*). Plant Growth Regul 37(2):127–133

Galic D, Dale A, Alward M (2014) Vegetative propagation of American chestnut. Acta Hortic 1019:99–103

Gandev S (2007) Budding and grafting of the walnut (Juglans regia L.) and their effectiveness in Bulgaria. Bulgarian J Agric Sci 13:683–689

Gandev S, Dzhuvinov V (2006) Performance of hypocotyl grafting of walnut under uncontrolled temperature conditions. Acta Hortic 705:351–353

Garner RJ (1988) The grafter's handbook, 5th edn. Oxford University Press, New York

Garner RJ, Bradley S (2013) The grafter's handbook. Chelsea Green Publishing

Georgi LL, Craddock JH, Bevins D, Kling R, Hebard FV (2013) Grafting gives legs to trees at Meadowview Research Farms, 30th Annual meeting. The American Chestnut Foundation., Abstract book, p 9

Gossard AC (1944) The rooting of pecan softwood cuttings under continuous mist. Proc Am Soc Hortic Sci 44:251–254

Grossman BC, Gold MA, Dey DC (2003) Restoration of hard mast species for wildlife in Missouri using precocious flowering oak in the Missouri river floodplain, USA. Agrofor Syst 59(1):3–10

Grossnickle SC, El-Kassaby YA (2016) Bareroot versus container stocktypes: a performance comparison. New For 47(1):1–51

Grossnickle SC, MacDonald JE (2018a) Seedling quality: history, application, and plant attributes. Forests 9:5

Grossnickle SC, MacDonald JE (2018b) Why seedlings grow: influence of plant attributes. New For 49(1):115–120

Hartman H, Kester D, Davies FT (1990) Plant propagation principles and practices, 5th edn. Prentice Hall Career & Technology, New Jersey, p 221

Hartmann C, Landgraf M, Bate M, Jackle H (1997) Kruppel target gene knockout participates in the proper innervation of a specific set of Drosophila larval muscles. EMBO J 16(17):5299–5309
Hartmann HT, Kester DE, Davies FT, Geneve RL (2002) Plant propagation: principles and practices. Prentice Hall Inc, New Jersey
Hartmann HT, Kester DE, Davies FT, Geneve RL (2007) Plant propagation: principles and practices, 7th edn. Prentice-Hall of India Pvt. Ltd., New Delhi, India
Hartmann HT, Kester DE, Davies FT, Geneve RL (2011) Plant propagation: principles and practices, 8th edn. Prentice Hall, New Jersey, p 915
Hassankhah A, Vahdati K, Lotfi M, Mirmasoumi M, Preece J, Assareh MH (2014) Effects of ventilation and sucrose concentrations on the growth and plantlet anatomy of micropropagated Persian walnut plants. Int J Hortic Sci Technol 2:111–120
Hazarika BN, Parthasarathy VA, Nagaraju V (2004) Influence of in-vitro preconditioning of Citrus sp. microshoots with sucrose on their ex vitro establishment. Indian J Hortic 61:29–31
Hodaj B, Rama P, Hodaj N (2014) Time and methods of walnut budding (Juglans regia L.). J Food Agric Environ 12(2):1350–1352
Hormaza JI, Wunsch A (2011) Pistacia. In: Kole C (ed) Wild crop relatives: genomic and breeding resources, Temperate fruits. Springer, Berlin, pp 119–128
Howes FN (1948) Nuts. Faber and Faber Limited, London
Huang H, Norton JD, Boyhan GE, Abrahams BR (1994) Graft compatibility among chestnut (Castanea) species. J Am Soc Hortic Sci 119:1127–1132
Huang Y, Wang Z, Zheng B, Huang J (2006) Effect of plant growth regulators on rooting capacity by lignified cuttings of *Carya illinoensis*. J Southwest For Coll 5:10
Hubert EG (1977) Vegetative propagation of the filbert (*Corylus avellana* L.) by means of budding and cuttings. Master of Science, Oregon State University
Janick J, Moore JN (1996) Hazelnuts, chapter 3. In: Maxin (ed) Fruit breeding, Nuts, vol III. Wiley, New York, p 177
Janick J, Paull RE (2008) The encyclopedia of fruit and nuts. Cambridge University Press, Cambridge, p 160
Jinks RL (1995) The effects of propagation environment on the rooting of leafy cuttings of ash (*Fraxinus excelsior* L.), sycamore (*Acer pseudoplatanus* L.) and sweet chestnut (*Castanea sativa* Mill.). New For 10:183–195
Kasmi M, Rama P, Hodaj B, Kukali E, Rabeta A (2013) Budding of walnut (*Juglans regia* L.). Albanian J Agric Sci 12:465–469
Key RN (1978) Prospects for vegetative propagation in the genus *Castanea*. Proceedings of American Chestnut Symposium, Morgantown, pp 10–16
Kuniyuki AH, Forde HI (1985) Walnut propagation. In: Ramos DE (ed) Walnut orchard management. University of California, Davis, CA, pp 38–45
Lagerstedt HB, Roberts WW (1972) Walnut grafting in Oregon problem and solutions. Ann. Rep. NNGA, Iowa State University
Landhausser SM, Rodriguez-Alvarez J, Marenholtz EH, Lieffers VJ (2012) Effect of stock type characteristics and time of planting on field performance of aspen (*Populus tremuloides* Michx.) seedlings on boreal reclamation sites. New For 43:679–693
Lee SH, Tewari RK, Hahn EJ, Paek KY (2007) Photon flux density and light quality induce changes in growth, stomatal development, photosynthesis and transpiration of *Withania somnifera* L. Dunal. Plantlets. Plant Cell Tissue Organ Cult 90:141–151
Letham DS (1974) Regulators of cell division in plant tissue. J Plant Physiol 32:66–70
Li J, Lian-fang L, Xiong X, Wang G, Liu H, Xi X (2013) Effects of parent tree age and the thickness of cuttings on rooting capability of Carya illinoensis. J Northwest For Univ 4:20
Littsle EL (1979) Check list of United States trees (Native and naturalizes). In: USDAM (ed) Agric. handbook. USDAM, Forest Services, Washington, DC, p 375
Lovelace W (2002) Root production method system. In: Dumroese RK, Riley LE, Landis TD, Technical Coordinators. National proceedings: Forest and Conservation Nursery

Associations-1999, 2000, and 2001. Proceedings RMRS-P24. USDA Forest Service, Rocky Mountain Research Station, Ogden, UT, pp 20–21

Maryam V, Mehrdad J, Majid S (2015) Effect of time, method of budding and type of scion on bud take of sour cherry scions onto Mahaleb rootstocks. Int J Agron Agric Res 6:233–239

Mauricio-de-Almeida (2020) Commercial viability of in-container budding and grafting of English Walnut (Juglans Regis). Thesis submitted to Jordan College of Agricultural Sciences and Technology California State University, Fresno

McCarthy B, Gilland K, Bauman J, Keiffer C (2010) Factors affecting performance of artificially regenerated American chestnut on reclaimed mine sites. J Am Soc Mining Reclam:582–597

McEachern GR (1973) The influence of propagation techniques, the rest phenomenon and juvenility on the propagation of pecan, Carya illinoensis (Wangenh.) K. Kpch, stem cuttings. Diss Abstr Int 34(3):947

Medina JP (1981) Studies of clonal propagation on pecans at Ica, Peru. Plant Propagator 27(2): 10–11

Merkle SA, Wetzstein HY, Sommer HE (1987) Somatic embryogenesis in pecan. Hortic Sci 22(1): 128–130

Miletic R, Mitrovic M, Rakicevic M (2008) Contrasting fruit properties of hazelnut cultivars grown on different rootstocks. Paper presented at 7th International Congress on Hazelnut, Viterbo

Mir MA, Shameem R, Pandit AH, Bhat KM, Wani WM, Jan A, Hassan S, Majid S, Hussain S (2016) Propagation of almond through budding under temperate conditions of north western Himalaya. Appl Biol Rese 18(3):299–304

Nacheva L (2012) Godishen doklad na Instituta po ovoshtarstvo-Plovdiv za 2012 godina. Instituta po ovoshtarstvo-Plovdiv

Nacheva L, Gandev S (2009) Transport and distribution of 14C-photoassimilates in walnut plants grafted by the hot callus method. Plant Sci 46:210–213

Najafian-Ashrafi E, Vahdati K, Ebrahimzadeh H, Mirmasoumi M, Lotfi N (2009) Optimization of walnut tissue culture medium using seed composition. Hortic Environ Biotechnol 50:1–6

Nas MN, Read PE (2001) Micropropagation of hybrid hazelnut: medium composition, physical state and iron source affect shoot morphogenesis, multiplication and explant vitality. Acta Hortic 556:252–257

Nas MN, Read PE (2004) A hypothesis for the development of a defined tissue culture medium for higher plants and micropropagation of hazelnuts. Sci Hortic 101:189–200

Nedev N, Vasilev V, Kavardzhikov L, Zdravkov K (1967) Oreh. Hristo G. Danov, Plovdiv

Nedev N, Vasilev V, Kavardzhikov L, Zdravkov K (1976) Propagation of walnut plant material. In: Nedev N (ed) Nut fruits. H. G. Danov, Plovdiv, pp 80–133

Nesbitt ML, Goff WD, Stein LA (2002) Effect of scionwood packing moisture and cut-end sealing on pecan graft success. HortTechnology 12(2):257–260

Ninic-Todorovic J, Cerovic S, Ognjanov V, Golosin B, Bijelic S, Jacimovic G, Kurjakov A (2008) Rootstocks of Corylus colurna L. for nursery production. Paper presented at 7th International Congress on Hazelnut, Viterbo

Obeidy AA, Smith MAL (1993) Organogenesis from mature pecan cotyledons and embryonic axes. HortScience 28:213–215

Olsen J, Smith D (2013) Growing hazelnuts in the pacific North West. Plant Propagation, Oregon State University Publication EM9075

Oraguzie NC, McNeil DL, Thomas MB (1998) Examination of graft failure in New Zealand chestnut (Castanea spp.) selections. Sci Hortic 76:89–103

Osterc G (2001) Phenomenon of physiological aging of woody plants as a factor cutting propagation. Sodobno Kmetijstvo 34:430–434

Osterc G, Fras ZM, Vodenik T, Luthar Z (2005) The propagation of chestnut (Castanea sativa Mill.) nodal explants. Acta Agriculturae Slovenica 82:411–418

Özkan Y, Edizer Y, Auca Y (2001) A study on propagation with patch budding of some walnut cultivars (*Juglans regia* L.). Acta Hortic 544:521–525

Pacholczak A, Szydło W, Łukaszewska A (2005) Effectiveness of foliar auxin application to stock plants in rooting of stem cuttings of ornamental shrubs. Propag Ornam Plants 5(2):100–106
Paillet F (2002) Chestnut: history and ecology of a transformed species. J Biogeogr 29:1517–1530
Pair JC, Khatamian H (1980) Propagation and growing of the Chinese pistachio. Proc Int Plant Propagators Soc 32:497–503
Palma AC, Laurance SG (2015) A review of the use of direct seeding and seedling plantings in restoration: what do we know and where should we go? Appl Veg Sci 18:561–568
Pandit AH, Kour A, Wani MS, Mir MA (2009) Genetic resources of chestnut in Kashmir valley. Acta Hortic 815:51–53
Payghamzadeh K, Kazemitabar SK (2010) In vitro germination of Pecan (Carya illinoensis) embryo. Biharean Biologist 4(1):37–43
Payghamzadeh K, Kazemitabar KS (2011) In vitro propagation of walnut-a review. Afr J Biotechnol 10(3):290–311
Pinchot CC, Clark SL, Schlarbaum SE, Saxton AM, Sung SS, Hebard FV (2015) Effects of temporal dynamics, nut weight and nut size on growth of American chestnut, Chinese chestnut and backcross generations in a commercial nursery. Forests 6(5):1537–1556
Pinghai D, Rongting X (1993) Effect of phenols on the survival of walnut grafting. Acta Hortic 311: 134–140
Pokorny FA, Sparks D (1967) Studies on air-layering pecans: effect of date of propagation, wounding and indole-3-butyric acid on rooting of air-layered pecan terminals, Carya illinoensis Koch. Cv. Stuart. Hortic Sci 2(2):50–51
Polat A, Ördek G (2006) Increasing of grafting success rate on walnut. Acta Hortic 722:253–256
Porebski S, Rzeźnicka B, Poniedzialek W (2002) Comparison of two methods of walnut grafting. J Fruit Ornam Plant Res 10:55–62
Rahemi A, Yadollahi A (2006) Rainfed almond orchards in Iran, ancient and new methods and the value of water harvesting techniques. Acta Hortic 726:449–453
Reddy PPN, Ray NR, Patel AD, Patel JS (2014) Effect of rooting media and IBA (indole butyric acid) levels on rooting and survival of air layering in fig (*Ficus carica* L.) cv. Poona under middle Gujarat agro-climatic conditions. Asian J Hortic 9(1):1–5
Reid W (2010) Propagating pecan and black walnut in Missouri. University of Missouri Center for Agroforestry, pp 1–12.
Reinert J, Bajaj YPS (1976) Applied and fundamental aspects of plant cell, tissue and organ culture. J Exp Bot 27:144–147
Rezaee R, Vahdati K, Grigoorian V, Valizadeh M (2008) Walnut grafting success and bleeding rate as affected by different grafting methods and seedling vigour. J Hortic Sci Biotechnol 83(1): 94–99
Rezaee R, Hassani D, Vahdati K (2014) Long term trials on topworking of walnut trees in Iran. Acta Hortic 1050:197–201
Rhodes HLJ (1968) Juvenility in relation to adventitious bud and root initiation in woody plants. Proc Int Plant Propagators Soc 18:252–254
Riel W, Sutter E (1996) Propagation. In: Almond production manual. University of California, USA, pp 1–2
Rodriguez APM, Wetzstein HY (1994) The effect of auxin type and concentration on pecan (Carya illinoensis) somatic embryo morphology and subsequent conversion into plants. Plant Cell Rep 13:607–611
Roots of Peace (2007) Vegetative propagation techniques. Afghanistan, Publication no. 2007-003-AFG.
Rose R, Campbell SJ, Landis TD (1990) Target seedling symposium: Proceedings, combined meeting of the Western Forest Nursery Associations, 1990 August 13–17, Roseburg, Oregon. Department of Agriculture, Forest Service, Rocky Mountain Forest and Range Experiment Station, Fort Collins, CO, p 286. Gen. Tech. Rep. RM-GTR-200
Rotondo-Walnuts (2004) Progress report for the period June to October 2004. Paper presented at IDC Board meeting, 25 November 2004, Rotondo Farm, Aliwal North, South Africa

Russell EWB (1987) Pre-blight distribution of castanea dentata (Marsh.) borkh. Bull Torrey Bot Club 114(2):183–190

Salimi S, Hoseinova S (2012) Selecting hazelnut (Corylus avellana L.) rootstocks for different climatic conditions of Iran. Crop Breed J 2(2):139–144

Serdar U (2009) Grafting in seeds, radicles, hypocotyls and epicotyls of chestnut. Acta Hortic 815: 199–204

Serdar U, Demirsoy H (2010) Graft compatibility in some Turkish chestnut genotypes (Castaneasativa Mill.). Acta Hortic 866:285–287

Serdar U, Soylu A (1999) Selection of chestnut (C. sativa Mill.) in Samsun vicinity. Acta Hortic 494:327–332

Serdar U, Kose B, Yilmaz F (2005) The structure of graft union in European chestnut using different grafting methods. Hortic Sci 40(5):1474–1477

Sharbat H, Mir MM, Bhat R, Wani SA, Shameem R (2016) Standardization of propagation techniques in Chestnut (*Castanea sativa* Mill.). Ecol Environ Conserv 22(2):911–915

Sharma RR (2002) Propagation of horticultural crops: principles and practices. Kalyani Publishers, New Delhi

Shreve LW (1974) Propagation of walnut, chestnut and pecan by rooted cuttings. In: Proceedings, Eighth Central States Forest Tree Improvement Conference, October 11–13, 1972, University of Missouri, Columbia, MO, pp 20–23

Skoog F, Armstrong DJ (1970) Cytokinin. Annu Rev Plant Physiol 21:359–384

Skousen JG, Dallaire K, Scagline-Mellor S, Monteleone A, Wilson-Kokes L, Joyce J, Jacobs DF (2018) Plantation performance of chestnut hybrids and progenitors on reclaimed Appalachian surface mines. New For 49(5):599–611

Solar A, Stampar F, Trost M, Barbo J, Avsec S (2001) Comparison of different propagation methods in walnut (*Juglans regia* L.) made in Slovenia. Acta Hortic 544:527–530

Spencer A (1980) The propagation of pecan stem cuttings. Hortus 27:28–32

Spetich M, Dey D, Johnson P, Graney D (2002) Competitive capacity of Quercus rubra L. planted in Arkansas' Boston Mountains. For Sci 48:504

Srivastava RP (1966) Research on horticultural crops at Chaubattia. Indian Hortic 10:9–11

Struve DK (2009) Tree establishment: a review of some of the factors affecting transplant survival and establishment. Arboric Urban For 35:10–13

Tilkat E, Onay A (2009) Direct shoot organogenesis from in vitro-derived mature leaf explants of pistachio. In Vitro Cell Dev Biol Plant 45:92–98

Tilkat E, Onay A, Yildirim H, Çetin Ozen H (2008) Micropropagation of mature male pistachio *Pistacia vera* L. J Hortic Sci Biotechnol 83(3):328–333

Tous JA, Romero A, Rovira M, Hermoso FJ (2009) Performance of "Negret" hazelnut cultivar grafted on 4 rootstocks in Catalonia (Spain). Acta Hortic 845:89–93

Ullah T, Muhammad W, Nabi G, Rehman N, Arshad M, Naeem N (2000) Bud take success of different almond varieties on peach rootstock. Pak J Biol Sci 3(11):180–185

Vahdati K, Zareie N (2006) Evaluation of side-stub and hypocotyle grafting efficiency for walnut propagation in Iran. Acta Hortic 705:347–350

Vahdati K, Leslie C, Zamani Z, McGranahan G (2004) Rooting and acclimatization of in-vitro grown shoots from three mature Persian walnut cultivars. Hortic Sci 39:324–327

Vahdati K, Bayat S, Ebrahimzadeh H, Jariteh M, Mirmasoumi M (2008) Effect of exogenous ABA on somatic embryo maturation and germination in Persian walnut (*Juglans regia* L.). Plant Cell Tissue Organ Cult 93(2):163–171

Vahdatia K, Aalifar M (2016) Development and extension of walnut propagation in Iran. Acta Hortic 1139:467–473

Van-Sambeek JW, Godsey LD, Walter WD, Garrett HE, Dwyer JP (2016) Field performance of Quercus bicolor established as repeatedly air-root-pruned container and bare root planting stock. Open J For 6(3):163–176

Vibha JB, Shekhawat NS, Mehandru P, Dinesh R (2014) Rapid multiplication of Dalbergia sissoo Roxb.: a timber yielding tree legume through axillary shoot proliferation and ex vitro rooting. Physiol Mol Biol Plants 20(1):81–87

Vieitez E, Vieitez MA (1976) Juvenility factors related to the rootability of chestnut cuttings. Acta Hortic 56:269–274

Vieitez AM, Ballester A, Garcia MT, Vieitez E (1980) Starch depletion and anatomical changes during the rooting of Castanea sativa Mill. cuttings. Sci Hortic 13:261–266

Vieitez J, Kingston IGD, Ballester A, Vieitez E (1987) Identification of two compounds correlated with lack of rooting capacity of chestnut cuttings. Tree Physiol 3:247–255

Vossen L (2000) Chestnut culture in California. Publication 8010, 2000. University of California, Oakland. Disponivalem: http://anrcatalog.ucdavis.edu. Acessoem

Wannan BS (2006) Analysis of generic relationships in *Anacardiaceae*. Blumea 51:165–195

Ward J, Gent M, Stephens G (2000) Effects of planting stock quality and browse protection-type on height growth of northern red oak and eastern white pine. Forest Ecol Manag 127:205–216

Wells L (2014) Budding and grafting of pecan. Univ Georgia Coop Ext Bull 1376:1–8

Wetzstein HY, Ault JR, Merkle SA (1989) Further characterization of somatic embryogenesis and plantlet regeneration in pecan, *Carya illinoensis*. Plant Sci 64:193–201

Wetzstein HY, Ault JR, Merkle SA (1990) Factors influencing somatic embryogenesis and plantlet regeneration in pecan, Carya illinoensis. Acta Hortic 280:69–73

Wetzstein HY, Rodriguez APM, Burns JA, Magner HN (1996) Carya illinoensis (pecan). In: Bajaj YPS (ed) Biotechnology in agriculture and forestry, vol 35. Springer, New York, pp 50–75

White J (2009) Patch budding pecans. New Mexico State University, College of Agricultural, Consumer and Environmental Sciences

Wilson E, Vitols K, Park A (2007) Root characteristics and growth potential of container and bare root seedlings of red oak (*Quercus rubra* L.) in Ontario, Canada. New For 34(2):163–176

Yates IE, Reilly CC (1990) Somatic embryogenesis and plant development in eight cultivars of pecan. HortScience 25(5):573–576

Zaczek JJ, Steiner KC, Bowersox TD (1995) Quality or quantity: stock choices for establishing planted northern red oak. In: Landis TD, Coords B, Luth F (eds) Forest Service. National Proceedings, forest and conservation nursery associations. Gen. Tech. Rep. PNW-GTR-365. U.S. Department of Agriculture, Forest Service, Pacific Northwest Research Station, Portland, OR, p 116

Zhang R, Peng FR, Le DL, Liu ZZ, He HY, Liang YW, Tan PP, Hao MZ, Li YR (2015) Evaluation of epicotyl grafting on 25- to 55-day-old pecan seedlings. HortTechnology 25(3):392–396

Zobayed SMA, Armstrong J, Armstrong W (2001) Leaf anatomy of in vitro tobacco and cauliflower plantlets as affected by different types of ventilation. Plant Sci 161(3):537–548

Chapter 6
Pollination Management

Sanjeev K. Banyal, Uday Raj Patial, and Ajay K. Banyal

Abstract Nut crops hold a major place in temperate zone fruit. These nut crops, namely walnut, chestnut, chilgoza nut, hazelnut, pecan nut, and pistachio are highly valued for their nutritive properties. Most of these nut crops offer deep root system and have been found to offer soil conservation properties. Although a lot of work has been done for the improvement of productivity in the nut fruit culture but still there are many barriers in the production of nut crops. The nut culture is limited to hilly tracts only and is totally dependent on the climatic potential of the locality. The plants exhibit different floral patterns, dichogamous behaviors, incompatibility systems which are even cultivar specific and ultimately affect the fruit setting by hindering the pollination. This chapter documents the information about floral biology, pollination patterns, pollination barriers, varietal combinations, climatic barriers and pollination management studies via natural or artificial ways in different nut crops.

Keywords Pollination · Varietal · Temperate nuts · Compatibility

1 Walnut

1.1 Floral Biology

Juglans is a monoecious genus, bearing unisexual flowers separately on a same tree. Since, most of the plantations are of seedling origin, they have long juvenile phase of 12–15 years, whereas grafted one are precocious and begun to fruit in 4–5 years. Though the staminate and pistillate flowers are present on the same plant, yet they bloom on the shoot of different shape and age (Wood 1934). The pistillate flowers

S. K. Banyal (✉) · A. K. Banyal
Department of Fruit Science, YSPUH&F College of Horticulture and Forestry, Neri, Hamirpur, Himachal Pradesh, India

U. R. Patial
Department of Fruit Science, PAU, Ludhiana, Punjab, India

M. M. Mir et al. (eds.), *Temperate Nuts*,
https://doi.org/10.1007/978-981-19-9497-5_6

emerge terminally in the pairs (nutlets) on the current season growth. Whereas, the staminate flowers emerge laterally from the buds present in the leaf axils of the twigs of old season growth. The staminate flowers develop on woody structure while females are born on twigs possessing a succulent growth. The pistillate flowers are rid of petals with ovary attached to a large reflexed stigma supported by the perianth of two pair of leaves. With the onset of the summers, the conical buds in the twigs begin to enlarge and catkins carrying densely packed immature flowers are observed hanging from the leaf axils and can be 3 in. long. However, the production of pistillate blooms and their development into fruits occur in a single season and they need not to pass from the dormancy period like staminate buds. Chandra and Tomar (2012) observed initiation of pistillate and staminate flowers in nine walnut cultivars from end-March to mid-April under Indian conditions. However, the floral initiation differs with cultivar and location. Hassankhah et al. (2017) studied the phenology of the three Persian walnut cultivars under Iran conditions. Bud break was recorded during mid-March in Jamal and Damavand while during first week of April in Chandler. The stigmatic receptivity also varied among cultivars as, first week of April in Damavand, second week of April in Jamal, and third week of April in Chandler. Staminate flower development is highly influenced by air temperature and usual temperature for catkin flowering, within the Romanian conditions was found between 14 and 15.9 °C (Cosmulescu et al. 2010).

1.2 Pollination

Almost all the genotypes under genus Juglans allow autogamy and allogamy. As the pendant catkins mature, anthers dehisce and the pollen is dispersed in the whole orchard. If there is no wind, the pollen is collected in the bucks of flowers present below having a cup shaped depression and blown away later when the winds arise. The stigmatic lobes begin to separate (up to 45°) with a sticky exudate over its surface acting a substrate for pollen germination and remains receptive for 7 days. The pollen germination begins after 2 days and fertilization takes place in a week (Krueger 2000).

1.2.1 Barriers in Pollination

1. Heterodichogamy: It is a rule in Persian walnut (Hassankhah et al. 2017). The chilling requirement of cultivars varies specifically for the male and female flowers. Hassani et al. (2014) observed higher chilling requirement of the terminal buds as compared to the male catkins in cultivar Jamal and Chandler, whereas an opposite scenario was noticed in cultivar "Damavand" revealing the protandrous and protogynous nature of the cultivars. Generally, the pistillate flowers begin to develop in May (Gao et al. 2012) and continue until blooming in the following year. Hence, the pollen shedding do not overlap with stigmatic receptivity (Golzari et al. 2016) resulting in poor setting of nuts.

2. Poor pollen viability: The pollen viability varies with the locality, season, climatic conditions, and cultivar. Bhat et al. (2016) recorded high pollen viability (91.4%) cultivar SKAU-W-0020 as compared to SKAU-W-0022 (87.40%) in Kashmir conditions. The pollen viability begins to decrease once it has been dehisced from the anthers, i.e., older the pollen, lower will be the viability (Wood 1934). Moreover, the pollen lost its viability within 2–3 h after exposure to sunlight at 25–32 °C.
3. Poor pollen distribution: The walnut pollen is very small (46 μm) in size (Smith et al. 1912) and is carried by wind. If plants are not placed at a proper distance, it will result in poor distribution of pollen within the orchard and ultimately affecting the yield.
4. Quantity of pistils produced: Regardless of satisfactory pollination factors, economic crops of walnuts cannot be produced until a large number of pistils develop on the tree. The production of pistillate flowers varies with the age of the plant. The young trees of cultivar Payne produce large number of pistils in their young age, but few pistils are observed on old trees. Whereas, in cultivar like Franquette plants produce few pistils in their young age and tends to develop more pistils as the plants get older (Wood 1934).

1.2.2 Pollination Management

Certain factors should be kept in mind for efficient pollination in a walnut orchard:

- Production of pollen grains in abundance ensuring efficient distribution within the orchard.
- Pollen should be light weight, so that it can be distributed easily by wind in the orchard.
- Prolonged maturity period of catkins, increasing pollen availability for a long duration.

1. Pollinizer density and placement in walnut orchards

   Since heterodichogamy is prevalent in all the cultivars, there is a difficulty in the synchronization of the blooming period of staminate and pistillate flowers on each tree. Cultivars with efficient synchronization (Tulare, Serr, and Sexton) seem to set fruit without pollinizer but cultivars with non-synchronous behavior (Chandler, Howard, Solano) have a pollinizer requirement. Therefore, a traditional recommendation of covering the orchard with 10% pollinizer has been followed. However, Krueger (2000) reported the problem of pistillate flower abscission in walnut due to excessive pollen loads on stigma under 10% pollinizer recommendation, so 2–3% pollinizer placement has been recommended.
2. Distance between pollinizers

   Pollen is carried by wind, and the direction or speed of the wind varies with region and season. The pollen grains remain viable up to a period of 24–48 h at 23–25 °C or maybe longer, depending upon the climatic conditions. Downwind fall-out from pollinizer trees in commercial orchards is thought to be around

Table 6.1 Pollinizer varieties for commercial walnut cultivars

Genotype	Pollinizer
CITH-W-1	CITH-W-2, CITH-W-3, CITH-W-4, CITH-W-5 CITH-W-6 CITH-W-7. CITH-W-8 CITH-W-9, CITH-W10, Hamdan, Sulaiman
CITH-W-2	CITH-W-3, CITH-W-4, CITH-W-5, CITH-W-6 CITH-W-7. CITH-W-8, CITH-W-9, CITH-W-10
CITH-W-3	CITH-W-2, CITH-W-4, CITH-W-5 CITH-W-6 CITH-W-7. CITH-W-7 Hamdan, Sulaiman, Opex Caulchery
CITH-W-4	CITH-W-1, CITH-W-2, CITH-W-3, CITH-W-5 CITH-W-6, Hamdan, Sulaiman
CITH-W-5	CITH-W-3, CITH-W-4, CITH-W-7, CITH-W-8, CITH-W10, Hamdan, Sulaiman, Opex Caulchery, Nugget Franquette
CITH-W-6	CITH-W-1, CITH-W-2, CITH-W-3, CITH-W-4, CITH-W-7, Hamdan, Sulaiman, Opex Caulchery, Nugget Franquette
CITH-W-7	CITH-W-1, CITH-W-2, CITH-W-3, CITH-W-4, CITH-W-5, CITH-W-6, CITH-W-8, Hamdan, Sulaiman Cheinova, Tutle
CITH-W-8	CITH-W-1, CITH-W-2, CITH-W-3, CITH-W-4, CITH-W-5, CITH-W-6, Hamdan, Sulaiman
CITH-W-9	CITH-W-1, CITH-W-2, CITH-W-3, CITH-W-4, CITH-W-5 CITH-W-6 CITH-W-7, Hamdan, Sulaiman
CITH-W-10	CITH-W-1, CITH-W-2, CITH-W-3, CITH-W-4, CITH-W-5 CITH-W-6 CITH-W-7. Opex Caulchery, Nugget Cheinova, Tutle

Source: Central Institute of Temperate Horticulture by S R Singh and co workers, named "Present Status and Future Strategies for Walnut Production in India". https://www.biotecharticles.com/PDF_Articles/Recent_advances_in_walnut_production_BA_4098.pdf

120–150 ft, but the pollen flow studies indicated significant contribution from outside of the orchard is also possible. Hence, there should be at least 250–300 ft distance between the rows of pollinizer.

3. Varietal compatibility

   For the improvised production of nuts, compatible pollinizer varieties should be placed within the orchard with the elite cultivars to ensure efficient pollination and obtain economic returns from the plants. Singh et al. (2012) have screen identified different pollinizer varieties for the elite walnut cultivars CITH-W-1 to 10 with the aim of maximizing pollination efficiency (Table 6.1).

   However, Hassani et al. (2012) observed Damavand to be a suitable pollinizer with longer pollen shading period (11–13 days) good synchronicity with different protandrous cultivars like Jamal, Serr, and other Iranian cultivars. Hassan et al. (2016) identified selection SKAU-W-0020 efficient to be used as a pollinizer with long pollen viability and better germination capacity under Indian conditions.

4. Artificial pollination

   The stigmatic receptivity period ranges from the slightly opening of lobes to appearance of brown stripes on its surface. The best period for carrying pollination is up to 3–5 days after the opening of split stigma. The pollen can be stored for 96 h under natural conditions but it loses its viability quickly to 33% after 4 h. The pollen viability is observed to be highest when the anthers shedding has reached 95%. Pollen viability could be reinforced at 4 °C (Liu et al. 2011).

Pollination can be performed by depositing a prepared load of pollen using the eraser end of pencil on to the stigmatic surface of flower, considered as the contact surface. Bhat et al. (2016) recorded higher fruit retention with artificial pollination (90.61%) as compared to open pollinated walnut plants (81.49%).

Hormonal Balance

Like climatic factors, the hormones have a major role in modifying the sex expression of the plants. The emergence of staminate flowers has been observed to be highly affiliated by the GA_3 formation in the leaf axils. Hassankhah et al. (2018) observed a significant increase in the male flowers and male:female ratio per shoot with the application of GA_3 @ 100 mL/L in Chandler walnut.

1.3 Influence of Climatic Factors

Many reports have revealed the significant effect climatic factors on the flowering behavior of walnut. There is fall in the flowering duration with the increase in the air temperature, thus ultimately shortening the pollination duration. Prevalence of spring frosts in the mountains lead to subsequent flower drop hence, late leafing/high chilling cultivars should be planted in mountain ranges with frequent spring frosts (Hassankhah et al. 2017). Cosmulescu et al. (2010) observed that sunshine has a major role in the production of viable pollen. Occurrence of rains and heavy fogs during the period of anther dehiscence sheds the pollen grains causing deterioration. Moreover, pollen grains are found to have lost their viability after having once been wet.

2 Chestnut

2.1 Floral Biology

Chestnut plants are monoecious in nature with flowering inflorescences carried by annual branches of the same tree. The floral buds begin to differentiate by mid-April and flowering commences during the mid-summer, attaining full bloom by end-June to mid-July (Bencat and Tokar 1978). Flowers are borne on spike structured catkins which arise from the axils of leaf on the current season's growth. Two types of catkins are observed on the plants, viz. (1) unisexual and (2) bisexual, respectively. The unisexual catkins carry staminate flowers only whereas the bisexual catkins consist of both pistillate and staminate flowers in a single inflorescence. The pistillate flowers are located in the base while staminate one's toward the tip in the bisexual inflorescence. However, the bisexual flowers tend to emerge near the apex whereas unisexual from the base of twigs and mature earlier than bisexual flowers (Miller et al. 2013). The catkins are observed to occur in a contiguous fashion in the center of the shoot and consists of vegetative buds at both the axial and distal parts of the shoot. The average length of catkins is 15–20 cm and are flexible or pendulous in appearance (Percival 1965), but it varies among cultivars and species (Mert and Soylu 2006). Based on the pollen viability and morphological length parameters of stamens, the European scientists have classified the catkins into four broad categories, namely (1) longistamine (7–9 mm), (2) mesostamine (4–5 mm), (3) brachystamine (1–3 mm), and (4) astamine (Bounous et al. 1992). Longistamine is considered as most suitable pollinizers as they produce abundant pollen and has long viability (Mert and Soylu 2006), whereas astamine types do not produce any pollen at all. There are one or more pistillate inflorescence (involucre) present at the base of the bisexual catkins. A single involucre generally consists of three pistillate flowers but sometimes it may range up to seven.

2.2 *Pollination*

Self-sterility is a norm in chestnut so it is necessary to plant two or three cultivars to ensure successful cross pollination (Bounous and Marinoni 2005). Since the plants are protandrous in nature, staminate flowers bloom a week earlier than the pistillate ones (Ayfer and Soylu 1993). The bisexual flowers bloom later than the unisexual flowers, and this phenomenon is known as duodichogamy (Clapper 1954). The pollination process continues up to a month, till the catkins turn brown, drop off and the styles darken. The involucre comprising three pistillate flowers grows spiny cupule (bur), fertilized ovaries stat growing and finally develops into nuts.

Pollinizer The production of abundant light pollen with less stigmatic nectar is a characteristic feature of anemophilous plants, but the chestnut flowers produce sticky and heavy (Chapa 1984) pollen which is a feature of entomophilus plants. So, both insects and wind play a major as pollinator. However, in case of staminate catkins, the pollination is possible with the mean of insects only. However, Hasegawa et al. (2015) observed higher efficient pollination and pollen dispersal to longer distances by bumble bees as compared to small bees, flies, and beetles. The fruit setting in chestnut is associated with beetle pollination syndrome because the beetles are majorly reported around the chestnut flowers (Larue et al. 2021).

2.2.1 Pollination Barriers

1. The catkins on the young plants do not produce abundant pollen resulting in inefficient pollination.

2. All three pistillate flowers within the involucre have a staggered time of stigma receptivity, so the staminate flowers should have a long pollen dehiscence period.
3. Plants are self-sterile in nature, and the highly selective stigma requires to ensure cross pollination to obtain an economic yield.
4. Pollen obtained from many interspecific hybrids is male sterile. The progenies of American and Chinese chestnut are mostly male sterile (Miller et al. 2013).
5. The presence of staminate catkins leads to non-availability of pollen to stigma resulting in no fertilization and formation of barren burs.
6. Adverse weather conditions at the time of stigma receptivity hampers anemophily.

2.2.2 Pollination Strategies

1. Varietal compatibility: Pistillate flowers of chestnut show incompatibility reaction with their own staminate flowers, and this selectivity of stigma for the pollen varies with cultivars and species. Hence it is necessary to select optimum pollinizer plant which can successfully fertilize the ovaries inside the involucre. Wang et al. (2012) identified some combinations with effective pollination as (Table 6.2).
2. Layout
   The pollinizer plants should be inter planted within the orchard or at the rows in the direction of wind flow, so as to ensure proper dispersal of pollen within the orchard.
3. Artificial pollination
   The light-colored catkins are considered as the best source of fresh pollen. Catkins can also be collected and kept at 68–70 °F till the anthers burst. The pollen is collected and stored in cool refrigerated conditions (Rutter 1990) as the warm humid conditions would not let pollen survive for more than 4 h (Schad et al. 1952). The pollen application is done on the receptive stigmatic surfaces with the help of camel's hairbrush and dusters.

2.2.3 Effect of Weather on Pollination

The pollination has been observed to be inefficient on the windy days. Successful pollination can be achieved on cloudy/sunny days in cold and hot weather without any hindrance. Higher success rate is obtained when flowers are pollinated in

Table 6.2 Effective pollinizers identified for chestnut cultivars

Pollinizer	Cultivar
Zunyu	Yanhong, Zipo, Duanci
Duanci	Zunyu,
Donglingmingzhu	Zunyu, Duanci

Source: Wang et al. (2012)

morning hours as compared to those in the afternoon (Clapper 1954). However, pollination was observed to be successful even on the rainy days also, though the success rates were not that high (Rutter 1990).

3 Hazelnut

3.1 Floral Biology

Hazelnut is a monoecious crop with very unusual flowering behavior, been a subject of various investigations. Distinct staminate and pistillate inflorescences can be observed on the leaf axils of 1 year old shoots on the same plant. The male inflorescence is catkin, containing 130–290 flowers. The female flowers are glomerulus, i.e., compound buds with vegetative part at the base and a fertile cluster at the top of the flower. These small pistillate flowers consist of an elongated pair of stigmatic styles having a minute ovary at the base (Germain 1994). The bud differentiation for male flowers begins in early May while rudimentary styles can be observed on the female flowers during late August after the formation of flowering primordia during mid-July. However, the pistillate flower formation is governed by three major factors, viz. (1) light incidence on 1 year old shoots, (2) origin, and (3) vigor. The emergence of pistillate flowers increases with the increase in the shoot length. The low vigor shoots (<15 cm) have a lower tendency to produce glomerulus, whereas vigorous shoots (>35 cm) are capable of producing large number of glomerulus (Baldwin 2015). Hazelnut flowers bloom in winter (end-December to February) when floral inflorescences break their dormancy after the completion of required chilling period. This chilling period varies with the cultivar and is considered to be higher for the pistillate flowers than those for catkins (Kavardzhikov 1980). The chilling requirement ranged from 100 to 860 h for catkins while 290 to 1550 h for the pistillate flowers (Mehlenbacher 1991).

3.2 Pollination

The peak period of pollination in hazelnut is from January to February and is totally anemophilous. However, Cristofori et al. (2018) reported pollination during mid-November in Northern hemisphere and June in the Southern hemisphere. The very minute pollen grains (25–40 μm) are carried by wind and spread the whole orchard. The climatic conditions prevailing at the time of the release of the pollen have a significant influence on its quality. The stigmatic styles emerge from the bud scales in the form of a red feathery tuft. Stigmas are receptive up to 2 months after the emergence of these red tufts till the weathering and darkening of surface, however the stigmatic receptivity was maximum during the initial 15 days after anthesis. A tiny bit of pollen grains are caught by stigma which reaches the stylar base after germination. The pollen growth takes place in 4–10 days. Pollen tube growth ceases when it reaches the apex of ovary, and the stylar tip gets blocked off with a heavy callose coat of irregular shape (Liu et al. 2014). These enclosed pollens are kept to rest for a period of 4 months, and development of ovules begins for the production of megaspores. The growth of these pollen tubes is resumed again during the formation of megaspores fertilization takes place by end-May or mid-June (i.e., 4–5 months after pollination). The growth rate of embryo is very slow during the first 3 weeks after fertilization (3–6% of the final volume). The shell hardening begins by the early-July, with hastened growth of embryo, attaining maturity by early-August and begun to fall down from plant by end-August to early-October.

3.2.1 Pollination Problems

The dichogamous behavior of hazelnut results in uneven flowering and is a major barrier in the pollination process. Majority of the cultivars are protandrous in nature while protogyny is very less frequent. However, this flowering pattern is related to the temperature differences preceding at the time of flowering. Protogyny is common at lower temperatures (<10 °C) while protandry is more frequent at higher (>10 °C) temperatures preceding the flowering (Baldwin 2015).

Novara et al. (2017) reported the prevalence of male sterility in hazelnut. The pollens obtained from the various cultivated cultivars like Barcelona and Giffoni were recorded with 50% or more sterility. Ascari et al. (2018) recorded highest pollen viability (50.1–57.1%) in wild hazelnut.

Moreover, the hazelnut is autosterile in nature, i.e., there is a prevalence of both self and cross-incompatibility in the diploid hazelnut ($2n = 22$), thus making allogamy a necessary criteria. This incompatibility reaction is sporophytic in nature and is governed by S gene/locus with multiple alleles. 25 S alleles have been identified till date (Ma et al. 2013). All these alleles exhibit an independent action in the pistil, whereas in case of pollen they act in a co-dominant or dominant manner. There is a strong linear dominance hierarchy present between the alleles. The incompatible pollens germinate successfully on the stigmatic surface but fails to penetrate inside the stigmatic cells.

Harsh weather conditions such as wind, rains, extreme cold adversely affect the pollination process (Ellena et al. 2012). The flowering period increases with the rise in temperature and shortens with low winter temperatures. As soon as temperature falls down to 0 °C, the development of catkins ceases down and there is no pollen shedding (Pinillos and Cuevas 2008).

3.2.2 Pollination Management

Effective pollination is must to obtain an economic yield. The keys to efficient pollination in filbert are:

Table 6.3 The compatibility reactions with different pollinizers for cultivar "Barcelona"

Source		S-alleles	
		S_1 allele	S_2 allele
(a) Non-producing cultivar	Barcelona	1	2
(b) Pollinizers	Butler	2	3
	Hall's Giant	5	15
	Ennis	1	11

1. Abundant supplies of the viable pollen.
2. Genetically compatible cultivars with synchronous flowering period.

However, there are certain factors to be kept in mind before selecting a pollinizer plant such as (1) compatibility with pistillate flower, (2) time of pollen shedding, and (3) the capacity of producing viable pollen.

1. Varietal compatibility
   Compatibility groups should be prepared after the identification S-alleles in each cultivar. Each cultivar possesses two S-alleles (S_xS_y) and both are expressed in pistillate flowers while in case of pollen alleles are expressed only if they are co-dominant in nature, i.e., of equal dominance (Olsen et al. 2013). Only the dominant allele will be expressed in the pollen if one allele is dominant over another. The S-alleles of pistillate flower of a cultivar must differ from the co-dominant or dominant pollen alleles to be successfully compatible (Table 6.3).
   The cultivar Barcelona has 2 S alleles but S_1 is dominant over S_2 so only it will be expressed. In Butler S_3 is the dominant allele which is different from the dominant allele of the Barcelona so it can be successfully compatible with Barcelona. Both the alleles of Hall's Giant are codominant, hence, two of them will be expressed and as they differ from the dominant allele (S_1) of non-producing cultivar, they can form a successful compatibility reaction with it. However, in case of Ennis, S_1 is the dominant allele which is similar to cultivar Barcelona, so it indicates that this interaction would not successful, i.e., they will be incompatible with each other (Baldwin 2015) (Table 6.4).
2. Standardize the planting and spacing of the pollinizers
   Since, wind is the pollination agent in the hazelnut orchards. It is necessary to place the pollinizer plants in desired number at optimum locations within the orchard. Pollen grains can be carried roughly to a distance of 80–100 ft within the orchard by wind. Pollinizer density varies from 3 to 30% around different filbert growing countries of the world. In Oregon, it has been standardized to grow 2–3 pollinizers in an orchard with different blooming time at a distance of 60 ft from other pollinizer plant to ensure the availability of the pollen in abundance (Olsen et al. 2000).
3. Artificial pollination
   Non-synchronous blooming of pistillate and staminate flowers is a barrier in pollination, hence pollen should be collected from catkins and stored in controlled conditions. Pistillate flowers are pollinated with the viable pollens of

Table 6.4 The possible pollinizer combinations used for different cultivars in kernel and in-shell market

Market	Cultivar	Pollinizer
Kernel	Clark	Hall's Giant Gem
	Lewis	Tonda di Giffoni Barcelona Hall's Giant Gem
	Willamette	Tonda di Giffoni Hall's Giant Gem
In-shell	Barcelona	Hall's Giant Gem Lewis VR4-31, VR11-27 VR20-11
	Ennis	Casina Hall's Giant Jemtegaard 5 VR20-11 VR23-18

Source: Olsen et al. (2000)

compatible pollinizer by the means of hand pollination, ventilators, dusting and spray pumps at favorable time. Ellena et al. (2012) observed a 37% increase in the yield by artificial pollination.

4. Spray pollination

   The dry air with high sunlight intensity during the opening of female adversely affects the pollen penetration and germination on the stigmatic surface. Hence a suspension media of pollen is prepared which comprised of 0.1% agar, 0.02% boric acid, and 10% sucrose (Sakamoto et al. 2009). It moisturizes the stigma and improves the pollen application rate (Ascari et al. 2018).

5. Installation of photoselective shade nets

   The relative humidity and air temperature at the time of blooming of female flowers have a significant effect on the pollination (Çetinbaş-Genç et al. 2019; Huo et al. 2014). In adverse conditions, the pollen tube growth gets arrested in the style leading to ovule abortion and formation of blank nuts. Hence photoselective shade nets are installed in the field which modifies the microclimate and improves the pollination process. Guastella et al. (2020) observed better pollen tube growth, low ovule abortion, and blank nuts inside the photosensitive nets as compared to open conditions.

4 Pecan Nut

4.1 Floral Biology

Pecan nut is a monoecious plant in which staminate and pistillate flowers are borne separately on the same tree and arranged into catkin and spikes, respectively. Compound buds in pecan nut enclose floral and mixed buds in separate bud scales but within a common outer scale covering. The mixed buds develop either into a vegetative shoot or in a single pistillate inflorescence, whereas floral buds develop into catkins (Wetzstein and Sparks 1986). The development of pistillate flower occurs few weeks before anthesis on the resumption of current season's growth in spring at the apical buds and very few on the lateral buds. Pistillate inflorescences consist of two to ten flowers, depending on the cultivar and vigor of the shoot (Sparks 1992). Each pistillate flower consists of a small yellow pubescent, 4 angled or lobed involucre enclosing single celled ovary and stigma finely divided.

Initiation and differentiation of staminate flower occur about 1 year in advance of anthesis when shoot growth begins in spring and expanding leaves are 2–3 weeks old. The catkin-like staminate inflorescence bears two or three catkins and the individual flower number in each catkin ranges from 72 to 123 (Fronza and Hamann 2016). Each staminate flower occurs in cluster of three, light green measures about 4–6 in. in length, with 5–6 stamens (Sparks 1992). Floral biology of 143 pecan seedling trees in Himachal Pradesh, India was studied by Kaushal and Sharma (2003) and they reported that catkin length ranged from 0 to 13 cm with 18.59% of coefficient of variation and number of male catkins per bud ranged from 0 to 3 with 9.42% coefficient of variation.

Both the protandrous and protogynous flowers differ in the development of bracted and imbricated floral primordia. In the former, the floral primordia is different in protandrous and protogynous cultivars; the floral apex, bracteoles, and anther primordial differentiate in June and July, whereas in the later this organogenesis occurs late, just before anthesis in March and April. The bilobed anthers develop in March in protandrous cultivars but in April in protogynous cultivars. Conner (2011) stored pollen samples at −80 °C and reported that pecan pollen can be stored for at least 8 years without a decrease in viability. Flowering time varies among the cultivars in northern climates (Sparks 2005). Low temperature exposure is essential for pistillate flower induction which takes place in late July or early August under USA conditions. However, Han et al. (2018) reported flowers development from April to May under Southeastern China conditions and also stated that the seasonality of flowering is largely controlled by temperature, with warmer years resulting in earlier flowering than cooler years. Wood (2000) indicated that flower maturity is earlier in older or larger trees and warmer springs accelerate development of catkin in relation to pistillate flowers.

4.2 Pollination

The mechanism of heterodichogamy, i.e., occurrence of both protandrous and protogynous individuals in pecan is well known (Adriance 1927). Pecan cultivars have been classified into two groups: protandrous and protogynous, depending upon the separation in time of pollen shed and pistil receptivity. In protandry or type I dichogamy pollen sheds before pistil receptivity and in protogyny or type II dichogamy pollen sheds after pistil receptivity. The behavior of these cultivars remains stable every year though the time of pollen shedding or stigma receptivity may vary from season to season. Pollen shedding for pecan tree as a whole ranged from about 4 to 14 days (Sparks 2000). Catkins produced on protandrous genotypes are short, thick, and compact as catkins growth of protandrous genotypes precedes shoot growth, whereas catkins produced on protogynous genotypes are longer and narrow as shoot growth precedes catkin growth and carry much more pollen to ensure pollination. Single catkin can produce enough pollen to pollinate flowers to produce around 50,000 pounds of average-sized pecans, and several thousand catkins are produced on an average bearing tree, thus further emphasizing how much pollen could be produce (Rohla 2016). High relative humidity and very low temperature hampered pollen dehiscence. Pollen emergence is visible within 3 h of pollination. Stigma of pecan is dry type. Kaushal and Sharma (2003) accessed behavior of 143 pecan cultivars and reported that duration of stigma receptivity ranged from 7 to 17 days.

The phenomenon of heterodichogamy promotes cross-pollination and self-pollination is rare. Dichogamy is a morphological mechanism developed by plant to prevent self-pollination which results in increased endogamy associated with small nuts and low kernel yield (Fronza et al. 2018). Complete dichogamy results in self-sterility, because there is no overlapping between periods of pollen shedding and stigma receptivity. However, in incomplete dichogamy self-pollination may occur when there is overlapping of pollen shed and pistil receptivity. While investigating the flowering behavior of 25 pecan cultivars, Maeda et al. (2006) observed protogyny in most cultivars and there was an overlapping period of several days during the blooming of male and female. Blooming season changes yearly and there was no correlation between the two characters. In general, complete dichogamy is observed in colder climate and incomplete in warm climate (Sparks 1992). The pollination of pecan is predominantly carried out by the wind (anemophilous). Wind pollination is effective up to 49 m between plants; however, if the spacing is larger there is reduction in fruiting (Conner 2010; Wells 2007). According to Baracuhy (1980), the period of pollen dispersal may range from 8 to 15 days, depending on the cultivar (Table 6.5).

Table 6.5 Classification of pecan nut cultivars on the basis of floral nature

Type I or Protandrous	Type II or Protogynous
• Caddo	• Burkett
• Cheyenne	• Candy
• Creek	• Elliot
• Desirable	• Forkert
• Jackson	• Kanza
• Oconee	• Apache
• Pawnee	• Tejas
• Western	• Nacono
	• Schley
	• Stuart
	• Sumner
	• Wichita

Source: Wood et al. (1997)

4.2.1 Problems in Pollination

The pecan nut tree has monoecious flowers, and the problems in pollination and fruit set are due to dichogamy. Lack of pollination, excessive self-pollination or xenia are major cause of crop loss and poor productivity (Wani et al. 2010). No pollination or possibly self-pollination results in fruit drop 14–45 days after pollination (Nesbitt 2018). Nuts arising from cross-pollination are usually larger and better filled than those arising from self-pollination, due to heterosis (Wolstenholme 1969). The problem of unfruitfulness occurs mostly in block-type orchards consisting of only one or two cultivars as there is no overlapping between periods of pollen shedding and stigma receptivity, but these can also occur at locations with a great number of different genotypes nearby. Sometimes pollination problems can also occur in orchards located in the regions having abundant wild pecan trees (Wood 2000).

Factors such as tree age/size, spring temperatures, and humidity influence the pollination by altering the maturity windows. Abnormally wet weather, strong dry winds or abnormally warm or cool springs during the flowering period reduce dissemination of pollen and hamper the pollination. Presence of high humidity adversely affects the growth, fruiting, and regularity bearing in plants by preventing pollination, increasing the incidence of disease on leaf and nuts (Ravindran et al. 2006). Spring freezes may damage the reproductive organ function and structure, affect pollen viability, pollen germination and adhesion, ovule viability, stigma receptivity, and reduce the effective pollination period and fertilization (Kaur et al. 2020).

4.2.2 Pollination Management

While planting an orchard, the pollination requirement of pecan should be kept in mind. For assuring proper pollination in the orchard, pollinizer should be planted with the main cultivar. Understanding about the phenological behavior and floral compatibility among cultivars is an essential factor in pollination. In general, at least 15% pollinizer trees in pecan orchard are recommended (Raseira 1990). Conner (2012) suggested that at least four cultivars including one producer and three pollinizers should be planted to start a pecan tree orchard. Wood (2000) reported highest crop set where pollinizer tree was adjacent to and not more than two trees from target tree. Wells (2007) indicated that main crop cultivars should generally be within about two rows of pollinizers to ensure cross-pollination. Thus, block widths exceeding about four rows between pollinizers are especially likely to exhibit serious problems in pollination. Scattered trees of off-type genotypes are potentially of major importance as backup orchard pollinizers. To ensure good quality nuts, the selection of suitable pollinizer for particular variety is necessary as pollination with different varieties significantly affects the nut setting of individual females (Mao et al. 2015).

Cultivar Moore, Texas Prolific, and San Saba have pollen available in time to pollinate the earliest flowers of any variety. Moneymaker and success usually require pollens from other varieties, however, Stuart, Burkett, Schley, and Delmas sometimes depend upon other varieties for pollination. Ajamgard et al. (2017) conducted a study in southwest of Iran to determine the best pollinizer for five pecan cultivars "GraTex," "10J," "Wichita 6J," "GraKing," "Choctaw" and reported that "Peruque," "GraKing" and "Stuart 2J" were the best pollinizers for five selected cultivars in southwest of Iran. Luo et al. (2016) indicated that cultivar Pawnee can be used as pollinizer for Mahan and Western cultivars and controlled or supplementary pollination had no significant influence on nut set, however, it increased nut yield of both the producer cultivars. Li et al. (2010) suggested that cultivar Western may pollinate cultivar Shawnee.

Controlled pollinations after pesticide sprays resulted in an inhibition of pollen germination and tube growth in pecan nut. Reilly and Wood (1996) evaluated the impact of standard foliar zinc and fungicide sprays on pollination of "Cheyenne" or "Desirable" orchards and found no detectable influence of these sprays on fertilization or fruit set.

4.3 Varietal Compatibility

See Table 6.6.

Table 6.6 Important pollinizers for different cultivars of pecan nut in Iran

Pecan variety	First recommend	Second recommend
GraTex	Graking + 10J	Graking + Comanche 4M
Choctaw	Peruque + Graking	Peruque + Stuart 2J
Wichita	6J Peruque + Graking	Peruque + Stuart 2J
GraKing	Peruque + Stuart 2J	Peruque + Mohawk
10J	Peruque + Mohawk	Peruque + Stuart 2J

Source: Ajamgard et al. (2017)

Table 6.7 Flowering periods of male and female trees of pistachio nut in Himachal Pradesh

Sr. No.	Inflorescence	No of seedling tree	Flowering time		Duration of flowering
			Start	End	
1.	Pistillate	16	10th Apr to 27th Apr	20th Apr to 7th May	9–12 days
2.	Staminate	15	4th Apr to 28th Apr	13th Apr to 8th May	9–12 days

Source: Thakur (2010)

5 Pistachio Nut

5.1 Floral Biology

The prevalence of dioecy in pistachio nut results in the presence of staminate and pistillate flowers on the separate plants (Afshari et al. 2008), thus making allogamy a necessary criterion. Both the male and female inflorescence consist of several hundred flowers generally blooming in April. During the initial 4–5 years of growth, floral buds are vegetative, but as the tree progresses into the productive phase, they develop into the floral buds. The flowers have a distinct apetalous appearance and do not produce any nectar, resulting in, very less or no insect attraction thereby, inhibiting entomophily in pistachio. Hence, the plants are totally anemophilous, i.e., transfer of pollen from staminate to pistillate flowers is essentially by the wind (Ak et al. 1996) (Table 6.7).

5.2 Pollination Problems

Pollination and fertilization are major steps in the pistachio culture, but there are certain barriers observed in the pollination process. The protandrous nature of the plants results in the non-synchronous behavior in the anthesis of the staminate and pistillate flowers. In case of the lack of winter cold, the male trees bloom earlier than the females and a huge amount of pollen get wasted because of non-receptivity of stigma giving rise of problem of empty nuts. In order to obtain economic yield, proper pollination is must for high fruit set, so, it is necessary to intersperse optimum number of male plats within the orchard whose flowering coincide with the pistillate plants.

5.3 Pollination Management

A ratio of 1:8 or 1:11 (male:female) should be maintained in the orchard (Kagka 1990). The position of the male plant should be kept either in the center of 8–11 female trees or they can be planted in a solid row along the direction of wind according to the site of the orchard to facilitate the proper pollen dispersal around the orchard.

Investigations had been carried in different pistachio growing countries for the selection of the male trees whose flowering coincide with their elite female cultivars. Ak et al. (1996) identified suitable male plants for seven different female varieties, namely Ohadi, Kirmizi, Siirt, Sefidi, Bilgen, Vahidi, and Mümtaz.

However, synchronization of flowering is not enough, there are many factors to be kept in mind before selecting a male tree such as (1) strong/upright growth, (2) synchronous flowering time, (3) long flowering duration, (4) more clusters, (5) big sized clusters, (6) high pollen production per cluster, (7) high pollen germination rate, (8) long pollen viability in vivo conditions, (9) regular bearing. Based on these characters, a study was conducted (Ak et al. 1996) on 24 male cultivars in Turkey for the 5 years and suitable male cultivars were determined for Kirmizi (5), Slirt (3), Ohadi (2), Rego (4), Bilgen (2), Vahidi (2), Sefidi (5) and Mumtaz (3) female types. Atli et al. (2005) identified Type 39 male for Kirmizi and Siirt, Type 10 for Halebi and Uzun; and Type 79 for Ohadi females in Turkey conditions.

The genus Pistacia contains about 11 species (*P. atlantica*, *P. khinjuk*, *P. terebinthus,* and *P. palaestina*) including *Pistacia vera* (cultured pistachio) and all these species can freely fertilize with each other (Acar and Kakani 2010). *P. terebinthus* has been identified as an efficient male plant because of its long flowering period coinciding with most female cultivars with high pollen viability (80.99%) and fruit set (14.48%) as compared to *P. vera* (Isfendiyaroglu et al. 2001). Since, most of the pistachio orchards are established on the spontaneous rootstocks, pollination performance is poor in orchards, hence budding or grafting of male cuttings of *P. terebinthus* can be done on the female plants of *P. vera* for facilitating better pollination within the orchard.

5.4 Supplementary Pollination

Since, the natural pollination is hampered by certain climatic or physiological barriers in pistachio orchards, artificial/supplementary pollination can also be practised. The clusters of male inflorescences are cut down when plants begin to bloom and kept at temperature range of 20–25 °C in a mesh for pollen shedding. The collected pollen is treated with $CaCl_2$ and stored at low temperature (−4 °C) in freezer for a period of 8–10 days, to let the pollen dry (Kuru 1994). As soon as the female plants begun to bloom, the pollen application is made on them via hand pollination, tractor mounted atomizer, or dusting with cotton bags. However, hand pollination had been observed to be a laborious and time-consuming task, nowadays it has been replaced by atomizers, dusters, electrostatic and spray pollination (Acar et al. 2001).

The pollen is mixed with wheat flour in a ratio of (2:98) before dusting for the fulfillment of large pollen requirements in the orchard. Acar and Eti (2008) recorded high fruit setting with the spray of 5–50% pollen mixture on pistillate flowers from a distance of 100 cm within 4 days of the anthesis. Fourteen to fifteen pollen grains are adequate for the pollination of a female flower. However, high pollen deposition on the stigma should be avoided as it decreases the reproductive success of female flowers by increasing the competition in pollen tube pathway, thereby resulting in excess flower abscission and poor fruit set (Young and Young 1992). The pollen deposition has been observed to be positively influenced with artificially charged pollen. Electrostatic pollination increased the yield as well as quality of pistachios (Vaknin et al. 2002). A significant increase has been recorded in the production of fruitlets per cluster (16.2%), yield (11.3%), and split percentage (18.6%) by the electrostatic pollination in pistachio orchards of California (Vaknin et al. 2001).

Another alternative technique of spray pollination can also be followed where the aqueous solutions of pollen grains have been applied on the stigma of pistillate flowers. The pollens are suspended in a media containing agar and sucrose in a standard concentration. This media is further supplemented with the boric acid for the enrichment of the pollen suspension (Hamid and Hajar 2016).

References

Acar I, Eti S (2008) Effect of pistil receptivity, pollen mixtures, and pollen application distances on fruit set of pistachios (*Pistacia vera*). N Z J Crop Hortic Sci 36(4):295–300

Acar I, Kakani VG (2010) The effects of temperature on in vitro pollen germination and pollen tube growth of Pistacia spp. Sci Hortic 125(4):569–572

Acar I, Ak BE, Kuzdere H (2001) An investigation on artificial pollination facilities in pistachios by using an atomizer. In: Proceedings of the XI GREMPA seminar on pistachios and almonds. Cahiers Options Mediterraneennes 56:145–148

Adriance GW (1927) A preliminary report on dichogamy in the pecan. J Am Soc Hortic Sci 24:95–97

Afshari H, Talaei A, Panahi B, Hokmabadi H (2008) Morphological and qualitative study of pistachio (Pistacia vera L.) pollen grains and effect of different temperatures on pomological traits. Aust J Crop Sci 1(3):108–114

Ajamgard F, Rahemi M, Vahdati K (2017) Determining the pollinizer for pecan cultivars. J Nuts 8(1):41–48

Ak BE, Acar I, Kaska N (1996) An investigation on the male determination for some female varieties throughout five years (1992–1996) grown at Ceylanpinar State farm in Sanliurfa conditions. In: Proceedings of the X. GREMPA seminar, pp 14–17

Ascari L, Guastella D, Sigwebela M, Engelbrecht G, Stubbs O, Hills D (2018) Artificial pollination on hazelnut in South Africa: preliminary data and perspectives. Acta Hortic 1226:141–148

Atli HS, Arpaci S, Uygur N (2005) Selection of pistachio pollinators. IV. International symposium on pistachios and almonds. Acta Hortic 726:417–422

Ayfer M, Soylu A (1993) Selection of chestnut cultivars (*Castanea sativa* Mill.) in Marmara region of Turkey. In: Proc. of the international congress on chestnut, pp 285–289

Baldwin BJ (2015) The growth and productivity of hazelnut cultivars (*Corylus avellana* L.) in Australia. Thesis

Baracuhy JBC (1980) Determinação do período de floração e viabilidade de pólen de diferentes cultivares de nogueira peca *Carya illinoensis* (Wang) K. Koch. Dissertação (Mestrado em Agronomia) - Curso de Pós-Graduação em Agronomia, Universidade Federal de Pelotas, Pelotas

Bencat F, Tokar F (1978) Phenological observation of the chestnut (Castanea sativa Mill.) on the experimental plot of Horne Lefantovce [Czechoslovakia]. Folia Dendrologica (Czechoslovakia) 4:49–89

Bhat K, Hassan S, Bhat MA, Mir MA, Kirmani SN (2016) Pollination studies in walnut (*Juglans regia* L.). Bioscan 11(4):2683–2686

Bounous G, Marinoni DT (2005) Chestnut: botany, horticulture, and utilization. Hortic Rev 31: 291–347

Bounous G, Bouchet M, Gourdon L (1992) Ricostituzione del castagneto a frutto tradizionale: interventi in Piemonte e nel Sud della Francia. Informatore Agrario 9:155–160

Çetinbaş-Genç A, Cai G, Vardar F, Ünal M (2019) Differential effects of low and high temperature stress on pollen germination and tube length of hazelnut (*Corylus avellana* L.) genotypes. Sci Hortic 255:61–69

Chandra P, Tomar CS (2012) Growth, flowering fruit set and yield in some cultivars/selections of walnut (Juglans regia). Indian J Agric Sci 82(5):402–404

Chapa J (1984) Pollinisation du chataignier. In: Pesson P, Louveaux J (eds) Pollinisation et production végétales. INRA, Paris, pp 187–194

Clapper RB (1954) Chestnut breeding, techniques and results. J Hered 45:107–114

Conner PJ (2010) Pecan pollination. Annual extension report. Horticulture Department, University of Georgia. 12p

Conner PJ (2011) Optimization of in-vitro pecan pollen germination. HortScience 46:571–576

Conner PJ (2012) Pollination charts revisited. Grower Pecan 23(4):34–39

Cosmulescu S, Baciu A, Botu M, Achim GH (2010) Environmental factors' influence on walnut flowering. Acta Hortic 861:83–88

Cristofori V, Pica AL, Silvestri C, Bizzarri S (2018) Phenology and yield evaluation of hazelnut cultivars in Latium region. Acta Hortic 1226:123–130

Ellena M, Sandoval P, Gonzalez A, Galdames R, Jequier J, Contreras M, Azocar G (2012) Preliminary results of supplementary pollination on hazelnut in south Chile. VIII Int Congr Hazelnut 1052:121–127

Fronza D, Hamann J (2016) Técnicas para o cultivo da nogueira-pecã. Universidade Federal de Santa Maria, Colégio Politécnico da UFSM, p 424

Fronza D, Hamann JJ, Both V, Anese RO, Meyer EA (2018) Pecan cultivation: general aspects. Ciência Rural 48

Gao Y, Liu H, Dong N, Pei D (2012) Temporal and spatial pattern of indole-3-acetic acid occurrence during walnut pistillate flower bud differentiation as revealed by immunohistochemistry. J Am Soc Hortic Sci 137:283–289

Germain E (1994) The reproduction of hazelnut (*Corylus avellana* L.): a review. Acta Hortic 351: 195–210

Golzari M, Hassani D, Rahemi M, Vahdati K (2016) Xenia and metaxenia in Persian walnut (*Juglans regia* L.). J Nuts 7(2):101–108
Guastella D, Sigwebela M, Suarez F, Stubbs O, Acevedo J, Engelbrecht G (2020) Effect of photo-selective shade nets on pollination process and nut development of *Corylus avellana* L. Front Plant Sci 11:1990
Hamid RK, Hajar Z (2016) Effects of artificial pollination using pollen suspension spray on nut and kernel quality of pistachio cultivar Owhadi. Int J Fruit Sci 16(2):171–181
Han M, Peng F, Marshall P (2018) Pecan phenology in southeastern China. Ann Appl Biol 172: 160–169
Hasegawa Y, Suyama Y, Seiwa K (2015) Variation in pollen-donor composition among pollinators in an entomophilous tree species, *Castanea crenata*, revealed by single-pollen genotyping. PLoS One 10(3):e0120393
Hassan S, Bhat KM, Rehman HU, Wani WM, Mir MA, Jan A (2016) Evaluation of some pollinizers and their suitability on productivity and quality of walnut (*Juglans regia*) cultivars. Appl Biol Res 18(1):39–45
Hassani D, Atefi J, Haghjooyan R, Dastjerdi R, Keshavarzi M, Mozaffari MR, Malmir A (2012) Damavand, a new walnut cultivar as a pollinizer for Iranian walnut cultivars and genotypes. Seed Plant Improv J 28(1):529–531
Hassani D, Dastjerdi R, Haghjooyan R, Soleimani A, Keshavarzi M, Atefi J, Mozaffari M, Rezaee R, Fahadan A, Rahmanian A (2014) Genetic improvement of Persian walnut (*Juglans regia* L.) in Iran. Acta Hortic 1050:95–102
Hassankhah A, Vahdati K, Rahemi M, Hassani D, Sarikhani Khorami S (2017) Persian walnut phenology: effect of chilling and heat requirements on budbreak and flowering date. Int J Hortic Sci Technol 4(2):259–271
Hassankhah A, Rahemi M, Mozafari MR, Vahdati K (2018) Flower development in walnut: altering the flowering pattern by Gibberellic acid application. Notulae Botanicae Horti Agrobotanici Cluj-Napoca 46(2):700–706
Huo H, Zhang R, Ma Q, Zhao T, Liang L, Wang G (2014) Stigma receptivity and characteristics of pollen tube growth of *Corylus kweichowensis*. For Res 27(3):403–409
Isfendiyaroglu M, Ozeker E, Misirli A, Saglam H (2001) Determination of pollinator characteristics of different Pistacia spp. in Manisa-Yunt mountain area. Cahiers Options Méditerranéennes 56: 267–270
Kagka N (1990) Pistachio research and development in the near east, north Africa and southern Europe. Nut production and industry in Europe, near east and north Africa. Reur Technical Ser 13:133–160
Kaur A, Ferguson L, Maness N, Carroll B, Reid W, Zhang L (2020) Spring freeze damage of pecan bloom: a review. Horticulturae 6(4):82
Kaushal R, Sharma SD (2003) Non-hierarchical Euclidean cluster analysis studies in seedling trees of pecan [*Carya illinoensis* (Wang) K. Koch.] growing in Himachal Pradesh. In: VII International symposium on temperate zone fruits in the tropics and subtropics-Part II 696:97–101
Kavardzhikov L (1980) Effect of air temperature on the phonological phase of flowering in filberts. 1. Effect of some temperature parameters on flowering [Corylus]. Gradinarska i lozarska nauka. Horticultural and Viticultural Science
Krueger WH (2000) Pollination of English walnuts: practices and problems. HortTechnology 10: 127–130
Kuru C (1994) Artificial pollination of pistachio trees under unsufficient pollination conditions. In: I International Symposium on Pistachio 419:121–124
Larue C, Austruy E, Basset G, Petit RJ (2021) Revisiting pollination mode in chestnut (*Castanea spp.*): an integrated approach. Bot Lett 168:1–25
Li X, Xu Y, Li Y, Liu Y, Zhai M (2010) The flowering phenology characteristic of five varieties of Carya illinoensis. J Jiangsu For Sci Technol 37(6):18–28
Liu H, Gao Y, Dong N, Xu H, Pei D (2011) Study on pollination activity of pistillate and staminate flowers in walnut excellent varieties. J Beijing For Univ 33(6):119–123

Liu J, Zhang H, Cheng Y, Wang J, Zhao Y, Geng W (2014) Comparison of ultrastructure, pollen tube growth pattern and starch content in developing and abortive ovaries during the progamic phase in hazel. Front Plant Sci 5:528
Luo X, Li Z, Sun Z, Wan X (2016) Analysis of pecan cultivars Mahan and Western in East China. Genet Mol Res 15(3)
Ma QH, Wang GX, Liang WJ, Chen X, Liang LS, Zhao TT (2013) Progress on pollen-stigma compatibility in Corylus (hazelnuts): a review. J For Res 24(3):397–402
Maeda TY, Hagiwara YS, Taniguchi M (2006) Investigation of varietal variations in pecan (*Carya illinoensis* Koch) cultivars for extending a business period of tourism farm in autumn. Hortic Res (Japan) 5(3):227–233
Mao MZ, Fu SL, Yao XH, Chang J, Shao WZ, Fu GL, Deng WP (2015) Effects of different pollen sources on pecan nut setting and its characteristics. J Anhui Agric Univ 42(4):504–510
Mehlenbacher SA (1991) Hazelnuts (Corylus). Genetic resources of temperate fruit and nut crops. Acta Hortic 290:791–838
Mert C, Soylu A (2006) Flower and stamen structures of male-fertile and male-sterile chestnut (*Castanea sativa* Mill.) cultivars. J Am Soc Hortic Sci 131:752–759
Miller A, Miller DD, Pijut PM (2013) How a flower becomes a chestnut. 103rd annual report of the Northern Nut Growers Association
Nesbitt ML (2018) Pecan varieties, pollination & spacing. Texas A&M AgriLife Extension Service
Novara C, Ascari L, LaMorgia V, Reale L, Genre A, Siniscalco C (2017) Viability and germinability in long term storage of *Corylus avellana* pollen. Sci Hortic 214:295
Olsen JL, Mehlenbacher SA, Azarenko AN (2000) Hazelnut pollination. HortTechnology 10(1): 113–115
Olsen JL, Mehlenbacher SA, McCluskey RL, Smith DC (2013) Growing hazelnuts in the Pacific Northwest: hazelnut varieties. Oregon State University
Percival M (1965) Floral biology. Pergamon Press, New York
Pinillos V, Cuevas J (2008) Artificial pollination in tree crop production. Hortic Rev 34:239–276
Raseira A (1990) A cultura da Nogueira-peca *Carya illinoinensis*. Comunicado Técnico, vol 63. Embrapa, Pelotas, pp 1–3
Ravindran C, Sharma M, Kher R (2006) Present status and problems of pecan nut (*Carya illinoinensis* (Wang.) K. Koch) cultivation in Jammu and Kashmir. In: XXVII International Horticultural Congress-IHC2006: International symposium on enhancing economic and environmental 772:179–182
Reilly CC, Wood BW (1996) No adverse effect of early season fungicide and zinc sprays on leaf area, fruit-set, and nut quality of pecan. HortScience 31:808–810
Rohla C (2016) Cross pollination is essential for pecan production. Center for Pecan and Specialty Agriculture, Noble Research Institute
Rutter PA (1990) Chestnut pollinator's guide. Badgersett Res Corporation Bull 1
Sakamoto D, Hayama H, Ito A, Kashimura Y, Moriguchi T, Nakamura Y (2009) Spray pollination as a labor-saving pollination system in Japanese pear (*Pyrus pyrifolia* (Burm.f.) Nakai): development of the suspension medium. Sci Hortic (Amsterdam) 119:280–285
Schad C, Solignat G, Miller D (1952) Recherches sur le chataignier a la station de Brive. Ann Amelior Plantes 3:376–458
Singh SR, Ahmed N, Mir JI, Mir A (2012) Present status and future strategies for walnut production in India
Smith CO, Hayes FM, Bioletti FT, Bryant HC, Quayle HJ, Thompson JI, Cruess WV (1912) Walnut culture in California: walnut blight. Calif Agric Exp Station 229
Sparks D (1992) Pecan cultivars - the orchard's foundation. Pecan Productions Innovations, Watkinsville, GA
Sparks D (2000) Pecan in warm climate. In: Erez A (ed) Temperate fruit crops in warm climates. Springer, Dordrecht, pp 381–403
Sparks D (2005) Adaptability of pecan as a species. HortScience 40:1175–1189

Thakur BS (2010) Pistachio nut cultivation in India. Advance training course on wild and underutilized fruits held from 22.07.2010 to 11.08.2010. Dr YS Parmar University of Horticulture and Forestry, Nauni, Solan (HP), pp 15–17

Vaknin Y, Gan-Mor S, Bechar A, Ronen B, Eisikowitch D (2001) Electrostatic pollination of pistachio (Pistacia vera L.): a novel technique of pollen supplementation in agriculture. Cahiers Options Méditerranéennes 56:53–57

Vaknin Y, Gan-Mor S, Bechar A, Ronen B, Eisikowitch D, Grunewaldt J (2002) Effects of supplementary pollination on cropping success and fruit quality in pistachio. Plant Breed 121(5):451–455

Wang Q, Su S, Zhao D, Kou YR (2012) Study on varieties combination suitable for mutual pollination of Yan Mountain Chestnut. Nat Resour 3(2):66–70

Wani IA, Bhat MY, Lone AA, Mir MY (2010) Unfruitfulness in fruit crops: causes and remedies. Afr J Agric Res 5(25):3581–3589

Wells L (2007) Pollination. Southeastern pecan grower's handbook, vol 1327. University of Georgia Cooperative Extension Publication, p 81

Wetzstein HY, Sparks D (1986) Flowering in pecan. Hortic Rev 8:217–255

Wolstenholme BN (1969) Effect of self and cross-pollination on fruit set and nut drop of the pecan at Pietermaritzburg. Agroplante 1:189–194

Wood MN (1934) Pollination and blooming habits of the Persian walnut in California, Technical Bulletin. United States Department of Agriculture, p 387

Wood BW (2000) Pollination characteristics of pecan trees and orchards. HortTechnology 10:120–126

Wood BW, Smith MW, Worley RE, Anderson PC, Thompson TT, Grauke LJ (1997) Reproductive and vegetative characteristics of pecan cultivars. HortScience 32(6):1028–1033

Young HJ, Young TP (1992) Alternative outcomes of natural and experimental high pollen loads. Ecology 73(2):639–647

Chapter 7
Mineral Nutrition

Aroosa Khalil, Mahrukh Mir, Mohammad Maqbool Mir, Umar Iqbal, Nowsheen Nazir, Munib Ur Rehman, Mahender K. Sharma, Ashaq H. Pandit, Rifat Bhat, and M. Amin Mir

Abstract In the age of precision farming, it is critical to manage mineral nutrition correctly in order to reduce agricultural inputs. As a result, efforts to increase nutrient use efficiency and protect the environment from redundant minerals, which can pose a serious risk of aquifer contamination, are critical. The best practices for managing the mineral nutrition of fruit trees are also necessary to promote fruit yield and quality which may have a significant impact on fruit tree management. The temperate nut crops like walnut, chestnut, hazelnut, and pecan nut are grown in traditional settings and not much attention is given toward their nutrition leading to reduced nut quality and overall yield.

Keywords Mineral nutrition · Analysis · Yield · Quality

1 Introduction

Thirteen essential nutrients in varying amounts are required by fruit trees for normal growth and production, these include nitrogen (N), phosphorus (P), potassium (K), calcium (Ca), magnesium (Mg), and sulfur (S) which are the macronutrients, required by trees in large amounts. Chlorine (Cl), iron (Fe), manganese (Mn), zinc (Zn), boron (B), copper (Cu) and molybdenum (Mo) are micronutrients that are required in smaller concentrations. In order to maximize growth and production while minimizing negative environmental effects, growers must learn how to manage these nutrients. The most important aspect is nutrient balance; when one of the essential elements is deficient, it can have a negative impact on plant processes,

A. Khalil (✉) · M. Mir · M. M. Mir · U. Iqbal · N. Nazir · Munib Ur Rehman · M. K. Sharma · A. H. Pandit · R. Bhat
Division of Fruit Science, FOH, Sher-e-Kashmir University of Agricultural Sciences and Technology of Kashmir, Srinagar, Jammu and Kashmir, India

M. Amin Mir
Ambri Apple Research Centre, Shopian, Sher-e-Kashmir University of Agricultural Sciences and Technology of Kashmir, Srinagar, Jammu and Kashmir, India

M. M. Mir et al. (eds.), *Temperate Nuts*,
https://doi.org/10.1007/978-981-19-9497-5_7

preventing optimal uptake, utilization, or distribution of other elements. An excess of any element, on the other hand, can be toxic to trees, reducing the availability of other nutrients in soil.

Aside from the wide range of nutrient requirements for plants, concentration of nutrients in individual plant organs is different during different times of the year. Effective nutrient management necessitates delivering the appropriate nutrient to the appropriate organ at the appropriate time. Some elements essential to trees need not be supplied in a regular fertilizer program due to three main reasons behind this (1) many elements are abundant (and available) in most soils to provide an adequate supply; (2) perennial trees recycle a substantial portion of some nutrients from the leaves back into the tree structure before leaf fall; and (3) for some nutrients, very little is removed in the crop, leaves, and prunings, most of the elements required by trees do not require regular fertilization. Visual identification of nutrient deficiencies and nutrient toxicities is common, but chemical analysis of leaf samples is recommended to identify the severity of the problem. Even if there are no obvious problems, leaf samples must be taken for evaluation of the effectiveness of the fertilization programme and to track nutrient level changes with time. In June or July, the nutrient levels are relatively stable and therefore this period is considered suitable for leaf nutrient sampling.

The nutrition of fruit trees is a complex system that connects plant demand with nutrient availability. The availability of nutrients is closely related to the mineral content of the soil, the development of the root system, and the soil environment (Tromp 1980, 2005; Wang et al. 2006). Nutrient uptake is actively regulated by the actual plant demand for growth and maintenance. Mineral nutrient management is governed by an internal nutrient cycle that allows remobilization and distribution from storage tissues until the trees are ready for further nutrient uptake in spring (Millard 1996; Neilsen and Neilsen 2003; White et al. 2015). The mobility of nutrients in xylem and phloem, which allows them to be redistributed from older organs to locations of active growth and development of new plant tissues, can also affect the final accumulation of a specific nutrient (Fernández et al. 2013; Socha and Eguerinot 2014; Briat et al. 2015; Kalcsits et al. 2017).

Fruit development causes a reduction in vegetative growth in parts such as roots, shoots, and leaves as crop load increases (Hansen 1971; Palmer 1992). Because specific organs frequently differ in their mineral composition, their development alters the demand for specific nutrients (Hansen 1971; Xia et al. 2009). As a result, the fruit load can significantly alter the macronutrient content of the leaves (Weinbaum et al. 1994; Samuoliene et al. 2016). Complex information about the variability of micronutrient content in fruit tree leaves in relation to actual growth intensity and fruit load is lacking (Mészáros et al. 2021). The nutrient availability to plants is influenced by number of factors like climate, plant species, soil type, nutrient status, moisture levels, oxygen content, humus content, pH and base saturation (Westwood 1978).

2 Almond

Optimizing plant nutrition in the nursery is the most practical way to achieve significant growth. The most important nutrient for plant growth and development is nitrogen (Marschner 2012). Different types of nitrogen have been shown to help fruit trees grow and perform better (Akhlaghi Amiri et al. 2016; Dehghanipoodeh et al. 2018; Rajaie and Motieallah 2018). Nitrogen is required abundantly for branch development, flowering, and fruiting in almond trees (Muhammad et al. 2015). The key time for nutrient application in case of almond is during nut growth—bloom to hull split; with boron and zinc needed mostly at bloom to early shoot growth, nitrogen needed in March–June and potassium and phosphorus during nut growth (Rehman et al. 2018a). Corrective measures should be adopted if nutrient concentrations fall below their critical values. Critical values and suggested range for essential nutrients are provided in Table 7.1. Foliar application in comparison to soil application is more effective during bloom (Rehman et al. 2018a). Application of foliar spray of boron (0.5–1 g/L) is done at early post-harvest (5% leaf fall) stage while zinc is applied in two forms, viz., $ZnSO_4$ (20–30 g/L) in dormant period and as ZnO (6 g/L) in mid-season. Among soil derived nutrients, potassium and nitrogen are removed in greatest quantities. Young vigorously growing trees are susceptible to potassium deficiency, especially in sandy soils. The leaves begin showing symptoms in late spring to early summer. In severe cases, the tip and sub-terminal margins of leaves on more vigorous shoots become necrotic and this often leads to rolling of leaves into a boat shape which is classically called "Viking's Prow". As potassium deficiency progresses, fruit bearing spurs often die and spur renewal is reduced (Ulrich 1952). Foliar sprays of potassium nitrate (KNO_3) sprayed thrice, 10 days apart from April to mid-May @ 10 g/L water is effective as a corrective measure (Brown and Uriu 1996).

Neilsen et al. (2001) observed that fertilization at the start of the growing season increased flowering and improved spur leaf and branch growth in the following growing season. In another study conducted by Muhammad et al. (2015) the highest almond yield was obtained by applying 466 g of nitrogen in four split doses, in late February (20%), mid-April (30%), late June another (30%) and finally in September

Table 7.1 Nutrient concentration in almond leaf

Nutrient	Critical value (CV)	Suggested range
Nitrogen	2.0%	2.2–2.5%
Phosphorus	0.14%	0.1–0.3%
Potassium	1.0%	1.4–2.0%
Calcium	1.9%	2.0–4.0%
Magnesium	0.25%	0.16–1.2%
Manganese	20 ppm	30–80 ppm
Boron	80 ppm	80–150 ppm
Zinc	15 ppm	15–20 ppm
Copper	4 ppm	6–10 ppm

Source: Doll (1996)

(20%) after harvest. The highest yield, number of fruits, leaf area, and mineral nutrient concentration was obtained upon applying 300 kg/ha of nitrogen (Zarata-Valdez et al. 2015). About 600 g nitrogen per tree must be given in two parts to achieve the highest yield and growth of almond 'Azar,' first dose should be applied in March (at the start of the growing season) and then other half, 45 days later in May (Rahnamoun 2002). According to studies, the highest almond yield was achieved when leaf dry matter contained nitrogen (2.7%), phosphorus (0.135%), potassium (1.25%), calcium (3.15%), magnesium (0.95%), and iron, zinc and manganese 100, 36 and 57 mg/kg, respectively (Rahnamoun 2002).

3 Walnut

Walnut obtains mineral nutrients from soil which normally meets all the requirements of the plant except in a few cases, when there occurs deficiency or toxicity of any element in the plant. Leaf sampling should be done to identify accurate nutritional requirements. Soil sampling is of limited utility in determining nutritional needs because nutritional deficiencies are not well correlated with soil analyses value (Proebsting and Davy 1946). Moreover, walnut has a very extensive root system trapping nutrients from deeper layers of the soil. Thus, leaf analysis provides accurate diagnosis of critical levels and nutritional requirements of walnut (Table 7.2). Masarat and Tomar (2008) demonstrated that N, P and K contents declined with the advancement of season basipetally on the shoot, while as Ca and Mg levels increased with age in basipetal direction. The least variation period of N, P, Ca and Mg contents accrued in median shoot leaflets from mid-July to mid-August and K contents showed the least variation during mid-September to mid-October.

Abaev et al. (1976) obtained growth and development of walnuts by supplying N, P_2O_5 and K_2O at 180:180:60 kg/ha each for getting the highest yield and net profit

Table 7.2 Critical nutrient levels of walnut leaves

Element	Deficient	Adequate	Toxic
Nitrogen	<2.1%	2.2–3.2%	–
Phosphorus	–	0.1–0.3%	–
Potassium	<0.9%	1.2%	–
Calcium	–	1.0%	–
Magnesium	–	0.3%	–
Sodium	–	–	0.1%
Chlorine	–	–	0.3%
Boron	<20 ppm	36–200 ppm	300 ppm
Copper	–	4 ppm	–
Manganese	–	20 ppm	–
Zinc	18 ppm	–	–

Source: Beutel et al. (1983)

(Thagusev and Caroev 1971; Kolov 1977). Liang et al. (2010) standardized the reasonable rate of fertilizer application for each walnut tree in southern Xinjiang (China). Each tree requires urea (3 kg), Re-SSP (3 kg) and potassium sulfate (105 kg) which resulted in yield increment up to 19.6–22.2 kg/tree. Beneficial effect of boron and zinc sprays on walnut was observed by Keshavarz et al. (2011). The results showed that B and Zn applications along with combinations had a significant effect on all traits except nut and kernel diameter, shell percent, husk thickness and pistillate flower abscission (PFA). Pollen germination, fruit set, vegetative growth, nut weight, kernel percent, nut and kernel length and chlorophyll index were highest when B and Zn were simultaneously sprayed at 174 and 1050 mg/L, respectively.

In non-competitive environments, Nicodemus et al. (2008) demonstrated that N fertilization can improve walnut growth and photosynthetic responses. NO_3 (as $NaNO_3$) or NH_4^+ (as $(NH_4)_2SO_4$) alone produced better physiological responses than a mixed N source (NH_4NO_3). The use of nitrification inhibitors to maintain NH_4^+ in mixed N-based fertilizers, should be taken into account in a field setting. Lotfi et al. (2009) investigated the differences in walnut cultivars' ability to accumulate essential elements involved in firmness of cell wall, ionic pumps, and transfer of ions through plasma and other cell membranes and also having a positive effect on salinity stress resistance. It was observed that increased salt stress levels increased calcium and potassium accumulations, especially in semi-sensitive and tolerant cultivar shoots. Furthermore, at all stress levels, differences in magnesium accumulation in root and shoot samples were significant and cultivar dependent.

4 Pecan Nut

Nutrients are required for pecan growth and fruit production. Pecan trees are slow to respond to fertilizers, so the effects of nutrients applied in 1 year will take 2 or 3 years to show up in tree growth and yield. As a result, pecan nutrition should be managed over a long period of time rather than just a few years. Before making fertilization decisions, it is important to understand some aspects of pecan nutrition. Even when leaf nitrogen levels are lower than the established threshold levels, yield does not decrease (Malstrom et al. 1983).

Nitrogen is the most common limiting factor in pecan growth and thus, profitability of the orchard. Nitrogen promotes tree growth, a higher kernel percentage, and overall tree health. Trees have two high-demand periods: approximately 3 weeks after budbreak, when the rapidly growing shoots and expanding leaves deplete nitrogen storage reserves, and late summer, when fruits begin the kernel filling process (Smith et al. 2007). The first application of nitrogen should be given during bud break in April, followed by applications in May and June. In June, 4.5 kg of actual nitrogen are required for every 45.35 kg of pecans planted per acre. The number of nitrogen applications should be proportional to the size of the crop. In response to growth, young trees should receive very small but frequent applications of nitrogen. Add more applications until early June if rapid growth occurs (Kumar

Table 7.3 Sufficiency levels for essential elements in pecan nut

Element	Sufficiency range (%)	Element	Sufficiency range (ppm)
N	2.30–3.00	Zn	60–150
P	0.14–0.30	Fe	50–300
K	1.00–2.50	Mn	100–2000
Mg	0.30–0.60	B	15–50
Ca	0.70–1.75	Cu	6–30
S	0.20–0.35	Ni	>2.5

Source: Smith et al. (2012)

et al. 2018). Dry land orchards should be fertilized with 34 kg of nitrogen per acre in late February or early March if leaf sample recommendations are not available, so that winter rains can help improve soil moisture and, in turn, nitrogen uptake (Smith 2005). Nitrogen applied via a drip or sprinkler irrigation system may increase the availability of nitrogen to the tree, lowering the amount of nitrogen needed to achieve the same response (Smith et al. 1979). Nitrogen is typically applied in four equal applications through the irrigation system, beginning around budbreak and then at 2-week intervals (Smith 2005).

The effects of N, P, and K fertilization on pecan seedling growth, nutrients, and photosynthetic properties were studied by Liu et al. (2021). The growth of pecan is found to be the most sensitive to nitrogen and the least sensitive to phosphorus. High concentrations of phosphorus, as well as iron, zinc, and copper, can prevent the pecan tree from absorbing nitrogen. Sparks (1988) observed increased nut growth as well as reduced leaf scorch and premature defoliation when 2.2 kg H_3PO_4 per tree was applied in the field. For maximum nut growth and minimum leaf scorch, the leaf phosphorus value was greater than 0.14% and 0.16%, respectively. The sufficiency range of different elements in pecan are given in Table 7.3. Phosphorus being relatively immobile in soil, a single planting application of 27–45 kg per acre of P_2O_5 can last for several years (Smith 2005).

Low winter temperatures (−7.2 °C) coupled with low potassium level severely injure or kill plants, even if they show no symptoms, and the yield from the damaged trees declines as well (Sharpe et al. 1952). Moreover, excessive Stage II fruit drop (also known as June drop) in pecan is frequently linked to a potassium deficiency. According to Wood et al. (2010) increasing leaf and fruit K concentrations through soil banding of potash over drip irrigation emitters increased fruit retention, increased in-shell nut yield, and improved nut quality by increasing kernel percentage. The most common way to get potassium is to apply muriate of potash to the soil. The rates should be determined by the amount of nitrogen applied and the amount of potassium in the leaves. In order to avoid leaf scorch, the N/K ratio should be kept at 2:1 for most varieties. It has been found that applying K as a banded application rather than a broadcast application has increased K absorption more effectively. On one side of the tree, apply 45 kg per acre of K_2O in a band about halfway between the trunk and canopy (Smith et al. 2012). Pecan trees are slow to absorb potassium from the soil, taking up to 5 years, therefore, by moving up to a depth of 70 cm, surface-applied potassium usually increases residual soil potassium (Worley et al. 1974; Smith et al. 1985).

According to Yuan et al. (2014), N, P, K, should be applied as primary base fertilizers during pollination and flowering while Ca and Mg as secondary base fertilizer flushing periods. In the production of nuts, trace elements are extremely important. Zinc is the most important trace element because most soils are deficient in zinc or have no zinc at all, so it must be added every year. Nuts fail to set and yield is low when zinc is absent or present in insufficient amounts. Apply $ZnSO_4$ (36% Zn) in water as the first leaf unfurls during budbreak and then at 2-week intervals until shoot growth ceases (Smith 2005). Foliar application of zinc should be done anytime particularly when leaf concentration falls below 60 ppm (Table 7.3) or when visible symptoms of zinc deficiency are present. When Zn is absorbed through roots, leaf Zn concentration is more uniform, thus 40 μg/g is adequate (Sparks 1982). However, soil-applied zinc is relatively ineffective in providing the Zn requirement of pecan trees grown in the high pH and calcareous soils (Smith et al. 1980).

5 Chestnut

Chestnut trees were traditionally grown in agro-forestry systems as either single trees, small orchards, or mixed stands, and seldom received any attention. However, now chestnut trees are being planted among fruit tree in orchards and other plantations (Martins et al. 2011). Chestnut (*Castanea sativa* Mill.) has become increasingly important for fruit production in some Mediterranean basin countries in recent years (Álvarez-Álvarez et al. 2010; Martins et al. 2011). Highly nutritious nuts and other by-products have contributed to the rising price of chestnuts (Borges et al. 2008; Echegaray et al. 2018), attracting farmers to chestnut farming. Producers have intensified cultivation processes to improve health of the tree along with nut production, compensating for reduced yield due to disease and pest infestation. Despite the fact that there have been only a few studies on chestnut fertilization, the ones that have been conducted have highlighted the critical role of boron (Portela et al. 2011, 2015; Arrobas et al. 2017), potassium (Arrobas et al. 2017) and nitrogen (Arrobas et al. 2018) in the nutrition of chestnut trees. On the other hand, the chestnut tree is greatly known for its ability to thrive in a variety of soil conditions (Bourgeois 1992).

Rodrigues et al. (2019) found the fertilizer Exactyon AG 18:5:13 (47% N encapsulated and 28.8% N as urea coated with ammonium sulfate) gave higher accumulated nut yield. Previous evidence for the importance of nitrogen fertilization (Arrobas et al. 2018) and boron (Portela et al. 2011, 2015) in chestnut tree has already been reported. Phosphorus and potassium seemed to have had little effect on the trees (Rodrigues et al. 2019). Some deficiencies, particularly calcium (Davis 1949; Ramalho et al. 1995) have a significant impact on growth of chestnut trees. Ringshake is a common wood failure with disastrous marketing consequences, despite interesting chemical and physical wood properties. This cohesion breakdown occurs in the middle lamella, a cell wall area rich in pectins (Chanson et al. 1989). These acid polysaccharides have been shown to be calcium-stabilized and involved in cell adhesion modification (Demarty et al. 1984; Jarvis 1984; Liners et al. 1994). This wood failure is related to calcium nutrition and its availability in soil.

6 Hazelnut

The number and amounts of mineral elements taken up by a fruit tree orchard from the soil each year can help with mineral fertilization planning. Despite its theoretical and practical importance, only a few experimental studies on hazelnuts have been conducted. The amounts of mineral nutrients taken up by hazelnuts from the soil were previously studied relative to the age of the orchards. Chaplin and Dixon (1979) applied nitrogen as urea to *Corylus avellana* at 0.68, 1.36 or 2.72 kg N per tree annually for a period of 7 years and found that increasing nitrogen significantly increased leaf nitrogen, yield and tree size. Leaf phosphorus level was reduced by raising nitrogen rates in all years. Soil pH after 7 years of nitrogen application significantly declined as nitrogen rate increased which increased availability of Mn. The studies conducted by Shrestha et al. (1986) with Barcelona hazelnut observed that boron spray was found to increase fruit set and decrease dry matter accumulation in the entire fruiting twig. In unsprayed twigs, prior to June, leaves constituted 80% of the total dry weight of fruiting twig. Later in the season, in early August, the dry weight was maximum in the nuts (66%) followed by in leaves (17%) and husks (13%). Leaves contained 87% of the total boron in fruiting twigs early in the season indicating that the leaf was the major sink for boron at that time. The interpretations of leaf tissue analysis are given in Table 7.4.

Roversi and Ughini (2006) studied how biennial bearing affected orchard macronutrient uptake so they could apply the right amount of mineral fertilizer. Two year trial was conducted in ‘Tonda Gentile delle Langhe’ (TGL) hazelnut orchards in

Table 7.4 Interpretation of leaf tissue results for hazelnut

Nutrient	Deficiency	Normal	Excess
Nitrogen	<1.80%	2.21–2.50%	>3.00%
Phosphorus	<0.10%	0.14–0.45%	>0.55%
Potassium	<0.50%	0.81–2.00%	>3.00%
Sulfur	<0.08%	0.13–0.20%	>0.50%
Calcium	<0.60%	1.01–2.50%	>3.00%
Magnesium	<0.18%	0.25–0.50%	>1.00%
Manganese	<20 ppm	26–650 ppm	>1000 ppm
Iron	<40 ppm	51–400 ppm	>500 ppm
Copper	<2 ppm	5–15 ppm	>100 ppm
Boron	<25 ppm	31–75 ppm	>100 ppm
Zinc	<10 ppm	16–60 ppm	>100 ppm

Source: USDA (2010)

Table 7.5 Uptake of different mineral elements (kg/ha) in two regions and its related decrease

	Montelupo			Rodello		
Elements	On-year	Off-year	Decrease (%)	On-year	Off-year	Decrease (%)
N	78.00	28.38	−63.6	91.26	39.32	−56.9
P	7.43	2.58	−65.3	8.00	2.52	−68.5
K	36.91	11.71	−68.3	42.82	12.85	−70.0
Ca	31.08	25.04	−19.4	32.92	17.80	−45.9
Mg	6.92	3.39	−51.0	6.57	2.57	−60.9

Source: Roversi and Ughini (2006)

Italy, viz., Montelupo and Rodello, with the first being 'on-year' and the second being a 'off-year'. It was concluded that the amount of mineral elements taken up by hazelnut from the soil was significantly lower in the 'off-year' compared to the on year. Furthermore, in both orchards, the percentage decrease between the 2 years was greater than 50%, with the exception of Ca (Table 7.5). The element with the highest uptake was nitrogen (91.26 kg) during the on-year while Mg had lowest uptake among all elements during both on and off year in both orchards. It was also depicted that the quantities of Phosphorus taken up yearly by the hazelnut are quite limited.

7 Pistachio Nut

Some soil analyses reveal sufficient nutrient available in the soil of pistachio growing areas but they may not be available to the leaves of the plant due to antagonism and lack of water. Antagonism means that if one element is abundant in the soil, the plants may be unable to absorb another. There are antagonistic relationships between the elements that are critical to the plant's survival. An excess of Ca, for example, leads to a Fe deficiency, while an excess of P leads to a Zn deficiency, and so on. Leaf and soil analysis should be performed, and element antagonism should be considered when creating proper fertilization programmes for plants (Ozbek 1977). Pistachio trees, on the other hand, struggle with salinity and alkalinity. Some elements cannot be translocated due to the high pH because availability of nutrients gets changed at critical pH levels (Westwood 1978; Ak and Parlakci 2006). The soil in the pistachio plantation area is largely deficient in nitrogen, phosphorus, potassium, and organic matter. Many orchards showed prominent signs of deficiency of phosphorus and zinc, fruit set is affected by zinc deficiency in particular, while as others were slightly lacking in nitrogen, iron, and manganese (Tekin et al. 1985).

In pistachio, nitrogen is the most important fertilizer element. Plants use nitrogen to make amino acids and nucleic acids, which are essential for all of the plant's functions. Pistachio shoot growth remains stunted due to nitrogen deficiency and have a reddish bark in more severe deficiencies. Petioles and midribs of N-deficient

Table 7.6 Nutritional range of healthy pistachio leaves

Element	Range
Nitrogen	2.5–2.9%
Phosphorus	1.0–2.0%
Potassium	0.14–0.17%
Calcium	1.3–4.0%
Magnesium	0.6–1.2%
Sodium	0.002–0.007%
Chlorine	0.1–0.30%
Manganese	30–80 ppm
Boron	55–230 ppm
Zinc	7–14 ppm

Source: Uriu and Crane (1977)

leaves turn red (Gonzales and Uriu 1985). Nut yield will decrease as N deficiency worsens. One of the pistachio nitrogen consumption studies was carried out by Weinbaum and Muraoka (1989) in which he found that at nitrogen levels below 1.8%, leaves turned yellow and fell while as, at 2.5% they remained intact until the end of vegetative period. The nutritional range of different elements in healthy pistachio leaves is given in Table 7.6.

Based on the findings, Baninasab et al. (2007) concluded that applying nitrogen to pistachio trees could be a useful method for reducing flower bud abscission and increasing crop in the following year. Nitrogen increases flower bud retention, leaf nitrogen level, current shoot and bud, photosynthesis rate, chlorophyll content and leaf area. Nitrogen was more effective in increasing flower bud retention in pistachio trees when applied at 0.3% in mid-June. Flower bud abscission decreased as leaf, shoot and flower bud nitrogen content and leaf area increased.

Pistachios require potassium in large quantities, therefore, 56.69 kg of potassium along with 90.72 kg of nitrogen are applied every year (Rehman et al. 2018b). Potassium is also required by plants for the formation of sugars and starch, protein synthesis, and cell division. It boosts the oil content of pistachios and makes them better resistant to cold. Potassium uptake is greatest during the nut-filling period, with very little during the spring vegetative flush and post-harvest. Leaf K status, nut yield, and quality have all been shown to improve with K fertilization (Zeng et al. 1998). Pistachio trees that are mature and heavy-bearing have been shown to absorb about 100 kg of K per acre during the growing season and 50 kg during the off-season (Brown et al. 1999). Karimi et al. (2012) found that foliar application of K_2SO_4 at 0.1% or 0.2% during the endospermic growth stage of seed and cotyledon appearance increased yield, splitting percentage, nut fresh mass and kernel dry mass, and decreased blank percentage.

Pistachios are frequently reported to have a zinc (Zn^{2+}) deficiency. Zn^{2+} fertilization is an effective way to correct Zn^{2+} deficiency and increase pistachio yield in Zn deficient soils. In pistachio grown under rainfed conditions, 2 g/L Zn application at 15-day intervals (before, during, and after blossom) would provide adequate Zn nutrition (Kizilgoz et al. 2010). At 50% leaf expansion, zinc (36% zinc sulphate) and boron are applied through foliar application (mid-April). Boron is applied in June or July through a drip system to orchards with chronically low tissue levels (Brar et al. 2015).

The hulls of some pistachio nuts are ruptured in the orchard prior to harvest, and these nuts are referred to as "early split." Even though only a small percentage of nuts split early, the kernels of these nuts are frequently decayed by fungi and contaminated with aflatoxin (Doster and Michailides 1995). Early splitting, which resulted in Aspergillus infection, is linked to unbalanced pistachio nutrition. Potassium, iron and phosphorus deficiency, as well as zinc toxicity, have all been linked to the occurrence of early splits in the past (Torabi and Malakouti 2000). Furthermore, Alipour and Panahi (2004) found a positive and significant correlation between leaf Cu content and early splitting, as well as a negative and significant correlation between leaf percentages of K and Ca. Hosseinifard and Panahi (2005) found Fe and Cu content in the leaf to be a significant factor influencing early splitting. These findings showed that a well-balanced mineral diet is critical for maintaining intact hulls that cover and protect the kernel from mould invasion.

In mature, alternate-bearing pistachio (*Pistacia vera*) trees, Rosecrance et al. (1996) investigated the relationships between NPK uptake, root growth and crop load during spring flush (mid-March to late May), nut filling (late May to early September), and post-harvest leaf senescence (late September to early December). Pistachio trees produce heavy fruit crop during the on-year and light crop during off-year. Primary uptake of these nutrients occurred during the nut filling stage in both on and off year. Comparing on and off year, N and K uptake increased by 35% and 112% during nut filling in the on year, respectively. In on-year trees, primary allocation of nutrients was to embryo development while in off-year it was to storage in perennial tissues. Minimum uptake of nutrients was observed between harvest and leaf senescence. The findings support the theory that sink demand regulates pistachio tree N, P, and K uptake and distribution.

8 Conclusion

Fruit yield and quality, both of which are important features of modern agriculture, require proper nutrition management of fruit tree species. In addition, fruit arboriculture is currently dealing with a decrease in fertile soils as a result of salinization events, as well as changes in global rainfall patterns as a result of climate change. By altering plant allometry, nutritional status, hydraulic features and water relations, photosynthetic process, and the nutritive and nutraceutical value of fruits, these events have a significant impact on the performance of fruit tree species. As a result of intensive production systems, various farming systems (biological, integrated, and conventional), and a plethora of new elite cultivars, nutrient management is becoming increasingly important. As a result, it's critical to learn how to properly manage the mineral nutrition of fruit tree species in the face of environmental changes and novel orchard management techniques in order to avoid both nutrient deficiency and excess. The above nutrient management discussion would only be worthy when backed up with valid methodologies of nutrient constraints diagnosis. The currently available leaf nutrient standards for different fruit crops have distinct limitations in terms of sampling index plant part(s) leaving very short time for remediation of identified constraints, application of diverse interpretation tools brings out varying limits of nutrient concentration and do we need to apply some correction factor when such nutrient standards are applied in fruits under fertigation programme.

9 Future Strategies

The traditional orchards growing temperate nut crops suffer a great deal of losses due to ill nutrient management. These orchards easily suffer from nutrient deficiencies thus, compromising the quality of the yield.

1. The growers should continuously implement soil testing and plant tissue analysis in their orchards to identify nutrient deficiency or toxicity.
2. These orchards have long been neglected in terms of nutritional requirements especially micronutrients. Problems arising due to deficiency of micronutrients like zinc and boron should be identified and care should be taken to rectify them.
3. A diligently designed fertilization schedule should be followed to avoid losses in plant health and nut quality and quantity.
4. The recommended doses of the fertilizers should be given regularly as prescribed to avoid future repercussions.
5. Precision tools like Geographical Information System (GIS) and Diagnosis and Recommendation Integrated System (DRIS) in mapping the fertility status of soil would provide the desired accuracy and effectiveness in fertilizer recommendations.
6. To improve nutrient use efficiency, tailoring the nutrient and water application as per crop phenological stages is a prerequisite.
7. Specific nutrient-signaling pathways, have made it feasible to enhance the uptake and use efficiency of applied nutrient. Analyzing signal transduction for various nutrients could provide some meaningful clues about the presence or absence of those signals to be later utilized in understanding the underlying principles of nutrient stress induced warning mechanism.
8. Incorporating fertigation would give a definite edge over conventionally used methods. A large number of studies have demonstrated much better NUE with fertigation using various bases of drip irrigation scheduling.
9. Use of organic fertilizers should be encouraged to prevent aquifer contamination with toxic pollutants.

References

Abaev MN, Lukin NF, Khvorova NE (1976) Diagnosis of mineral nutrition and estimation of requirements of young nursery trees of walnut. Agrokhimiia 12:113–116

Ak BE, Parlakci H (2006) The low yield reasons and solution of pistachio nut in Turkey. Nucis 13: 37–40

Akhlaghi Amiri N, Asadi Kangarshahi A, Arzani K, Barzegar M (2016) Calyx biochemical changes and possibility of reducing Thomson orange june drop by nutrition elements and growth regulators. Int J Hortic Sci Technol 3(2):179–186

Alipour H, Panahi B (2004) Effect of organic fertilizer on pistachio nut quality and its contamination to Aflatoxin. In: VII Soil Science Congress of Iran, Rasht, Iran

Álvarez-Álvarez P, Díaz-Varela E, Cámara-Obregón A, Afif-Khouri E (2010) Relating growth and nutrition to site factors in young chestnut plantations established on agricultural and forest land in northern Spain. Agrofor Syst 79:291–301

Arrobas M, Afonso S, Ferreira IQ, Moutinho-Pereira JM, Correia CM, Rodrigues MA (2017) Liming and application of nitrogen, phosphorus, potassium and boron on a young plantation of chestnut. Turk J Agric For 41:441–451

Arrobas M, Afonso S, Rodrigues MA (2018) Diagnosing the nutritional condition of chestnut groves by soil and leaf analyses. Sci Hortic 228:113–121

Baninasab B, Rahemi M, Javanshah A (2007) Effects of time of foliar application of nitrogen and its concentrations on the flower bud retention in pistachio trees. Int J Soil Sci 2(1):40–47

Beutel J, Uriu K, Lilleland O (1983) Leaf analysis for California deciduous fruits. In: Reisenauer HM (ed) Soil and plant tissue testing in California. University of California Bull 1879, Oakland, pp 11–14

Borges O, Gonçalves B, Soeiro JL (2008) Nutritional quality of chestnut (*Castanea sativa* Mill.) cultivars from Portugal. Food Chem 106:976–984

Bourgeois C (1992) Le châtaignier: un arbre, un bois. Institut Pour le Développement Forestier, Paris

Brar GS, Doll D, Ferguson L, Fichtner E, Kallsen CE, Beede RH, Klonsky K, Tumber KP, Anderson N, Stewart D (2015) Sample costs to establish and produce pistachios. University of California Cooperative Extension, Davis

Briat JF, Dubos C, Gaymard F (2015) Iron nutrition, biomass production, and plant product quality. Trends Plant Sci 20:33–40

Brown PH, Uriu K (1996) Nutrition deficiencies and toxicities: diagnosis and correcting imbalances. In: Micke WC (ed) Almond production manual. UCANR Publication, pp 179–188

Brown P, Zhang Q, Huang Z, Holtz B, Hornung C (1999) Agronomic and economic responses of mature 'Kerman' pistachio trees to potassium applications in California. California Pistachio Ind. Annual Report, pp 84–85

Chanson B, Leban JM, Thibaut B (1989) La roulure du châtaignier (*Castanea sativa* Mill.). Forêt Méditerranéenne XI:15–34

Chaplin MH, Dixon AR (1979) The development of standard ranges for leaf nitrogen in the Filbert (*Corylus avellana* L.). J Am Soc Hortic Sci 104:710–712

Davis DE (1949) Some effects of calcium deficiency on the anatomy of *Pinus tadea*. Am J Bot 36: 276–282

Dehghanipoodeh S, Askari MA, Talaei AR, Babalar M (2018) Evaluation the preharvest application of iron and nitrogen on some qualitative characteristics of two apple cultivars during cold storage. Int J Hortic Sci Technol 5(1):31–42

Demarty M, Morvan C, Thellier M (1984) Calcium and the cell wall. Plant Cell Environ 7:441–448

Doll D (1996) Almond nutrients and fertilization. Fruit and nut research and information. University of California. http://fruitsandnus.ucdavis.edu/almondpages/AlmondNutrientsFertilization/

Doster MA, Michailides TJ (1995) The relationship between date of hull splitting and decay of pistachio nuts by Aspergillus species. Plant Dis 79:766–769

Echegaray N, Gomez B, Barba FJ, Franco D, Estevez M, Carballo J, Marszałek K, Lorenzo JM (2018) Chestnuts and by-products as source of natural antioxidants in meat and meat products: a review. Trends Food Sci Technol 82:110–121

Fernández V, Sotiropoulos T, Brown PH (2013) Foliar fertilization: scientific principles and field practices. Paris, France, International Fertilizer Industry Association (IFA)

Gonzales V, Uriu K (1985) Nutritional deficiency symptoms in pistachio. California Pistachio Ind. Annual Report, pp 69–73

Hansen P (1971) The effect of cropping on uptake, contents and distribution of nutrients in apple trees. Tidsskr Planteavl 75:615–625

Hosseinifard J, Panahi B (2005) The effect of different mineral nutrients on early splitting in pistachio. IV Int Symp Pistachios Almonds 726:325–328

Jarvis MC (1984) Structure and properties of pectin gels in plant cell walls. Plant Cell Environ 7: 153–164

Kalcsits L, Van Der Heijden G, Reid M, Mullin K (2017) Calcium absorption during fruit development in 'honeycrisp' apple measured using 44ca as a stable isotope tracer. HortScience 52:1804–1809

Karimi HR, Sevandi Nasab S, Roosta HR (2012) The effect of salicylic acid and potassium on some characteristics nut and physiological parameters of pistachio trees cv. Owhadi. J Nuts 3(3): 21–26

Keshavarz K, Vahdati K, Samar M, Azadegan B, Brown PH (2011) Foliar application of zinc and boron improves walnut vegetation and reproductive growth. Hortic Technol 21(2):181–186

Kizilgoz I, Sakin E, Aslan N (2010) The effects of zinc fertilisation on the yield of pistachio (Pistacia vera L.) grown under rainfed conditions. Afr J Agric Res 5(24):3427–3430

Kolov OV (1977) Effect of mineral fertilizers on the growth of walnut seedlings. Agris 1:61–62

Kumar A, Rehman MU, Rather GH, Angmo T, Sau S (2018) Pecan nut. In: Ghosh SN, Kumar A, Rehman MU, Sharma RR (eds) Temperate fruit crops theory to practical, vol II. New India Publishing Agency, New Delhi, pp 711–784

Liang Z, Zou Y, Zhang J (2010) Effect of N, P and K fertilizers on walnut trees in southern Xinjiang. Xinjiang Agric Sci 47(5):958–963

Liners F, Gaspar T, Van Cutsem P (1994) Acetyl- and methyl-esterification of pectins of friable and compact sugar beet calli: consequences for inter cellular adhesion. Planta 192:545–556

Liu J, Chen W, Wang H, Peng F, Chen M, Liu S, Chu G (2021) Effects of NPK fertilization on photosynthetic characteristics and nutrients of pecan at the seedling stage. J Soil Sci Plant Nutr 21(3):2425–2435

Lotfi N, Vahdati K, Kholdebarin B, Ashrafi EN (2009) Germination, mineral composition, and ion uptake in walnut under salinity conditions. HortScience 44(5):1352–1357

Malstrom HL, Fenn LB, Riley T (1983) Nitrogen fertilization of pecans in Far West Texas. Pecan Q 17:5–13

Marschner H (2012) Mineral nutrition of higher plants. Academic, London, pp 135–190

Martins A, Marques G, Borges O, Portela E, Lousada J, Raimundo F, Madeira M (2011) Management of chestnut plantations for a multi-functional land use under Mediterranean conditions: effects on productivity and sustainability. Agrofor Syst 81:175–189

Masarat J, Tomar CS (2008) Standardization of leaf positioning and sampling time for macronutrient estimation in walnut (*Juglans regia*). Indian J Agric Sci 78(6):531–533

Mészáros M, Hnátková H, Čonka P, Náměstek J (2021) Linking mineral nutrition and fruit quality to growth intensity and crop load in apple. Agronomy 11(3):506

Millard P (1996) Ecophysiology of the internal cycling of nitrogen for tree growth. J Plant Nutr Soil Sci 159:1–10

Muhammad S, Sanden BL, Lampinen BD, Saa S, Siddiqui MI, Smart DR, Olivos A, Shackel KA, DeJong T, Brown PH (2015) Seasonal changes in nutrient content and concentrations in a mature deciduous tree species: studies in almond (*Prunus dulcis*). Eur J Agron 65:52–68

Neilsen GH, Neilsen D (2003) Nutritional requirements of apple. In: Ferree DC, Warrington IJ (eds) Apples: botany, production and uses. CAB International, Wernstein, Germany, pp 267–302

Neilsen D, Millard P, Herbert LC, Nelsen GH, Hogue EJ, Parchomchuk P, Zebarth BJ (2001) Remobilization and uptake of N by newly planted apple (*Malus domestica*) trees in response to irrigation method and timing of N application. Tree Physiol 21:513–521

Nicodemus MA, Salifu FK, Jacobs DF (2008) Growth, nutrition, and photosynthetic response of black walnut to varying nitrogen sources and rates. J Plant Nutr 31(11):1917–1936

Ozbek S (1977) Genel Meyvecilik Çukurova Ziraat Fakültesi Yayınları, 111. Ders Kitabı 6:386

Palmer JW (1992) Effects of varying crop load on photosynthesis, dry matter production and partitioning of Crispin/M.27 apple trees. Tree Physiol 11:19–33

Portela E, Ferreira-Cardoso JV, Louzada JL (2011) Boron application on a chestnut orchard: effect on yield and quality of nuts. J Plant Nutr 34:1245–1253

Portela E, Ferreira-Cardoso J, Louzada J, Gomes-Laranjo J (2015) Assessment of boron application in chestnuts: nut yield and quality. J Plant Nutr 38:973–987

Proebsting EL, Davy S (1946) In: Childers NF (ed) Fruit Nutrition. Rutgers State University, New Brunswick, NJ

Rahnamoun H (2002) The effects of nitrogen, phosphorus and potassium on almond trees. Agricultural Research, Education and Extension Organization. Bulletin No.706. Jahad-e-Keshavarzi Press, Tabriz. 29 pp. (In Persian)

Rajaie M, Motieallah ME (2018) Ameliorative effect of ammonium sulfate on salt tolerance and ion homeostasis in lemon (*C. limon*) Seedlings. Int J Hortic Sci Technol 5:111–122

Ramalho JC, Rebelo MC, Santos ME, Antunes ML, Nunes MA (1995) Effects of calcium deficiency on *Coffea arabica*. Nutrient changes and correlation of calcium levels with some photosynthetic parameters. Plant Soil 172:87–96

Rehman MU, Rather GH, Gull Y, Mir MM, Iqbal U (2018a) Almond. In: Ghosh SN, Kumar A, Rehman MU, Sharma RR (eds) Temperate fruit crops theory to practical, vol II. New India Publishing Agency, New Delhi, pp 711–784

Rehman MU, Rather GHM, Mir M, Iqbal U, Gull Y, Parray EA, Qayoom S (2018b) Pistachio. In: Ghosh SN, Kumar A, Rehman MU, Sharma RR (eds) Temperate fruit crops theory to practical, vol II. New India Publishing Agency, New Delhi, pp 711–784

Rodrigues MÂ, Grade V, Barroso V, Pereira A, Cassol LC, Arrobas M (2019) Chestnut response to organo-mineral and controlled-release fertilizers in rainfed growing conditions. J Soil Sci Plant Nutr 20(2):380–391

Rosecrance RC, Weinbaum SA, Brown PH (1996) Assessment of nitrogen, phosphorus, and potassium uptake capacity and root growth in mature alternate-bearing pistachio (Pistacia vera) trees. Tree Physiol 16(11/12):949–956

Roversi A, Ughini V (2006) Effect of alternate bearing on the mineral nutrition in hazelnut. V Int Symp Mineral Nutr Fruit Plants 721:77–82

Samuoliene G, Viškelien A, Sirtautas R, Kviklys D (2016) Relationships between apple tree rootstock, crop-load, plant nutritional status and yield. Sci Hortic 211:167–173

Sharpe RH, Blackmon GH, Gammon N (1952) Relation of potash and phosphate fertilization to cold injury of Moore pecans. Proc Southeastern Pecan Growers Assoc 45:81–85

Shrestha GK, Thompson MM, Righetti TL (1986) Foliar applied boron increases fruit set in Barcelona hazelnut. J Am Soc Hortic Sci 112:412–416

Smith MW (2005) Understanding alternate bearing. Pecan South 38(10):32–37

Smith MW, Kenworthy AL, Bedford CL (1979) The response of fruit trees to injection of nitrogen through a trickle irrigation system. J Am Soc Hortic Sci 104:311–313

Smith MW, Storey JB, Westfall PN, Anderson WB (1980) Zinc and sulphur in pecan leaflets as affected by application of sulphur and zinc to calcareous soils. Hortic Sci 15:77–78

Smith MW, Ager PL, Endicott DSW (1985) Effect of nitrogen and potassium on yield, growth, and leaf elemental concentration of pecan. J Am Soc Hortic Sci 110:446–450

Smith MW, Wood B, Raun WR (2007) Recovery and partitioning of nitrogen from early spring and midsummer applications to pecan trees. J Am Soc Hortic Sci 132:758–763

Smith MW, Charles TR, William DG (2012) Pecan leaf elemental sufficiency ranges and fertilizer recommendations. HortTechnology 22(5):594–599

Socha AL, Eguerinot ML (2014) Mneuvering manganese: the role of transporter gene family members in manganese uptake and mobilization in plants. Front Plant Sci 5:106

Sparks D (1982) Zinc concentration in pecan leaflets associated with zinc deficiency symptoms. Hortic Sci 17:670–671

Sparks D (1988) Growth and nutritional status of pecan in response to phosphorus. J Am Soc Hortic Sci 113:850–859

Tekin H, Genc C, Kuru C, Akkok F (1985) Investigations in the determinations of nutrient contents of pistachio nuts. (In Turkish). Bahce 14(1–2):47–57

Thagusev NA, Caroev BJ (1971) Manuring walnut orchards. Sadovodstvo 4:30–31

Torabi M, Malakouti MJ (2000) Effect of nutrition on pistachio contamination to Aflatoxin. The First Congress of Pistachio Aflatoxin, Kerman, Iran

Tromp J (1980) Mineral absorption and distribution in young apple trees under various environmental conditions. In: Atkinson D, Jackson JE, Sharples RO, Waller WM (eds) Mineral nutrition of fruit trees. Butterworths, London, UK, pp 173–182

Tromp J (2005) Mineral nutrition. In: Tromp J, Webster AD, Wertheim SJ (eds) Fundamentals of temperate zone tree fruit production. Backhuys Publishers, Leiden, Poland, pp 55–64

Ulrich A (1952) Physiological basis for assessing the nutritional requirement of plants. Annu Rev Plant Physiol 3:207–228

Uriu K, Crane JC (1977) Mineral element changes in pistachio leaves. J Am Soc Hortic Sci 102: 155–158

USDA (2010) USDA Agricultural Research Service, Hazelnut Clonal Germplasm Repository, Corvallis, GRIN database. www.ars.usda.gov/Main/docs.htm?docid=1035

Wang H, Inukai Y, Yamauchi A (2006) Root development and nutrient uptake. Crit Rev Plant Sci 25:279–301

Weinbaum SA, Muraoka TT (1989) Nitrogen usage and fertilizer nitrogen recovery by mature pistachio trees. California Pistachio Industry. In: Annual Report Crop Year, 1988–1989, Univ. of California Davis CA, pp 84–86

Weinbaum SA, Niederholzer FJA, Ponchner S, Rosecrance RC, Carlson RM, Whittlesey AC, Muraoka TT (1994) Nutrient uptake by cropping and defruited field-grown 'Frendi' prune trees. J Am Soc Hortic Sci 119:925–938

Westwood MN (1978) Temperate zone pomology. W.H. Freeman and Company, San Francisco, p 404

White AC, Rogers A, Rees M, Osborne CP (2015) How can we make plants grow faster? A source–sink perspective on growth rate. J Exp Bot 67:31–45

Wood BW, Wells L, Funderburke F (2010) Influence of elevating tree potassium on fruit drop and yield of pecan. Hortic Sci 45(6):911–917

Worley RE, Harmon SA, Carter RL (1974) Effect of repeated N, P, K and lime applications on soil pH, P and K under old and young pecan trees. J Am Soc Hortic Sci 99:57–62

Xia G, Cheng L, Lakso A, Goffinet M (2009) Effects of nitrogen supply on source-sink balance and fruit size of 'gala' apple trees. J Am Soc Hortic Sci 134:126–133

Yuan ZQ, Yang XY, Ling H, Shen C, Zhao WM, Huang JQ, Wang ZJ (2014) Dynamic changes of leaves' main mineral elements in pecan 'Mahan's growth period. Acta Botanica Boreali Occidentalia Sinica 34(7):1443–1449

Zarata-Valdez JL, Muhammad S, Saa S, Lampinen BD, Brown PH (2015) Light interception, leaf nitrogen and yield prediction in almonds: a case study. Eur J Agron 66:1–7

Zeng DQ, Brown PH, Rosecrance RC (1998) The effects of alternate bearing, soil moisture and gypsum on potassium nutrition of pistachio (*Pistacia vera* L.). Acta Hortic 470:412–420

Chapter 8
Plant Water Relations and Irrigation

Munib Ur Rehman, Yasmeen Gull, Mohammad Maqbool Mir, Umar Iqbal, Tashi Angmo, Mehvish Hanief, Romana Mahmood, Gh. Hassan Rather, G. I. Hassan, and S. A. Banday

Abstract Water is one of the most important inputs in agricultural production. Plants require both carbon dioxide and water for photosynthesis. With the exchange of gases, substantial amount of water is lost through the stomata via transpiration. Carbon assimilation will be maximal as long water is not a limiting factor. Reduction in amount of energy available to the plant will impact vegetative growth as well as fruit and kernel development. A standard, science-based approach for scheduling irrigation is to calculate the evapotranspiration (ET) of the crop. The amount of water thus used by a particular species (ET_c) will keep varying with the factors influencing ET such as increase in leaf area, weather, time of the day, etc. At the same time, availability and use efficiency of irrigation water will also depend on many other factors like soil, topography, method of application, and climatic conditions. Nut crops are mostly grown on marginal lands as rainfed crops, but proper irrigation scheduling has been found to enhance the yields. Though it has been suggested that irrigation should exceed 100% of ET_c to avoid water stress, but considering future climatic scenarios major restrictions in irrigation may be faced due to less water availability for the purpose, necessitating to increase water use efficiency. Adopting regulated deficit irrigation (RDI) strategies where water application during

---

Munib Ur Rehman (✉) · M. M. Mir · U. Iqbal · T. Angmo · M. Hanief · R. Mahmood
Division of Fruit Science, FOH, Sher-e-Kashmir University of Agricultural Sciences and Technology of Kashmir, Srinagar, Jammu and Kashmir, India

Y. Gull
Department of Physics, GDC, Kulgam, Jammu and Kashmir, India

G. H. Rather
Ambri Apple Research Centre, Shopian, Sher-e-Kashmir University of Agricultural Sciences and Technology of Kashmir, Srinagar, Jammu and Kashmir, India

G. I. Hassan
Research Centre for Residue and Quality Analysis, FOH, Sher-e-Kashmir University of Agricultural Sciences and Technology of Kashmir, Srinagar, Jammu and Kashmir, India

S. A. Banday
Krishi Vigyan Kendra, Sher-e-Kashmir University of Agricultural Sciences and Technology of Kashmir, Ganderbal, Jammu and Kashmir, India

M. M. Mir et al. (eds.), *Temperate Nuts*,
https://doi.org/10.1007/978-981-19-9497-5_8

non-sensitive growth periods is reduced by 20–30% can have minimal impact on yield and quality, with increased water use efficiency.

Keywords Crop coefficient · Deficit · Almond · Walnut · Pistachio · Pecan nut · Hazelnut

1 Introduction

Water is critical for almost all processes and the greatest component of trees by mass. Nearly about 70% of the fresh water is being consumed by agriculture sector, of which 40% comes through irrigation (FAO 2017), making agriculture sector as most important user of water with a heavy impact on ecosystems' water share. Withdrawal of water for agriculture is leading to falling groundwater table, dried-up rivers, desertification, salinated soil, and polluted waterways. From hydrological models and volumetric-based calculations, global groundwater depletion due to increasing global freshwater withdrawals, primarily associated with the expansion of irrigated agriculture in drylands, has been estimated to be ~100 and ~300 km^3/year (Bierkens and Wada 2019).

Water availability for agricultural production is becoming a major limiting factor in many areas, which is due to increased demand and competition for its use in various growing sectors like industrial, agriculture, rural, and urban. Demand for irrigation water for crop production is expected to increase further by 30% compared to current consumption, which may cause further imbalance between demand and availability (IPCC 2021). In areas where water is abundant, quality of water may be a concern due to salinization and contamination by pesticides, nitrates, etc. Availability of this limited resource is also governed by the growing imbalance in the economic conditions between urban and rural areas in and among the countries. Taking all these important scenarios including climate change into consideration, major restrictions will be faced, curtailing irrigation of traditional crops (García-Tejero et al. 2011).

Water management in modern agriculture thus turns out to be an important aspect to ensure water consumption necessities of plants without much reduction in yield and quality. The basic requirement to achieve this is to estimate the water status of plants and of the soil (Gerhards et al. 2016; Buitrago et al. 2016), with the adoption of irrigation technologies and management practices that contribute to less water use. As the water loss through transpiration exceeds the amount of easily accessible soil water, the plant exhibits stress responses, with the closure of stomata and ultimately reduction of photosynthesis and production of carbohydrates. The quantity of energy available to the plant will get reduced, hampering vegetative growth and fruit and kernel development (Doll and Shackel 2015). Vegetative growth is rapid just after leafing-out during spring and is necessary to establish fruiting positions and carbohydrate reserves for future yields. As the growth starts, water requirement is very low and reaches to maximum level with the full expansion of leaves at around 4–5 weeks after bloom. Any deficit in irrigation here will affect proper canopy development and nut load which is directly related to canopy growth and size (Lampinen et al. 2007).

Fruit and kernel development in temperate nut crops is a three-stage process. Stage I is characterized with a rapid growth of the hull, shell, and integuments. The stage ends with the maximum external dimensions, which is about 2 months after bloom. During this period, severe tree water stress rarely occurs, due to stored soil moisture and moderate temperatures, but if it does, reduced size of the nuts and kernels can be observed due to reduction in cell division and subsequently the number. Stage I is followed by another stage (Stage II), also called as lag phase, in which endocarp hardening takes place, with little or no visible growth in dimensions. The final growth period is the Stage III, where kernel begins to accumulate weight at a continuous rate until harvest. The stage is supposed to be crucial, as starvation at this stage will lead to reduction in kernel dry weight, shriveled and textured kernels, or in severe case blank nuts. In case of pistachio, water shortage at nut filling stage results in high percentage of non-splits (nuts with shells that do not split open).

2 Irrigational Demand

The quantity of water actually consumed by the plant (ET) or reaching at the field plot is considerably less than the water released at the irrigation source. Same is the case with the rains received over the year or growing period with the seepage and percolation losses in the field. The irrigation efficiency, which is a ratio of usage to supply, usually between 60% and 70%, varies depending on soil, climate, crop, cropping sequences and intensity, crop calendar (the time and length of growing period), and application method. The practicality of an estimation of demand requires appropriate techniques largely depending upon the precise prediction made of an area under different crops. A standard and relatively simple, science-based approach to assess irrigation requirements of an orchard is to calculate the evapotranspiration (ET). Plants behave like wicks, drawing out water from soil to atmosphere, which is lost to atmosphere through stomata by the process of evapotranspiration. Water flow can be resisted with the closure of these stomata.

2.1 Almond

Among other factors, water requirement of almond will depend on the cultivar, rootstock, or tree density and has been reported to range between 900 and 1350 mm annually (Golhamer and Girona 2012). Production function for the crop reaches maximum yield of 4000 kg/ha (depending on the season) at around 1250 mm of irrigation. Maximum yield declined by 14% as the irrigation rate was reduced from 1350 to 1000 mm. Unit kernel weight also showed a progressive decline from 1.2 to 1.0 g when irrigation amount was brought down from 1350 to 1000 mm (Goldhamer and Fereres 2016). Prgomet et al. (2020), in line with the above observation, recorded no loss in yield when water application was cut down by 35% of ET_c during nut filling stage.

Crop water requirement (K_c) of almond orchards studied during last several decades indicates a progressive increasing trend with time, which has been attributed to the changes associated with intensification of almond cultivation system including higher planting density, minimal pruning, better nutrient management, extra crop load and most importantly the change in irrigation pattern in recent years, replacing surface irrigation with most frequent soil-wetting drip or micro-sprinkler irrigation systems increasing the surface evaporation (Tejero et al. 2018).

While determining orchard water requirement, it is important to consider year-to-year variations in K_c, as its value changes radically with the attainment of growth from young to maturity. K_c adjustment becomes difficult in situations as ground surface is covered by tree canopy completely (Espadafor et al. 2015).

The relationship between the ET_c of mature trees and area shaded by young almond trees was given by Fereres et al. (1982) for the shade values within 8–65%. Beyond 65%, ET_c of mature trees is assumed to be 100%.

$$ET_c\ (\%) = -0.0194S2 + 2.8119S - 1.0080$$

2.2 Pecan Nut

Irrigation is vital for survival and rapid growth of young trees and for consistent production of quality nuts on bearing trees. Pecan trees need an ample supply of soil water for good growth and nut production. Almost all growth and productivity parameters in pecan have benefited from irrigation. Supplemental water treatments significantly increased kernel yield, nut weight, diameter, oil content, and appearance (Heaton et al. 1982). Irrigating bearing trees helps in improving nut size and nut fill. Generally before nut shell formation, less water availability reduces nut size and after shell hardening, kernel fill is inhibited. Pecan trees do not like wet or saturated soils, so do not over water; however, irrigation should be sufficient to thoroughly soak the soil.

Trees at early stage of plantation need immediate irrigation and thereafter careful application is required, as excess water may cause roots to rot. Young pecan trees will require 40–60 L of water weekly by irrigation or rain. Even though pecan nut tree has a deep-tap root system, but to maintain optimum moisture levels, irrigation of top 1 m will be adequate. Ideally, regular watering of plants at least at weekly intervals during mid-March to September, especially in mid-summer is required. Stein et al. (1989) reported increased yield, nut size, trunk diameter, and kernel percentage due to summer and fall irrigation. Interestingly, they reported the need of water even before shuck opening as late season irrigation reduced stick-tights and viviparous nuts.

2.3 *Pistachio*

While pistachio trees are less thirsty than other nut trees and able to survive periods of drought, irrigation is essential for commercial production. Soil moisture monitoring devices should be used to ascertain frequency and duration of irrigation schedules. Pistachio trees are more tolerant to saline conditions than other nut trees. For best quality yields and tree health, trees should never be subjected to moisture stress during the growing season. This is particularly important during nut fill in late summer because inadequate soil moisture at this stage can cause a high percentage of non-splits (nuts with shells that do not split open). All ecophysiological parameters like stomatal conductance, transpiration, net CO_2 uptake are influenced by irrigation and must be restored to 50% ET_c (crop evapotranspiration) (de Palma et al. 1997). Higher levels of irrigation lower photosynthetic rates and decrease fruit/inflorescence, which may be due to root sensitivity to soil oxygen depletion in saturated soils (Monastra et al. 1995; de Palma and Novello 1998). Irrigating pistachio trees with 120 mm water at 20–30 days interval has been found to double the yields during both on and off years (Kanber et al. 1993). Currently irrigation water usage ranges from 9 ML/ha in some regions down to as 2 ML/ha in other orchards.

2.4 *Hazelnut*

Hazelnut (*Corylus avellana* L.) is nutritionally one of the most important nut crops in the world, mostly cultivated in areas with the evenly distributed annual rainfall between 800 and 1000 mm, throughout the year (Cristofori et al. 2014). However, recently its cultivation has expanded significantly in areas particularly in southern hemisphere (FAOSTAT 2016), where rains are scarce during the growing season (Stolpe and Undurraga 2016; Garreaud et al. 2017; Deitch et al. 2017).

3 Evapotranspiration

There are several procedures and variety of instruments with a well-defined accuracy which can be used for measuring evapotranspiration (ET). Crop ET (ET_c) is normally computed from weather data using FAO56 method (Allen et al. 1998). The method is simple but while deriving crop coefficient (K_c) values, accuracy in measurements and computations of field observations or prediction of ground cover and height are of particular importance (Allen et al. 2011). ET_c can be calculated from the product of K_c and reference grass evapotranspiration (ET_o). Here K_c represents the main distinguishing characteristics integrated together in terms of the energy balance from the reference grass and is calculated adopting the

FAO56 method by estimating ET_c (mm/day) and grass reference evapotranspiration ET_o (mm/day).

$$K_c = {}^{ET_c}\!/_{ET_o}$$

The daily ET_o depends upon the energy available for evaporation resulting from the energy balance at the surface and is computed with the PM-ET_o equation obtained by fixing the corresponding aerodynamic and surface resistance parameters of the reference grass in the Penman–Monteith combination equation (Pereira et al. 1999). The energy available for the evaporation process or latent heat flux (λE) will be the difference in net radiation flux (R_n), soil heat flux (G), and sensible heat flux (H). In arid regions, K_c values observed in relation to grass reference range from 1.2 to 1.4, and in sub-humid regions, the upper limit is 1.2. Higher ET values than the upper limits might result due to errors committed in ET measurements like weather data (ET_o) calculation or data processing procedure (Pereira et al. 2021a). In absence of advection, the resulting upper limit of K_c in case of orchard crops is close to 1.2 as they do not attain full canopy cover. Hence, daily grass reference evapotranspiration can be obtained as follows:

$$ET_o = \frac{0.408\Delta(R_n - G) + \gamma \frac{900}{T+273} u_2 (e_s - e_a)}{\Delta + \gamma(1 + 0.3u_2)}$$

where

Δ—slope of the saturation vapour pressure-temperature curve at mean air temperature (kPa/°C)
$(R_n - G)$—available energy at the vegetated surface (MJ/m^2 day)
Γ—psychometric constant (kPa/°C)
T—mean daily air temperature (°C)
u_2—mean daily wind speed (m/s) at 2 m height
$(e_s - e_a)$—vapour pressure deficit (VPD) of the atmosphere (kPa)

The PM-ET_o equation represents evaporative demands of the atmosphere, considering only vertical fluxes of heat and vapour and not the advective heat energy fluxes. As each crop possesses a specific characteristic distinguishing it from others, accordingly K_c for every crop varies. Two approaches are considered for accessing K_c: one includes multi-day effects of plant transpiration in addition to soil evaporation, calculated as the ratios between crop transpiration (T_c) or soil evaporation (E_s) and ET_o; and the other is the sum of the basal crop coefficient (K_{cb}) and the soil evaporation coefficient (K_e) as given by Allen et al. (1998).

$$K_c = K_{cb} + K_e \text{ with } K_{cb} = T_c/ET_o \text{ and } K_e = E_s/ET_o.$$

Under water stress conditions, ET_c is replaced by actual crop ET ($ET_{c\ act}$), and accordingly K_c by actual $K_{c\ act}$ or using the dual approach, by $K_sK_{cb} + K_e$.

Table 8.1 Indicative standard values for single and basal crop coefficients relative to the mid- and end-season for deciduous temperate nut crops

				Indicative standard values (±10%) of K_c and K_{cb}			
Crop	Density	f_c	Plant height (m)	$K_{c\ mid}$	$K_{cb\ mid}$	$K_{c\ end}$	$K_{cb\ end}$
Almond	Young	0.15–0.30	1.0–2.0	0.40	0.35	0.35	0.25
	Low	0.30–0.40	4.0–5.0	0.45	0.40	0.40	0.30
	Medium	0.40–0.50	4.0–5.0	0.65	0.60	0.50	0.40
	High	0.50–0.60	4.0–5.0	0.90	0.85	0.65	0.60
	Very high	>0.60	4.0–5.5	1.05	1.00	0.75	0.70
Hazelnut	Young	0.15–0.30	1.0–2.0	0.45	0.40	0.40	0.20
	Low	0.30–0.45	4.0–5.0	0.75	0.70	0.55	0.45
	High	0.45–0.60	1.5–2.0	0.95	0.90	0.65	0.55
Pecan	Young	0.25–0.30	1.0–5.0	0.55	0.50	0.50	0.35
	Low	0.30–0.60	5.0–10.0	0.75	0.70	0.55	0.45
	Medium	0.60–0.70	7.0–11.5	0.85	0.80	0.60	0.55
	High	>0.70	10.0–13.5	1.00	0.95	0.65	0.60
Pistachio	Young, low	0.15–0.30	1.5–2.0	0.45	0.40	0.35	0.25
	Medium	0.30–0.50		0.80	0.75	0.55	0.50
	High	>0.50	2.0–3.0	0.95	0.90	0.65	0.60
Walnut	Young, low	<0.30	1.0–2.0	0.50	0.45	0.45	0.25
	Medium	0.45–0.75	3.5–7.0	0.85	0.80	0.50	0.40
	High	0.75–0.85	3.5–7.0	0.95	0.90	0.55	0.50
	Very high	>0.85	3.5–7.0	1.05	1.00	0.60	0.55

$$ET_{c\ act} = K_s ET_c = K_s K_c ET_o = (K_s K_{cb} + K_e) ET_o$$

where K_s (values running from 0 to 1) is the stress coefficient, depending upon the sufficiency of available soil water to maintain the crop ET rate.

The standard K_c and K_{cb} values for tree crops vary with the crop age, crop management, training system, fraction of ground cover and height (Pereira et al. 2021b, c). A standard K_c and K_{cb} on the basis of estimation of crop density or the fraction of ground shaded or the fraction of intercepted photosynthetic active radiation is required to be established. A most common value obtained at a particular growth stage can be attributed to each crop (Pereira et al. 2021b, c). Also, crop coefficient can be derived from f_c (fraction of ground covered by vegetation) and h (Pereira et al. 2020).

Indicative updated values (Table 8.1) of the crop coefficients ($K_{c\ mid}$, $K_{c\ end}$, $K_{cb\ mid}$, and $K_{cb\ end}$) of some nut crops have been obtained by Rallo et al. (2021) from the observed values and those proposed in the referred studies. With a limited soil evaporation due to small fraction of soil wetting under drip irrigation system and shadowing of soil by canopies $K_{c\ mid}$ for all trees was obtained by adding 0.05 to $K_{cb\ mid}$. However, according to the f_c value and precipitation occurrence at the end of the season, the additive factor to calculate $K_{c\ end}$ varied in the range of 0.05–0.40.

The water requirement estimated by determining ET_c is to be sufficed either by irrigation or rainfall. Any reserve water within the soil from winter rainfall, capillary rise, etc. is also to be included. Thus equation for total amount of irrigation applied during a season would be:

$$\text{Water applied} = ET_c - (\text{In– season rainfall}) - (\text{Stored soil moisture})$$

Indicative f_c and plant height values from Me et al. (2004), Olsen (2013), Lampinen (2014), Fischbach (2017), Kallsen and Fanucchi (2008), Simmons et al. (2007), Samani et al. (2011), Taylor et al. (2017), Stevens et al. (2012), Espadafor et al. (2015), García-Tejero et al. (2015), Bellvert et al. (2018), and López-López et al. (2018).

4 Different Approaches of Demand Management

Selection and design of the irrigation system depend upon site topography and soil type. The choice of irrigation system, in turn, influences final land levelling and tillage requirements. Though choosing a flood or furrow irrigation system minimizes initial capital cost but properly designed and managed micro-irrigation systems like micro-sprinkler, surface drip, subsurface drip, and sprinkler are advantageous in reducing physical soil limitations.

Alternate furrow irrigation: Instead of every furrow, alternate furrow irrigation can reduce the water use by about 33% (Sepaskhah and Ahmadi 2010). Water saving mostly occurs on the lower part of the field. Transpiration rate is found to be reduced as compared to conventional furrow irrigation system without any effect on photosynthetic rate (Du et al. 2013).

Efficient irrigation method: Water use efficiency can be enhanced by adapting more efficient systems of irrigation. Sprinkler and drip irrigation systems can save non-effective water loss. However, the most important factor influencing the adaption of any water conservation method aimed at increasing irrigation efficiency will depend on its water productivity (WP). WP is defined as the ratio of biological yield or economic output to crop ET, which may not always be economical and will depend on the product (Fereres et al. 2003).

Deficit irrigation: Irrigation requirement can be reduced with a minimal reduction in yields by withholding irrigation at less-sensitive stages of plant growth. In a research conducted at ICARDA, yield reductions of only 10–15% were recorded by curtailing 50% of full supplemental irrigation requirement (Zhang and Oweis 1999).

5 Deficit Irrigation (DI) Strategies

Exposing trees to a moderate degree of water stress can have a positive impact on various factors like checking the excessive vegetative growth, reduction in fungal diseases, and ease of harvest in addition to water saving (Goldhamer et al. 2006; Teviotdale et al. 2001). Regulated deficit Irrigation (RDI) strategies can be adopted, where water application during non-sensitive growth periods is reduced by 20–30% with minimal impact on yield and quality (Cabonell-Barrachina et al. 2015; Romero et al. 2015). Moreover, the RDI strategy using irrigation cut-off (ICO) has been suggested to be applied in many tree crops (Trentacoste et al. 2015; Ahumada-Orellana et al. 2018). Under this strategy, irrigation during less-sensitive phenological periods is completely suppressed till a predefined water threshold is reached. This way the water status and gas exchange are maintained within an optimal range, so as not to cause much reduction in stomatal conductance (g_s) and net carbon dioxide assimilation rate (A_n) by monitoring stem water potential (Ψ_{stem}) (Cifre et al. 2005; Galindo et al. 2018).

Even though, fruit yield and quality considerably depend upon the irrigation (Egea et al. 2009), deficit irrigation of pistachio trees in April and May result in substantial increase in the formation of early-split nuts in late summer (Doster et al. 2001). For almond, there are contradictory reports related to its response to deficit irrigation strategies (Girona et al. 2005; Goldhamer et al. 2006; Egea et al. 2010). Torrecillas et al. (1996) considered almond among drought-tolerant crop because of its xeromorphic properties. Franco et al. (2000) and Castel and Fereres (1982) considered almond resistant to drought and salinity as it can withstand water stress. However, it is also being established that one of the reasons for low productivity and poor nut quality in almond is due to its cultivation confined traditionally under marginal areas associated with limited irrigation conditions (CAPDR 2016; Goldhamer and Fereres 2016).

6 Crop Water Stress Index

The crop water stress index (CWSI) for measuring water stress in plants, proposed by U.S. Water Conservation Laboratory, Phoenix, Ariza, is a non-invasive direct method based on canopy temperature. CWSI of a particular plant or orchard can be estimated by determining real-time canopy temperature measurements using infrared thermometers, and the absolute humidity status (Idso et al. 1981). Value for CWSI runs from 0 to 1, with 0 indicating no water stress, while as 1 indicating severe water stress. In well-irrigated conditions, CWSI was reported to remain below 0.2 throughout the season (Testi et al. 2008).

$$\mathrm{CWSI} = (T_c - T_a - D_2)/(D_1 - D_2)$$

where

T_c—average canopy temperature (°C)
T_a—air temperature (°C)
D_2 and D_1—predicted respective minimum and maximum values of (T_c − T_a) at a specific VPD (kPa)

When soil water is not limiting the plant transpiration ability, leaf temperature can be predicted at a specific vapour pressure deficit (VPD). As and when the soil moisture is depleted, with the closing of stomata less amount of water will flow out of the leaf, resulting in increased leaf or canopy temperature.

CWSI is also defined on the basis of variations in environmental conditions (Jackson et al. 1981):

$$\mathrm{CWSI} = 1 - E/E_p$$

where

E—actual evaporation rate of leaf canopy
E_p—potential evaporation rate of the leaf canopy

7 Water Quality/Salt Stress

Agriculture production is severely affected by salt stress in many parts of the world especially in arid and semi-arid regions. This salinity stress is affecting irrigated lands as well. Worldwide, 20% of irrigated land has got effected due to secondary salinization lead by poor irrigation management (Glick et al. 2007). The characteristic feature of saline soils is the electrical conductivity (EC) exceeding 4 ds/m (approx. 40 mM NaCl) at 25 °C with exchangeable sodium of 15% in the root zone saturation extract (Shrivastava and Kumar 2014). Increased soil salt content affects plants negatively by inducing osmotic stress, ion toxicity (sodium, chloride), and nutritional disorders (Yeo et al. 1991; Marschner 1995). Among temperate nut crops, walnut shows more sensitivity towards soil or irrigation water salinity (Table 8.2). The tree growth rate and yield have been observed to decline by 18–21% with the

Table 8.2 Tolerance rating of nut crops to salt stress (Grieve et al. 2012)

Crop	Rating
Almond	Sensitive
Pecan	Moderately sensitive
Pistachio	Moderately tolerant
Hazelnut	Moderately tolerant
Walnut	Sensitive

increase in root zone salinity from 1 to 1.5 dS/m (Fulton et al. 1988). Salt tolerance determined by a genetic system can vary within the same species. Paradox and *J. hindsii* rootstocks are comparatively less sensitive to salt than *J. regia*. Rootstock comparisons for salt sensitivity are primarily based on their salt translocation capacity to the scion part and signs of leaf toxicity. Chlorine content of 0.3%, sodium 0.1%, and boron 30 ppm in leaves of walnut are considered as excessive levels (Caprile and Grattan 2011). Almond also considered as salt-sensitive crop has been observed to show little tolerance than walnut with no deleterious effect up to 3 dS/m (Dejampour et al. 2012). However, as the salinity is increased from 0.8 to 4.26 dS/m, production rate was observed to be reduced drastically by about 46% (Franco et al. 2000). For almond, Empyrean (*Prunus persica* × *davidiana*) as a root stock shows superiority in salt tolerance by exclusion of sodium ions from the root vasculature (Shao et al. 2021). Pistachio considered as relatively tolerant to salt tress also shows variation in exhibiting tolerance among cultivars. Among the three pistachio rootstocks (Badami-e-Zarand, Sarakhs, and Ghazvini) tested, all showed decline in most of the growth parameters with the increase in salt content above 75 mM NaCl. However, 'Ghazvini' rootstock was found more resistant than others on the basis of Na accumulation in leaves and stem (Hokmabadi et al. 2005; Tavallali et al. 2008). It was also proposed that $CaSO_4$ at 50–100 mM concentrations can ameliorate the negative effects of salinity on plant dry matter and chlorophyll content. A reduction of 80% in relative dry weight of pistachio plants was recorded with 13 dS/m of saline irrigation water (Sepaskhah and Karimi-Goghari 2005). In Pecan, the growth was suppressed at 0.5% of salt stress, though being suggested for cultivation in soils with high levels of salinity (Sheng et al. 2014). Hazelnut is also considered resistant at 100–150 mM NaCl (Zhang et al. 2014).

8 Physiological Response to Water Stress

To avoid water stress under drought conditions, trees involve different resistance processes. The first important response by plants to minimize water loss under mild-to-moderate water stress conditions is by reduction in stomatal conductance (g_s) (Hsiao 1990; Torrecillas et al. 1996). This influences carbon dioxide absorptions and finally photosynthesis and growth rate. Down the line. this reduction in g_s is triggered by sharp decrease in leaf water potential (Ψ_{leaf}) or stem water potential (Ψ_{stem}).

Though response being cultivar specific, almond in general is not considered as an isohydric crop; that is, when subjected to water stress, it has a limited capability to regulate g_s (Egea et al. 2011; Eichi 2013). Spinelli et al. (2016) observed notable contrast in terms of Ψ_{leaf} in Guava cultivar of almond trees when subjected to two different irrigation treatments. But under severe and prolonged water stress conditions, stomatal conductance was regulated by leaf area reduction promoted by leaf senescence. Hence increasing water use efficiency (WUE) together by maintaining acceptable levels of g_s by lowering crop water potential and at the same time by upholding carbon assimilation and photosynthetic rate at optimum level (Rouhi et al. 2007). Transpiration rate was not affected until 0.5 MPa but tended to decrease when the stress level exceeded 0.5 MPa of deficit stem water potential (DSWP) (Dhillon et al. 2019). Nevertheless, at Ψ_{stem} values ranging between −0.5 and −2.0 MPa, the relationship between g_s values ($R2 = 0.74$) and photosynthetic rate (A) ($R2 = 0.61$) was found highly significant (Spinelli et al. 2016). Pattern may vary as optimum range of predawn and midday Ψ_{leaf}, g_s and photosynthetic rate for every cultivar may not be same under proper irrigation (Gomes-Laranjo et al. 2006). Walnut crop showed decrease in transpiration rate at low plant water stress (PWS) of less than −0.5 MPa of DSWP (Dhillon et al. 2014). Investigations in pistachio led by Guerrero et al. (2005) by subjecting trees to regulated deficit irrigation (RDI) at 60% of ET_c during the three stages of fruit development, with full irrigation at third stage, i.e. fruit development, observed lowest value of −1.8 MPa of midday Ψ_{stem} in RDI tress, with the greatest reduction of 60% in gas exchange over control (−1.1 MPa at 100% ET_c), suggesting that at fruit development stage third irrigation should exceed that of 100% ET_c. Othman et al. (2014) observed more than 50% decline in photosynthesis and gas exchange rate in pecan, with midday Ψ_{stem} ranging between −1.5 and −2.0 MPa. However, they recommended maintaining of −0.90 MPa midday Ψ_{stem}.

9 Improving Irrigation Efficiency (E_a)

In most of the nut orchards, commonly practiced irrigation systems are surface or gravity systems, having efficiencies of 60–70%. This in other way means that 30–40% of irrigation water gets wasted or lost from the conveyance system by surface run-off, evaporation from exposed irrigation channels, and by deep percolation of water. Efficiencies of gravity irrigation systems can however be enhanced to a great extent by adapting certain modification which can reduce conveyance losses like change in rate or duration of water application, length or slope of the field, using zero slope or level basins In case of sprinkler or drip irrigation systems, irrigation rates and water distribution patterns are more governed by irrigation system rather than soil type. Within the orchards, surface irrigation systems like flood and furrow and pressurized systems, including sprinkler, drip and micro-sprinkler, are the commonly used irrigation systems with varying efficiencies. Pressurized systems are usually more efficient (Table 8.3) due to the reduction of both run-off, water loss through deep percolation and relatively small area of wetted soil. As an example, an irrigation system that only wets 20% of the orchard floor can hold 2/3rd of the water of a system that wets 30% of an orchard floor. Efficiency can further be increased using subsurface drip system with an approximate water saving of 25% compared to surface drip system (Sedaghati et al. 2012). Sprinkler or drip systems are not necessarily the only efficient systems of irrigation. Efficiency of 80–90% can be obtained with surface irrigation systems as well as long the system is finely designed and managed (Smith et al. 2005).

Table 8.3 Application efficiency (E_a) of various irrigation systems

Irrigation method	Average efficiency (%)
Surface method	
Graded furrow	65
Level furrow	80
Graded border	65
Level basins	85
Sprinkler	
Periodic move	75
Side roll	75
Moving big gun	65
Centre pivot	
Impact heads w/end gun	80
Spray heads wo/end gun	90
Low energy precision application	95
Lateral move	
Spray heads w/hose feed	90
Spray heads w/canal feed	85
Micro-irrigation	
Trickle	85
Subsurface drip	90
Micro-spray	85
Water table control	
Surface ditch	65
Subsurface drain lines	75

Source: Howell 2003

Table 8.4 Estimates of available water for various soil types

Type of soil	Range	Average
Coarse (sand/loam sand)	5–8.3	6.25
Sandy (loam sand/sandy loam/loam)	8.3–12.5	10.4
Medium (loam/silty clay loam)	10.4–18.3	12.5
Fine (silty loam/silty clay loam/clay loam/silty clay)	14.2–20	16.6

10 Irrigation Scheduling

An important aspect of irrigation water management is proper irrigation scheduling—that is, correct decisions about when and how much to irrigate. This will depend on various factors like type of soil, flow rate, emitter spacing, etc. Various methods—including those that use evapotranspiration (ET) estimates or plant monitoring techniques—are effective means of irrigation scheduling. Likewise, measuring soil moisture offers many benefits. This practice has merit, but it is only an estimate of soil moisture and sampling from deeper depths may be time consuming. Table 8.4 includes the estimates of available water for various soil types.

Water conservation by improving irrigation scheduling is of practical importance in situation where run-off and/or deep percolation from the irrigated fields cannot be reused. Irrigation scheduling is based on soil water measurements or by using climatological methods for estimation of daily evaporation rates, evaporation pans, or lysimeter. Infrared thermometers used to access crop canopy temperature remotely can provide a reliable measure of plant water status for scheduling irrigation. Also the timing of irrigation is an important aspect for increasing crop water use efficiency by way of reducing leaching losses and enhancing crop yield per unit of evapotranspiration.

11 Measures for Water Conservation

Water conservation can be done by adopting the following principles:

11.1 Addition of Organic Matter

To increase the availability of soil water for plants, it is necessary to enhance water storage capacity within the soil profile. The long-recognized practice to increase the water holding capacity of a soil is to increase soil's organic matter (OM), as it effectively improves the soil chemical and physical properties. Hudson (1994), observed more than twofold increase in available water content in soils of all texture groups (sand, silt loam, and silty clay loam) with 0.5–3% increase in organic matter content. Soil organic matter results in efficient use of water by slow water release, facilitating proper crop growth and thus increasing yield and water productivity. This increased retention and availability of water with the organic matter amendments may suffice for approximately 35 water-stress days, contributing as much as 80–90 mm to transpiration (Ankenbauer and Loheide 2016).

11.2 Tillage and Subsoiling

Surface tillage breaks the soil crust leading to increased infiltration and storage of water in the root zone area. But at the same time, tillage exposes more soil surface area towards drying and may not be suitable practice for surface feeding crops. For deep rooted tree crops, it is helpful under water stressed conditions as it reduces the subsurface moisture loss due to breakage in soil capillarity. Tree crops with active deep roots also reduce drainage losses by extracting moisture from deeper zones. In most cases, where in root extension is inhibited or unsuitable for extension due to the presence of hard layer (plow pan) down the ground, subsoiling or deep tillage facilitates extension of roots for more soil water extraction.

11.3 Reducing Evaporation

Evaporation losses from fields can be reduced by weed control. Weeds draw out subsoil moisture and loose it to atmosphere through transpiration. Tillage in some instances may expose the moist soil and increase the evaporative surface area. But in case of heavy soils, tillage is primarily needed to reduce the evaporative losses by plugging the cracks developed when fallow. Sandy soils not showing cracks are "self-mulching" and need not to be tilled. Covering the field during fallow periods with crop residues and stubble reduces evaporation losses from the soil by insulating the evaporative soil surface from high temperatures and wind velocities. These dry land farming techniques are reducing evaporation losses by about half the annual precipitation in the Northern Great Plains of the United States (Hansen et al. 2012). Thus for precipitation of 38 cm/year, about 19 cm of water per year can be conserved by adapting these dryland farming techniques. Further, soil evaporation losses can be reduced by efficient use of irrigation like in drip irrigation systems, where the extent of wet areas is limited to the shaded areas near the plants.

11.4 Mulching

Soil moisture retention can be enhanced by using crop residues as mulch during the summer fallow. Mulch forms an insulating cover over the surface reducing water loss due to evaporation by 35–50% (Sauer et al. 1996). The material used for the purpose is easily accessible, low cost, and environment friendly without contaminating soil. The residue on decomposition may also add up the organic content of the soil.

11.5 Chemical Amendments

Chemical makeup of soil or water can be improved to enhance water infiltration capacity by adding chemical amendments to either soil or water. The working principle behind these chemical amendments is that they increase the total salt concentration of the soil water with the reduction in sodium adsorption ratio (SAR). Aggregate stability is enhanced making soil resistant to depression and pore blockage as well as soil crusting gets decreased. Materials usually used to improve water penetration can be categorized into four classes as below:

1. *Salts*: Amendments containing salts or any fertilizer salt that increase the electrolyte concentration of soil solution help in reducing the aggregate breakdown in soils ultimately influencing the water penetration in soils. Usefulness of salt amendments may, however, depend upon the total salinity and SAR values of the irrigation water or soil.

2. *Calcium materials*: Electrolyte concentration as well as soluble calcium content of soil or water can be increased by adding calcium salts. Effectiveness of the calcium amendment used may depend upon its solubility. High solubility of calcium materials like calcium chloride ($CaCl_2$) and calcium nitrate ($Ca(NO_3)_2$) makes them convenient for direct application with irrigation water but are more expensive than gypsum. Water solubility of less soluble/insoluble calcium amendments like dolomite and lime can be increased by reducing pH of the solute (pH less than 7.0). Lime and dolomite when added to acidic soils are raise soil pH.
3. *Acids and acid-forming materials*: Acids or acid-forming amendments react with soil-lime to form Ca-salt (gypsum). Availability of exchangeable Ca (gypsum) increases on dissolving with irrigation water and reduces the soil pH on applying in sufficient quantity. Commonly used acid or acid-forming amendments include sulphuric acid H_2SO_4 products, soil sulphur, ammonium polysulphide, and calcium polysulphide. Since these materials (with the exception of polysulphide) all contain sulphur (S) or H_2SO_4 but no calcium, they supply exchangeable calcium indirectly, by dissolving lime that is inherently in the soil.
4. *Soil conditioners*: Soil amendments of this group can be categorized into two types: organic polymers or surfactants. Organic polymers are mainly water-soluble polyacrylamides (PAMs) and polysaccharides other than crystal-like, cross-linked PAMs that expand when exposed to water. Effectiveness of these long chain organic molecules on water infiltration highly depends on polymer properties such as molecular weight, structure and electrical charge, polymer concentration and salinity of the irrigation water (Liu et al. 2018). The use of anionic polymers improved infiltration in solutions of 0.7 dS/m and reduced infiltration to solution of 0.05 dS/m (Helalia and Letey 1988).These conditions provide resistance against the disruptive forces of droplet impact by binding soil aggregates together. Organic polymers applied on furrows can effectively prevent carrying out of sediments and its loading. The most effective results on soil aggregate stability were obtained on recently tilled soils, when sprayed at the rate of about 4 kg/ha followed with an application of gypsum and irrigated through sprinkler system (Kumar and Saha 2011).

   Surfactants in another class of soil conditioners are mostly effective in soils containing high percentage of organic matter like turf soils, forest soils, and burned range lands. The functioning of these amendments is by reducing the surface tension of water.

12 Conclusion

Increased demand and competition for water use in various growing sectors are making it a scarcest resource for the crop production to feed the growing world population. Water management in modern agriculture thus turns out to be an important aspect to ensure water consumption necessities of plants without much reduction in yield and quality. Presently in most of the nut orchards, commonly practiced irrigation systems are surface or gravity systems, where 30–40% water is lost from the conveyance system. Efficiencies of irrigation systems can however be enhanced to a great extent by adapting certain modification. The basic requirement to achieve this is to estimate the water status of plants and of the soil with the adoption of irrigation technologies and management practices (scheduling in terms of time and quantity) that contribute to less water use. Deficit irrigation strategies can be adopted with minimal impact on yield and quality of nuts so as to build an integrative system to control water use and actively regulate the plant's growth and productivity.

13 Future Strategies

With near about 6 million ha of land area under temperate nuts, it is vital to consider water saving techniques keeping in view the current water scarcity and changing climatic conditions. Research in a broader perspective of better production and resource conservation is required in areas of impact of effective ground cover and mulches on crop coefficient. Precision in irrigation management can be achieved by proper modelling of crop coefficient as well as soil water balance models, incorporated with remote sensing. While aiming at real-time irrigation scheduling, forecasted weather data may be required. In addition, crop production functions need to be designed to solve the problem of allocation of limited water resources in crops competing for limited available water, for which knowledge of the crop's physiological behaviour is crucial in order to manage water stress and avoid excessive deficit.

References

Ahumada-Orellana LE, Ortega-Farias S, Searles PS (2018) Olive oil quality response to irrigation cut-off strategies in a super-high density orchard. Agric Water Manag 202:81–88

Allen RG, Pereira LS, Raes D, Smith M (1998) Crop evaporation. In: Guidelines for computing crop water requirements, FAO Irrig. Drain. Paper 56. FAO, Rome, Italy. 300pp

Allen RG, Pereira LS, Howell TA, Jensen ME (2011) Evapotranspiration information reporting: I. Factors governing measurement accuracy. Agric Water Manag 98:899–920

Ankenbauer KJ, Loheide SP (2016) The effects of soil organic matter on soil water retention and plant water use in a meadow of the Sierra Nevada, CA. Hydrol Processes 31(4):891–901

Bellvert J, Adeline K, Baram S, Pierce L, Sanden BL, Smart DR (2018) Monitoring crop evapotranspiration and crop coefficients over an almond and pistachio orchard throughout remote sensing. Remote Sens 10:200

Bierkens MFP, Wada Y (2019) Non-renewable groundwater use and groundwater depletion: a review. Environ Res Lett 14(6):063002

Buitrago S, Orlando F, Forni E, De Gregorio T, Falzoi S, Boni C, Pisetta M, Confalonieri R (2016) Development and evaluation of new modeling solutions to simulate hazelnut (Corylus avellana L.) growth and development. Ecol Model 329:86–99

Cabonell-Barrachina AA, Memmi H, Noguera-Artiaga L, Gijón-López MDC, Ciapa R, Pérez-López D (2015) Quality attributes of pistachio nuts as affected by rootstock and deficit irrigation. J Sci Food Agric 95(14):2866–2873

CAPDR (2016) Characterización del sector de la almendra en Andalucía. Secretaría General de Agricultura y Alimentación, Consejería de Agricultura, Pesca y Desarrollo Rural, Junta de Andalucía, p 34

Caprile J, Grattan S (2011) Evaluating the salt and boron tolerance of English walnut rootstock. http://cekings.ucdavis.edu/files/66721.pdf

Castel JR, Fereres E (1982) Responses of young almond trees to two drought periods in the field. J Hortic Sci Biotechnol 57:175–187

Cifre J, Bota J, Escalona JM, Medrano H, Flexas J (2005) Physiological tools for irrigation scheduling in grapevine (Vitis vinifera L.). An open gate to improve water-use efficiency. Agric Ecosyst Environ 106:159–170

Cristofori V, Pica AL, Silvestri C, Bizzarri S (2014) Phenology and yield evaluation of hazelnut cultivars in Latium region. Acta Hortic 1226:123–130

de Palma L, Novella V (1998) Effect of drip irrigation on leaf gas exchanges and stem water potential in pistachio cv. Larnak. Acta Hortic 470:317–323

de Palma L, Novello V, Monastra F (1997) First observations on physiological effects of irrigation in pistachio. Acta Hortic 449:545–550

Deitch MJ, Sapundjieff M, Feirer S (2017) Characterizing precipitation variability and trends in the world's Mediterranean-climate areas. Water 9(4):259

Dejampour J, Aliasgarzad N, Zainalabedini M, Niya MR, Hervan EM (2012) Evaluation of salt tolerance in almond (Prunus dulcis (L) Batsch) rootstocks. Afr J Biotechnol 11(56): 11907–11912

Dhillon R, Udompetaikul V, Rojo F, Roach J, Upadhyaya S, Slaughter D, Lampinen B, Shackel K (2014) Detection of plant water stress using leaf temperature and microclimatic measurements in almond, walnut and grape crops. Am Soc Agric Biol Eng 57(1):297–304

Dhillon R, Rojo F, Upadhyaya SK, Roach J, Coates R, Delwiche M (2019) Prediction of plant water status in almond and walnut trees using a continuous leaf monitoring system. Precis Agric 20: 723–745

Doll D, Shackel K (2015) Drought tip: drought management for California almonds. UC Agriculture and Natural Resources. ANR Publication. https://escholarship.org/uc/item/5zn49824

Doster MA, Michailides TJ, Goldhamer DA, Morgan DP (2001) Insufficient spring irrigation increases abnormal splitting of pistachio nuts. Calif Agric 55(3):28–31

Du T, Kang S, Yan B, Zhang J (2013) Alternate furrow irrigation: a practical way to improve grape quality and water use efficiency in arid Northwest China. J Integr Agric 12(3):509–519

Egea G, González-Real MM, Baille A, Nortes PA, Sánchez-Bel P, Domingo R (2009) The effect of contrasted deficit irrigation strategies on the fruit growth and kernel quality of mature almond trees. Agric Water Manag 96:1605–1614

Egea G, Nortes PA, González-Real MM, Baille A, Domingo R (2010) Agronomic response and water productivity of almond trees under contrasted deficit irrigation regimes. Agric Water Manag 97:171–181

Egea G, Dodd IC, González-Real MM, Domingo R, Baille A (2011) Partial rootzone drying improves almond tree leaf-level water use efficiency and afternoon water status compared with regulated deficit irrigation. Funct Plant Biol 38:372–385

Eichi VR (2013) Water use efficiency in almond (*Prunus dulcis* (Mill) D.A. Web). Ph.D. thesis, School of Agriculture, Food and Wine, Faculty of science, Univ. of Adelaide, p 77

Espadafor M, Orgaz F, Testi L, Lorite IJ, Villalobos FJ (2015) Transpiration of young almond trees in relation to intercepted radiation. Irrig Sci 33:265–275

FAO (2017) Water for sustainable food and agriculture. Food and Agriculture Organization of the United Nations, Rome, pp 1–21

FAOSTAT (2016) Food and Agriculture Organization of the United Nations. https://www.fao.org/faostat/en/#data/QC

Fereres E, Martinich DA, Aldrich TM, Castel JR, Holzapfel E, Schulbach H (1982) Drip irrigation saves money in young almond orchards. Calif Agric 36:12–13

Fereres E, Goldhamer DA, Parsons LR (2003) Irrigation water management of horticultural crops. HortScience 38(5):1036–1042

Fischbach J (2017) A production and economic model for hedgerow hazelnut production in the midwestern United States. Upper midwest hazelnut development initiative. University of Wisconsin, p 15

Franco JA, Abrisqueta JM, Hernansaez A, Moreno F (2000) Water balance in a young almond orchard under drip irrigation with water of low quality. Agric Water Manag 43:75–98

Fulton AE, Oster J, Nanson B (1988) Salinity management of walnut. In: Ramos DE (ed) Walnut production manual. Division of Agriculture and Natural Resources, University of California, pp 54–65

Galindo A, Collado-González J, Griñán I, Corell M, Centeno A, Martín-Palomo MJ, Girón IF, Rodríguez P, Cruz ZN, Memmi H, Carbonell-Barrachina AA, Hernández F, Torrecillas A, Moriana A, Pérez-López D (2018) Deficit irrigation and emerging fruit crops as a strategy to save water in Mediterranean semiarid agrosystems. Agric Water Manag 202:311–324

García-Tejero IF, Durán-Zuazo VH, Jiménez-Bocanegra JA, Muriel JL (2011) Improved water use efficiency by deficit irrigation programs: implications for saving water in citrus orchards. Sci Hortic 124:278–282

García-Tejero IF, Hernández A, Rodríguez VM, Ponce JR, Ramos V, Muriel JL, Durán-Zuazo VH (2015) Estimating almond crop coefficients and physiological response to water stress in semiarid environments (SW Spain). J Argic Sci Technol 17:1255–1266

Garreaud RD, Alvarez-Garreton C, Barichivich J, Boisier JP, Christie D, Galleguillos M, LeQuesne C, McPhee J, Zambrano-Bigiarini M (2017) The 2010–2015 megadrought in central Chile: impacts on regional hydroclimate and vegetation. Hydrol Earth Syst Sci 21:6307–6327

Gerhards M, Rock G, Schlerf M, Udelhoven T (2016) Water stress detection in potato plants using leaf temperature, emissivity and reflectance. Int J Appl Earth Obs Geoinf 53:27–39

Girona J, Mata M, Marsal J (2005) Regulated deficit irrigation during the kernel-filling period and optimal irrigation rates in almond. Agric Water Manag 75:152–167

Glick BR, Cheng Z, Czarny J, Duan J (2007) Promotion of plant growth by ACC deaminase-producing soil bacteria. Eur J Plant Pathol 119:329–339

Goldhamer DA, Fereres E (2016) Establishing an almond water production function for California using long-term yield response to variable irrigation. Irrig Sci 35:169–179

Goldhamer DA, Viveros M, Salinas M (2006) Regulated deficit irrigation in almond: effects of variation in applied water and stress timing on yield and yield components. Irrig Sci 24:101–114

Golhamer DA, Girona J (2012) Crop yield response to water: almond. In: Steduto P, Hsiao TC, Fereres E, Raes D (eds) FAO Irrigation and drainage paper No. 66. Food and Agriculture Organization of the United Nations, Rome, pp 358–373

Gomes-Laranjo J, Coutinho JP, Galhano V, Cordeiro V (2006) Responses of five almond cultivars to irrigation: photosynthesis and leaf water potential. Agric Water Manag 83:261–265

Grieve CM, Grattan SR, Maas EV (2012) Plant salt tolerance, Chapter 13. In: Wallender WW, Tanji KK (eds) ASCE manual and reports on engineering practice No. 71 Agricultural salinity assessment and management, 2nd edn. ASCE, Reston, VA, pp 405–459

Guerrero J, Moriana A, Perez-Lopez D, Couceiro JF, Olmedilla N, Gijon MC (2005) Regulated deficit irrigation and the recovery of water relations in pistachio trees. Tree Physiol 26:87–92

Hansen NC, Allen BL, Baumhardt RL, Lyon DJ (2012) Research achievements and adoption of no-till, dryland cropping in the semi-arid U.S. Great Plains. Field Crops Res 132:196–203

Heaton EK, Daniell JW, Moon LC (1982) Effect of drip irrigation of pecan quality and relationship of selected quality parameters. Food Sci 47(4):1272–1275

Helalia AM, Letey J (1988) Polymer type and water quality effects on soil dispersion. Soil Sci Soc Am J 52(1):243–246

Hokmabadi H, Arzani K, Grierson PF (2005) Growth, chemical composition and carbon isotope discrimination of pistachio (*Pistacia vera* L.) rootstock seedlings in response to salinity. Aust J Agric Res 56:135–144

Howell TA (2003) Irrigation efficiency. Encyclopedia of water science. Marcel Dekker. https://doi.org/10.1081/E-EWS120010252

Hsiao TC (1990) Measurements of plant water status. In: Steward BA, Nielsen DR (eds) Irrigation of agricultural crops, Agronomy monograph No. 30. Am. Soc. Agro., Madison, WI, pp 243–279

Hudson BD (1994) Soil organic matter and available water capacity. J Soil Water Conserv 49(2): 189–194

Idso SB, Jackson RD, Pinter PJ, Reginato RJ, Hatfield JL (1981) Normalizing the stress-degree-day parameter for environmental variability. J Agric Meteorol 24:45–55

IPCC (2021) IPCC (Intergovernmental Panel on Climate change), 2021. The physical science basis. Sixth assessment report, WGI. Cambridge University Press, Cambridge, UK

Jackson RD, Idso SB, Reginato RJ, Pinter PJ Jr (1981) Canopy temperature as a crop water stress indicator. Water Resour Res 17:1133–1138

Kallsen C, Fanucchi C (2008) Design of efficient pistachio orchards. In: University of California Cooperative Extension. Pistachio Short Course, p 39

Kanber R, Yazar A, Önder S, Köksal H (1993) Irrigation response of pistachio (Pistacia vera L.). Irrig Sci 14:7–14

Kumar A, Saha A (2011) Effect of polyacrylamide and gypsum on surface runoff, sediment yield and nutrient losses from steep slopes. Agric Water Manag 98(6):999–1004

Lampinen B (2014) Optimum walnut canopies: spacing and managing orchards for both early and mature production. Quad County Walnut Institute. https://ucanr.edu/sites/CEStanislausCo/files/188306.pdf

Lampinen B, Dejong T, Weinbaum S, Metcalf S, Negron C, Viveros M, McIlvane J, Ravid N, Baker B (2007) Spur dynamics and almond productivity. Modesto: 2007–2008. Annual Research Report to the Almond Board of California. UC Drought Management website https://ucmangedrought.ucdavis.edu/PDF/Lampinen%20et%20al%202007.pdf

Liu D, Edraki M, Berry L (2018) Investigating the settling behavior of saline tailing suspensions using kaolinite, bentonite and illite clay minerals. Powder Technol 326:228–236

López-López M, Espadafor M, Testi L, Lorite IJ, Orgaz F, Fereres E (2018) Water requirements of mature almond trees in response to atmospheric demand. Irrig Sci 36:271–280

Marschner H (1995) Mineral nutrition of higher plants. Ann Bot 78(4):527–528

Me G, Valentini M, Caviglione M, Lovisolo C (2004) Effect of shade on flowering and yield for two different hazelnut training systems. Acta Hortic 686:187–192

Monastra F, Insero O, Tamponi G (1995) The present situation of some underutilized fruit crops in Italy. In: Llácer G, Aksoy U, Mars M (eds) Underutilized fruit crops in the Mediterranean region. CIHEAM, Zaragoza, pp 39–46

Olsen J (2013) Growing hazelnuts in the Pacific Northwest: Orchard Design. Oregon State University Extension Service, p 4

Othman Y, VanLeeuwen D, Heerema R, St. Hilaire R (2014) Midday stem water potential values needed to maintain photosynthesis and leaf gas exchange established for pecan. Am Soc Horticult Sci 139(5):537–546

Pereira LS, Perrier A, Allen RG, Alves I (1999) Evapotranspiration: review of concepts and future trends. J Irrig Drain Eng 125(2):45–51

Pereira LS, Paredes P, Melton F, Johnson L, Wang T, Lopez-Urrea R, Cancela JJ, Allen R (2020) Prediction of crop coefficients from fraction of ground cover and height. Background and validation using ground and remote sensing data. Agric Water Manag 241:106197

Pereira LS, Paredes P, López-Urrea R, Hunsaker DJ, Mota M, Mohammadi S (2021a) Standard single and basal crop coefficients for vegetable crops, an update of FAO56 crop water requirements approach. Agric Water Manag 241:106197

Pereira LS, Paredes P, Hunsaker DJ, López-Urrea R, Mohammadi S (2021b) Standard single and basal crop coefficients for field crops. Updates and advances to the FAO56 crop water requirements method. Agric Water Manag 243:106466

Pereira LS, Paredes P, Melton F, Johnson L, Mota M, Wang T (2021c) Prediction of crop coefficients from fraction of ground cover and height. Practical application to vegetable, field and fruit crops with focus on parameterization. Agric Water Manag 241:106197

Prgomet I, Pascual-Seva N, Morais MC, Aires A, Barreales D, Ribeiro AC, Silva AP, Barros AIRNA, Goncalves B (2020) Physiological and biochemical performance of almond trees under deficit irrigation. Sci Hortic 261:108990

Rallo G, Paco TA, Paredes P, Puig-Sirera A, Massai R, Provenzano G, Pereira LS (2021) Updated single and dual crop coefficients for tree and vine fruit crops. Agric Water Manag 250:106645

Romero P, Muñoz RG, Fernández-Fernández JI, del Amor FM, Martínez-Cutillas A, García-García J (2015) Improvement of yield and grape and wine composition in field-grown Monastrell grapevines by partial root zone irrigation, in comparison with regulated deficit irrigation. Agric Water Manag 149:55–73

Rouhi V, Samson R, Lemeur R, Van Damme P (2007) Photosynthetic gas exchange characteristics in three different almond species during drought stress and subsequent recovery. Environ Exp Bot 59:117–129

Samani Z, Bawazir S, Skaggs R, Longworth J, Pi A (2011) A simple irrigation scheduling approach for pecans. Agric Water Manag 98:661–664

Sauer TJ, Hatfield JL, Prueger JH (1996) Corn residue age and placement effects on evaporation and soil thermal regime. Soil Sci Soc Am J 60:1558–1564

Sedaghati N, Hosseinifard SJ, Mohammadi MAA (2012) Comparing effects of surface and subsurface drip irrigation systems on growth and yield on mature pistachio trees. J Water Soil 26(3):575–585

Sepaskhah AR, Ahmadi SH (2010) A review on partial root-zone drying irrigation. Int J Plant Prod 4(4):241–258

Sepaskhah AR, Karimi-Goghari S (2005) Shallow groundwater contribution to pistachio water use. Agric Water Manag 72(1):69–80

Shao Y, Cheng Y, Pang H, Chang M, He F, Wang M, Davis DJ, Zhang S, Betz O, Fleck C, Dai T, Madahhosseini S, Wilkop T, Jernstedt J, Drakakaki G (2021) Investigation of salt tolerance mechanisms across a root developmental gradient in almond rootstocks. Front Plant Sci 11: 595055

Sheng J, Pujuan Z, Haijun Z, YaHui C, GuangQin L (2014) Effects of application of ALA on growth and development of pecan seedlings under salt stress. J Yangzhou Univ 35(3):90–94

Shrivastava P, Kumar R (2014) Soil salinity: a serious environmental issue and plant growth promoting bacteria as one of the tools for its alleviation. Saudi J Biol Sci 22(2):123–131

Simmons LJ, Wang J, Sammis TW, Miller DR (2007) An evaluation of two inexpensive energy-balance techniques for measuring water use in flood-irrigated pecans (Carya illinoinensis). Agric Water Manag 88:181–191

Smith RJ, Raine SR, Minakevich J (2005) Irrigation application efficiency and deep drainage potential under surface irrigated cotton. Agric Water Manag 71(2):117–130

Spinelli GM, Snyder RL, Sanden BL, Shackel KA (2016) Water stress causes stomatal closure but does not reduce canopy evapotranspiration in almond. Agric Water Manag 168:11–22

Stein LA, McEachern GR, Storey JB (1989) Summer and fall moisture stress and irrigation scheduling influence pecan growth and production. HortScience 24:607–611

Stevens RM, Ewenz CM, Grigson G, Conner SM (2012) Water use by an irrigated almond orchard. Irrig Sci 30:189–200

Stolpe N, Undurraga P (2016) Long term climatic trends in Chile and effects on soil moisture and temperature regimes. Chilean J Agric Res 76(4):487–496

Tavallali V, Rahemi M, Panahi B (2008) Calcium induces salinity tolerance in pistachio rootstock. Fruits 63(5):285–296

Taylor NJ, Annandale JG, Ibraimo NA, Steyn JM, Gush MB (2017) Are simple empirical crop coefficient approaches for determining pecan water use readily transferrable across a wide range of conditions? Acta Hortic 1150:9–14

Tejero IFG, Moriana A, Pleguezuelo CRR, Zuazo VHD, Egea G (2018) Sustainable deficit-irrigation management in almonds (*Prunus dulcis* L.): different strategies to access the crop water status. In: Tejero IFG, Moriana A, Pleguezuelo CRR, Zuazo VHD (eds) Water scarcity and sustainable agriculture in semiarid environment: tools, strategies and challenges for woody crops. Academic, pp 271–298

Testi L, Goldhamer DA, Iniesta F, Salinas M (2008) Crop water stress index is a sensitive water stress indicator in pistachio trees. Irrig Sci 26:395–405

Teviotdale B, Goldhamer D, Viveros M (2001) Effects of deficit irrigation on hullrot disease of almond trees caused by *Monilinia fructicola* and *Rhizopus stolonifer*. Plant Dis 85:399–403

Torrecillas A, Alarcón JJ, Domingo R, Planes J, Sánchez-Blanco MJ (1996) Strategies for drought resistance in leaves of two almond cultivars. Plant Sci 118:135–143

Trentacoste ER, Puertas CM, Sadras VO (2015) Effect of irrigation and tree density on vegetative growth, oil yield and water use efficiency in young olive orchard under arid conditions in Mendoza, Argentina. Irrig Sci 33:429–440

Yeo AR, Lee KS, Izard P, Boursier PJ, Flowers TJ (1991) Short- and long-term effects of salinity on leaf growth in rice (Oryza sativa L.). J Exp Bot 42(240):881–889

Zhang H, Oweis T (1999) Water-yield relations and optimal irrigation scheduling of wheat in the Mediterranean region. Agric Water Manag 38(3):195–211

Zhang L, Jia ZG, Ma QH, Zhang G, Wang GX (2014) Effects of saline-alkali stress on electrical impedance spectroscopy parameters and ion contents in shoots of Ping'ou hybrid hazelnut. J Appl Ecol 25(11):3131–3138

Chapter 9
Canopy Management

Aroosa Khalil, Mahrukh Mir, Safura Nabi, Mohammad Maqbool Mir, Umar Iqbal, Nowsheen Nazir, Shafat Ahmad Banday, Rifat Bhat, Saba Q. Khan, and Tajamaul F. Wani

Abstract Management of canopy design is a fundamental component in the plant kingdom for improving crop production by increasing the amount of dry matter partitioning towards the reproductive period. The most essential fruit plant management practice is canopy architecture. Managing a canopy helps in developing a strong framework of tree that will be able to bear heavy crop loads, at the same time improving fruit production and quality in the long haul. Fruit tree training and pruning are specialized horticultural procedures. The trees should be trained in such a way that enough air and light can enter the foliage, allowing for proper colouring and the growth of high-quality fruit. Adequate attention must be paid to training young plants to an acceptable training method. The temperate nut crops grow into large trees with a spreading canopy if allowed to grow uninterrupted. Traditionally in crops like walnut, pecan, chestnut, hazelnut, and chilgoza, no pruning is done, and there is little concept of a well-defined training system. Large trees can be seen growing outrageously in orchards. Such tall trees become unmanageable and suffer a great deal in crop yield and quality due to reduced interception of light. Therefore, they should be trained very early in their life cycle and then pruned annually to ease handling and managing the crop.

Keywords Training · Pruning · Temperate nuts · Framework · Management · Bearing habit

The canopy of a fruit tree refers to a tree's physical components, which include the trunk, branches, shoots, and leaves (Vandana et al. 2017). The number, length, and direction of the stem, branches, and shoots determine the canopy architecture,

---

A. Khalil (✉) · M. Mir · S. Nabi · M. M. Mir · U. Iqbal · N. Nazir · R. Bhat · S. Q. Khan · T. F. Wani
Division of Fruit Science, FOH, Sher-e-Kashmir University of Agricultural Sciences and Technology of Kashmir, Srinagar, Jammu and Kashmir, India

S. A. Banday
Krishi Vigyan Kendra, Sher-e-Kashmir University of Agricultural Sciences and Technology of Kashmir, Ganderbal, Jammu and Kashmir, India

M. M. Mir et al. (eds.), *Temperate Nuts*,
https://doi.org/10.1007/978-981-19-9497-5_9

Table 9.1 Different types of training systems recommended for temperate nut crops

S. No.	Crop	Training systems	References
1.	Almond	Cordon system, palmette system, vase system, hedge system	Almond Report: Australia tour of Spain (2014)
		Super high-density system	Casanova-Gascón et al. (2019)
2.	Walnut	Free vertical system, modified leader system	Germain et al. (1995)
3.	Pecan nut	Central leader system	McEachern and Stein (1986)
4.	Hazelnut	Open centre system, vertical axis system, V-shaped hedge system	Germain and Sarraquigne (1997)
		Free vase system, monocone system, ipsilon system, hedge row system	Tous et al. (1994)
		Double hedge row system	Me et al. (2001)

whereas the height and the spread of the canopy determine the volume (Pathak 2009). The number of leaves and their size comprise the canopy density. The canopy density of a tree, along with the tree architecture and its photosynthetic efficiency, determines its fruiting potential to a large extent (Kallow et al. 2005).

Canopy management is defined as the manipulation of the tree structure for maximum fruit production. The development of the tree structure and its maintenance in terms of shape, size, and direction of the branches lead to a better light interception and optimize the production and quality of fruits. Green leaves utilize the sunlight for the production of carbohydrates and sugars, which are needed by the buds, flowers, and fruits for their growth and development. Canopy management primarily consists of training and pruning to develop a proper framework of the tree, which is responsible for better light interception. Temperate fruits and nuts have lower productivity in comparison to other fruits, one of the reasons being improper canopy management as it directly influences productivity and quality (Singh 2008).

The temperate nut crops grow into large trees with a spreading canopy if allowed to grow uninterrupted. Traditionally in crops like walnut, pecan, chestnut, hazelnut, and chilgoza, no pruning is done, and there is little concept of a well-defined training system. Large trees can be seen as old as 50 years or greater, growing outrageously in orchards. Such tall trees become unmanageable and suffer a great deal in crop yield and quality due to reduced interception of light. Larger canopies are bound to create a shadow effect which would reduce the amount of light falling on leaves for photosynthesis. Temperate and tropical fruit crops only take up less than 70% of the radiated light, and hence, canopy regulation should be prioritized to maximize interception of light and yield quality and quantity (Whiley et al. 2013) (Table 9.1).

1 Almond

Almond (*Prunus dulcis*) is a part of the Rosaceae family, which includes apple, pear, cherry, peach, plum, apricot, and strawberry. It originated in Central Asia. Flowers are produced laterally on spurs or small lateral branches or sometimes on long stalks.

It grows to a height of 3–4.5 m and a diameter of 30–40 cm or more. Fruiting begins in 3–4-year-old grafted trees and reaches a peak in 6–10 years. In contrast to its short-lived cousin, the peach, almond trees can yield for 50 years or more. Thinning is unnecessary; for standard cropping, a significant proportion of blooms must set fruit. In mature orchards, the main crop grows on spurs (short, proleptic shoots) that can live for many years and produce fruits alone or in groups of five (Lampinen et al. 2011).

1.1 Canopy Management in Almond

An almond tree is productive for more than 50 years and thus is likely to develop a sufficiently large trunk width, trunk height, and root depth. All this leads to shading, causing lowered light interception and compromised yield out. A change in traditional cultivation of almonds from widely spaced trees in an orchard to a new system of closely spaced trees in modern high-density orchards is being adopted using dwarfing rootstocks, chemical growth inhibitors, and other orchard management practices to reduce the canopy size to accommodate a greater number of plants per unit area. Availability of size controlling rootstock has made high-density orcharding possible. In high-density orchards, the closer spacing of trees is achieved by developing the tree canopy into a conical form rather than a globular form, as can be seen in a traditional orchard. The practice of canopy management must be started as early as possible when the trees are still in the nursery.

The various factors governing canopy management in almonds are as under:

- Training
- Pruning
- Use of size controlling rootstocks

1.2 Bearing Habit

The almond, like other Prunus species, forms flower buds laterally on both spurs and stronger current-year stems. The number of flower buds generated at a node might vary both within and between cultivars. The number of blooms produced by each spur is determined by the previous year's spur leaf area and whether the spur produced fruit in the previous year. Figure 9.1 shows the spurs borne on 2- and 3-year-old wood. Spurs that produced fruit in 1 year rarely blossomed or produced fruit in the following year (Lampinen et al. 2011). Understanding the relative impact of blossom number and relative fruit set on almond tree output in commercial orchards is crucial for guiding efforts to boost orchard productivity and supporting growers in identifying the most profitable almond crop management options (Tombesi et al. 2016).

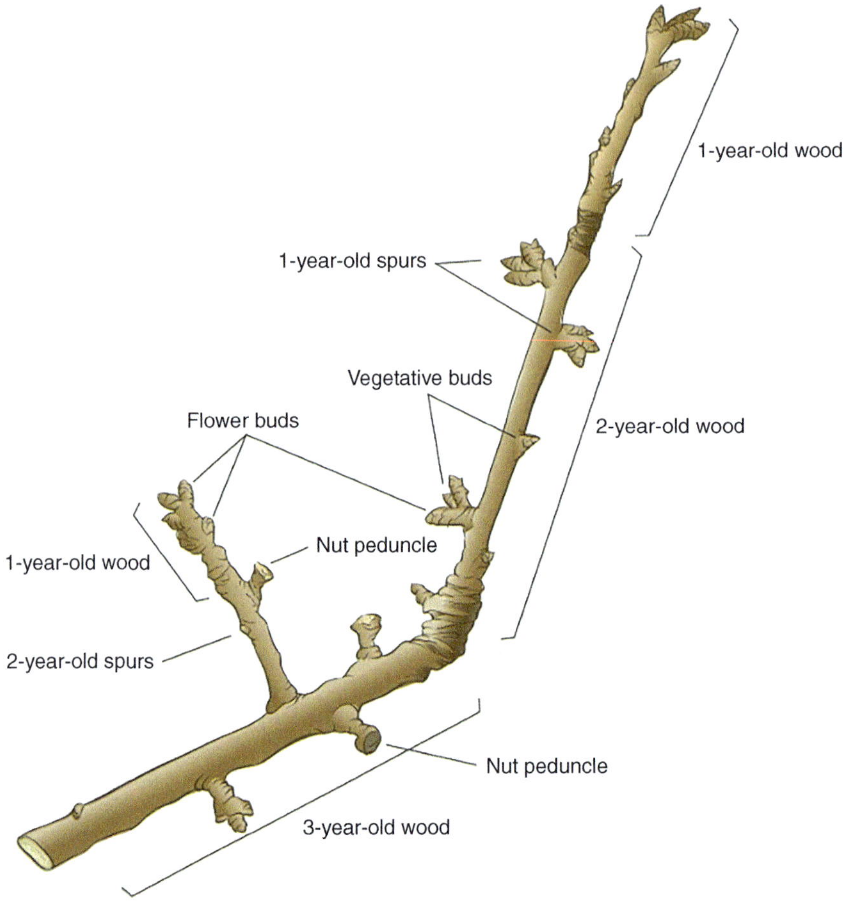

Fig. 9.1 Almond bearing habit. (Source: Tombesi et al. 2016)

1.3 Training

The newly planted trees should be clipped at a height of 36 in. above the ground. Plants should be staked and irrigated right away. When the trees begin to grow, a regular fertilization and irrigation programme is initiated. Almond trees are trained according to modified central leader system (Fig. 9.2). Other systems like cordon system, palmette system (Fig. 9.3), standard vase (vase shape with regular pruning) (Fig. 9.4), minimal vase (vase shape, but with very little pruning), minimal central leader, intensive central leader, hedge (minimal prune), hedge (heavier prune), etc. have also been used. The central leader system with the fewest prunes appears to be the most natural and balanced. The vase systems are outperformed by the single leader and hedge systems. The single leader and hedge systems would also make an over-the-row trunk shake and capture harvesting strategy more feasible. These

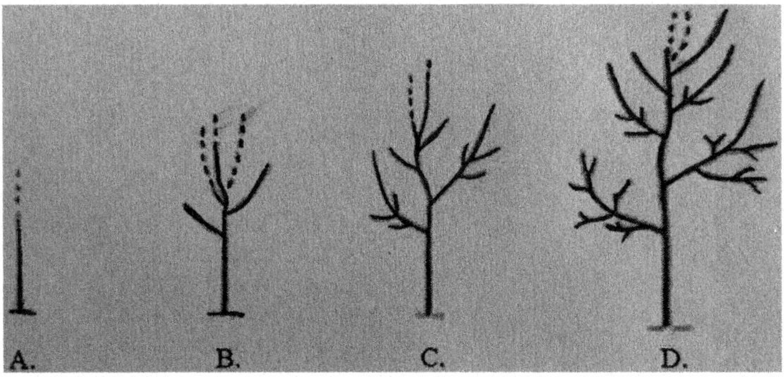

Fig. 9.2 Modified leader system. (**a**) Year 1 of training, (**b**) Year 2 of training, (**c**) Year 3 of training, (**d**) Year 4 of training. (Source: https://www.groworganic.com/)

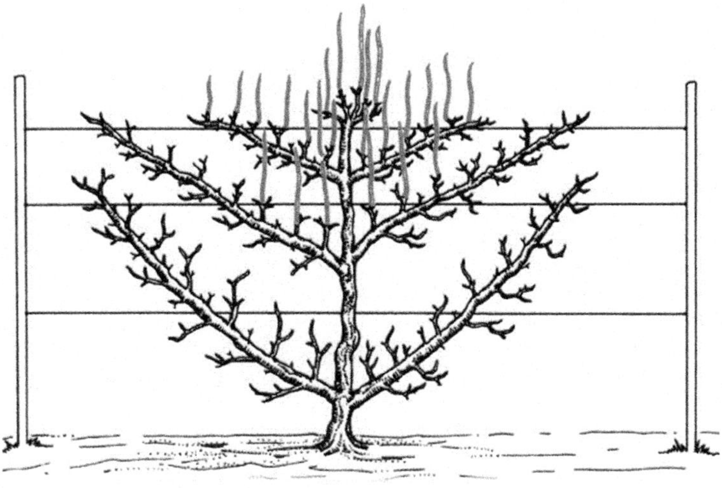

Fig. 9.3 Palmette system. (Source: https://catalog.extension.oregonstate.edu/)

training techniques differ in terms of initial investment, establishment and maintenance expenses, early versus mature yield potential, picking efficiencies, and so on. Figures 9.5, 9.6, and 9.7 show pruned vase and unpruned vase; minimal prune central leader and heavier hedge prune; minimal hedge prune and intensive prune central leader in an Australian almond orchard (Almond Report: Australia tour of Spain 2014).

A super high-density training system is being adopted in Spain. As a result of which in just a few years, the almond business in Spain has progressed from a marginal crop farmed in poor and arid soils to an alternative product to the conventional ones of deciduous fruit, cereals, and citrus. Figure 9.8 illustrates the lateral and frontal view of super-high-density orchard. Because of this shift in production management, the almond tree now has a 20–25% better profitability than cereals

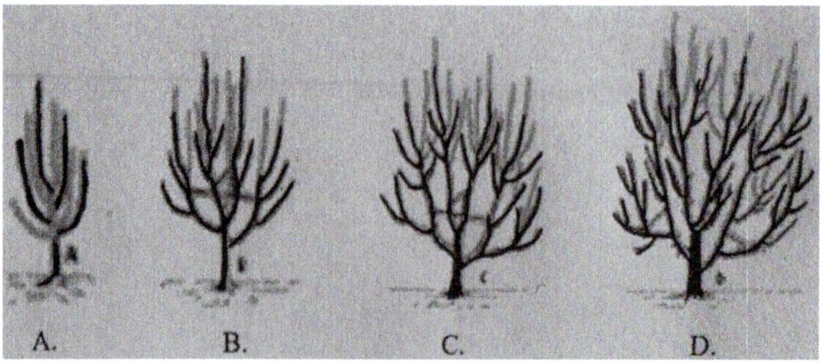

Fig. 9.4 Standard vase system. (**a**) Year 1 of training, (**b**) Year 2 of training, (**c**) Year 3 of training, (**d**) Year 4 of training. (Source: https://www.groworganic.com/)

Fig. 9.5 Pruned vase and unpruned vase. (Source: Almond Report: Australia tour of Spain 2014)

or olive trees on the same land. At the same time, almond consumption has climbed significantly, as global demand has grown at a pace of 5% per year (Casanova-Gascón et al. 2019).

1.4 Pruning

Bearing trees are typically trimmed once a year to foster new growth by removing 10–15% of the older, less productive wood. However, trimming is less detailed than with many other fruit species. In many circumstances, removing a few older branches with diameters of 2–4 cm will result in the desired quantity of regrowth following pruning (Micke and Kester 1997). Each system's pruning mechanism is unique. However, adjustments are required based on the growth tendencies of different cultivars. Pruning almond trees should resume in the second growing

Fig. 9.6 Minimal prune central leader and heavier hedge prune. (Source: Almond Report: Australia tour of Spain 2014)

Fig. 9.7 Minimal hedge prune and intensive prune central leader. (Source: Almond Report: Australia tour of Spain 2014)

season when the tree is dormant. The tree will most likely have multiple lateral branches at this point. Two laterals should be marked per branch to remain and become secondary scaffolds. A secondary scaffold will emerge from a primary scaffold limb in the shape of a 'Y'. Remove any lower branches that may interfere with irrigation or spraying. To allow more air and light penetration, prune any shoots or branches growing through the tree's centre. Excess water sprouts (sucker growth) should also be removed at this time. Also, while pruning second-year almond trees, eliminate narrow angled secondary branches. In the third and fourth years, the tree will have primary, secondary, and tertiaries that are permitted to live and grow. They make up the robust scaffold. In the third and fourth years, trees now have primaries

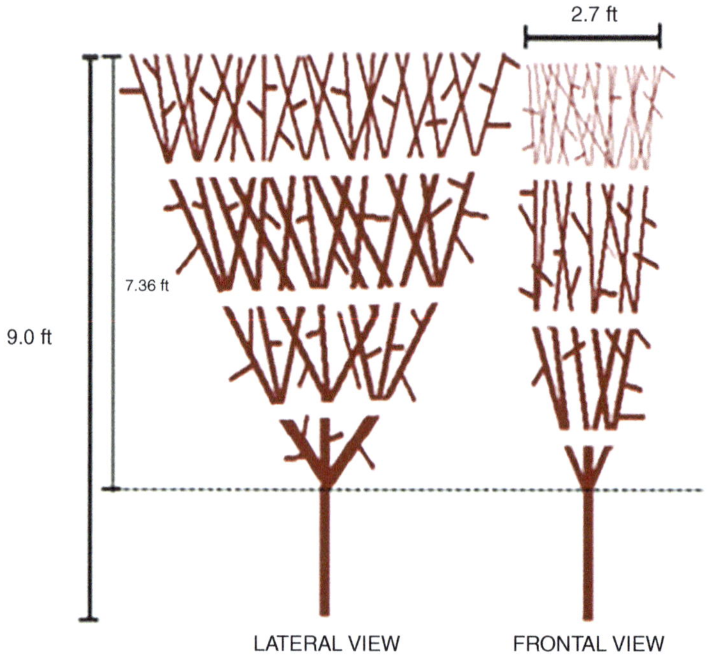

Fig. 9.8 Super high-density orchard. (Source: https://www.agromillora.com/)

that branch to secondaries and tertiaries. Allow these to develop. Implement the pruning tactics used in the second-year dormant pruning at a level or two higher in the tree. The goal is to fill the canopy's top edge while leaving a fairly open middle that allows sunlight to penetrate. Pruning is more about maintenance than it is about building structure or retarding size. This includes the removal of limbs that are damaged, dead, or diseased, as well as those that cross over the existing scaffolding to prevent air or light circulation through the canopy from being obstructed.

1.5 Use of Size Controlling Rootstocks

Almond size management with other *Prunus* sp. rootstocks has historically proved unsuccessful. Commercial almond rootstocks Root-PAC 40 and Root-PAC 20 rarely reduce scion vigour (Rubio-Cabetas 2016). Several diploid plum clones were chosen from the local 'Pollizo de Murcia' population, and three rootstocks were released: 'Adesoto' (Moreno et al. 1995) from EEAD-Zaragoza, and 'Monpol' and 'Montizo' from CITA-Zaragoza (Felipe et al. 1990, 1997a, b). The primary selection goal of these rootstocks was to introduce vigour lowering and waterlogging tolerance. It is recommended to employ selections that are currently in advanced commercial stages, such as 'Adesoto' and 'Montizo'. The trees are smaller, enabling for semi-intensive irrigated almond farming while yet providing adequate anchoring. Because of their low drought tolerance, plum rootstocks should only be used for irrigated crops (Rubio-Cabetas 2016). Other almond rootstocks under investigation include Replant-PAC (Pinochet 2010), Root-PAC-40, and Root-PAC-20 (Gasic and Preece 2014), which exhibit 40% and 20% less vigour than 'Garnem', respectively. Peach seedlings have recently been reported in the United States as almond rootstocks 'Controler' (Gasic and Preece 2014), generated from an open-pollinated *P. persica* HBOK series (Harrow Blood × Okinawa), both unpatented, with 60% vigour control compared to 'Nemaguard' (DeJong et al. 2004). The rootstock P.S.A5 is more dwarfing than Rubira and P.S.A6 (Rubio-Cabetas 2016).

2 Walnut

Juglans regia L., commonly called as the Persian or English walnut, is a member of the genus *Juglans* of the family Juglandaceae. It is native to Eurasia, ranging from the Near East through Central Asia, the Himalayas, and Western China. Juglans is a genus of roughly 20 deciduous trees in the Juglandaceae family. The eastern North American black walnut (*Juglans nigra*) and the English, or Persian walnut (*Juglans regia*), are both valuable timber trees that produce edible walnuts. The eastern North American butternut (*J. cinerea*) also produces an edible nutlike seed. The walnut tree attains a height of 12–18 m and has a broad spreading top and a thick, hefty stem.

2.1 Canopy Management in Walnut

Walnut trees have been genetically engineered to grow swiftly upwards and outwards in order to out-compete neighbouring plants for sunshine and survival. As a result, the tree effectively shades out competitors and then reproduces by blossoming and distributing seed, allowing a new generation to form. This pattern, while helpful in preserving the species' existence, is far from the perfect package for a nut orchard. Instead, we favour compact trees that begin nut production early and generate high yields without shadowing each other. There is a favourable relationship between sunlight interception in the orchard and walnut yield per hectare. The output of an umbrella-shaped walnut tree is restricted by the shade it casts on itself when left unpruned and separated from other trees. Light levels can fall to less than 30% at a distance of more than 1.5 m from the edge of the canopy, resulting in little or no nut output. The canopy volume to surface area ratio is also affected by the tree's shape. A well-planned orchard should be able to intercept 70–80% of the sunlight that falls on the ground. Because tree height influences sunlight interception, the maximum tree height in summer should not exceed 80% of the row width.

The various factors governing canopy management in walnut are as under:

- Training
- Pruning
- Use of sizc controlling rootstocks
- Use of growth regulator

2.2 Bearing Habit

In walnut, the fruits are borne on 1 year old wood which is produced either terminally or both terminally and laterally. On this basis, the cultivars are classified into two types: terminal bearer and lateral bearer. Annual shoot branching density and blooming buds' position are used to describe fruiting behaviour. There are three sorts of fruit-bearing habits, according to Germain (1990, 1992): terminal, intermediate, and lateral. Terminal fruit-bearing kinds have flower buds solely on the terminal or subterminal sections of annual shoots growing on 3-year-old stems. Intermediate bearers generate female flowers mostly on terminal and subterminal buds on yearly shoots when implanted on 2-year-old branches. Flowering buds can be seen along 1-year-old branches of lateral fruit-bearing varieties. The terminal and subterminal buds are induced by female flowering, as well as the majority of axillary buds, to form on the current growth stalk.

2.3 Training

In terminal bearing type of walnut (Hartley and Franquette), no heading back is carried out but thinning is done because of proliferative vegetative growth due to lack of early fruiting. Whereas, in the lateral bearing type of walnut (Vina and Chico), both heading back and thinning out are carried out to promote shoot growth which gets discouraged due to early fruiting (Rathore 1991). Walnut trees can be trained either individually on goblet or open centre, modified central leader system, or in hedgerow by oblique palmette, free vertical flattened axis, the sloping axis, etc. The modified central leader system outperforms the goblet system, but the free vertical axis is the best training system. The modified central leader and the free vertical system both allow for a 30% and 70% reduction in pruning time when compared to the goblet, respectively. The first yield appears substantially earlier on trees established in the free vertical axis system, and to a lesser extent on trees trained in the modified central leader system. The cumulative yields of free vertical axis and modified central leader trees are 46% and 10% greater, respectively, than open-centre trees. This advantage is maintained for up to 13 years after planting. The various types of training systems employed have little effect on fruit quality (Germain et al. 1995).

2.4 Pruning

Pruning is a dwarfing procedure, and the apparent vigour caused by harsh cutting of shoots is simply the tree trying to regrow what was cut off. Avoiding heading cuts (cutting shoots by 25–50%) and leaving the leaders alone are the quickest approach to grow a tree. Skilled canopy training is directing tree development into a canopy in order to produce a robust framework, optimize light interception, and maximize nut output. A suitable method to achieve this is by removal of unwanted growth by pinching out or pruning shoot tips before the tree expends a lot of energy and resources on unwanted growth. Bracing or tying the shoot at the correct angle is an efficient approach to teach the tree without sacrificing growth if the shoot location is correct but the angle is incorrect. Pruning of the lateral bearers consists of heading back numerous new shoots to decrease the fruiting and encourage shoot growth in the periphery of the tree while that of terminal bearers involve heading back of the main branches and thinning out the competition. In mature trees, there is a problem of shading which results in reduction of photosynthetic activity which weakens the growth of spur. Therefore, before overcrowding becomes a serious problem, selective branches should be thinned out in top side portions of the tree. No more than 25% of the branch should be removed so that sufficient leaf area is retained for food production (Stoke 1969). The pruning is carried out in the dormant season but early spring pruning is preferable. Delay of pruning operation beyond this results in excessive bleeding. Pruning a young-developing orchard on an annual or biennial basis is costly and time consuming. The use of a hedger instead of manual pruning allows for faster and less expensive pruning without affecting tree growth or yield. Various hedging strategies were tested in a 13-year-old mature walnut (Chandler) orchard for three years. According to recent study, some hedging techniques are more advantageous than others. Although not statistically significant, the hedging treatments are starting to outperform the non-hedged/non-trimmed treatment, which has not been pruned in 3 years (Olson et al. 2001) (Figs. 9.9 and 9.10).

2.5 Use of Size Controlling Rootstocks

In close plantings, reducing the size of *J. nigra* trees can result in higher yields per hectare and lower labour expenses. Size reduction can be accomplished by selecting smaller-sized trees or by using dwarfing rootstocks. *Juglans ailantifolia* interstocks have been shown to inhibit *J. regia* scion growth (Tourjee 1998). Traditionally, walnut had two rootstock options: Northern California black walnut (*J. hindsii*) seedling and paradox seedling. Three clonal paradox rootstocks have become available in recent years: Vlach, RX1, and VX211. RX1 has smaller trees and is less vigorous than VX211 (Buchner et al. 2012).

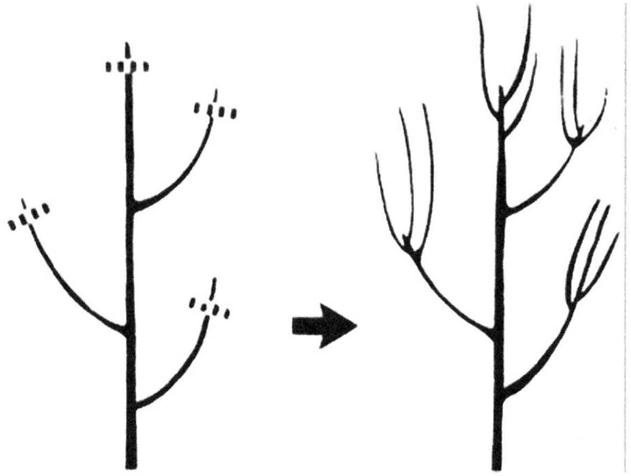

Fig. 9.9 Heading back in walnut. (Source: http://cecontracosta.ucanr.edu/)

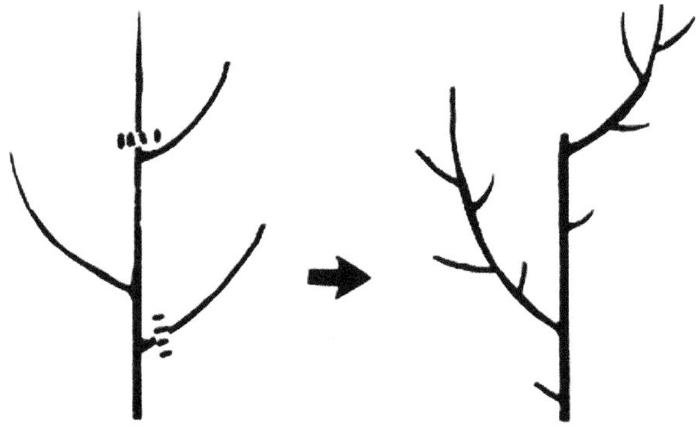

Fig. 9.10 Thinning out in walnut. (Source: http://cecontracosta.ucanr.edu/)

2.6 *Use of Growth Regulators*

Although producing walnuts requires fewer chemical compounds than growing other fruit species, advisers and producers have contemplated using some of these tools under particular conditions in order to improve nut yield and quality. During the winter, paclobutrazol is administered to the root zone of 'Serr'. Its application influences growth of shoot, bud break, time of flowering, and fruit load. Bud break in treated plants begins 5–7 days earlier than control trees. Male flower production is accelerated, and the lower pollen burden influences female flowering, resulting in increased fruit output. Depending on the dosage, length of shoot is drastically reduced per tree, and growth of shoot ceases early in the season. The leaves of treated trees become dark green and curlier. Sunlight can enter the canopy more easily. Following that, less wood is eliminated, less pruning is needed, and more fruiting sites emerge across the canopy. However, the effect of a treatment lasts for two seasons, and residues are low enough to encourage additional research into the benefits of using it on young vigorous unproductive trees or after harsh pruning

(Lemus 2009). The use of hydrogen cyanamide can improve earliness and bud set quality. When compared to control trees, cyanamide treatments resulted in faster and higher lateral bud break. The best effects were observed when the chemical was sprayed around the mid-July, consistently throughout numerous seasons (Lemus 2009). Thus, to prevent pistillate flowers abscission in low chill locations or when low chilling accumulation winters occur, 2% hydrogen cyanamide plus surfactant or mineral oil is used to improve and speed up bud break, accelerate productive shoot development, and diminish matching between male and female blooming (Lemus 2009). Amino ethoxyvinyl glycine (AVG) treatment at the start of flowering overcomes PFA and reliably enhances fruit set (Lemus 2009). Six cultivars, namely Franquette, Payne, Serr, Hartley, Ashley, and Tehama were given a foliar spray of dikegulac sodium @ 500, 1000, or 2000 ppm in May, June, and July, respectively, resulting in increase in tree compactness without affecting set or quality of nut (Martin et al. 1980).

3 Chestnut

Chestnut (*Castanea* sp.) a genus of seven deciduous trees in the family Fagaceae is native to temperate region of the Northern Hemisphere. The edible nuts are found in the bur-like fruits, and various species are grown as decorative and timber trees. When mature, most chestnut species are towering trees with wrinkled bark. The majority of male flowers are borne in tall upright catkins, whileas, female flowers are borne singly or in clusters at the foot of short male catkins. The spiny bur, depending on the species, surrounds one to seven nuts and splits when mature. The seeds lose viability quickly and usually germinate soon after falling to the ground in autumn. The fruit is surrounded with a spiny cupule that measures 5–10 cm in diameter. Burrs are found in clusters and can contain up to seven nuts depending on the species.

3.1 Canopy Management in Chestnut

As far the growing conditions allow, chestnut trees grow up to 45–120 cm or more in a year. There are dwarfing root stocks in fruiting trees. Because chestnut trees lack dwarfing root stocks, they will naturally want to grow tall and rapidly. Furthermore, chestnut trees, like other fruiting trees, will yield more nuts per tree or per acre if the plants are managed in ways that enhance nut production.

The various factors governing canopy management in chestnut are as under:

- Training
- Pruning
- Use of growth regulators

3.2 Bearing Habit

The chestnut fruit is produced terminally on spurs or short laterals. Chestnut is a monoecious tree that has male and female inflorescences on the same tree. During the late summer, flower buds form on stalk growth above the growing burs. New shoots sprout from these buds in the next spring, with catkins blooming midway along the shoot. Chestnuts have two kinds of catkins: those with just male, pollen-producing flowers (staminate catkins) and the other with both male and female inflorescence (bisexual catkins). The first few (basal) catkins are staminate catkins that produce pollen approximately 10 weeks following bud break. The last (most distal) catkin to form during the current season's development is bisexual, with one to three pistillate inflorescences at the basal end of a catkin. Each bur contains three ovaries (involucre); if all three ovaries are pollinated, then three nuts sprout in the bur.

3.3 Training

The central leader system is the most widely used training system in chestnut. A tree training system called the Eurovase system has been created in Europe. On peach trees, this system, also known as open-centre (Fig. 9.11) is used. The tree will grow to be quite short, with scaffold formed by only two or three branches. All the branches point in different directions that form the tree canopy. Given the association between sunshine on the canopy and nut yield, this training approach appears promising. Chestnut cultivars such as Colossal and Qing may function well with this strategy. The Y-top forming training system is comparable to the canopy system in trees, with a focus on sunshine exposure. The key distinction is that the Y-top employs four branches in a two-layer structure. Each branch covers around 90°. This system is more suited for chestnut since it spreads the load across more branches and allows for a larger vertical area, which chestnut trees prefer to develop naturally. Precoce Migoule, Okie, and Marival are some chestnut cultivars that may perform better with this training strategy.

The trees are trained for maximum light interception into the canopy. The open-centre and the Eurovase system both are known to serve this objective. In the beginning, the tree is allowed to grow up to a height of 1.5 m as a single stem from the ground level. The scaffold branches are maintained at suitable positions. The chestnut tree naturally attains a low headed framework if left unpruned and comes into bearing sooner. One of the practices followed is that the trees are left unpruned till they come into bearing and then afterwards only the lower most branches are removed every year till the tree is subsequently headed (Jindal and Karkara 1991).

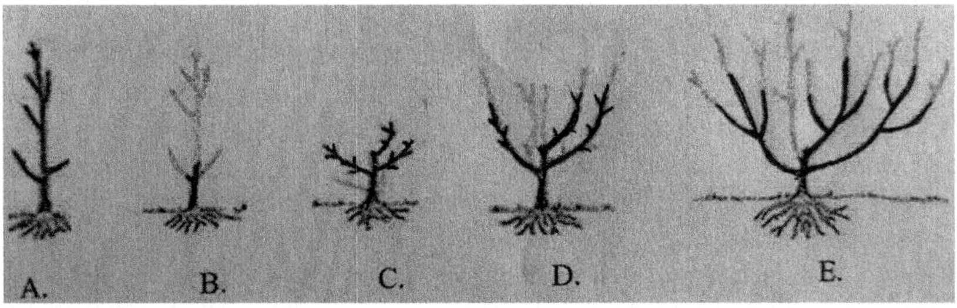

Fig. 9.11 Open centre or vase system of training. (**a**) Year 1 training. (**b**) Year 2 of training. (**c**) Year 3 of training. (**d**) Year 4 of training. (**e**) Year 5 of training. (Source: https://www.groworganic.com/)

3.4 *Pruning*

Winter pruning, also known as dormant pruning, is performed on chestnut trees in the middle of winter (late Dec to late Feb). Dormant pruning has no influence on the tree's vitality as the energy reserves are kept in the main trunk and roots. Summer pruning has an effect on tree vigour, so it is best to limit summer training/pruning to removing new growth. Summer trimming is done between the time the tree buds out up to the end of July. Any new growth after August 1st will not have enough time to harden off before winter. Except when branches break, no pruning should be done from August to December end. If the chestnut tree's structure is no longer balanced, adjustments to the tree's structure must be performed during dormant pruning. A balanced tree is less likely to be blown over in heavy winds. When planning to rebalance a chestnut tree, consider opening up the tree with fewer central branches. This has two effects on the tree. The first being, it exposes the tree to more sunshine, which helps to keep it healthy. The second advantage of open system is that it has a lower wind profile, which reduces the likelihood of the wind pushing the tree down. After about 3–5 years of training, thinning operations are carried out in which limbs are selectively removed from the canopy's perimeter, especially those that are growing close together or beyond the ideal canopy size. Branches with narrow attachment angles should also be removed. Branches should be cut back to their point of emergence or to the laterals that approximately have one-third the diameter of the limb being removed. Trees with thin bark are susceptible to sunscald. At no time should one remove more than 30% of the overall foliage.

3.4.1 Japanese Pruning

Dr. Hitoshi Araki, Senior Researcher at Hyogo Research Institute, Japan discovered that chestnuts require 35% relative sun radiation only to bear fruit. This is illustrated in Fig. 9.12. Dr. Araki endeavoured to design pruning procedures that would alleviate this strain, as well as to demonstrate the effectiveness of pruning in an orchard linked to his Research Institute (Hall 2000).

Dr. Araki implemented two methods to make trimming easier: limiting trees to a maximum height of 3.5–4 m and reducing the space between the canopy's edge and its centre. He accomplished this by using an elliptical tree design. Figure 9.13 depicts the shift in tree shape. Trees in alternate rows are clipped to achieve an elliptical shape, while those in between have at least a 2 m gap between tree canopies. After 5 or 6 years, the intermediate trees are removed, leaving elliptical trees spaced 4 × 8 m apart. Pruning to obtain the shape proposed by Dr. Araki should begin in the first year and continue on a regular basis throughout the tree's life. The initial goal should be to promote a leader and two primary branches that are strategically placed. Figure 9.14 illustrates dehorning of the main leader and the gap filled by dehorning of the main branch. The goal in year 3 should be to encourage the tree to expand out (Fig. 9.15). This entails thinning outwardly developing branches. Leave decent limbs if they're pointing in the proper direction, but be prepared to cut them off if they're not. The long-term shape of the tree should be the most important concern during pruning. Only two good scaffold branches are required for an elliptical tree. It is better to leave the central leader in place as these develop, but this can be replaced later if necessary (Hall 2000).

It is preferable not to thin out older trees all at once. Cutting back gradually allows the tree to continue producing without interruption. Cutting back too much can be counterproductive since it leaves too little growth for blossoming. Once the trees are under control, annual reduction pruning helps to preserve their productivity by allowing light to enter the canopy. You should aim for a fifth to a third of last year's growth—a half is too high. It is critical to keep adequate space between trees in order to allow for adequate light penetration. Figure 9.16 shows how this can be accomplished in a heavily planted orchard that shall be thinned later.

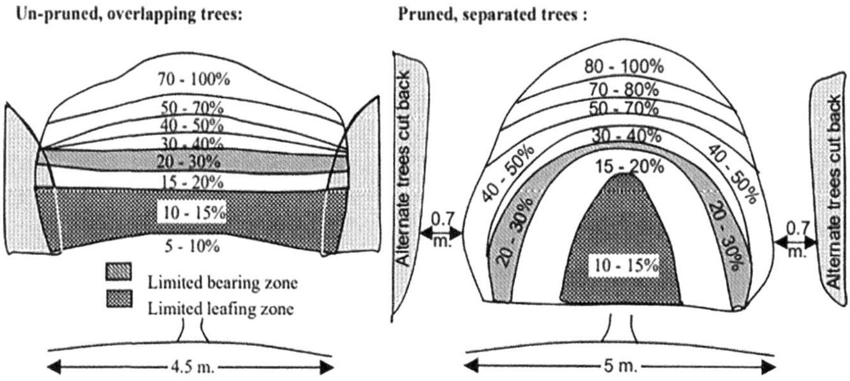

Fig. 9.12 Measurements of relative solar radiation within the tree canopy. (Source: The Western Chestnut Growers Assn., Inc. Spring 2000)

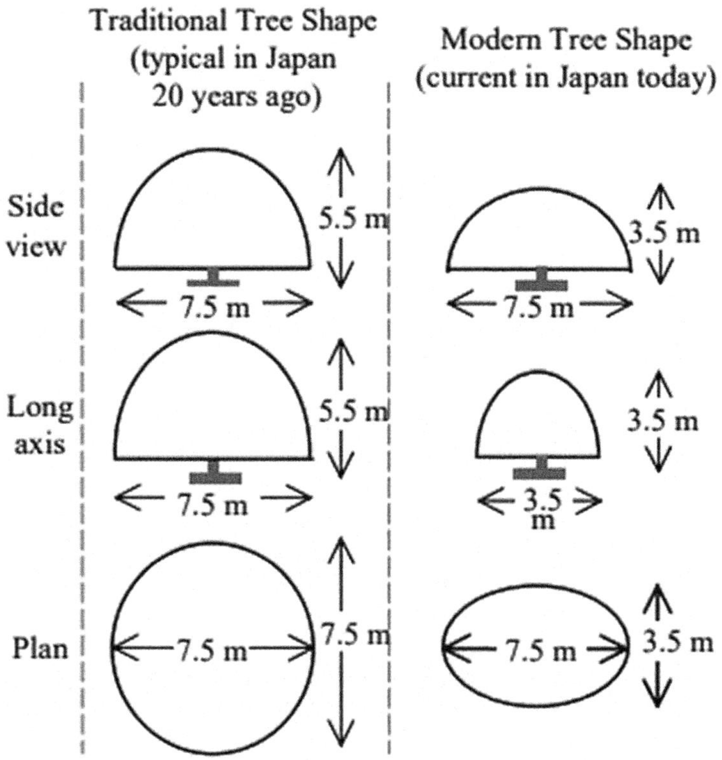

Fig. 9.13 Re-shaping of chestnut trees in Japan. (Source: The Western Chestnut Growers Assn., Inc. Spring 2000)

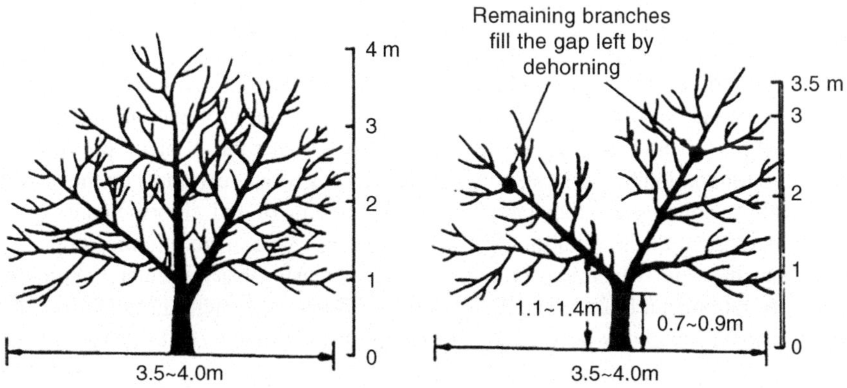

Fig. 9.14 De-horning the central leader. (Source: The Western Chestnut Growers Assn., Inc. Spring 2000)

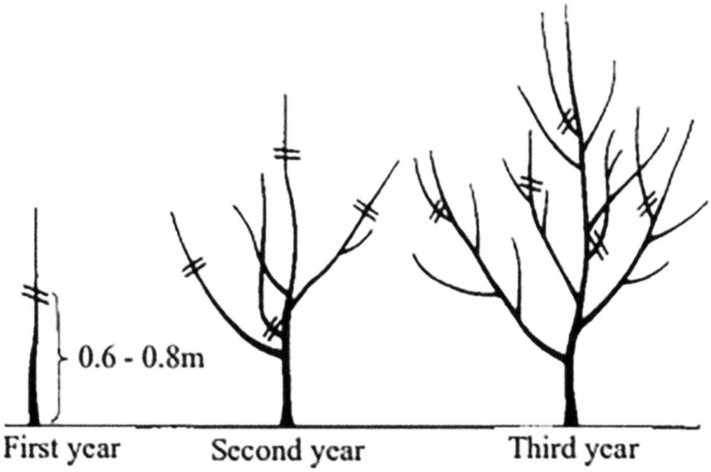

Fig. 9.15 Pruning a young tree in three successive years. (Source: The Western Chestnut Growers Assn., Inc. Spring 2000)

3.4.2 Pruning of Senescent Orchards

When confronted with trees up to 10 m tall, Dr. Araki's method was to severely prune them to a height of roughly 1.5 m so that further trimming of regrowth could keep the trees at a reasonable height of 3.5–4 m. Staged cutback, as shown in Fig. 9.17, may be appropriate for older trees that still retain lower limbs.

3.5 Use of Growth Regulators

The effects of paclobutrazol (PP333) and chlormequat (CCC) on the growth of chestnut young branches and leaves, as well as carbon-nitrogen metabolism of leaves, were investigated by Zhang et al. (2019), with the goal of providing a scientific foundation for understanding the effects of growth retardants on the growth of chestnut saplings and their application. Paclobutrazol and chlormequat significantly decreased the longitudinal growth of young chestnut tree mother branches, fruit branches, and vegetative branches while increasing basal diameter. Spraying chlormequat on fruit branches had the greatest influence on length growth. Spraying chlormequat also greatly boosted the growth of the base diameter of the fruit branch.

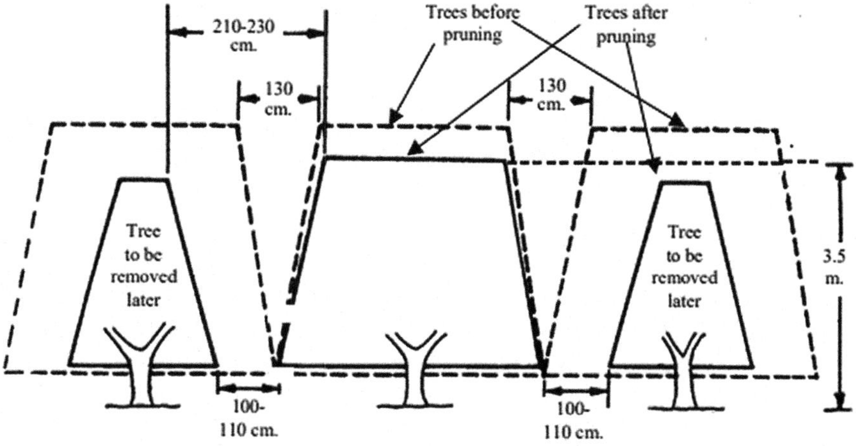

Fig. 9.16 Pruning to achieve inter-tree spacing. (Source: The Western Chestnut Growers Assn., Inc. Spring 2000)

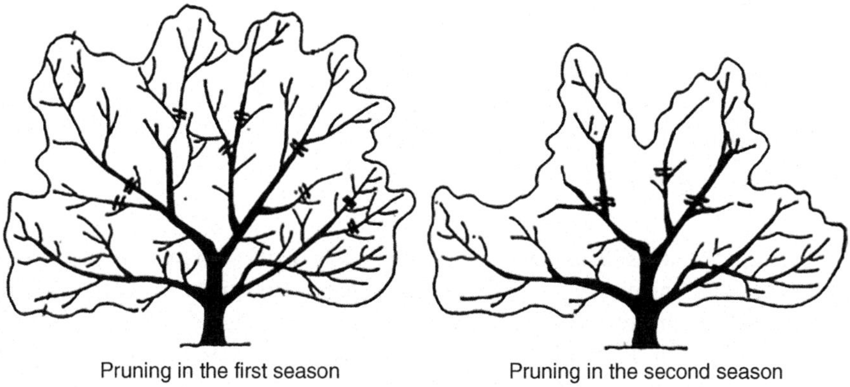

Fig. 9.17 Progressive pruning of senescent trees. (Source: The Western Chestnut Growers Assn., Inc. Spring 2000)

4 Pecan Nut

Pecan (*Carya illinoinensis*), tree of the Juglandaceae family indigenous to temperate North America. *Carya*, sometimes called hickory, is a commercially important genus in the walnut family, Juglandaceae, with around 18 deciduous tree species suitable to temperate and subtropical parts of the Northern Hemisphere. On occasion, it grows up to a height of about 50 m and a trunk diameter of 2 m. The male flowers are called catkins, while the female flowers form compact clusters at the terminals of the branches. When the fleshy hulls of the short-clustered fruits dry, break along suture lines, and divide into four roughly equal parts, gradually releasing the nuts. Their shape ranges from long and cylindrical with a sharp tip to short and roundish.

4.1 *Canopy Management in Pecan Nut*

If tree canopies invade and generate excessive orchard shadowing, the profitability of pecan orchard companies would eventually collapse. This worsening of the canopy light environment, as well as the associated 'sunlight stress', often increases alternate bearing intensity (Pearce and Dobersek-Urbanc 1967), which is likely the most economically significant biological challenge confronting commercial pecan companies. In a pecan orchard, production is directly proportional to light interception (Loomis and Gerakis 1975). Tree growth and invasion necessitate two or three separate temporal phases of tree orchard thinning by removal—leaving few trees per unit area and excessive inter-tree spacing for much of the orchard's existence.

The various factors governing canopy management in pecan nut are as under:

- Training
- Pruning
- Use of growth regulators

4.2 *Bearing Habit*

The staminate and pistillate flowers borne separately on the same tree and organized into catkins and spikes, respectively. The pecan has compound buds enclosing floral and mixed buds in separate bud scales but within a common outer scale covering (Sparks 1986). On developing, floral buds produce catkins but mixed buds grow either into a vegetative shoot or in a single pistillate inflorescence. The catkins are borne on current season's shoot and not on 1-year-old wood (Wetzstein and Sparks 1984).

4.3 *Training*

The initial 5 years of a pecan tree's life are crucial for the development of the canopy framework. To create tree which is medium-sized, durable, and wind-resistant tree, use a central leader or modified central leader system with well-spaced and broadly slanted scaffolds growing spirally (Heerema 2015). Heading back, branch selection, tip trimming, and pinching are all effective strategies for establishing new pecan trees and ensuring adequate training throughout the first few years.

Remove the top one-third to half of the previous season's growth when planting dormant nursery trees. This normally yields a whip that stands 96–105 cm tall. This type of pruning promotes vigorous, upright branch growth right below the head-back point the following spring, from which a allowing strong central leader is chosen a year later. The strongest, most robust branch should be picked to continue the central leader development in each dormant season during the first 4 or 5 years following planting. It is critical but challenging to train a young grafted pecan tree to a central leader. Young pecan trees are difficult to train due to their slow growth. If the tree is not healthy and growing quickly, it will not react to training. Training and removing shoots from slow-growing trees can weaken the tree and cause more harm than good (McEachern and Stein 1986).

To train the tree as per central leader system of training, give a heading cut to prune 1/3 to 1/2 of the selected central leader shoot in the first year. Choose one sprout at the top of the tree to be the central leader in May or June. This should be the strongest and fastest growing sprout. Remove any more shoots that are within 15 cm of the cut back spot. This promotes the maximal growth of the chosen central leader shoot. Allow side shoots on the trunk's lowest half to form the traditional trashy trunk. In the first dormant season, any 30–45 cm long side shoots or branches should be pruned at the tip by removing or pinching the branch's growth point. Figure 9.18 illustrates the cut back and training in the first year of training (McEachern and Stein 1986).

In the second year of tree training, select and leave the strongest shoot at the cut backpoint in May or June after growth has begun. This chosen shoot will be the central leader. Remove all other shoots that are forming at the clipped back location. During the second dormant season, any 30–80 cm long side shoots is to be tip pruned. Figure 9.19 shows the pruning cuts to be made during the second year of tree training. In the third year of tree training, select the final central leader shot at the top of the tree at the cut back point in May or June, as rapid growth begins. Remove all other shoots from the tree's top 12 in. This only allows for one central leader. The tree should have only one central leader and a huge trashy trunk during the first 4 years. Never cut back side shoots more than one-third of the way and leave them till they are 1 in. in diameter. Figures 9.20 and 9.21 illustrate the cut backs that should be made in the third and fourth year of training. In the third dormant season, the side branches with a diameter of 2.5 cm and which are less than 1.2 m above the ground should be totally removed. Prune smaller side shoots and branches at the tip. Similar pruning is done in the fourth dormant season. If a tree has two upright trunks (leaders), it is critical to remove one of the two trunks as early as possible in the tree's existence.

There are up to three buds at each node along the central leader: primary, secondary, and tertiary buds. The primary bud being the largest amongst the buds is easy to spot. The primary, secondary, and tertiary buds shrink in size and have wider angles relative to the shoot axis, respectively. When the primary bud is permitted to grow, the secondary and tertiary buds normally do not. Primary buds typically form narrower crotch angles in their shoots than secondary or tertiary buds, resulting in weaker scaffolds. Because primary buds have a natural predisposition to

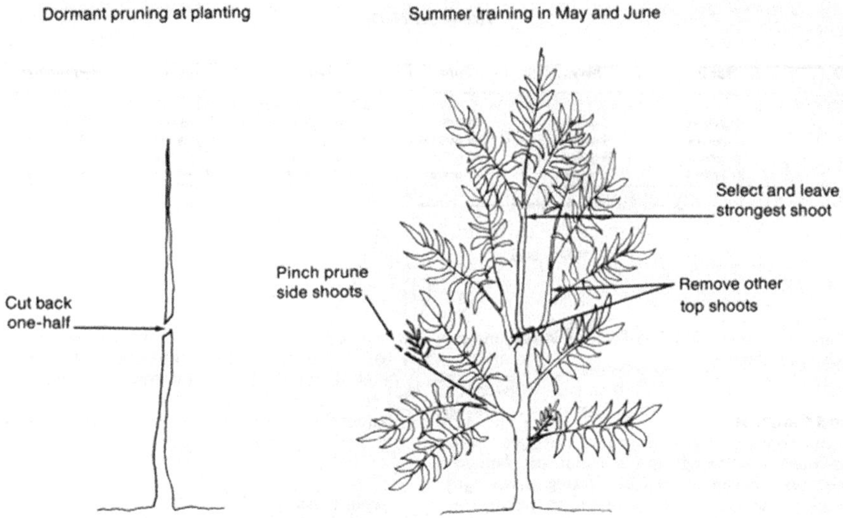

Fig. 9.18 Cut back and select pecan tree training the first year. (Source: McEachern and Stein 1986)

Fig. 9.19 Cut back and select pecan tree training the second year. (Source: McEachern and Stein 1986)

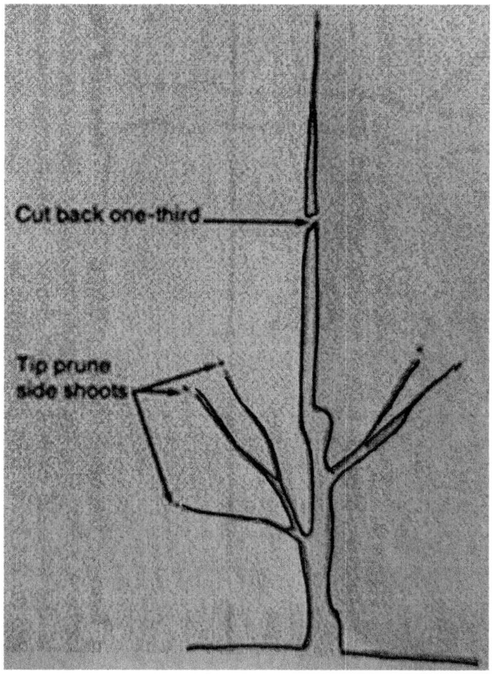

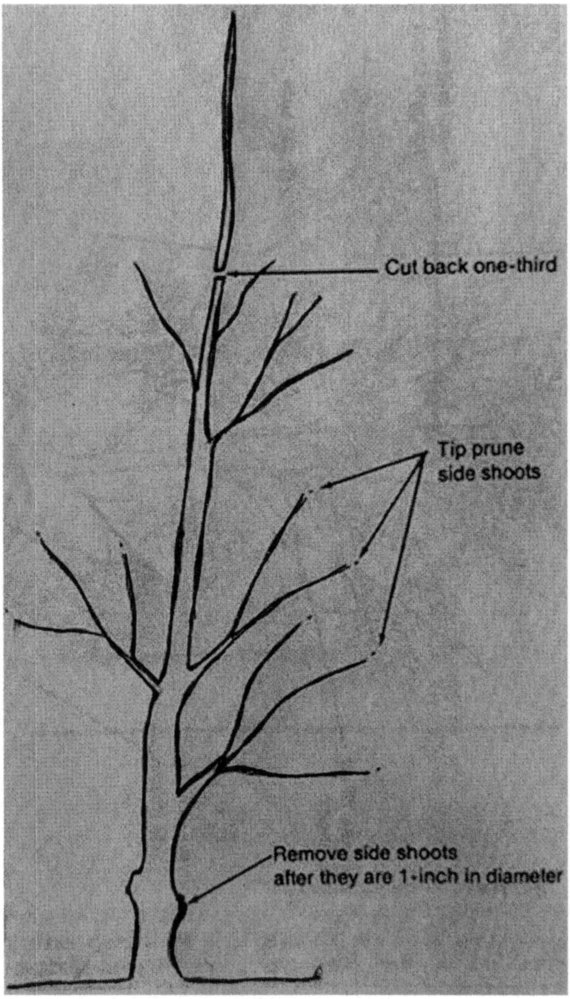

Fig. 9.20 Cut back and select dormant pruning in January and February of pecan trees in the third and fourth years

promote straight, upright growth, the central leader should always come from one. The dormant primary buds along the central leader, on the other hand, may be pinched off to force the secondary buds to form stronger, more widely angled shoots. This may make it easier to choose appropriate scaffold branches in the future.

4.4 Pruning

The training and pruning type and intensity of that should be performed are heavily influenced by the tree's intended use. The bearing habit of a pecan tree is that it is a terminal bearer. So only light or moderate pruning is required. If the trees are pruned

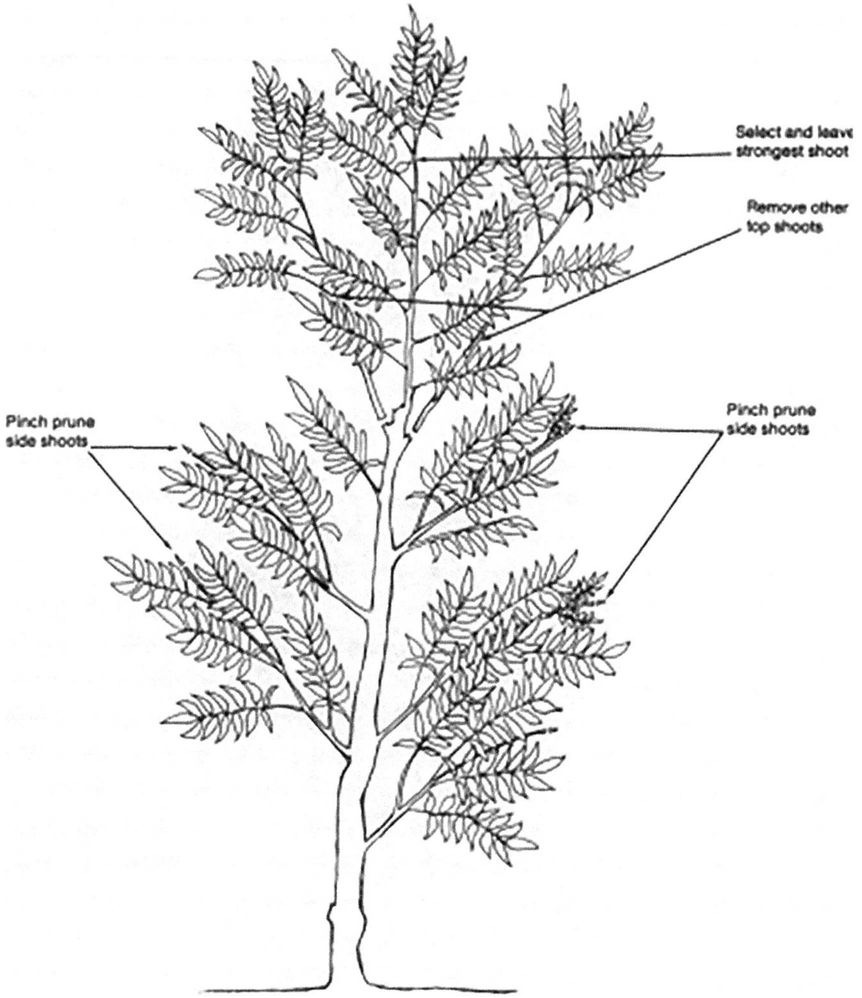

Fig. 9.21 Cut back and select summer training of pecan tree in May and June of the third and fourth years. (Source: McEachern and Stein 1986)

heavily, it results in excessive vigorous growth leading to no production for several years (Hanna 1987). Pruning mature trees is costly, but it is required. The majority of pruning procedures are performed to maintain tree size and prevent shading. Mature trees are pruned to reduce shadowing, correct limb angles, remove dead limbs, and lower branches to make room for orchard cultural operations. Some of the hedging strategies are as follows:

4.4.1 Hedging Using Mechanical Means

Hedging is a short-term treatment that must be repeated after three or four seasons, depending on the pruning strategy used. Hedging yield increases may become smaller with each subsequent hedging operation. Light pruning is preferable to heavy pruning on an annual basis (but only on branches 1″ to 1.5″ in diameter). Light mechanical pruning is also advised in the winter prior to the season when a heavy yield is expected (White 2000).

4.4.2 Mature Tree Cut Back

In cases where trees were planted too close together and allowed to grow together and crowd, gardeners may try to fix the situation by harshly pruning the trees. This is frequently referred to as dehorning or pollarding. This pruning method is ineffective in restoring the trees to full output. Trees are either severely or moderately dehorned (White 2000).

4.4.3 Pruning for Correction

Narrow crotches should be avoided at all costs. To improve sunlight penetration, whole branches should be frequently removed from the heart of the tree. Branches that are in each other's path should also be removed on a regular basis to prevent limb darkening. Suckers should be cut to the nearest crotch to prevent regrowth. This approach will postpone orchard crowding, but it will eventually be essential to thin the orchard or begin mechanical hedging. Lower scaffold limbs that hinder the tree shaker from being clamped during harvest must be removed. Limbs that are low enough due to the tree's heavy crop to be run over or damaged by tractor wheels should be removed. Although dead wood can be removed, most farmers just shake it away at harvest (White 2000).

4.5 Growth Regulators

Controlling tree growth is a key issue in high-density pecan orchards due to the strong vegetative tendency of relatively young pecan trees and the lack of dwarfing rootstocks or cultivars. Paclobutrazol, an efficient inhibitor of gibberellin production, outperforms standard and typically ineffective pruning or hedging approaches for controlling pecan tree size (Sparks 1979). However, cultivar responses vary based on tree size and age, as well as the balance of vegetative and reproductive growth (Gash and David 1989; Wood 1988; Worley et al. 1996).

Shoots can be transformed into fruiting shoots by reducing their vegetative growth. Terminal growth was slowed following a single paclobutrazol (PBZ) therapy for thc first year. The growth regulator PBZ also enhanced the proportion of shoots that were short or extremely short (Zhu and Stafne 2019).

With the arrival of gibberellin biosynthesis inhibitors, the potential to limit vegetative growth of pecan was re-examined. Paclobutrazol administered as a soil drench significantly restricted growth of both 3 and 10 year old pecan trees for 3–4 years after treatment (Wood 1988). There appears to be a cultivar reaction as well as a response differential based on tree age. In another study, three gibberellin biosynthesis inhibitors were tested for their relative effectiveness in regulating vegetative development, and the order of efficacy was discovered to be uniconazole > paclobutrazol > flurprimidol (Davis et al. 1991).

5 Hazelnut

Hazelnut (*Corylus avellana*) is often known as filbert, cobnut, or hazel. It belongs to the family Betulaceae. The plants are found in the northern temperate zone. The European filbert, or common hazel (*Corylus avellana*), produces an oil that is used in food goods, perfumes, and soaps; the tree also produces a reddish white soft timber.

Hazelnuts are deciduous trees. Monoecious blooms (male and female flowers on the same plant) appear on bare branches from late winter to early spring (March–April) before leaf emergence. Male flowers are somewhat spectacular, pale yellow-grey, and appear in sessile drooping catkins (each to 5–7.5 cm long). Just above the male catkins, inconspicuous female flowers with red stigmas bloom. The medium green leaves (up to 10 cm long) are double serrate, elliptic to ovate to orbicular, round to cordate at the base, and generally hairy. The fruit consists of a hard edible brown nut (up to 7.5–10 cm long) surrounded in a leafy, hairy, light green husk. Nuts appear in clusters of 1–4 at terminal ends and are wrapped in ragged husks on half of their length. The husk (involucral tube) that surrounds the nut extends at least 2.5 cm beyond the nut to form a beak. In late August and September, nuts ripen.

5.1 Canopy Management in Hazelnut

The critical factor in ensuring success for more constant annual production in a hazelnut crop is light management. Reduced light levels caused by canopy shadowing decrease yield, nut quality, and female flower and catkin density the following year (Azarenko et al. 1997). Inadequate light infiltration into the canopy could be one of the causes of low productivity. In the dense canopies of hazelnut orchards, photosynthetically active radiation (PAR) attenuates rapidly with depth. In most orchards, limited light entry lowers floral bud growth, fruit set, and fruit quality (Jackson and Palmer 1977; Snelgar et al. 1992).

The various factors governing canopy management in hazelnut are as under:

- Training
- Pruning

5.2 *Training*

Hazelnut orchards are established in Piedmont (Northwestern Italy) using various training systems. The usage of 'bush' with light pruning is traditional in the original areas. New training systems such as the 'double hedgerow' (Fig. 9.22) and the 'free vase' were compared and found that the double hedge system provided higher yields per tree and per hectare (Me et al. 2001). When the monocone, ipsilon, free vase, and hedgerow systems were compared in Spain (Figs. 9.23 and 9.24), the hedgerow system generated the most suckers but cropped the earliest (Tous et al. 1994). Hedge training technologies appear to provide good yields and crop early. Other countries have devised new training systems to boost output and better accommodate cultural and technical norms. The vase with a single trunk is a training device extensively employed the United States and France. Sucker management, cultural practises, and mechanical harvesting can all be carried out efficiently in this system (Lagerstedt and Painter 1973; Olsen 2002). In France, yields per hectare did not differ between the open centre vase, vertical axis, and V-hedge systems (Fig. 9.25), with the exception of the V hedge, which provided better yields in the initial years after planting (Germain and Sarraquigne 1997). When compared to the traditional bush system planted at high densities, the V-hedge system performed well in Italy when managed without irrigation, particularly for early cropping (Romisondo et al. 1984).

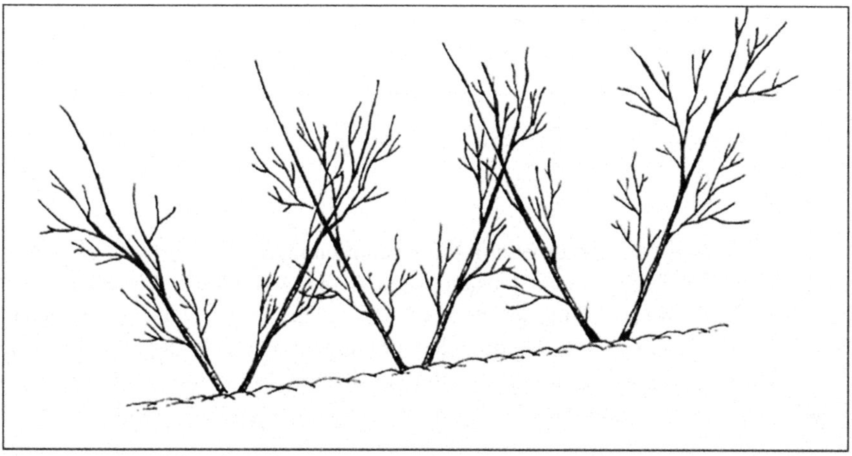

Fig. 9.22 Double hedge or hedgerow system. (Source: Me et al. 2001)

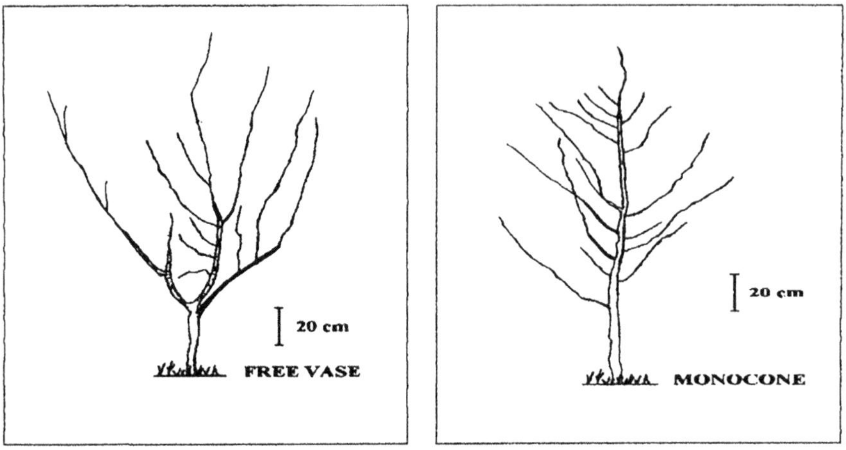

Fig. 9.23 Comparison between free vase and monocone training systems in hazelnut. (Source: Tous et al. 1994)

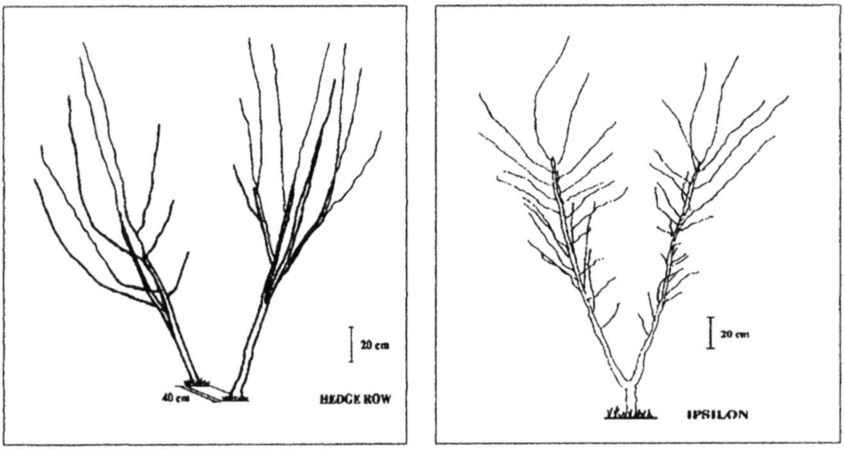

Fig. 9.24 Comparison between hedge row and ipsilon training systems in hazelnut. (Source: Tous et al. 1994)

The 'tree shape' exhibited better light penetration than the other two forms, most likely due to the higher plant density, whereas the 'free-vase' system was more shadowed. Nonetheless, 'bush' systems are closer to the spontaneous growth behaviour of *Corylus avellana* and so require less pruning than 'tree-shape' and 'free vase' systems, which require more precise pruning. In fact, trees with single trunk age faster, particularly in cultivars with low or medium tree vigour (Valentini et al. 2008).

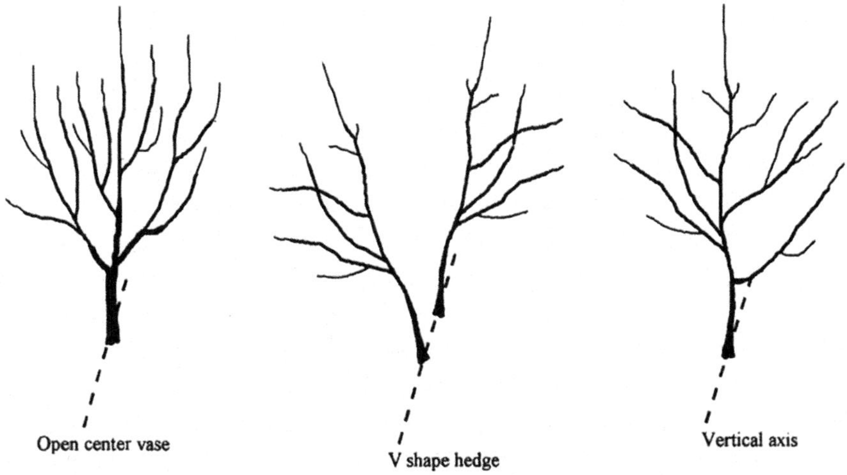

Fig. 9.25 Comparison between open centre, V-shape hedge and vertical axis training systems in hazelnut. (Source: Germain and Sarraquigne 1997)

5.3 Pruning

The state of the tree is the best predictor of the need for pruning. Remove around half of the fruiting wood from each pruned tree. Catkins indicate the location of the fruiting wood in a tree. Remove as much of the moss-covered, poorly growing wood as possible while leaving broad scaffold branches on which to develop new fruiting wood. Shorten the low laterals and remove the middle branches. This makes moving equipment in the orchard easier. Suckers that grow up through the canopy should be removed unless they can be used to fill a gap in the tree's canopy. In general, thin the tree so that sunlight can reach all areas of it.

Many mature hazelnut orchards require height reduction. Overgrown orchards create several difficulties. It is difficult to attain complete spray coverage at the treetops. Too-tall trees also shade off much of the sides of the trees, limiting the crop to the tops of the trees. The top third of the tree produces all of the nuts, while the other two-thirds serve only as a support framework. This is not a particularly efficient tree shape. The tree's centre third should also be fruitful. The bottom part of an orchard will always just support wood in older orchards.

6 Pistachio Nut

Pistachio nut *Pistacia vera* L. belongs to the family Anacardiaceae. It is the only economically edible nut among the 11 species of genus *Pistacia*, all of which exude turpentine or mastic. The pistachio tree originated in Asia Minor (now Turkey), Iran,

Syria, Lebanon, the Caucasus region of southern Russia, and Afghanistan. The pistachio tree is dioecious, meaning that male and female blossoms grow on different trees. The pistachio tree has 13 principal branches, each with one terminal blossom and 5–19 lateral flowers. The flowers have up to five sepals and are apetalous. Male flowers feature five tiny stamens, whereas females have a single superior tricarpellate ovary. On 1-year-old wood, pistachios grow in grape-like clusters. Pistachio trees are small to medium in size and can grow to a height of 12 m, but they are often smaller throughout the cultivation phase (Kashaninejad and Tabil 2011).

6.1 Canopy Management in Pistachio Nut

Pistachios have significant apical dominance in their vegetative growth habit, which means that vigorous current season shoots emerge only from buds closest to the terminal of 1-year-old branches. Because of the terminal bud's hormonal regulation, vegetative buds further down the branch are capable of growing long, robust shoots, but only transient spur growth. This characteristic becomes more obvious as the tree ages. As a result, branches can lengthen for years without lateral branching, causing the tree to produce fruit further distant from its central axis. Apical dominance in pistachio promotes tree crowding, loss of tree structure, and hinders bloom bud formation on lower fruit wood due to a lack of light. The eventual reduction in growth on the canopy's sides encourages further growth at the tree's top, where light is not a constraint. Despite its prolific growth, the upper canopy eventually outgrows the reach of hand pruning crews. As a result, one of the key purposes of trimming bearing pistachios is to confine the trees to their allotted space while still boosting light penetration for viable fruit wood and nut production across the tree.

The various factors governing canopy management in pistachio are as under:

- Training
- Pruning

6.2 Bearing Habit

Pistachios grow fruit laterally on 1-year-old wood. Botanically, an individual flower bud (inflorescence) is a panicle with an extended central axis and lateral branches that yield up to 200 individual flowers. The structure that supports the blossoms is referred to as the 'rachis' by growers, while the phrase 'cluster' refers to the rachis and the nuts it finally supports. Dormant, 1-year-old fruit wood is represented by both preformed (spur) and neoformed (whip) development. Spurs can form on 1-year-old structural branches with a large diameter in young trees where apical dominance is not as strong. Spurs are frequently generated in the first flush (preformed growth) of whips if the buds did not differentiate into flower buds the previous season. Flower buds are frequently generated in the preformed growth of long, whip-like growth (first flush). If a 1-year-old whip is left untipped, the lateral buds are all capable of producing spur growth, which could lead to the development of fruit buds in the following year (Beede et al. 2005).

6.3 Training

A trimmed tree may have the ideal shape, but it will always grow at a slower rate than an unpruned tree. The most stimulating training is during the dormant season (winter). This is a critical concept for pistachios because they develop slowly and require dormant pruning to stimulate growth in specific branches. The most dwarfing effect on trees is caused by in-season training (tipping). This is due to the removal of segments of the tree that the tree expended energy to produce.

In-season tipping, also known as summer pruning, reduces the number of leaves and consequently the amount of growth chemicals produced by photosynthesis. As a result, producers should not put off in-season tipping when training young trees because it removes excess shoot growth and foliage. In-season branching tipping should only remove the amount of shoot growth that can be physically pinched off. Avoid removing more than 6 in. (15 cm) from a single shoot. Heading cuts (removal of only a portion of a limb) stimulate more vigour and rapid growth than thinning cuts (removal of an entire limb at its origin). On young trees, heading cuts are most commonly used to guide growth, force branching, and produce long shoots with rapid growth. Thinning cuts are used to manage tree shape, eliminate undesirable limbs, and maintain the desired level of tree vigour. Thinning cuts also remove fewer buds, resulting in a less stimulating influence on subsequent growth (Beede et al. 2005).

The first dormant season generally consists of reducing rootstock growth from successfully budded trees and selectively removing dormant shoots from unbudded trees. Shoots higher than 1 m are pruned at 1.1 m to force the growth of the major scaffolds. Dormant heading has been shown to force more laterals on pistachios than in-season heading. The main shoot hooked up the stake in the previous season can be headed at 90 cm in the spring to force laterals. This is known as scaffold budding, and it helps to improve orchard uniformity. During the second growing season, if the trees have developed sufficiently by late June to mid-July, they are trained up the stake and topped at 1.1 m (Beede et al. 2005).

During the second dormant season, secondary branches on trees are trimmed to 2–3 branches originating from each primary. It is better to have branches spread 8–10 cm down the primary, rather than retaining limbs exactly opposite one another. Following selection, each is led to a length of 28–33 cm and, if necessary, tied up. Trees are thinned to three or four well-distributed scaffolds before being headed to 28–33 cm and knotted to produce an upright vase shape (Beede et al. 2005).

During the third dormant season, pruning trees with developed tertiary branches is similar to pruning trees with established secondary branches. Two to three branches are chosen for placement and strength per secondary and directed to 46–61 cm. Long shoots with a considerable girth are usually cut in half, and minor limb removal is done in the middle of thick trees (Beede et al. 2005).

6.4 Pruning

It is critical that the grower understands what type of pruning is desired and is able to communicate the objectives clearly and simply. The desired goal can be achieved by dividing the instructions into four steps. The first step is to eliminate any branches that are damaged, excessively low, cross over the centre, or overlap one another. These are the obvious cuts that can be made by less skilled team members. Second, thinning cuts must be performed near the canopy's edge with the goal of 'pushing' the bearing branches upright. This is performed by removing branches with an angle of less than 45°. Branches must be pruned so that those that remain point upward rather than out. Because apical dominance causes most new growth to occur on the canopy edge, few thinning cuts are required in the core of pistachios. The next stage is to execute selective heading cuts in weak spots closer to the tree axis. These should promote dormant buds to sprout into new shoots that can be used to regenerate fruitwood. The final stage is to tip the 1-year-old whips produced mostly at the canopy's top with pole pruners. Shorter growth can also be tipped elsewhere in order to generate as many fresh spur growths as possible. It is critical to distinguish between floral and vegetative buds in order to keep at least one of the latter on the tipped 1-year-old wood (Beede et al. 2005).

7 Chilgoza

Chilgoza, often known as pine nuts, is a type of nut that grows on a pine tree (*Pinus Gerardiana*). *Pinus pinea*, *Pinus koraiensis*, *Pinus sibirica*, and *Pinus gerardiana* are the most often consumed species in Europe. These edible nuts with a crisp, nutty, and buttery flavour are also known as pignoli, pinyon, pinon nuts, cedar nuts, etc. and are primarily grown in the Western Himalayan region. The major producers are China, North Korea, the Russian Federation, Pakistan, and Afghanistan; in the Mediterranean (*Pinus pinea*), the top producing countries are Italy, Turkey, Spain, and Portugal. Despite the fact that there are over 30 different species of pine nuts, only three are consumed. Chilgoza is protected by a hard shell, which should be removed before eating. Always preserve the pine nuts in their shell for a longer shelf life. Pines are softwoods; however, they can be classified as soft pines or hard pines commercially. Soft pines, such as white, sugar, and pion pines, have comparatively soft wood, needles in bundles of five (less usually, one to four), stalked cones with scales devoid of prickles, and little resin. Male pollen-bearing cones are coated with many fertile scales, each having two pollen sacs. Female cones with two ovules are borne on the same tree and contain many spirally arranged bracts (modified leaves), each of which is positioned beneath a scale with two ovules (potential seeds). In some pines, the scale that contains the nutlike seed may stretch to resemble a wing for airborne distribution.

7.1 Canopy Management in Chilgoza

Chilgoza trees are predominantly found at altitudes ranging from 2000 to 3350 m in the western forests of the Himalayan region. It is distributed in patches around the arid mountainous areas of Pakistan, Afghanistan, and India. The Chilgoza can be found in Kashmir's Kishtwar and Himachal Pradesh's Pangi, Bharmour, and Kinnaur. The Chilgoza forests in Kinnaur are the largest, comprising roughly 2000 hectares.

Pinus gerardiana is the botanical name for the Chilgoza pine, which grows to a height of 25 m and has distinctive three-needle leaves. Chilgoza pines are medium-sized trees with a wide range of heights (Bhattacharyya et al. 1988). The average height of mature trees has been reported to range from 5 to 27 m (Alam 2011; Bhattacharyya et al. 1988; Saeed and Thanos 2006). Chilgoza pine trees, like other pines, are monoecious, with male and female reproductive organs located in different parts of the tree. According to these investigations, the average height of chilgoza pine might be around 18 m. Numerous environmental factors determine girth, and trees as tall as 4.5 m and as wide as 2.4 m (Stebbing 1906).

The canopy of chilgoza pine trees depends upon the environment it grows in. The canopy may be open, wide, and deep with erect and long branches in open situations or it may be shallow and narrow in dense forest in which the branches are short and horizontal resulting in a compact canopy structure (Gupta and Sharma 1975). Chilgoza pine trees have a long lifespan. According to Yadav (2009), the oldest chilgoza pine tree was found in Kinnaur, Himachal Pradesh, India and was 1086 years old. Being an evergreen tree, Chilgoza does not require much pruning and training.

8 Conclusion

Significant evolution and transition in tree canopy morphologies have happened in the previous few decades, resulting in novel production systems. To harvest a large number of high-quality fruits, canopy management must be improved through manipulation of plant population, plant architect, use of correct scion and rootstock combinations, training pruning system, and use of growth regulators. Canopy management is also used to reduce inside tree shade, increase warmth in the inner portion of the tree, and reduce insect pests and disease shelter points.

9 Future Strategies

Tree design or canopy management, particularly size control, has emerged as a priority for lowering production costs while enhancing yield and quality. Light interception is influenced by canopy design and form, resulting in higher monitory returns for fruit growers. As a result, early height control and tree canopy management are crucial approaches that should be used in fruit crops to increase grower returns. As already discussed, the management of plant architecture has to begin early in the life cycle of a tree. The selection of appropriate rootstock combined with other horticultural practices can be adopted to regulate the growth and development of a bearing tree to ensure a good quality crop. Traditionally, not much importance has been given to maintaining the canopy structure in the orchards of nut crops leading to overgrown trees having more wood than fruiting branches which also increases shading causing a decline in the size and quality of the nuts. However, adopting proper management practices like training, pruning, and the use of PGRs can successfully help to control tree size and spread of canopy. This would allow for sufficient light interception equally throughout the tree and thus uniform cropping. The growers should prioritize this at beginning of the orchard establishment as it becomes increasingly difficult to control tree vigour with the passing years.

References

Alam M (2011) Trees and shrubs of Afghanistan: a dendrological guide. Rossolis, Muséebotanique, Lausanne

Anonymous (2014) Australian almond industry study tour of Spain. Almond Board of Australia (ABA), Horticulture Australia Ltd

Azarenko AN, McCluskey RL, Hampson CR (1997) Time of shading influences yield, nut quality, and flowering. IV International symposium on hazelnut. Acta Hortic 445:179–184

Beede RH, Ferguson L, Wylie C, Fanucchi C (2005) Planting and training young trees. In: Ferguson L (ed) Dans: pistachio production manual. University of California, Davis, pp 57–59

Bhattacharyya A, LaMarche VC Jr, Telewski FW (1988) Dendrochronological reconnaissance of the conifers of northwest India. Tree-Ring Bull 48:21–30

Buchner R, DeBuse C, Lampinen B, McGranahan G, Leslie C, Ramos D, Metcalf S, Hackett W, Gilles C, Spanfelner F (2012) Walnut clonal paradox rootstock experiments in Solano and Tehama counties. Walnut research reports. University of California

Casanova-Gascón J, Figueras-Panillo M, Iglesias-Castellarnau I, Martín-Ramos P (2019) Comparison of SHD and open-center training systems in almond tree orchards cv. 'Soleta'. Agronomy 9(12):874

Davis TD, Curry EA, Steffens GL (1991) Chemical regulation of vegetative growth. Crit Rev Plant Sci 10(2):151–188

DeJong T, Johnson R, Doyle J, Weibel A, Solari L, Basile B, Marsal J, Ramming D, Bryla D (2004) Growth, yield and physiological behaviour of size-controlling peach rootstocks developed in California. Acta Hortic 658(2):449–456

Felipe AJ, Blasco AB, Gella R (1990) 'MONTIZO': Un clonselecto de "pollizo de Murcia". Acta Hortic 1:113–118

Felipe AJ, Gómez-Aparisi J, Socíasi Company R (1997a) The almond x peach hybrid rootstocks breeding program at Zaragoza (Spain). Acta Hortic 451(1):259–262
Felipe AJ, Carrera M, Gómez-Aparisi J (1997b) 'Montizo' and 'Monpol', two new plum rootstocks for peaches. Acta Hortic 451:273–276
Gash D, David I (1989) Paclobutrazol effect on growth and cropping of pecan trees. Acta Hortic 239:301–304
Gasic K, Preece J (2014) Register of new fruit and nut cultivars list 47. HortScience 49(4):396
Germain E (1990) Inheritance of late leafing and lateral bud fruitfulness in walnut (*Juglans regia* L.), phenotypic correlations among some traits of the trees. Acta Hortic 284:125–134
Germain E (1992) Le noyer. In: Gallais A, Bannerot H (eds) Améloration des especes végétales cultivées, objectifs et critères de sélection. INRA, Paris, pp 125–134
Germain E, Sarraquigne JP (1997) Hazelnut training systems: comparison between three systems used on three varieties. Acta Hortic 445:237–245
Germain E, Lespinasse JM, Reynet P, Bayol M (1995) Orchard training of lateral fruit-bearing walnut varieties assessment of trials carried out in France. III Int Walnut Congr 442:313–320
Gupta BN, Sharma KK (1975) The chilgoza pine, an important nut pine of Himalayas. Wans Year Book 1:21–32
Hall P (2000) Pruning chestnuts for improved productivity, vol 2(2). The Western Chestnut Growers Assn., Inc.
Hanna JD (1987) In: Rom RC, Carlson RF (eds) Rootstocks for fruit crops. Wiley, New York, pp 401–410
Heerema R (2015) Training young pecan trees. College of Agricultural, Consumer and Environmental Sciences, New Mexico State University
Jackson JE, Palmer JW (1977) Effects of shade on the growth and cropping of apple trees. II. Effects on consumption of yield. J Hortic Sci 52:253–266
Jindal KK, Karkara BK (1991) Chestnuts. In: Mitra SK, Bose TK, Rathore DS (eds) Temperate fruits. Horticulture and Allied Publishers, Calcutta, India, pp 498–518
Kallow G, Reddy BMC, Singh G, Lal B (2005) Rejuvenation of old and senile orchard. CISH, Lucknow, p 40
Kashaninejad M, Tabil LG (2011) Pistachio (Pistacia vera L.). In: Yahia EM (ed) Postharvest biology and technology of tropical and subtropical fruits. Woodhead Publishing, pp 218–247e
Lagerstedt HB, Painter JH (1973) A comparison of filbert training to tree and bush forms. HortScience 8:390–391
Lampinen BD, Tombesi S, Metcalf S, DeJong TM (2011) Spur behaviour in almond trees: relationships between previous year spur leaf area, fruit bearing and mortality. Tree Physiol 31:700–706
Lemus G (2009) Innovative methods of walnut production in South America. VI Int Walnut Symp 861:191–198
Loomis RS, Gerakis PA (1975) Productivity of agricultural ecosystems. In: Cooper JP (ed) Photosynthesis and productivity in different environments. Cambridge University Press, London, pp 145–172
Martin G, LaVine P, Sibbett GS, Nishijima C (1980) Chemical "pruning" of walnut trees. Calif Agric 34(10):16–17
McEachern GR, Stein LA (1986) Planting and establishing pecan trees. Bull Texas Agric Ext Serv 1545
Me G, Valentini N, Miaja ML (2001) Comparison of two training systems in hazelnut. Acta Hortic 556:321–326
Micke WC, Kester DE (1997) Almond growing in California. II Int Symp Pistachios Almonds 470: 21–28
Moreno MA, Tabuenca MC, Cambra R (1995) 'Adesoto 101', a plum rootstock for peaches and other stone fruits. HortScience 30:1314–1315
Olsen J (2002) Growing hazelnuts in the Pacific Northwest. Oregon State University, Corvallis, OR. Extension Service, EC1219

Olson WH, Ramos DE, Micke WC, Yeager J, Shawareb N (2001) Walnut training and hedging for early production and profit. Acta Hortic 544:437–442

Pathak RK (2009) Canopy management in minor fruit crop. Summer school on management of canopy architecture for higher productivity in subtropical fruits. CIAH, Lucknow, India, pp 42–48

Pearce SC, Dobersek-Urbanc S (1967) The management of irregularity in growth and cropping. J Hortic Sci 42:295–305

Pinochet J (2010) 'Replantpac' (Rootpac R), a plum-almond hybrid rootstock for replant situations. HortScience 45(2):299–301

Rathore DS (1991) Walnuts. In: Mitra SK, Bose TK, Rathore DS (eds) Temperate fruits. Horticulture and Allied Publishers, Calcutta, India, pp 377–414

Romisondo P, Manzo P, Tombesi A (1984) Sceltadelle cultivar. Aspetti della tecnica colturale e loro reflessis ulla qualita delle produzioni. Atti del Convegno Internazionale sul Nocciuolo, Avellino, Italy, pp 61–78

Rubio-Cabetas JM (2016) Almond rootstocks: overview. In: Kodad O, López-Francos A, Rovira M, Socias i Company R (eds) XVI GREMPA meeting on almonds and pistachios. CIHEAM, Zaragoza, pp 133–143

Saeed M, Thanos CA (2006) The effect of seed coat removal on seed germination of *Pinus gerardiana* Wallich ex D. Don. chilgoza pine. J Appl Emerg Sci 1(1):174–177

Singh J (2008) Basic horticulture. In: Training and pruning in orchard, vol 5. Kalyan Publishers, Ludhiana, Punjab, pp 183–193

Snelgar WP, Manson PJ, Martin PJ (1992) Influence of time of shading on flowering and yield of kiwifruit vines. J Hortic Sci 67:481–487

Sparks D (1979) Physiology-site, growth, flowering, fruiting, and nutrition. In: Jaynes RA (ed) Nut tree culture in North America. Northern Nut Growers Assn, Hamden, CT, pp 211–239

Sparks D (1986) Growth and nutrition of pecan seedlings from potassium phosphate foliar sprays. HortScience 21(3):451–453

Stebbing EP (1906) Notes on the chilgoza forests of Zhob and the Takht-I Suliman. Government of India Printing Office, Calcutta

Stoke HF (1969) In: Jaynes RA (ed) Handbook of North American nut trees. The W. F. Humphrey Press Inc., Geneva, pp 39–41

Tombesi S, Lampinen B, Metcalf S, DeJong T (2016) Yield in almond is related more to the abundance of flowers than the relative number of flowers that set fruit. Calif Agric 71(2):68–74

Tourjee KR (1998) Missouri eastern black walnut breeding program. In: Jones JE, Mueller R, Van Sambeek JW (eds) Nut production handbook for eastern black walnut. Republic, MO, Southwest Missouri Resource Conservation & Development, pp 90–96

Tous J, Romero A, Rovira M, Clave J (1994) Comparison of different training systems on hazelnut. Acta Hortic 351:455–461

Valentini N, Caviglione M, Ponso A, Lovisolo C, Me G (2008) Physiological aspects of hazelnut trees grown in different training systems. VII Int Congr Hazelnut 845:233–238

Vandana U, Solanki SPS, Lamo K (2017) Canopy management: way to develop fruit tree architecture. Biomol Rep:1–3

Wetzstein HY, Sparks D (1984) The morphology of staminate flower differentiation in pecan. J Am Soc Hortic Sci 109(2):245–252

Whiley AW, Wolstenholme BN, Faber BA (2013) Crop management. In: Schaffer B, Wolstenholme BN, Whiley AW (eds) The avocado. Botany, production and uses. CABI, Wallingford, UK, pp 342–379

White JM (2000) Pruning mature pecan trees. Esteban Herrera, Extension Horticulturist, New Mexico State University

Wood BW (1988) Paclobutrazol suppresses shoot growth and influences nut quality and yield of young pecan trees. J Am Soc Hortic Sci 113:374–377

Worley RE, Mullinix BG, Daniel JW (1996) Selective limb pruning, tree removal, and paclobutrazol growth retardant for crowding pecan trees. Sci Hortic 67:79–85
Yadav RR (2009) Tree ring imprints of long-term changes in climate in western Himalaya, India. J Biosci 34(5):699–707
Zhang Y, Guo S, Sun C (2019) Effects of growth retardants on branches growth, leaves development and physiology of chestnut. J Northwest A & F Univ Nat Sci Ed 47(9):79–89
Zhu H, Stafne ET (2019) Influence of paclobutrazol on shoot growth and flowering in a high-density pecan orchard. HortTechnology 29(2):210–212

Chapter 10
Biotechnological Interventions for Improvement of Temperate Nuts

Vishal Sharma, Jagveer Singh, and Gurupkar Singh Sidhu

Abstract Nuts are dry edible fruits or seeds that have a high fat, protein, fibre, and oil content. Commercial propagation and genetic improvement of temperate nuts are difficult by conventional methods. Future is foreseen for temperate nut breeding initiatives, notably based on modern biotechnological approaches. The propagation of temperate nut trees has also improved significantly. Over the last 30 years, many forms of grafting and tissue culture (micropropagation or somatic embryogenesis) procedures for propagation have been extensively explored, with the most effective approaches being commercialized. Commercial grafted and micropropagation plants of walnut, pistachio, and other nut crops are available from a number of authorized nurseries. Professional advisors are currently advising the establishment of modernized orchards of temperate nut crops utilizing novel cultivars and rootstocks. Recent advances in molecular biology, i.e. genetic transformation, CRISPR-Cas9, next-generation sequencing (NGS) techniques, high-throughput genotyping, and genomics-based approaches like genome-wide association studies (GWAS) and genomic selection, have opened up new avenues for temperate nut tree breeding efficiency. Nut crop breeders have successfully used modern molecular and genomic techniques such as molecular markers, genetic transformations, and high-throughput genotyping to investigate the genetic basis of desired characteristics and, as a result, develop novel varieties and rootstocks over the last few decades.

Keywords Micropropagation · Biotechnological · Somatic embryogenesis · Nuts · Cryopreservation · Molecular · Genomics

V. Sharma · G. S. Sidhu (✉)
School of Agricultural Biotechnology, Punjab Agricultural University, Ludhiana, Punjab, India
e-mail: gurupkar-soab@pau.edu

J. Singh
School of Agricultural Biotechnology, Punjab Agricultural University, Ludhiana, Punjab, India

Department of Fruit Science, College of Horticulture & Forestry, Acharya Narendra Deva University of Agriculture & Technology, Ayodhya, India

M. M. Mir et al. (eds.), *Temperate Nuts*,
https://doi.org/10.1007/978-981-19-9497-5_10

1 Introduction

A nut is a fruit that consists of a hard or strong nutshell that protects an edible kernel. A broad range of dry seeds are referred to as nuts in common language and in the culinary sense, but 'nut' indicates that the seed's shell does not open to release the seed in a botanical context. Nuts and seeds are high in protein, with some being comparable to meat, fish, and chicken. They are consumed raw, roasted, or pressed for oil, and also used in baking and cooking. Walnut (*Juglans regia*) is the most important temperate nut crop produced in India. The other temperate nuts grown in India include the European chestnut (*Castanea sativa*), which grows at 1600–2300 m above sea level and the ornamental horse chestnut (*Aesculus indica*) both grown from seedlings and the Filbert or European hazelnut (*Corylus avellana*) locally called 'Bindak' or 'Bhotia badam'.Wild forms of the latter are found growing in the Western Himalayas from 1800 to 3300 m in partially shaded gullies. The Turkish hazelnut (*Corylus colurna*) also grows wild in the Himalayas. *Pistacia vera* L. is a member of the genus *Pistacia,* in the family *Anacardiaceae*, order *Sapindales*, tribe Rhoae (Stevens 2008). Chestnut trees were designated as *Fagus castanea* L. until 1754, when Miller proposed the genus *Castanea*. Chestnut belongs to the family *Fagaceae*. Pecan nut (*Carya illinoinensis*) is also an important and less explored temperate nut which belongs to the *Juglandaceae* family. All the above-mentioned temperate nut crops are rich in proteins, fibre, carbohydrates, fats and have many other nutritional properties. The most important role in consumption of these temperate nuts is in various cardiovascular diseases.

The poor rate of multiplication and high cost of plants are two of the most significant disadvantages of traditional vegetative propagation via grafting and cutting (Maggs 1975; Joley 1979; Holtz et al. 1995). Furthermore, traditional propagation methods are too sluggish to satisfy the demands of large-scale plantings. Micropropagation is a technique for propagating cultivars on their own roots, producing chosen rootstock clones, and creating genetically modified plants. These technologies might provide a cost-effective way to produce huge volumes of superior planting material all year long, with no seasonal limitations. Tissue culture methods appear to be a significant tool for increasing plantlet output. Many efforts have gone into developing in vitro multiplication and preservation methods for various temperate nuts. Using biotechnological interventions, one can resolve the breeding issues associated with the improvement and propagation of temperate nuts. Genetic resources can be conserved for long interval of time using techniques like cryopreservation. Genome editing is the recent and most advanced technique which can be employed for the alteration of different genes responsible for different traits like various biotic and abiotic stresses as well as for nutritional enhancement in temperate nut crops. Using genomic tools, different marker systems can be developed for molecular studies in these nuts. Differential gene expression studies can provide in depth knowledge of different genes responsible for various metabolic processes in the temperate nuts.

2 Important Temperate Nuts

*2.1 Chestnut (*Castanea sativa*)*

Chestnut trees were designated as *Fagus castanea* L. until 1754, when Miller proposed the genus *Castanea*, which comprises up to 13 species of chestnut trees and chinkapins native to Asia, North America, and Europe, with greatest diversity in South-east Asia and eastern USA (Fig. 10.1a). Chestnut trees are deciduous, with alternate, simple leaves 20–30 cm long and 3–6 cm wide, oblong to lanceolate, with a serrate margin and an acute to acuminate apex. The exceptional nutritional value of chestnuts has long been recognized; ancient Greeks and pre-Roman tribes venerated the chestnut tree and considered chestnuts to be superior to almonds, hazelnuts, and walnuts (Merkle et al. 2020).

*2.2 Chilgoza Pine (*Pinus gerardiana*)*

Pinus gerardiana Wall. ex D. Don, often known as 'Chilgoza' or 'neoza pine', is one of the world's most promising trees with ecological and commercial potential (Fig. 10.1b). This tree is critical to the economic development and livelihood of those living near the forest. It has a very limited range and is only found in the highlands of eastern Afghanistan, portions of Pakistan, and a few isolated locations in the arid inner Himalayas (Kant et al. 2006).

Fig. 10.1 Some important temperate nut crops. (**a**) Chestnut; (**b**) Chilgoza pine; (**c**) Hazelnut; (**d**) Pecan nut; (**e**) Pistachio nut; and (**f**) Walnut

2.3 *Hazelnut (***Corylus mandshurica***)*

There are 20 species of hazelnut in the genus *Corylus* L., which belongs to the Betulaceae family and yield edible nuts (Fig. 10.1c). Hazelnuts are frequently utilized in food processing for the production of confectionary items such as chocolate, cookies, and hazelnut oil. Hazelnuts have a high overall fatty acid content (60% of the hazelnut kernel), with oleic acid accounting for the majority of the fatty acids (80%). Fatty acids are significant nutritional components for humans (Li et al. 2021).

2.4 *Pecan Nut (***Carya illinoensis***)*

The most significant species in the *Carya* genus is the pecan (*Carya illinoensis*) (Fig. 10.1d). It is grown primarily for its oil and protein-rich nut, as well as for its high-quality wood. Pecan is traditionally propagated by budding or grafting onto open pollinated rootstocks. The pecan is a native of North America, which is the world's top producer. It is grown in Australia, Brazil, Canada, Mexico, Israel, and South Africa (Casales et al. 2018).

2.5 *Pistachio (***Pistacia vera** *L.)*

Pistachio trees (*Pistacia vera* L.) are widely cultivated in Europe, North Africa, Middle East, China, and California (Onay et al. 2004a). *Pistacia vera* is a member of the Anacardiaceae (cashew) family and belongs to the *Pistacia* genus (Fig. 10.1e). *Pistacia vera* is the most commercially significant of the 11 species in the genus *Pistacia* since it yields fruit of sufficient size to be sold (Al-Safadi and Elias 2010).

2.6 *Walnut (***Juglans regia** *L.)*

The English walnut (*Juglans regia* L.), often known as the Persian walnut, is the most widely produced and economically farmed walnut species (Fig. 10.1f). *Juglans regia* is a Central Asian tree that grows as a wild or semi-cultivated tree throughout a wide range of habitats, from south-eastern Europe and the Caucasus to Turkey and Iran, to the former Soviet Union's southern areas to China and the eastern Himalayas (Brown et al. 2020).

3 Plant Tissue Culture of Temperate Nuts

Plant tissue culture is a technique which involves different strategies for the improvement of various plant species. These strategies are classified as in vitro propagation, somatic embryogenesis, organ culture, in vitro mutagenesis, in vitro shoot tip grafting, cryopreservation for long-term storage, etc. Plant tissue culture techniques are employed by different co-workers for the improvement of some commercially important temperate nut crops (Table 10.1).

3.1 In Vitro Propagation

Chestnut cuttings are very difficult to root (Vieitez 1974); early studies focused on plant tissue culture as a propagation technique. Several studies, primarily using seedling material, have been published on the beginning and maintenance of shoot proliferation. Viéitez et al. (1986) reviewed the chestnut tissue culture. By utilizing stump sprouts as explants, these researchers were able to micropropagate mature plants from this refractory species (Vieitez et al. 1983).

Vieitez and Vieitez (1980) promoted shoot proliferation on MS media with 3 months old seedlings and nodal explants. Blayde (1966) medium, Heller's (1953) medium + 1 mM NH_4NO_3, and Lepoivre's (Quoirin and Lepoivre 1977) medium with 0.1–0.5 mg/L (0.4–2.2 pM) BA were determined to be appropriate after testing a variety of macro- and micro-nutrient basal media. Although there has been little success with mature explants, there are still many issues. Browning of explant has been seen throughout the establishing process. After surface sterilization and before transfer to medium, explants were soaked in sterile water to alleviate this problem. Auxin dips before transfer to hormone-free media were used to accomplish rooting in adult material, although only to a limited extent. Furthermore, cultivars differ significantly in their responses to shoot proliferation and roots (Jacquiot 1950; Vieitez and Vieitez 1980; Vieitez et al. 1983; Chevre et al. 1983).

The use of basal explants rather than shoot tips improved the multiplication of Hazelnut shoots (Anderson 1983). Perez et al. (1985) discovered that growing seedling shoot and cotyledonary segments on Cheng's media in Hazelnut resulted in optimal shoot initiation.

Pecans are mostly cultivated from seedlings and natural trees in the southern United States and Mexico. Attempts to grow Pecans in vitro have had a very less success rate. Seedlings were used to create cultures, with the pre-treatment of the seedlings proved to be very important (Hansen and Lazarte 1984). Explants from etiolated stock plants did not proliferate, whereas explants from stock plants maintained in a greenhouse on 16-h photoperiods, proliferated satisfactorily. WPM was used to promote shoot growth in nodal explants (Wood 1982; Hansen and Lazarte 1984).

Table 10.1 Overview of different plant tissue culture methods employed for improvement of important temperate nut crops reported or reviewed by various authors

Sr. No.	Name of nut	Plant tissue culture method	Reported or reviewed by
1	Chestnut	In vitro propagation	Vieitez and Vieitez (1980), Vieitez et al. (1983), Viéitez et al. (1986)
2		In vitro regeneration	Chevre et al. (1983)
3		Callus culture	Jacquiot (1950), Vieitez et al. (1978)
4		In vitro shoot tip grafting	Fernandez-Lorenzo and Fernandez-Lopez (2005)
5		Cryopreservation	Pence (1990, 1992)
6		Micropropagation, micrografting, organogenesis	Merkle et al. (2020)
7	Hazelnut	Direct regeneration	Anderson (1983), Perez et al. (1985)
8		Callus culture	Kai et al. (1984)
9		Somatic embryogenesis	Merkle et al. (1987), Wetzstein et al. (1989)
10		Cryopreservation	Pence (1990), Gonzalez-Benito and Perez (1994), Normah et al. (1994)
11	Pecan nut	*Micropropagation*	Wood (1982)
12		Cryopreservation	Pence (1990)
13		Somatic embryogenesis, somaclonal variations	Vendrame and Wetzstein (2020)
14	Pistachio nut	In vitro regeneration	Benmahioul et al. (2016)
15		Shoot culture, embryo culture	Al-Safadi and Elias (2010)
16		In vitro propagation, somatic embryogenesis	Onay and Jeffree (2000), Onay (2000)
17		Micropropagation	Benmahioul (2017)
18		*Direct regeneration*	Tilkat et al. (2013)
19		In vitro shoot tip necrosis	Kermani et al. (2017)
20		Micropropagation	Delijam et al. (2016)
21		In vitro micrografting	Onay et al. (2003)
22		Somatic embryogenesis	Ghadirzadeh-Khorzoghi et al. (2019)
23		Tissue culture propagation	Pour et al. (2019)
24		In vitro propagation	Gabr and Hassanen (2012)
25		In vitro selection—salt tolerance	Raoufi et al. (2021)
26		Direct shoot organogenesis	Tilkat and Onay (2009)
27		In vitro shoot tip grafting	Abousalim and Mantell (1992), Onay et al. (2003, 2004a, 2016), Can et al. (2006), Garcia et al. (2012), Marin et al. (2016)
28		Micropropagation, micrografting, somatic embryogenesis, organogenesis, somaclonal variations, cryopreservation, genetic transformation	Onay et al. (2020)
29	Walnut	Direct and indirect regeneration	Payghamzadeh and Kazemitabar (2011)

(continued)

Table 10.1 (continued)

Sr. No.	Name of nut	Plant tissue culture method	Reported or reviewed by
30		Somatic embryogenesis	Tulecke and McGranahan (1985), Tulecke et al. (1988)
31		Endosperm culture	Tulecke et al. (1988)
32		Somaclonal variation	Yajima et al. (1997)
33		Callus culture	Cummins and Ashby (1969)
34		Cryopreservation	Brison et al. (1991), de Boucaud et al. (1991, 1994)
35		Direct and indirect regeneration, somatic embryogenesis, genetic transformation, cryopreservation	Brown et al. (2020)

In Italy and Spain, a technology for micropropagating the rootstock *Pistacia integerrima* has been devised, and the plants produced are being evaluated in the field (Martinelli 1988). Micropropagation of *P. atlantica* and numerous cultivars of *P. vera* has also progressed. After being transplanted to soil in a greenhouse, rooted shoots of *P. vera* cultivars failed to develop (Martinelli 1988).

In vitro propagation on walnut was precisely and thoroughly reviewed by Payghamzadeh and Kazemitabar (2011). In the review, they reported the in vitro studies on different *Juglans* spp., commonly used sterilization techniques in explant sterilization, commonly used hormones, mediums, and explant for micropropagation, different solidifying agents and different culture conditions used for incubation of cultures by various researchers. In vitro propagation methods including sterilizing agents, different media concentration, plant growth regulator combinations, and solidifying agents were thoroughly reviewed by different authors in pistachio, pecan nut, walnut, and chestnut (Onay et al. 2020; Merkle et al. 2020; Vendrame and Wetzstein 2020; Brown et al. 2020).

3.2 *Somatic Embryogenesis*

In chestnuts, cambial tissues and cotyledons have been used for developing callus (Jacquiot 1950; Vieitez et al. 1978). Kai et al. (1984) discovered that IBA at 0.1 mg/L (0.5 pM) was preferable to IAA or NAA, with greater auxin concentrations causing callus in hazelnuts. Merkle et al. (1987) described somatic embryogenesis development from pre-mature zygotic embryos. Using a series of media enhanced the rate of embryogenic response substantially (Wetzstein et al. 1989). Somatic embryogenesis of pistachio nut from female flowers was done by Onay et al. (2004b, 2007).

Early research on *Juglans* spp. focused on root initiation using callus as a way to solve difficulties with rooting of cutting. Using an auxin-free media, Jacquiot (1951) does not obtained callus on *J. regia* (English walnut) explants. Later, using White

(1963) medium containing NAA and kinetin, callus was generated and sustained for a brief period from stem tissue of *J. nigra* (black walnut), but no differentiation occurred (Cummins and Ashby 1969). More success has been achieved with the hybrid rootstock 'Paradox' (*J. hindsii* × *J. regia*). Driver and Kuniyuki (1984) discovered that WPM or B-5 media performed better than MS or Cheng (1978) media in a variety of tests. Somatic embryogenesis may also yield walnut plants, and such plants have been grown in the field (Tulecke and McGranahan 1985).

Further, somatic embryogenesis and in vitro regeneration of pistachio, pecan nut, walnut, and chestnut were thoroughly reviewed by different authors in which they covered all the previous reports (Onay et al. 2020; Merkle et al. 2020; Vendrame and Wetzstein 2020; Brown et al. 2020).

3.3 Organ Culture

Organ culture is a method of plant tissue culture in which an organ of a plant is excised and cultured on different media for development of complete plant. Organ culture includes endosperm, ovary, anther, pollen, and embryo culture. A very few work has been done on organ culture when we talk about temperate nut crops.

American chestnut embryogenic culture induction has utilized immature seeds as explants; and embryogenic cultures of American chestnut were induced on a modified woody plant medium (WPM) (Lloyd and Mccown 1980).

The embryogenic process was utilized to produce triploids ($3n = 48$) from walnut endosperm (Tulecke et al. 1988). The cultivars 'Payne', 'Early Ehrhardt', and 'Manregian' generated embryogenic cultures on a regular basis. At University of California, Davis, triploids from 'Manregian' endosperm are kept in the *Juglans* germplasm collection.

3.4 In Vitro Mutagenesis

A tetraploid 'Mitsuru' walnut which was developed through colchicine treatment was compared to a diploid 'Mitsuru' walnut, and pollen properties like size, germination percentage, and fertility were examined (Yajima et al. 1997). The Shinano cultivar 'Mitsuru' was created by crossing *J. regia* var. Orientis Kitamura (Teuchi walnut) with the Persian walnut (*J. regia* L.).

3.5 In Vitro Shoot Tip Grafting

Micrografting of in vitro shoot tips of mature European chestnut on in vitro shoots of juvenile origin resulted in higher shoot multiplication rates when the elongated scions were excised and cultured on multiplication medium. Shoot multiplication only increased for one of the two tested mature clones, and in vitro rooting was not improved (Fernandez-Lorenzo and Fernandez-Lopez 2005).

Micrografting of pistachio was investigated in vitro and ex vitro by Abousalim and Mantell (1992). Different micrografting techniques have been reported in pistachio by different authors (Abousalim and Mantell 1992; Onay et al. 2003, 2004a; Can et al. 2006; Garcia et al. 2012; Marin et al. 2016).

3.6 *Cryopreservation*

Cryopreservation is presently the sole option for ensuring the safe and long-term protection of genetic resources of many species (Engelmann 2000). Most pecan nut embryonic axes dried to 5–10% moisture content before being subjected to liquid nitrogen sprouted or partially germinated, with some callus after thawing as reported by Pence (1990). Fresh seed had the best in vitro growth and development, which worsened as the seed aged, going from shoots to callus with no growth.

Initial experiments revealed that drying chestnut embryonic axes to around 8% moisture before freezing resulted in callus formation following recovery (Pence 1990). Before liquid nitrogen treatment, chestnut embryonic axes dried to 20–30% moisture exhibited enhanced survival and some shoot development (Pence 1992).

In early studies, Pence (1990) revealed that cryopreserved embryonic axes and control axes of hazelnut seeds produced only callus. Maximum recovery (85%) was achieved when embryonic axes from fresh seed of *Corylus avellana* L. 'Morell' were frozen at 12% moisture content, whereas 'Butler' needed 11% moisture to achieve 50% recovery (Gonzalez-Benito and Perez 1994). Whole seeds of *C. avellana* 'Barcelona' did not survive in liquid nitrogen exposure after desiccation pre-treatment; however, embryonic axes were extracted from the thawed seeds and regrown in culture (Normah et al. 1994). Axes from stratified Barcelona seed improved shoot growth in both the control and liquid nitrogen exposure treatments. Axes from stratified seed that had been maintained and dried to 8% moisture showed 85% viability and 70% shoot development when cryopreserved, while only 30% of unstratified axes produced shoots (Reed et al. 1994). Embryonic axes from seeds of *Corylus colurna* L., *Corylus americana* Marsh., and *Corylus sieboldiana* var. Mandschurica (Maxim.) *C. Schneider* was stored in liquid nitrogen using this method at NCGR-Corvallis and the National Seed Storage Laboratory in Ft. Collins, CO. The regrowth of the thawed axes was between 75% and 80% for all of these species.

Dried axes of Juglans seeds (5% moisture) germinated or partially germinated in vitro after cryopreservation, producing shoots and/or roots (Pence 1990). Cryoprotectant comprising 5 M 1,2-propanediol and 20% sucrose resulted in 75–91% *Juglans* embryonic axis survival and regeneration (de Boucaud et al. 1991). Slow freezing of walnut shoot tips produced in vitro was also done by de Boucaud and Brison (1995). The combination of a modified PVS2 treatment and gradual freezing

(0.5 °C/min) of shoot tips resulted in a 34% survival rate (Brison et al. 1991). With isolated walnut somatic embryos, encapsulation-dehydration and slow freezing techniques were successfully done (de Boucaud et al. 1994).

Furthermore, cryopreservation of pistachio nut, pecan nut, walnut, and chestnut was thoroughly reviewed by different authors in which they reported different cryopreservation techniques like desiccation and vitrification (Onay et al. 2020; Merkle et al. 2020; Vendrame and Wetzstein 2020; Brown et al. 2020).

4 Genetics/Breeding

Chestnut are diploid with $2n = 2x = 24$, and the different species hybridize readily. Chestnuts are self-sterile and usually protandrous (Dermen and Diller 1962; Jaynes 1962, 1964; Li 1981; Stout 1926). *Castanea* displays high levels of genetic variability within and among species as a result of their wide geographical ranges, outcrossing habit and long history of cultivation (Merkle et al. 2020).

Pecan nut is a diploid ($2n = 2x = 32$) deciduous tree within the genus *Carya*. There are 20–25 species of large trees in the genus *Carya* (Reed and Davidson 1954; Sparks 1992). The leaves are alternate and compound, varying in size, e.g. 'Cheyenne' has small leaves, while 'Mahan' has large leaves. Leaflets are usually 5–10 cm in length and vary with 9–17 leaflets per leaf. Leaf colour is variable, from yellow-green of 'Desirable' to the extremely dark leaves of 'Pawnee'. Pecan is a monoecious species with staminate (male) and pistillate (female) flowers on the same tree. Pistillate flowers are borne in terminal spikes on new shoot growth, while pendulous staminate inflorescences are borne on the base of the shoot and along the length of the supporting 1-year-old wood. Cultivar selection either from dooryard seedlings or from seed orchards has been the major source of commercially important cultivars. Cultivar selection by breeding was initiated in the early twentieth century, mainly by programmes conducted by state or federal research institutions. A more detailed review and update about the current status of breeding programmes in pecan have been reported by Thompson and Conner (2012) and Vendrame and Wetzstein (2020).

Pistacia vera L. is a member of the genus *Pistacia*, in the family *Anacardiaceae*, order Sapindales (Stevens 2008). Genetics and breeding of *P. vera* occur in Iran (Rahemi and Tavallali 2007). The breeding potential of different *Pistacia* spp. genotypes is being studied (Arpaci et al. 2013) at the Pistachio Research Institute, Gaziantep in Turkey. More details about breeding, genetics, botany, and history of pistachio are compiled by Onay et al. (2020). RFLPs, non-coding chloroplast DNA, random amplified polymorphic DNA markers (RAPDs), amplified fragment length polymorphisms (AFLPs), inter-simple sequence repeats (ISSRs), sequence-related amplified polymorphisms (SRAPs), simple sequence repeats (SSRs), sequence characterized amplified region (SCARs), and single nucleotide polymorphisms (SNPs) have all been used to investigate the genetic relationships of *Pistacia vera* cultivars (Sheikhi et al. 2019).

The English or Persian walnut (*Juglans regia* L.) is the most extensively grown walnut species in the world. The *Juglans* genus is divided into four parts. Walnut ($2n = 2x = 32$) breeding began by selecting better nuts and seedlings by accident (Brown et al. 2020). Steady advances in walnut genomics have resulted in the development of genetic and physical maps (Luo et al. 2015), and genome assembly by Martinez-Garcia et al. (2016) and a reference genome and gene resource for six *Juglans* species by Stevens et al. (2018). Genome of *J. regia, J. microcarpa, J. hindsii* and other *Juglans* have development of *Juglans* genomic resources has been reviewed recently (Bernard et al. 2018) and will facilitate future efforts in marker-assisted selection (MAS), gene cloning and functional genomics. More details about breeding, genetics, botany, and history of pistachio are compiled by Merkle et al. (2020), Vendrame and Wetzstein (2020), Onay et al. (2020) and Brown et al. (2020). Breeding objectives for different nut fruits, viz. pistachio, chestnut, pecan, and walnut have been also discussed.

5 Genetic Diversity

For the study of genetic diversity, several molecular markers are being used. RAPD are ideal for DNA fingerprinting since they are quick and simple to evaluate. The ISSR is also effective since it does not require any previous sequence (Adams et al. 2003). ISSR analysis was performed to elucidate the genetic diversity of *Pinus gerardiana* (Chilgoza) in Himachal Pradesh (Kant et al. 2006; Ginwal et al. 2009) and in Zhob, Balochistan, Pakistan (Gul et al. 2021).

Morphological studies in pistachio nut were performed by Zohary (1952). Several studies have reported diversity at the intraspecific level using RAPDs, AFLP. SSR markers indicated the genetic diversity of cultivated pistachio (Khadivi et al. 2018; Choulak et al. 2019). Iran is one of the main diversity centres and origin of pistachio in the world. More recently, different markers such as ISSR, inter-retrotransposon amplified polymorphism (IRAP), and retrotransposon microsatellite amplified polymorphism (REMAP) were used for genetic diversity analysis in pistachio nut (Pourian et al. 2019; Labdelli et al. 2020).

Several studies have revealed diversity at the intraspecific level using RAPDs (Karimi et al. 2012), AFLPs, RFLPs, ISSRs, and SSRs demonstrated the geographic distribution of pistachio with RAPD markers; there were two distinct clusters: Mediterranean and Iranian-Caspian. Vahdati et al. (2019) published a review for the diversification and history of fingerprinting techniques used in Walnut.

Molecular and morphological markers, viz. RFLPs, RAPD, ISSR and AFLP, SNP, and SSRs for chestnut, chilgoza, hazelnut, pecan nut, pistachio, and walnut species have been valuable tools for cultivar identification, the study of genetic structure, the construction of genetic maps and taxonomic studies were compiled by Merkle et al. (2020), Vendrame and Wetzstein (2020), Onay et al. (2020) and Brown et al. (2020).

6 Molecular Breeding

The extended juvenility phase, which delays the observation of reproductive characteristics in perennial tree species, is a severe breeding restriction. Disease resistance/susceptibility characteristics, which require consistent techniques and duplicates, are another challenging feature to identify. In these situations, molecular breeding can help speed up the process and increase accuracy.

Because of the present availability of molecular and genomic techniques, researchers have been able to answer a number of significant problems in the field of chestnut breeding, as compiled by Hill Craddock and Taylor Perkins (2019). In *Castanea mollissima* breeding projects in China, quantitative trait locus (QTL) mapping has been used to aid in marker-assisted selection (MAS), with the primary features being nut quality and harvest date, as well as resistance to *Phytophthora* root rot and Blight (Hill Craddock and Taylor Perkins 2019). Botta et al. (2019) produced hazelnut molecular linkage and association maps, as well as MAS.

Through MAS, pecan nut breeding attempts to enhance yield, quality, earliness, disease and pest tolerance, and other important agronomic characteristics (Gantait et al. 2019). QTLs were developed for shelling percentage by Luo et al. (2017). Pistachio scion and rootstock breeding operations might benefit from marker-assisted selection (MAS) of female offspring. Due to a lack of morphological indicators, identifying sex in pistachio seedlings has proven challenging (Sheikhi et al. 2019). To date, one genetic linkage map for pistachio was constructed using a cross between pistachio and monoecious *P. atlantica* (Turkeli and Kafkas 2013). Kafkas et al. (2015) reported that pistachio has a ZZ/ZW sex determination system. RAPD marker (OPO08945) was ineffective for sex determination in other *Pistacia* spp. (Hormaza et al. 1994). Other studies on sex determination were done by Kafkas et al. (2001) using RAPD markers, sequenced characterized amplified region (SCAR) marker by Yakubov et al. (2005) and Esfandiyari et al. (2012). Kafkas et al. (2015) used restriction site associated DNA sequencing (RAD sequencing) to distinguish sex determination.

Isozymes have been overtaken by restriction fragment length polymorphisms (RFLPs), random amplified polymorphic DNAs (RAPDs), amplified fragment length polymorphisms (AFLPs), and simple sequence repeats as more informative and robust DNA markers with higher polymorphism rates (SSRs). Vahdati et al. (2019) used RFLPs, RAPDs, AFLPs, and SSRs in a sequential order to identify walnut germplasm, evaluate its genetic diversity, and assess the uniformity and stability of cultivars. Genome-wide molecular marker data are used widely in breeding programmes for both trait mapping and prediction (Aradhya et al. 2001). Most efforts in walnut till date have focused on mapping qualitative traits. For example, Woeste et al. (1996a, b, 2002) found a sequence characterized amplified region (SCAR) marker associated with tolerance of cherry leaf roll virus among a backcross population of English × Black walnut.

7 Gene Cloning

MnSOD (manganese superoxide dismutase) was effectively cloned and produced in *E. coli* to see if pistachio MnSOD might induce allergy responses in atopic people (Noorbakhsh et al. 2010). The results showed that this protein might trigger some cross-reactions to IgE antibodies and thus could be considered as a pan allergen. Nucleotide binding site-leucine–rich repeat (NBS-LRR) regions of known disease resistance (R) genes were also cloned in pistachio by Bahramnejad (2014). *LEAFY* homologous gene (*PcBLFY*) of *P. chinensis* was cloned and used for breeding cultivars with shorter juvenile phase and earlier flowering (Zhang et al. 2012).

The National Centre for Biotechnology Information (NCBI) has the walnut genome data (Martinez-Garcia et al. 2016) which will be used for walnut improvement. Several genes involved in biosynthesis of tannin from shikimic acid (Muir et al. 2011), melanin biosynthesis from tyrosine (Escobar et al. 2008), and flavonoids from phenylalanine (Beritogoli et al. 2002) have been identified.

8 Functional Genomics

Jazi et al. (2015) developed a protocol for RNA isolation from pistachio that could be used for downstream studies, including real-time PCR and RNA-seq analysis. Several reference genes were studied, including ACT, EF1, -TUB, -TUB, GAPDH, CYP2, UBQ10, and 18S rRNA, to determine the best collection of reference genes for transcript level normalization under various circumstances, such as cold, drought, and salt (Jazi et al. 2016). Under all abiotic treatments, *EF1α* was the better reference gene in all samples. *β-TUB* was the second most stable gene in samples subjected to cold and drought treatments, whereas ACT held the same position in saline-treated samples.

Next-Generation Sequencing (NGS) also ushered in the current era of genotyping-by-sequencing (GBS) (Elshire et al. 2011) and its application in genetics and breeding with forest trees compiled by Neale and Kremer (2011), Alexander and Lawson (2014), and Nelson et al. (2014b). The Forest Health Initiative (Nelson et al. 2014a) funded a study to sequence and assemble the genome of the Chinese chestnut 'Vanuxem' (794 Mb). Three significant blight resistance QTLs were discovered and sequenced by Staton et al. (2015). Furthermore, a cyto-molecular method based on fluorescent in situ hybridization is being utilized to confirm chromosome-level assemblies and uncover structural rearrangements across the genomes of various genotypes and species of *Castanea* (Islam-Faridi et al. 2016; Staton et al. 2015).

Cloning and analysis of a walnut shikimate dehydrogenase (SkDH) were also responsible for synthesis of gallic acid, a key intermediate in the synthesis of hydrolysable tannins (HTs) (Muir et al. 2011; Martinez-Garcia et al. 2016; Escobar et al. 2008; Araji et al. 2014). Walnut transgenic lines suppressed for the expression of JrPPO1 is a biosynthetic enzyme responsible for the conversion of tyrosine to melanin and other hydroquinones and is expressed in all tissues. The walnut genome sequence revealed the presence of a second closely linked JrPPO2 gene, expressed only in walnut roots using RNAi (Araji et al. 2014; Martinez-Garcia et al. 2016; Stevens et al. 2018; Beritogoli et al. 2002).

9 Transcriptomes and Gene Discovery

The first genes for *Castanea* species were obtained by purifying and sequencing proteins or selecting and sequencing individual cDNA clones. Connors et al. (2002) reported the cystatin gene expressed in the stems of American chestnut. Schafleitner and Wilhelm (2002) identified genes responsible for chestnut blight fungus (*C. parasitica* (Murr.) Barr). The Genomic database for the *Fagaceae* Project (https://hardwoodgenomics.org/) will be used for identification of disease resistance genes in *Castanea* and other *Fagaceae* species.

Barakat et al. (2009) and Barakat et al. (2012) reported on the transcriptome from American and Chinese chestnut as two have variation for fungal infection. Serrazina et al. (2015) compared the root transcriptomes of the ink disease-susceptible species, European chestnut, and the resistant species, Japanese chestnut, after *P. cinnamomi* inoculation. The transcriptome data sets are publicly available on the Hardwood Genomics Project website (http://hardwoodgenomics.org/). Santos et al. (2017) and Serrazina et al. (2015) reported expression profiling of candidate genes potentially involved in defence response to *P. cinnamomi* in European and Japanese chestnuts. The *Cast_Gnk2* like gene was found to best discriminate between susceptible and resistant genotypes. The mode of action of ginkbilobin-2 (Gnk2) protein may be to chemically prevent pathogen growth through a plant specific cysteine rich motif DUF26, which belongs to cysteine rich receptor like kinases (CRKs) protein family (Miyakawa et al. 2014), which activates HR-related actin dependent cell death (Gao et al. 2016).

10 Future Perspective

This chapter discusses existing nut crop breeding techniques as well as future prospects for advancement. This chapter also gives an overview of current research and approaches that might be employed in the future. Biotechnological interventions will open up new avenues for the in vitro propagation and genetic improvement of difficult to grow temperate nut crops, in future. Advanced molecular breeding, genomics, proteomics, and metabolomics (omics) techniques will lead to improvement of commercially important temperate nut crops by developing different genetic platforms. Using these approaches, one can enhance the yield of the globally demanding temperate nut crops with the enhanced resistance to different biotic and abiotic stresses and nutritional enrichment. Biotechnology in combination with commercial breeding will reduce the longer period of varietal development. Overall, these tools will substantially benefit future efforts to discover key genes, as well as breeders' efforts to create information relevant for genetic engineering methods and marker-assisted breeding initiatives.

11 Conclusion

Nuts are important dietary source for human beings which are rich in fats, proteins, essential oils, and fatty acids. Biotechnology plays a tremendous role in the improvement and quality enhancement in various temperate nut crops. Plant tissue culture methods like micropropagation, somatic embryogenesis, direct or indirect regeneration can develop large number of planting material of important temperate nut crops in short interval of time. Callus cultures and cell suspension cultures of nut crops can produce some important secondary metabolites using large vessels (Bioreactors). Using modern biotechnological approaches, e.g. genetic transformation, RNAi and CRISPR, various genes responsible for many diseases can be silenced lead to improvement in production and quality of temperate nut crops.

References

Abousalim A, Mantell SH (1992) Micrografting of pistachio (*Pistacia vera* L. cv. Mateur). Plant Cell Tissue Org Cult 29:231–234

Adams RP, Schwarzbach AE, Pandey RN (2003) The concordance of terpenoid, ISSR and RAPD markers, and ITS sequence data sets among genotypes: an example from Juniperus. Biochem Syst Ecol 31:375–387. https://doi.org/10.1016/S0305-1978(02)00157-6

Alexander LW, Lawson SW (2014) Genomics in hardwood tree improvement: applications of a growing resource. In: Benkeblia N (ed) Omics technologies and crop improvement. CRC Press, Boca Raton, FL, pp 45–104

Al-Safadi B, Elias R (2010) Enhancement of Pistachio (*Pistacia vera* L.) propagation *in vitro* utilizing the shoot and embryo culture techniques. Int J Fruit Sci 10:96–108

Anderson WC (1983) Micropropagation of filberts, *Corylus auellana.* Comb. Proc Int Plant Propag Soc 33:132–137

Aradhya MK, Potter D, Woeste K, Simon CJ (2001) An integrated high-density AFLP-SSR map of Persian walnut. Hort Sci 36:536

Araji S, Grammer TA, Gertzen R, Anderson SD, Mikulic-Petkovsek M (2014) Novel roles for the polyphenol oxidase enzyme in secondary metabolism and the regulation of cell death in walnut (*Juglans regia*). Plant Physiol 164:1191–1203

Arpaci S, Açar I, Karadag S, Atl HS, Bilgel L (2013) Sulu Koşullarda Antepfıstığı İçin Uygun Anaç ve Dikim Aralıklarının Belirlenmesi. TAGEM Project Report. Gaziantep, Turkey, p 335

Bahramnejad B (2014) Identification and characterization of a NBS–LRR class resistance gene analog in *Pistacia atlantica* subsp. kurdica. Mol Cell Biol Res Commun 3:175–185

Barakat A, DiLoreto DS, Zhang Y, Smith C, Baier K (2009) Comparison of the transcriptomes of American chestnut (*Castanea dentata*) and Chinese chestnut (*Castanea mollissima*) in response to the chestnut blight infection. BMC Plant Biol 9:51

Barakat A, Staton M, Cheng CH, Park J, Buang N (2012) Chestnut resistance to the blight disease: insights from transcriptome analysis. BMC Plant Biol 12:38

Benmahioul B (2017) Factors affecting *in vitro* micropropagation of Pistachio (*Pistacia vera* L.). Agric For J 1(1):56–61

Benmahioul B, Kaïd-Harche M, Daguin F (2016) *In vitro* regeneration of *Pistacia vera* L. from nodal explants. J For Sci 62(5):198–203

Beritogoli I, Magel E, Abdel-Latif A, Charpentier JP, Jay-Allemand C, Breton C (2002) Expression of genes encoding chalcone synthase, flavanone 3-hydroxylase and dihydroflavonol 4-reductase correlates with flavanol accumulation during heartwood formation in *Juglans nigra*. Tree Physiol 22:291–300

Bernard A, Lheureux F, Dirlewanger E (2018) Walnut: past and future of genetic improvement. Tree Genet Genomes 14:1. https://doi.org/10.1007/s11295-017-1214-0

Blayde DF (1966) Interaction of kinetin and various inhibitors in the growth of soybean tissues. Physiol Plant 18:748–753

Botta R, Molnar TJ, Erdogan V, Valentini N, Marinoni DT, Mehlenbacher SA (2019) Hazelnut (*Corylus* spp.) breeding. In: Jameel M, Khayri AI, Jain SM, Johnson DV (eds) Advances in plant breeding strategies: nut and beverage crops, vol 4. Springer, pp 157–220. https://doi.org/10.1007/978-3-030-23112-5

Brison M, Paulus V, de Boucaud MT (1991) Cryopreservation of walnut and plum shoot tips. Cryobiology 28:738

Brown PJ, Leslie CA, Dandekar A (2020) *Juglans regia* walnut. In: Litz RE, Pliego-Alfaro F, Hormaza JI (eds) Biotechnology of fruit and nut crops, 2nd edn. CAB International, pp 246–257

Can C, Ozaslan M, Toremen H, Sarpkaya K, Iskender E (2006) *In vitro* micrografting of Pistachio, *Pistacia vera* L. var. Siirt, on wild pistachio rootstocks. J Cell Mol Biol 5:25–31

Casales FG, Watt EV, Coetzer GM (2018) Propagation of pecan (*Carya illinoensis*): a review. Afr J Biotechnol 17(18):586–605

Cheng TY (1978) Clonal propagation of woody plant species through tissue culture techniques. Proc Int Plant Propag Soc 28:139–158

Chevre AM, Gill SS, Mouras A, Salesses G (1983) *In vitro* vegetative multiplication of chestnut. J Hortic Sci 58:23–29

Choulak S, Chatti K, Chatti R, Guenni K, Salhi-Hannachi A, Said K, Chatti N (2019) Microsatellite (SSR) markers reveal genetic diversity and population structure in Tunisian pistachio. Fruits 74(2):73–81

Connors BJ, Laun NP, Maynard CA, Powell WA (2002) Molecular characterization of a gene encoding a cystatin expressed in the stems of American chestnut (*Castanea dentata*). Planta 215:510–514

Cummins JN, Ashby WC (1969) Aseptic culture of *Juglans nigra* L. stem tissues. FOX Sci 15:102–103

de Boucaud MT, Brison M (1995) Cryopreservation of germplasm of Walnut (*Juglans* species). In: Bajaj YPS (ed) Biotechnology in agriculture and forestry, vol 32. Berlin, Springer, pp 129–147

de Boucaud M, Brison M, Ledoux C, Germain E, Lutz A (1991) Cryopreservation of embryonic axes of recalcitrant seed: *Juglans regia* L. cv. Franquette. Cryo-Letters 12:163–166

de Boucaud MT, Brison M, Negrier P (1994) Cryopreservation of walnut somatic embryos. Cryo-Letters 15:151–160

Delijam MA, Garoosi G, Nezami-Alanagh E, Hosseini R (2016) Improving *Pistacia vera* micropropagation: with emphasis on the efficiency of minerals, vitamins and PGRs. J Plant Mol Breed 4(1):43–54

Dermen H, Diller JD (1962) Colchiploidy of chestnuts. For Sci 8:43–50

Driver JA, Kuniyuki AH (1984) *In vitro* propagation of Paradox walnut rootstock. Hort Sci 19:507–509

Elshire RJ, Glaubitz JC, Sun Q, Poland JA, Kawamoto K (2011) A robust, simple genotyping-by-sequencing (GBS) approach for high diversity species. PLoS One 6:e19379

Engelmann F (2000) Importance of cryopreservation for the conservation of plant genetic resources. In: Engelmann F, Takagi H (eds) Cryopreservation of tropical plant germplasm: current research progress and application. Japan International Research Center for Agricultural Sciences, Tsuba International Plant Genetic Resources Institute, Rome, pp 8–20

Escobar MA, Shilling A, Higgins P, Uratsu SL, Dandekar AM (2008) Characterization of polyphenol oxidase from walnut. J Am Soc Hortic Sci 133:852–858

Esfandiyari B, Davarynejad GH, Shahriari F, Kiani M, Mathe A (2012) Data to the sex determination in *Pistacia* species using molecular markers. Euphytica 185(2):227–231

Fernandez-Lorenzo JL, Fernandez-Lopez MJ (2005) Reinvigoration of mature *Castanea sativa* by serial micrografting onto a juvenile clone. Acta Hortic 693:293–298

Gabr MF, Hassanen SA (2012) *In vitro* propagation of *Pistacia vera* L. rootstock. Catrina 7(1): 25–31

Gantait S, Panigrahi S, Patel IC, Labrooy C, Rathnakumar AL, Yasin JK (2019) Peanut (*Arachis hypogaea* L.) breeding. In: Jameel M, Khayri AI, Jain SM, Johnson DV (eds) Advances in plant breeding strategies: nut and beverage crops, vol 4. Springer, pp 253–300. https://doi.org/10.1007/978-3-030-23112-5

Gao N, Wadhwani P, Mühlhäuser P, Liu Q, Riemann M (2016) An antifungal protein from *Ginkgo biloba*. Protoplasma 253:1159–1174

Garcia E, Imbroda I, Lorente P, Marín JA, Arbeloa A (2012) Micropropagation and *in vitro* grafting techniques to assist the selection of a Pistachio rootstock from a population of terebinth (*Pistacia terebinthus* L.) collected in the SE of Spain. Acta Hortic 961:245–252

Ghadirzadeh-Khorzoghi E, Jahanbakhshian-Davaran Z, Seyedi SM (2019) Direct somatic embryogenesis of drought resistance Pistachio (*Pistacia vera* L.) and expression analysis of somatic embryogenesis-related genes. S Afr J Bot 121:558–567

Ginwal HS, Bisht P, Chauhan P, Maurya SS, Barthwal S (2009) Genetic relationship between three Himalayan pines of Indian occurrence. Indian J Genet Plant Breed 69(1):81–83

Gonzalez-Benito ME, Perez C (1994) Cryopreservation of embryonic axes of two cultivars of hazelnut (*Corylus avellana* L.). Cryo-Letters 15:41–46

Gul J, Saeed S, Ahmed A, Leghari SK, Basit A, Rehman A, Khan MZ (2021) Genetic diversity and morphological variation of *Pinus gerardiana* along the environmental gradient from Zhob, Balochistan, Pakistan. Nusantara Biosci 13:121–128

Hansen KC, Lazarte JE (1984) *In vitro* propagation of pecan seedlings. Hort Sci 19:237–239

Heller R (1953) Recherches sur la nutrition minerale des tissus vegetaux cultives *in vitro*. Ann Sci Nat Bot Biol Veg 14:1–223

Hill Craddock J, Taylor Perkins M (2019) Chestnut (*Castanea* spp. Miller) breeding. In: Jameel M, Khayri AI, Jain SM, Johnson DV (eds) Advances in plant breeding strategies: nut and beverage crops, vol 4. Springer, pp 105–156. https://doi.org/10.1007/978-3-030-23112-5

Holtz B, Ferguson L, Allen GE (1995) Rootstock production and budding. In: Ferguson L (ed) Pistachio production. University of California, Davis, pp 54–56

Hormaza JI, Dollo L, Polito VS (1994) Identification of a RAPD marker linked to sex determination in *Pistacia vera* using bulked segregant analysis. Theor Appl Genet 89:9–13

Islam-Faridi N, Majid MA, Zhebentyayeva T, Georgi LL, Fan S (2016) FISH confirmation of a reciprocal translocation in chestnut. Cytogenet Genome Res 148:144

Jacquiot C (1950) Sur la culture *in vitro* de tissu cambial de chltaignier (*Castanea uesca* Gaertn). Acad Sci Ser D 231:1080–1081

Jacquiot C (1951) Sur les facteurs determinant la reprise de l'activite du cambium chez diverses essences forestieres. Acad Sci Ser D 232:2350–2352

Jaynes RA (1962) Chestnut chromosomes. For Sci 8:372–377

Jaynes RA (1964) Interspecific crosses in the genus *Castanea*. Silvae Genet 13:146–154

Jazi MM, Rajaei S, Seyedi SM (2015) Isolation of high quality RNA from pistachio (*Pistacia vera* L.) and other woody plants high in secondary metabolites. Physiol Mol Biol Plants 21:597–603

Jazi MM, Ghadirzadeh KE, Botanga C, Seyedi SM (2016) Identification of reference genes for quantitative gene expression studies in a non-model tree pistachio (*Pistacia vera* L.). PLoS One 11:e0157467

Joley LE (1979) Pistachios. In: Jaynes RA (ed) Nut tree culture in North America. Northern Nut Growers Association, Handen, CT, pp 163–174

Kafkas S, Çetiner MS, Perl-Treves R (2001) Development of sex-associated RAPD markers in wild *Pistacia* species. J Hortic Sci Technol 76:242–246

Kafkas S, Khodaeiaminjan M, Güney M, Kafkas E (2015) Identification of sex-linked SNP markers using RAD sequencing suggests ZW/ZZ sex determination in *Pistacia vera* L. BMC Genomics 16:98–108

Kai AH, Salesses G, Mouras A (1984) Multiplication *in vitro* dunoisetier (*Corylus auellana* L.). Agro 4:399–402

Kant A, Pattanayak D, Chakrabarti SK, Sharma R, Thakur M, Sharma DR (2006) RAPD analysis of genetic variability in *Pinus gerardiana* Wall. in Kinnaur (Himachal Pradesh). Indian J Biotechnol 5:62–67

Karimi HM, Abadi MHH, Kohbanani AM (2012) Genetic diversity of *Pistacia khinjuk* stocks using RAPD markers and leaf morphological characters. Plant Syst Evol 298:963–968

Kermani SA, Hokmabadi H, Jahromi MG (2017) The evaluation of the effect of multiwall carbon nano tube (MWCNT) on *in vitro* proliferation and shoot tip necrosis of Pistachio rootstock UCB-1 (*Pistacia integrima* × *P. atlantica*). J Nuts 8(1):49–59

Khadivi A, Esmaeili A, Mardani N (2018) Genetic diversity of cultivated pistachio as revealed by microsatellite molecular markers. Biotechnol Biotechnol Equip 32:602–609

Labdelli A, Herrán RDL, Arafeh R, Resentini F, Trainotti L, Halis Y, Adda A, Merah O (2020) Genetic variation in damaged populations of *Pistacia atlantica*. Plants 9:1541. https://doi.org/10.3390/plants9111541

Li MQ (1981) Study on the kariotype and Giemsa C band type of common *Camellia oleosa* and *Castanea mollissima*. For Sci Technol 4:9–11

Li Y, Sun P, Lu Z, Chen J, Wang Z, Du X, Zheng Z, Wu Y, Hu H, Yang J, Ma J, Liu J, Yang Y (2021) The *Corylus mandshurica* genome provides insights into the evolution of *Betulaceae* genomes and hazelnut breeding. Hortic Res 8:54

Lloyd G, Mccown B (1980) Commercially feasible micropropagation of mountain laurel, *Kalmia latifolia*. Comb Proc Int Plant Propag Soc 30:421–427

Luo MC, You FM, Li P, Wang JR, Zhu T (2015) Synteny analysis in Rosids with a walnut physical map reveals slow genome evolution in long-lived woody perennials. BMC Genomics 16:707

Luo H, Xu Z, Li Z (2017) Development of SSR markers and identification of major quantitative trait loci controlling shelling percentage in cultivated peanut (*Arachis hypogaea* L.). Theor Appl Genet 130:1635–1648

Maggs DH (1975) Prospects for pistachio nut growing in Australia. West Aust Nutgrowing Soc 1: 47–63

Marin JA, Garcia E, Lorente P, Andreu P, Arbeloa A (2016) A novel approach for propagation of recalcitrant pistachio cultivars that side steps rooting by *ex vitro* grafting of tissue cultured shoot tips. Plant Cell Tissue Org Cult 1:191–200

Martinelli A (1988) Use of *in vitro* techniques for selection and cloning of different *Pistacia* species. Acta Hortic 227:436–437

Martinez-Garcia PJ, Crepeau MW, Puiu D, Gonzalez-Ibeas D, Whalen J (2016) The walnut (*Juglans regia*) genome sequence reveals diversity in genes coding for the biosynthesis of non-structural polyphenols. Plant J 87:507–532

Merkle SA, Wetzstein HY, Sommer HE (1987) Somatic embryogenesis in pecan. Hort Sci 22(1): 128–130

Merkle SA, Viéitez FJ, Corredoira E, Carlson JE (2020) *Castanea* spp. chestnut. In: Litz RE, Pliego-Alfaro F, Hormaza JI (eds) Biotechnology of fruit and nut crops, 2nd edn. CAB International, pp 206–237

Miyakawa T, Hatano K, Miyauchi Y, Suwa Y, Sawano Y, Tanokura M (2014) A secreted protein with plant-specific cysteine-rich motif functions as a mannose-binding lectin that exhibits antifungal activity. Plant Physiol 166:766–778

Muir RM, Ibanez AM, Uratsu SL, Ingham ES, Leslie CA (2011) Mechanism of gallic acid biosynthesis in bacteria (*Escherichia coli*) and walnut (*Juglans regia*). Plant Mol Biol 75: 555–565

Neale DB, Kremer A (2011) Forest tree genomics: growing resources and applications. Nat Rev Genet 12:111–122

Nelson CD, Powell WA, Maynard CA, Baier KM, Newhouse A (2014a) The Forest Health Initiative, American chestnut (*Castanea dentata*) as a model for forest tree restoration: biological research program. In: Double ML, MacDonald WL (eds) Proceedings of the Vth international chestnut symposium, Acta Horticulturae. ISHS, Leuven, Belgium, pp 179–189

Nelson CD, Powell WA, Merkle SA, Carlson JE, Hebard FV (2014b) Biotechnology of trees: chestnut. In: Ramawat K, Merillon JM, Ahuja MR (eds) Tree biotechnology. CRC Press, Boca Raton, FL, pp 3–35

Noorbakhsh R, Mortazavi SA, Shahidi F, Maleki SJ, Nasirah LR (2010) Influence of processing on the allergenic properties of pistachio nut assessed in vitro. J Agric Food Chem 58:10231–10235

Normah MN, Reed BM, Yu X (1994) Seed storage and cryoexposure behavior in hazelnut (*Corylus avellana* L. cv. Barcelona). Cryo-Letters 15:315–322

Onay A (2000) Micropropagation of pistachio from mature trees. Plant Cell Tissue Organ Cult 60: 159–162

Onay A, Jeffree CE (2000) Somatic embryogenesis in Pistachio (*Pistacia vera* L.). In: Jain SM, Gupta PK, Newton RJ (eds) Somatic embryogenesis in woody plants, vol 6. Kluwer Academic Publishers, pp 361–390

Onay A, Pirinc V, Adiyaman F, Isikalan C, Tilkat E, Basaran D (2003) *In vivo* and *in vitro* micrografting of pistachio, *Pistacia vera* L. cv. "Siirt". Turk J Biol 27:95–100

Onay A, Pirinc V, Adiyanan F, Isikalan C, Tilkat E, Basaran D (2004a) *In vivo* and *in vitro* micrografting of mature pistachio (*Pistacia vera* var. Siirt). Plant Cell Tissue Organ Cult 77: 215–219

Onay A, Pirinç V, Tilkat E, Aktürk Z, Yildirim H (2004b) Somatic embryogenesis of pistachio from female flowers. J Hortic Sci Biotechnol 79:960–964

Onay A, Tilkat E, Yildirim H, Süzerer V (2007) Indirect somatic embryogenesis from mature embryo cultures of pistachio, *Pistacia vera* L. Propag Ornam Plants 7:68–74

Onay A, Tilkat E, Süzerer V, Karakaş-Metin Ö, Özden-Çiftçi Y (2016) Rejuvenation of mature lentisk by micrografting and evaluation of genetic stability. Turk J Biol 40:781–796

Onay A, Çiftçi YO, Koç HA, Süzerer V, Tilkat E (2020) *Pistacia vera* Pistachio. In: Litz RE, Pliego-Alfaro F, Hormaza JI (eds) Biotechnology of fruit and nut crops, 2nd edn. CAB International, pp 44–64

Payghamzadeh K, Kazemitabar SK (2011) *In vitro* propagation of walnut - a review. Afr J Biotechnol 10(3):290–311

Pence VC (1990) Cryostorage of embryo axes of several large-seeded temperate tree species. Cryobiology 27:212–218

Pence VC (1992) Desiccation and the survival of *Aesculus*, *Castanea* and *Quercus* embryo axes through cryopreservation. Cryobiology 29:391–399

Perez C, Rodriguez R, Tames RS (1985) "*In vitro*" filbert (*Corylus avellana* L.) micropropagation from shoot and cotyledonary node segments. Plant Cell Rep 4:137–139

Pour MM, Saberi-Riseh R, Mohammadinejad R, Hosseini A (2019) Nano-encapsulation of plant growth-promoting *Rhizobacteria* and their metabolites using alginate-silica nanoparticles and carbon nanotube improves UCB1 Pistachio micropropagation. J Microbiol Biotechnol 29(7): 1096–1103

Pourian MA, Bakhshi D, Aalami A, Hokmabadi H (2019) Assessment of genetic relationships among cultivated and wild pistachios (*Pistacia vera* L.) using molecular markers. J Hortic Res 27(1):37–46

Quoirin M, Lepoivre P (1977) Etude de milieux adaptes aux cultures *in vitro* de *Prunus*. Acta Hortic 78:437–442

Rahemi M, Tavallali V (2007) Effects of rootstock on Iranian pistachio scion cultivars. Fruits 62: 317–323

Raoufi A, Salehi H, Rahemi M, Shekafandeh A, Khalili S (2021) *In vitro* screening: the best method for salt tolerance selection among Pistachio rootstocks. J Saudi Soc Agric Sci 20:146–154

Reed CA, Davidson J (1954) The improved nut trees of North America and how to grow them. The Devin-Adair Company, New York, pp 128–158

Reed BM, Normah MN, Yu X (1994) Stratification is necessary for successful cryopreservation of axes from stored hazelnut seed. Cryo-Letters 15:377–384

Santos C, Duarte S, Tedesco S, Fevereiro P, Costa RL (2017) Expression profiling of *Castanea* genes during resistant and susceptible interactions with the oomycete pathogen *Phytophthora cinnamomi* reveal possible mechanisms of immunity. Front Plant Sci 8:1–12

Schafleitner R, Wilhelm E (2002) Isolation of wound-responsive genes from chestnut (*Castanea sativa*) microstems by mRNA display and their differential expression upon wounding and infection with the chestnut blight fungus (*Chryphonectria parasitica*). Physiol Mol Plant Pathol 61:339–348

Serrazina S, Santos C, Machado H, Pesquita C, Vicentini R (2015) *Castanea* root transcriptome in response to *Phytophthora cinnamomi* challenge. Tree Genet Genomes 11:6

Sheikhi A, Arab MM, Brown PJ, Ferguson L, Akbari M (2019) Pistachio (*Pistacia* spp.) breeding. In: Jameel M, Khayri AI, Jain SM, Johnson DV (eds) Advances in plant breeding strategies: nut and beverage crops, vol 4. Springer, pp 353–400. https://doi.org/10.1007/978-3-030-23112-5

Sparks D (1992) Pecan cultivars: the Orchard's Foundation. Pecan Production Innovations, Watkinsville, GA

Staton M, Best T, Khodwekar S, Owusu S, Xu T (2015) Preliminary genomic characterization of ten hardwood tree species from multiplexed low coverage whole genome sequencing. PLoS One 10(12):e0145031

Stevens PF (2008) Angiosperm phylogeny, Version 9. http://www.mobot.org/MOBOT/research/Apweb

Stevens KA, Woeste K, Chakraborty S, Crepeau MW, Leslie CA (2018) Genomic variation among and within six *Juglans* species. G3 8:2153–2165. https://doi.org/10.1534/g3.118.200030

Stout AB (1926) Why are chestnuts self-fruitless? J N Y Bot Garden 27:154–158

Thompson TE, Conner PJ (2012) Pecan. In: Badenes ML, Byrne DH (eds) Fruit breeding. Springer, New York, pp 771–802

Tilkat E, Onay A (2009) Direct shoot organogenesis from *in vitro* derived mature leaf explants of pistachio. In Vitro Cell Dev Biol Plants 45:92–98

Tilkat E, Süzerer V, Akdemir H, Tilkat EA, Çiftçi YO, Onay A (2013) A rapid and effective protocol for surface sterilization and *in vitro* culture initiation of adult male pistachio (*Pistacia vera* L. cv. "Atlı"). Acad J Sci Res 1(8):134–141

Tulecke W, McGranahan G (1985) Somatic embryogenesis and plant regeneration from cotyledons of walnut, *Juglans regia* L. Plant Sci 40:57–63

Tulecke W, McGranahan GH, Ahmadi H (1988) Regeneration by somatic embryogenesis of triploid plants from endosperm of walnut, *Juglans regia* L. cv. Manregian. Plant Cell Rep 7(5):301–304

Turkeli Y, Kafkas S (2013) First genetic linkage map in pistachio constructed using an interspecific cross between *Pistacia vera* L. and monoecious *Pistacia atlantica* Desf. Sci Hortic 151:30–37

Vahdati K, Arab MM, Sarikhani S, Sadat-Hosseini M, Leslie CA, Brown PJ (2019) Advances in Persian walnut (*Juglans regia* L.) breeding strategies. In: Jameel M, Khayri AI, Jain SM, Johnson DV (eds) Advances in plant breeding strategies: nut and beverage crops, vol 4. Springer, pp 401–474. https://doi.org/10.1007/978-3-030-23112-5

Vendrame W, Wetzstein H (2020) *Carya illinoinensis* pecan. In: Litz RE, Pliego-Alfaro F, Hormaza JI (eds) Biotechnology of fruit and nut crops, 2nd edn. CAB International, pp 238–245

Vieitez E (1974) Vegetative propagation of chestnut. N Z J For Sci 4:242–252
Vieitez AM, Vieitez ML (1980) Culture of chestnut shoots from buds in vitro. J Hortic Sci 55:83–84
Vieitez AM, Gonzalez ML, Vieitez E (1978) *In vitro* culture of cotyledon tissue of *Castanea satiua* Mill. Sci Hortic 8:243–237
Vieitez AM, Ballester A, Vieitez ML, Vieitez E (1983) *In vitro* plantlet regeneration of mature chestnut. J Hortic Sci 58:475–463
Viéitez AM, Viéitez ML, Viéitez E (1986) Chestnut (*Castanea* spp.). In: Bajaj YPS (ed) Biotechnology in agriculture and forestry, Trees 1, vol I. Springer, Berlin, pp 393–414
Wetzstein HY, Ault JR, Merkle SA (1989) Further characterization of somatic embryogenesis and plantlet regeneration in pecan, *Carya illinoensis*. Plant Sci 64:193–201
White P (1963) The cultivation of animal and plant cells. Ronald Press, New York
Woeste K, McGranahan GH, Bernatzky R (1996a) Randomly amplified polymorphic DNA loci from a walnut backcross [(*Juglans hindsii* × *J. regia*) × *J. regia*]. J Am Soc Hortic Sci 121:358–361
Woeste K, McGranahan GH, Bernatzky R (1996b) The identification and characterization of a genetic marker linked to hypersensitivity to the cherry leafroll virus in walnut. Mol Breed 2: 261–266
Woeste K, Burns R, Rhodes OE, Michler C (2002) Thirty polymorphic nuclear microsatellite loci from black walnut. J Hered 93:58–60
Wood BW (1982) *In vitro* proliferation of pecan shoots. Hort Sci 17:890–891
Yajima M, Nakamura H, Takahashi K, Watanabe Y, Saito S, Yokozawa Y (1997) Somatic chromosome numbers of colchicine-treated Shinano walnut (*Juglans regia* L.) and its F-1 seedlings. J Jpn Soc Hortic Sci 65:677–683
Yakubov B, Barazani O, Golan-Goldhirsh A (2005) Combination of SCAR primers and touchdown-PCR for sex identification in *Pistacia vera* L. Sci Hortic 103:473–478
Zhang P, Liu J, Zheng B, Mao F, Tang M (2012) Cloning and sequence analysis of LEAFY homologue from *Pistacia chinensis* Bunge. In: International conference on biomedical engineering and biotechnology. IEEE, Macao, China, pp 104–110
Zohary M (1952) A monographical study of the genus *Pistacia*. Palestine J Bot 5:187–228

Chapter 11
Organic Approaches in Temperate Nuts

M. H. Chesti, Hujjat Ul Baligah, Zahoor Ahmad Baba, Umar Iqbal, Mohammad Maqbool Mir, Inayat M. Khan, Shakeel A. Mir, Irfan A. Bisati, Syed Andleeba, Tabasum N. Qadri, and Zaffar Mahdi

Abstract Nut crops form an important component of diet and serve as a healthy alterative for meat especially in places where there is lesser availability of resources for animal rearing. In addition to energy and nutrients, nut crops are a rich source of monounsaturated fatty acids, polyunsaturated fatty acids, mono acylglycerols, tocopherols, tocotrienols, phytosterols, phytostanols, squalene, terpenoids, sphingolipids, and essential oils. In response to multiple health benefits of nut crops and their increasing market demand, their cultivation is gaining momentum globally. However, due to increased diet consciousness among consumers, due attention is given to quality and safety of food consumed. Consumers now prefer organically grown food over conventionally grown food with a belief that organic foods have negligible chemical residues and pose lesser health risk. This has led to shift from conventional cultivation to organic cultivation of all food crops including nuts. Different strategies either separately or in combination are employed for successful organic cultivation of nuts. This includes the use of organic fertilisers, mulches, biocontrol agents to reduce/eliminate the use of agrochemicals. This

M. H. Chesti (✉) · H. U. Baligah · I. M. Khan · S. A. Mir · S. Andleeba
Division of Soil Science and Agricultural Chemistry, FOA, Sher-e-Kashmir University of Agricultural Sciences and Technology of Kashmir, Wadura, Jammu and Kashmir, India
e-mail: mhchesti@gmail.com

Z. A. Baba
Division of Basic Sciences and Humanities, Sher-e-Kashmir University of Agricultural Sciences and Technology of Kashmir, Wadura, Jammu and Kashmir, India

U. Iqbal · M. M. Mir · I. A. Bisati
Division of Fruit Science, FOH, Sher-e-Kashmir University of Agricultural Sciences and Technology of Kashmir, Srinagar, Jammu and Kashmir, India

T. N. Qadri
S.P. College, Srinagar, Jammu and Kashmir, India

Z. Mahdi
Division of Basic Science & Humanities, FOH, Sher-e-Kashmir University of Agricultural Sciences and Technology of Kashmir, Srinagar, Jammu and Kashmir, India

M. M. Mir et al. (eds.), *Temperate Nuts*,
https://doi.org/10.1007/978-981-19-9497-5_11

chapter aims at understanding various organic strategies that can be employed for organic cultivation of temperate nuts in different parts of the world.

Keywords Nuts · Health benefits · Organic cultivation · Conventional cultivation

1 Introduction

Nuts are dried fruits that develop hard outer wall on maturity and are consumed in seeded form. Some of the examples of nut crops are almond (*Prunus* spp.), Brazil nut (*Bertholletia excelsa*), cashew (*Anacardium occidentale*), hazelnut (*Corylus avellana*), macadamia (*Macadamia* spp.), pecan (*Carya illinoinensis*), pine nut (*Pinus* spp.), pistachio (*Pistacia vera*), and walnut (*Juglans regia*). Among these nuts, walnut, hazelnut, pecan, pistachio, chestnut and chilgoza pine are examples of temperate nuts. Nut consumption is believed to be an important component of human food supply. Human beings started consumption of nuts millennia ago when man was a food gatherer. The nut crops are believed to be source of protein especially for vegetarians or at places where there is lesser availability of meat. These are consumed in both raw and processed form. These are rich in energy contributed by 45–75% lipids, 10–25% proteins (Sathe et al. 2005). Nuts are also rich in micronutrients such as calcium, magnesium, potassium and iron (Segura et al. 2006). These also comprise monounsaturated fatty acids, polyunsaturated fatty acids, mono acylglycerols, tocopherols, phytosterols, phytostanols, terpenoids, and essential oils, etc. (Shahidi and Miraliakbari 2005; Miraliakbari and Shahidi 2008). Most of the nuts such as almonds, Brazil nuts, cashews, chestnut, hazelnuts, macadamias, pecans, pine nuts, pistachios and walnuts have been scientifically studied and reported to possess high medicinal values (Ros 2010). Nuts are reported to improve favourable lipid profiles (Griel and Kris-Etherton 2006), help fight cancer (González and Salas-Salvadó 2006), check type-2 diabetes (Tapsell et al. 2004) reduce inflammation (Jiang et al. 2006), etc. Some of the important health benefits of temperate nuts are enlisted in Table 11.1. In 2017, *Global Burden of Disease Study* stated that insufficient intake of nuts is the fourth dietary risk factor for non-communicable disease and account for 2% deaths globally (Afshin et al. 2019). The dieticians and people advocating sustainable consumption now advice replacing animal protein with nut products keeping in view environmental concerns related to animal production (Willett et al. 2019). In present era, nuts, their oils, flours and other edible products are popular worldwide. In addition to their culinary uses, nuts have many industrial uses, e.g. tannins from the bark can be used for tanning and dyes. Some nut trees having hard wood or have good rot resistance have good scope for wood working, veneering, fence making and other uses requiring ground contact. Nut cultivation due to its immense dietary and industrial potential contributes greatly to local and national economies with production and consumption gaining momentum globally (De Souza et al. 2017). According to statistical yearbook of nuts and dried fruits (Nuts 2019), the production of nuts had increased by 24% compared to the average of the last 10 years.

Table 11.1 Health benefits of some of commonly consumed nuts in world

Nut	Origin	Benefits
1. Cashew nut *Anacardium occidentale*	Brazil	• Lower cholesterol level in blood • Control diabetes • Reduce coronary heart disease risk • Promote healthy bone development • Prevent of high blood pressure Creates an antibacterial, vesicant and anthelmintic property
2. Hazel nut *Corylus avellana*	Asia	• Reduce weight gain • Protect against cell damage • Lower cholesterol • Improve insulin sensitivity • Support heart health • Reduce inflammation • Improve sperm count
3. Macadamia *Macadamia* spp.	Australia	• Lower heart disease risk • Improve metabolic syndrome and diabetes • Prevent cancer • Protect the brain • Prevent weight gain
4. Pecans *Carya illinoinensis*	America and Mexico	• Heart Healthy • Improves Digestion • Helps to reduce weight • Acts as anti-inflammatory
5. Walnut *Juglans regia*	Persia and North America	• Good for the brain and heart • Mood-boosting • Supports weight loss • Supports a healthy digestive system • Help Manage Type 2 Diabetes • May Help Lower Blood Pressure
6. Pistachio *Pistacia vera*	Persia	• Lowers blood sugar levels • Promotes a healthy heart • Boosts the immune system • Improves hemoglobin • Improves vision • Makes digestion better • Prevents premature ageing • Imparts UV protection

With the increase in importance of nuts in international markets, their cultivation shifted from traditional to intensive pattern with immense use of agrochemicals that encompass pesticides, insecticides, fertilisers, etc. The introduction of agrochemicals, although with many positive effects on agricultural production, has led to many foreseen dangers in terms of un-natural substances entering agricultural produce. The use of chemicals in conventional cultivation harms producers as well as consumers in addition to deleterious effect on environment. For instance, in a study, it was found that foods grown from conventional agricultural are high in nitrogen, increasing its carcinogenicity, posing severe threat to living world (Savino et al. 2006). Moreover, in many countries such as France, the extensive use of fertilisers has increased nitrate content in water bodies making water unfit for use (Schroder et al. 2004; Camargo and Alonso 2006). The nitrates when present in soils in excessive amount enter plants and hence food chain via crops and finally ingested by humans causing severe problems such as nitrate poisoning (Taylor et al. 2002). Other fertilisers such as phosphates and super phosphates that are produced from phosphoric acid pollute the environment with heavy metals such as arsenic and cadmium which affect even the non-target species including humans. Pesticide residue in food is another alarming food safety issue arising due to conventional agriculture. A screening conducted in Germany detected 361 active pesticide substances in edible items with table grapes showing traces of almost 23 pesticide substances and 10–12 in fruits such as apple, pear, etc. (Johansson et al. 2014). Several studies show that the use of pesticide and the presence of their residues in the food may contribute to the development of cancer, Parkinson's disease, and endocrine related disorders (Ryan et al. 2013; Landauo-Ossondo et al. 2009). Organic cultivation is the answer to this issue which prevents health of a million producers per annum otherwise affected due to usage of pesticides (Soares et al. 2013). It is an approach that prevents synthetic chemicals, hormones, antibiotic agents, genetic engineering, and irradiation in plant and animal husbandry (Baydas et al. 2021). Sustainability encloses agricultural, economic, environmental and legal spheres. To check the safety of organic foods over inorganic food, urinary bio-monitoring was performed on children which showed that two specific metabolites malathion and chlorpyrifos were reduced to undetectable levels after 5-day substitution of inorganic food with organic food (Lu et al. 2006). Furthermore, children who consumed exclusively organic produce showed no measurable pesticide metabolites (Lu et al. 2001). Such findings led to increased diet consciousness and outlook towards choosing food changed from nutritional quality to food safety. Consumers now believe that organic foods are better than inorganic foods in terms of nutritional quality and safety, more beneficial for health and tastier, and they are willing to pay more for it (Hansen et al. 2018; Asif et al. 2018). In recent years, the increased demand for organic foods has led to a rise in organic food production and organic farming (Maggio et al. 2013). Almost 80 billion Euros of organic food was produced by 2.7 million producers on 57.8 million hectares of land as per a report by FiBL and IFOAM (2018). Organic agriculture is practiced in fruits, vegetables, cereals, oilseeds, nuts and plantations crops. Very limited data is available on organically cultivated nuts and organic Fruit and Nut Farming Market was estimated to be valued at $19.1 billion in 2021.The growth rate of the Global Organic Fruit and Nut Farming Market is 10.2%, with an estimated value of $28.2 billion expected by 2025 (Research and Markets 2021). Forty-six countries reported organic nut production, which increased by 180% from 2004 to 2006 (Willer et al. 2010). Since then, there is a constant increase in organic cultivation of nuts. Different techniques have been applied on nut crops to make their organic cultivation successful. This chapter gives a broader account of different organic practices used in organic cultivation of temperate nuts.

2 Organic Fertiliser Application of Temperate Nuts

The history of fertiliser use is at least as old as the agricultural history. It began with the use of organic manures and gradually shifted to chemical fertilisation. Nowadays, trend is shifting back to minimise the use of chemical fertilisers in all crops including fruits and nuts not just for integrated fruit productions but also to ensure environment sustainability (Reganold et al. 2001). Various studies have been conducted to compare effect organic fertiliser application on temperate nut crops. In a study, pecan trees were fertilised with NPK fertilisers (CF) and biosolids including a treatment with rhizospheric fungi *Pisolithus tinctorius* + *Scleroderma* sp. and *Trichoderma* sp. for 3 years. After studying the effect of treatments, it was found that bio solids increased the growth of Pecan trees by 9.5%. The yield and quality of nuts produced was same in both cases, i.e. biosolids were as efficient as conventional NPK fertilisers with mycorrhizal inoculation favouring the growth and increasing fructification in fertilised pecans (Rivero et al. 2009). The development of pecans was also compared after the application of organic amendment vermicompost, with inorganic solid fertiliser in two doses including a control. Plant growth was same in first year of study but during second year, the highest growth in diameter was detected in plants where either an organic amendment or high rate of a synthetic fertiliser was applied. However, in third year organic amendment showed much improved growth compared to synthetic fertiliser. Giuffré et al. (2011) also compared compost, vermicompost liquid fertiliser with control in pecans and reported an increment in soil microbial population, microbial biomass carbon, total carbon and phosphorus in soil which in turn improved nut production and plant diameter. In similar kind of study, animal wastes (blood powder + eggshell powder combination) were used as fertilisers to quantify their effect on various growth and quality parameters in young walnut seedlings. The animal wastes helped to mitigate drought by improving the flux of available nutrients to plants in addition to improving water holding capacity to plants that in turn improved the dry matter accumulation and branching architecture of plants. It also improved nutrient metabolism and photosynthesis resulting in high leaf water use efficiency and dry matter production in young walnut trees (Zhang et al. 2020). Similarly, poultry litter was applied on young seedlings and 35-year-old walnut orchard which resulted in improved height and leaf nitrogen of younger seedlings in first year and increased nuts in second year. Age old orchard did not show any significant response towards application of poultry litter (Ponder Jr et al. 2005). Bilgin et al. (2021) applied microbial fertilisers to walnut trees both as foliar and soil application. Soil application of microbial strain improved the shoot length, nut weight and overall yield. This increase was attributed to increased hormone activity in plants. Several endomycorrhizal fungi are reported to have a potential of drought mitigation in walnut trees. To exemplify, two mycorrhiza species *Glomus mosseae* and *Glomus etunicatum* were inoculated on walnut saplings and drought stress was induced to saplings for a period of 20 days. *Mycorrhiza* improved macro as well as micronutrient uptake in stressed plants and inoculated plants had a better drought tolerance towards drought stress (Behrooz et al. 2019).

Effect of organic fertiliser has also been studied on Hazelnuts. A comparative study was made between hazelnuts grown under organic and mineral fertilisation. It was found that concentration and peak intensity values of Ca, Fe, Mn, P, Mg, Zn, Cl, Na, Br, Rb, F, K and Se elements were higher in hazelnuts grown under organic farming regime but Al, Cr and Ni levels were found to be higher in samples grown under conventional farming regime concluding that organic hazelnuts are likely to have higher nutritional mineral content (Akbaba et al. 2011). In another study, hazelnut husk compost, cattle manure, and mineral fertilisers were applied on hazelnut trees to study their effect on yield, quality and soil parameters. It was found that husk compost improved soil biological activity and concentration of potassium and phosphorus in soil. There was a considerable improvement in yield as well as quality parameter of hazelnuts (Ozenc and Caliskan 2000). Application of mycorrhizal fungi on hazelnuts has also shown positive results in improving their growth attributes. For instance, micro propagated plants of hazelnut were treated with mycorrhizal fungi to obtain better propagation in controlled conditions. Inoculated plants attained greater biomass compared to un-inoculated plants paving the way for the use of fungi for improvement in growth of hazelnuts (Mirabelli et al. 2007).

Pistachio nut which is mostly grown in Iran on low organic matter soils have also responded fairly well to application of organic treatments. In a study, pistachio was treated with municipal solid waste compost, cow manure and humic acids (Razavi Nasab et al. 2019). Foliar application of humic acid significantly improved total chlorophyll by 29% compared to control. Humic acids also interacted with organic matter to influence leaf number, and leaf surface area. Similarly, time of humic acid application influenced stem diameter, leakage and chlorophyll a while as time of organic waste application influenced total chlorophyll, chlorophyll a, chlorophyll b, and carotenoid contents of pistachio leaves. The improvement in physiological characters was attributed to slow release of nutrients and their contiguous storage in organic matter. Hence, combination of municipal solid waste compost and cow manure with humic acids is a good alternative to organically supply nutrients and improve plant quality parameters in pistachio nut. Nadi et al. (2011) used vermicompost synthesised from different organic wastes such as cow dung, pistachio waste cotton residue and date waste enriched with iron sulphate to grow pistachio nuts. This vermicompost when mixed with loamy soil increased the seed germination, leaf area index shoot dry weight and stem diameter of plants and resulted in increase in chlorophyll and iron content in leaves. Not only different organic manure, microbial fertilisers have also been studied for their use in pistachio nuts, e.g. a study was effectuated under controlled conditions where microbial fertilisers and organic manure were applied pistachio plants and effects on growth parameters were recorded. The combined application of microbial fertilisers and organic manure increased leaf dry weight, stem dry weight and shoot dry weight compared to control. Microbial fertilisers also increased the uptake of micro and macronutrients in seedlings with a positive impact on soil properties (Jamalifard et al. 2017). Microbial fertilisers may also help to mitigate drought in pistachio plants which was confirmed in one of the studies conducted by Salmani et al. (2020) by treating pistachio nuts with effective microorganisms and subjecting seedlings to salinity stress. After 3 months, various physiological parameters of plants were studied which revealed that in presence of effective microorganisms, the salinity could not decrease relative leaf water content, macro and micronutrients chlorophyll a and b to lethal levels. It also prevented over secretion of proline and prevented over accumulation of sodium in plants which effectively reduced the damages of salinity stress to plants (Fazely-Salmani et al. 2020). All the studies discussed establish that there are multiple alternatives available for shifting from mineral fertilisers to organic for supplying nutrients to nut crops, however, further studies need to be done to establish location specific practice keeping in view the economics and profitability of crop.

3 Biocontrol Strategies in Temperate Nuts

Organic farming that aims at production of food without the application of harmful chemical fertilisers is prone to many diseases and pests. There is a need to cope up with these disease and pests to prevent economic losses to crop and prevent any kind of environmental degradation. Hence, emphasis is laid on application of biocontrol agents to control pests and diseases instead of chemical pesticides (Ram et al. 2018). Biocontrol agents may be precisely defined as any living entities, e.g. microbes or substances derived from living entities, e.g. botanicals that potentially reduce the incidence of disease or pests in a plant. In organic farming, biocontrol agents play a curative role to bring down the soaring demand of chemical pesticides. The biocontrol agents may include various microorganisms such as *Trichoderma* or antimicrobial botanicals such as neem (*Azadirachta indica*), garlic (*Allium sativum*), eucalyptus (*Eucalyptus globules*), turmeric (*Curcuma longa*), tobacco (*Nicotiana tabacum*) and ginger (*Zingiber officinale*). Various biocontrol strategies have been tested on various crops, namely fruits, vegetables ornamentals, etc. which shall be discussed to chalk out plan for shifting nut cultivation from conventional to organic.

3.1 Biocontrol of Diseases in Temperate Nuts

Walnut which is one of the important nut crops of temperate region is attacked by several disease-causing micro-organisms that lead to potential damage to tree as well as its produce. One of the diseases that occur on a destructive scale is the black spot of walnut (*X. campestris* pv. *Juglandis*). Researchers tried to control this disease using actinomycetes, strain A217 a biocontrol agent. Under controlled conditions, a fermentation broth actinomycetes strain A217 was used against black spot of walnut using acupuncture inoculation and streptomycin was used as a positive control. The control effect of actinomycetes broth and streptomycin was 79.33% and 93.33% on leaves, respectively. The control effect of actinomycetes A217 fermentation broth on walnut black spot disease in branches was 81.52% and for streptomycin 78.42% (He et al. 2020). Another disease of walnut, 'walnut bacterial blight', caused due to *Xanthomonas arboricola* pv. *Juglandis* is mostly controlled chemically by application of various copper-based fungicides from bud break up to August. However, while trying to search for organic alternative for the control of this disease, Ozaktan et al. (2012) tried to control this disease using bacterial antagonists mostly isolated from phylloplane of healthy walnut trees and tested against Xaj. *Pantoea vegans* strain C9/1. Among the selected strains, 60% had a potential to significantly reduce the bacterial blight by 41–77% on immature nuts. The effective strains were further tested on walnut seedlings and four putative antagonists effectively controlled bacterial blight on walnut leaves reducing the attack blight up to 82%. The strains that successfully controlled the disease were designated as *Pseudomonas fluorescence* which needs a further investigation to check the optimum levels that must be applied to achieve better control. Fungal diseases of walnut have also been controlled organically by application of fungal antagonists. To exemplify, a study was conducted out to control anthracnose of walnut (*Colletotrichum gloeosporioides*) using an antifungal agent *Bacillus velezensis*. It decreased the diseases severity of anthracnose in field by 1.3-fold to 6.9-fold. The decrease in spore germination and mycellial growth was 99.3% and 33.6% which was attributed to secretion of chitinase, protease, and β-l,3-glucanase by *Bacillus velezensis*. In addition to disease control, the bacterium successfully increased nutrient uptake, enhanced chlorophyll content and consequently biomass by 1.5-fold to 2.0-fold (Choub et al. 2021).

Pecans have also responded well to biological control of diseases. Low environmental risk fungicides have been tried as an alternative to conventional fungicides for controlling pecan scab caused by *Fusicladium effusum*. The metabolites were obtained from the nematode symbiont *Photorhabdus luminsens*. After separating by fractionating approach and identifying by, the metabolite was identified as *trans*-cinnamic acid (TCA) which was tested for its potential to combat *Fusicladium effusum*. TCA was effective against disease at 148–200+ μg/mL. In liquid media, TCA arrested all growth of *F. effusum* at a concentration even as low as 64 μg/mL (Bock et al. 2019). In another study conducted in the USA, pecan scab caused due to *Venturia effuse* was controlled using organic fungicides biocontrol agents *Bacillus subtilis* and extract of giant Knotweed *Reynoutria sachalinensis*. The biocontrol agents reduced severity on foliage with little or no difference between (Bock et al. 2014). Several biocontrol agents have been isolated for treatment of diseases occurring in hazelnuts. Bacterial strains isolated from rhizosphere and phyllosphere of hazelnut to test for their potential against die back of hazelnut. Among the isolates, two strains *Pseudomonas avellanae* and *Pseudomonas syringae* pv. *coryli* successfully inhabited growth of dieback of hazelnuts (Gentili et al. 2008). Biocontrol agents have also been tested to control various diseases in pistachio nuts. Die back and canker caused due to *Paecilomyces formosus* was controlled biologically by testing 50 strains of streptomycetes in vitro and selecting two strains BH4-1 and BH4-3 for biocontrol in green house. The strains showed promising results in control of disease reducing the diameter of lesions almost as equally as chemical fungicide (Torabi et al. 2019). Previously, Zarei Jalalabadi et al. (2015) investigated biocontrol of *P. variotii* by different *Trichoderma* species in vitro. Biological control of *Xanthomonas translucens*, the causal agent of this disease in Australia was carried out via: *Bacillus subtilis* and *Pseudomonas* sp. (Salowi 2010). *Trichoderma harzianum* has also been used to treat wilt disease of pistachio nut which is one of the most devastating diseases of pistachio nut caused due to *Verticillium dahliae*. The biocontrol agents have been tested in both green house and laboratory conditions and 20 isolates were obtained that had a capability of inhibiting mycelia growth that showed promising results. Among these, 5 isolates were selected to be tested in green house among which 2 isolated Tr8 and Tr19 showed an inhibition of 94.94% and 88.15% (Fotoohiyan et al. 2017). Panicle and shoot blight of pistachio nuts caused due to *Botryosphaeria dothidea* affects leaves, panicle rachises, fruits, shoots and buds and always threatens growers by its yield destroying frequency. The disease in conventional farming is controlled by application of almost 5–6 sprays of fungicide in a year. The disease is alternatively controlled by the application of microbial fungicide *Paenibacillus lentimorbus* obtained from healthy pistachio leaves. This microbial fungicide stopped the germination of pycnidiospore as well as development of lesions on pistachio leaves and showed a good resistance to fungicides such as Azoxystrobin, Benomyl, Tebuconazole, Propiconazole, or Trifloxystrobin at their highest concentration (Chen et al. 2003). Not only pistachio tree, several biocontrol agents have been isolated from pistachio growing soils that have antagonistic activity against pistachio plant pathogens. An antibiotic streptomycetes was isolated from pistachio soil to test its inhabiting effect on *Phythopthora drechsileri* that causes root rot and gummosis in pistachio. The biocontrol agent inhibited the growth of fungal mycelium in vivo and in vitro increasing the number of healthy plants considerably (Shahidi-Bonjar et al. 2006). The efficiency of biocontrol agents against different pathogens in walnut, pecans, pistachio nut and hazelnut pave a way for shifting from conventional cultivation to organic cultivation in other temperate nuts such as pine nut, chestnut, etc.

3.2 *Biocontrol of Pests in Temperate Nuts*

Insects are the most abundant and diverse tiny creatures but quite versatile and major competitors with humans for resources generated by agriculture (Oerke and Dehne 2004). These cause considerable losses to crops globally and cause a great loss to yield threatening food security and sustainability. According to a survey conducted by FAO, almost 40% of worlds agricultural crops are lost to pests each year that cost $70 billion dollars annually to countries (Gullino et al. 2021). To overcome these losses, a large quantity of pesticides is applied, and global market value is expected to increase from US$50.62 billion in 2017 to US$68.82 billion by the end of 2025 as estimated by fortune business industries (2020). This is an alarming issue because this not only increases the cost of cultivation but also affects air, water and soil in addition to increase in pest resistance. This will have a disastrous effect on biodiversity, ecology and human health (Sexton et al. 2007). Efforts are being constantly made to reduce application of pesticides with some biocontrol agents called bio pesticides. Biopesticides have shown positive results in many crops including nut crops. Among nut crops, biocontrol has been tested against walnut aphid, *Chromaphis juglandicola* that is present in large numbers during spring season and reduces tree vigor, nut quality causing huge economic losses. The trial for biological control of this pest dates back to 1969 when parasitic wasp *Trioxys pallidus* from Iran led to a dramatic success in its biological control. This was followed by another experiment where a predatory bug *Orius sauteri* was tested as a biocontrol agent against two walnut aphids, i.e. dusky-veined aphid and walnut aphid species that coexist in adaxial and abaxial side of leaves. The efficiency of biocontrol agent was as high as 77% for *P. juglandis* and 80% for *C. juglandicola*. However, the efficiency declined when both aphid species were present on the same leaf (Wang et al. 2021). In some cases, biological control of pests in temperate nuts has been successfully achieved using entomopathogenic nematodes. With pecans, Shapiro-Ilan et al. (2005) tested the use of entomopathogenic nematodes (*S. carpocapsae*, *S. feltiae*, and *H. bacteriophora*) to control weevil in pecans (*Curculio caryae*). However, the experiment was partially successful as only 35% of larvae were infected that too up to certain age. Pecan weevil has also been organically managed using microbial insecticide Grandevo (based on *Chromobacterium subtsugae*) which worked as efficiently as chemical insecticide. Later it was tested on Pecan aphid complex black pecan aphid, black margined aphid and yellow pecan aphid. This microbial insecticide successfully controlled aphid complex showing much effect against black pecan aphid and in addition maintained high population of natural enemies (Oliveira-Hofman et al. 2021). Studies were further continued and control of chestnut pests, *Curculio elephas* and *Cydia splendana* in laboratory using three species of entomopathogenic nematodes *Steinernema feltiae*, *S. weiseri* and *Heterorhabditis bacteriophora*, were investigated. In addition to virulence, the effect of temperature on the development of infection and death of host was studied. The cold-adapted *S. weiseri* and *S. feltiae* were the most virulent species at 10 and 15°C, whereas the warm-adapted *H. bacteriophora* was the most effective at 20 and 25°C. In soil pot experiments conducted at 15°C, *S. weiseri* was the most virulent species against *C. elephas* and *C. splendana*. Asian Chestnut Gall Wasp (ACGW) that drastically reduces the yield of this nut (up to 80%) by causing flower abortion has also been controlled using organic practices. Multiple tests were carried to explore natural parasitoids on gall wasps from different countries however most of them showed low parasitism. The population of ACGW was regulated by a complex of 11 hymenopteran parasitoid species (Bailey et al. 2009; Zhang 2009), with *T. sinensis Kamijo* showing the most promising result due to its univoltine, highly host-specific and phenologically adapted to the life cycle of the ACGW (Kamijo 1982; Moriya et al. 2003).

Preventive biological measures such as trap crops are adopted to control pests in nut crops that include pistachio nut. Pistachio nut which is the main crop of California's central valley suffers attack from hemipteran pests in large numbers contributing to nut damage. The incidence of attack is mostly in organic production, hence needs efficient preventive and curative measures. Therefore, in a 2-year-old study, trap crops that included alfalfa, vetch, mustard and radish were sown between rows. After collecting samples from ground cover and tree canopy, small bugs were abundant in tree canopy and large bugs in trap crops. Trap crops also harboured more numbered of natural enemies which suggested that population of hemipteran pests can be maintained below thresh hold level to reduce economic losses to pistachio nuts (Stahl et al. 2021). All the studies conducted suggest that there is great diversity of micro and macro organisms available that may act as biocontrol agents to combat pest attack on temperate nuts. These can be further studied for time of release, optimum population to achieve better control and phonological stage of pests when biocontrol is most effective.

4 Mulching

Mulching is an effective strategy for soil management as it improves the tillage quality by increasing the soil's biological activity, regulating the soil temperature, controlling weed growth, reducing soil evaporation, and preventing nutrient leaching. It also provides benefits for growth, annual and perennial yield, and the health of plants, and its low cost makes it a widespread technique. It is one of the important techniques used in organic cultivation of many crops including the nuts. Applying organic mulches to base of trees has so far been as effective as chemical herbicides or mechanical weed control. These mulches have more impact on physical condition of soils. They release nutrients, organic acids and increases organic matter of soil causing better aggregation and increased retention of nutrients. Organic mulches around walnut trees are reported to increase growth by almost 89% (Van Sambeek and Garrett 2004). In an experiment conducted by Dong et al. (2021), three types of green manures cum mulch, namely orchid, hairy vetch, rattail and fescue were used on walnut trees. After recording different growth parameters, it was observed that height, ground diameter, root diameter and root density increased compared to no treatment. Organic mulch applied during establishment of pecan orchard at three different depths was compared with un-mulched herbicide treatment and a common Bermuda grass sod. While measuring the trunk cross sectional area (TCSA) after 3 years, trees supplied with organic mulch had TCSA 60–70% more than trees supplied with chemical herbicide and sod. It was inferred that mulches can be used to improve growth of pecan trees (Foshee et al. 1996). Similarly, mulching showed a positive effect on growth of chestnut trees in an experiment laid down by Tian and Li (2022) in which fragment branches, chestnut shell, and involucres were grinded into 3–10 cm pieces, forming lingo cellulose mulch which was applied to cover the soil surrounding Chinese chestnut trees. This mulch improved soil moisture and soil carbon. Moreover, it successfully inhabited weed growth and helped to combat chestnut blight which provides an alternative strategy for the management of chestnut and other perennial trees. In another experiment, brassica cover crops and hazelnut husk were used as weed management strategy in hazelnut (*Corylus avellana*) orchard. The cover crops reduced weed density, weed dry weight and diversity of weed species compared to control in addition to showing positive effect on growth of hazelnut trees (Mennan and Ngouajio 2012).

5 Future Strategies

The use of organic practices such as mulching, bio control agents, and PGPRs restricts the amount of pesticides, weedicides and other agrochemicals used in conventional cultivation. Although many studies have so far been effectuated to establish concrete organic practices for nut crops, most of the studies are carried out under controlled conditions especially the ones related to biocontrol practices. Therefore, there is a need to further explore the practices beyond lab and green house in field conditions. In addition, there is a need to establish proper application rates, application time of different biocontrol agents and study the environmental factors that influence their efficiency. Efforts must be taken to increase the efficiency of organic production which will make a way towards sustainability, food security and resilient environment for common benefit.

6 Conclusions

Nut crops hold an important place in organic markets and land under nut crops is increasing continuously. Increasing research in organic farming is opening new horizons towards the cultivation of organic nut crops and helps to overcome pest, fertility and yield challenges in diverse environment making the cultivation more sustainable. The use of organic cultivation practices for growing temperate nuts seems to have many positive effects on the environment in general and soil in particular. Organic management favours the growth of bacterial communities and enhances enzyme activity of soil which influences all nutrient turnover cycles. Organic cultivation of nuts improves the quality of cultivated nuts increasing their consumer demand and market value.

References

Afshin A, Sur PJ, Fay KA, Cornaby L, Ferrara G, Salama JS, Murray CJ (2019) Health effects of dietary risks in 195 countries, 1990–2017: a systematic analysis for the Global Burden of Disease Study 2017. Lancet 393(10184):1958–1972

Akbaba U, Şahin Y, Türkez H (2011) Comparison of element contents in hazelnuts grown under organic and conventional farming regimes for human nutrition and health. Fresenuis Environ Bull 20(7):1655–1660

Asif M, Xuhui W, Nasiri A, Ayyub S (2018) Determinant factors influencing organic food purchase intention and the moderating role of awareness: a comparative analysis. Food Qual Prefer 63: 144–150

Bailey R, Schönrogge K, Cook JM, Melika G, Csóka G, Thuróczy C, Stone GN (2009) Host niches and defensive extended phenotypes structure parasitoid wasp communities. PLoS Biol 7(8): 1000179

Baydas A, Yalman F, Bayat M (2021) Consumer attitude towards organic food: determinants of healthy behaviour. Market Manag Innov:96–111. https://doi.org/10.21272/mmi.2021.1-08

Behrooz A, Vahdati K, Rejali F, Lotfi M, Leslie C, Behrooz A (2019) Effects of endomycorrhizal fungi and drought stress on nutrient acquisition of Walnut (*Juglans regia* L). J Res Biol 9(2): 2644–2655

Bilgin NA, Yagmur B, Misirli A (2021) Preliminary study on investigation of the use of effective microorganism applications in walnut. Int J Agric Environ Food Sci 5(3):329–334

Bock CH, Shapiro-Ilan DI, Wedge DE, Cantrell CL (2014) Identification of the antifungal compound, trans-cinnamic acid, produced by *Photorhabdus luminescens*, a potential biopesticide against pecan scab. J Pest Sci 87(1):155–162

Bock CH, Hotchkiss MW, Shapiro-Ilan DI, Brock JH, Brenneman TB, Wilkins B, Mizell RF (2019) A comparison of organic fungicides: alternatives for reducing scab on pecan. Org Agric 9(3):305–314

Camargo JA, Alonso Á (2006) Ecological and toxicological effects of inorganic nitrogen pollution in aquatic ecosystems: a global assessment. Environ Int 32(6):831–849

Chen WQ, Morgan DP, Felts D, Michailides TJ (2003) Antagonism of *Paenibacillus lentimorbus* to *Botryosphaeria dothidea* and biological control of panicle and shoot blight of pistachio. Plant Dis 87(4):359–365

Choub V, Ajuna HB, Won SJ, Moon JH, Choi SI, Maung CEH, Ahn YS (2021) Antifungal activity of Bacillus velezensis CE 100 against anthracnose disease (Colletotrichum gloeosporioides) and growth promotion of walnut (Juglans regia L.) trees. Int J Mol Sci 22(19):10438

De Souza RGM, Schincaglia RM, Pimentel GD, Mota JF (2017) Nuts and human health outcomes: a systematic review. Nutrients 9(12):1311

Dong N, Hu G, Zhang Y, Qi J, Chen Y, Hao Y (2021) Effects of green-manure and tillage management on soil microbial community composition, nutrients and tree growth in a walnut orchard. Sci Rep 11(1):1–13

Fazely-Salmani AL, Davarynejad GH, Abedi B (2020) Effect of effective microorganisms on physiological and biochemical responses of UCB1 pistachio under salinity stress. Res Pomol 5(1):71–80

FİBL and IFOAM Organics International (2018) The world of organic agriculture, statistics and emerging trends 2018. https://shop.fibl.org/CHen/mwdownloads/download/link/id/1093/?ref=1

Fortune Business Insights (2020) Crop protection chemicals market size, share and industry analysis by product type (herbicides, insecticides, fungicides), origin (synthetic chemicals, bio-based) and regional forecast 2018–2025

Foshee WG, Goff WD, Tilt KM, Williams JD, Bannon JS, Witt JB (1996) Organic mulches increase growth of young pecan trees. HortScience 31(5):811–812

Fotoohiyan Z, Rezaee S, Bonjar GHS, Mohammadi AH, Moradi M (2017) Biocontrol potential of *Trichoderma harzianum* in controlling wilt disease of pistachio caused by Verticillium dahliae. J Plant Protect Res 57:185–193

Gentili A, Mariotti E, Vincenzi A, Mazzaglia A, Heydari A, Schaad NW, Balestra GM (2008) Dieback (Moria) of hazelnut: isolation and characterization of two potential biocontrol agents. J Plant Pathol 90:383–386

Giuffré L, Romaniuk R, Ríos RP, Zubillaga MM, Rastrelli L (2011) Sustainable management in pecan cultivation in Argentina. Emirates J Food Agric 23(3):243–248

González CA, Salas-Salvadó J (2006) The potential of nuts in the prevention of cancer. Br J Nutr 96(2):S87–S94

Griel AE, Kris-Etherton PM (2006) Tree nuts and the lipid profile: a review of clinical studies. Br J Nutr 96(2):S68–S78
Gullino M, Albales R, Al-jboory I, Angelotti F, Chakraborty S, Garrett K, Stephenson T (2021) Scientific review of the impact of climate change on plant pests: a global challenge to prevent and mitigate plant pest risks in agriculture, forestry and ecosystems. Embrapa Semiárido-Livro Técnico
Hansen T, Sorensen MI, Eriksen MLR (2018) How the interplay between consumer motivations and values influences organic food identity and behavior. Food Policy 74:39–52
He H, Hao X, Zhou W, Shi N, Feng J, Han L (2020) Identification of antimicrobial metabolites produced by a potential biocontrol Actinomycete strain A217. J Appl Microbiol 128(4): 1143–1152
Jamalifard A, Abbaszadeh Dahaji P, Akhgar A (2017) The effect of organic and microbial fertilizers on growth and nutrition of pistachio seedlings. J Pistachio Sci Technol 2(4):41–60
Jiang R, Jacobs DR Jr, Mayer-Davis E, Szklo M, Herrington D, Jenny NS, Barr RG (2006) Nut and seed consumption and inflammatory markers in the multi-ethnic study of atherosclerosis. Am J Epidemiol 163(3):222–231
Johansson E, Hussain A, Kuktaite R, Andersson SC, Olsson ME (2014) Contribution of organically grown crops to human health. Int J Environ Res Public Health 11(4):3870–3893
Kamijo K (1982) Two new species of Torymus (*Hymenoptera, Torymidae*) reared from Dryocosmus kuriphilus (*Hymenoptera, Cynipidae*) in China and Korea. Kontyu 50:505–510
Landau-Ossondo M, Rabia N, Jos-Pelage J, Marquet LM, Isidore Y, Saint-Aimé C, ARTAC International Research Group on Pesticides (2009) Why pesticides could be a common cause of prostate and breast cancers in the French Caribbean Island, Martinique. An overview on key mechanisms of pesticide-induced cancer. Biomed Pharmacother 63(6):383–395
Lu C, Knutson DE, Fisker-Andersen J, Fenske RA (2001) Biological monitoring survey of organophosphorus pesticide exposure among pre-school children in the Seattle metropolitan area. Environ Health Perspect 109(3):299–303
Lu C, Toepel K, Irish R, Fenske RA, Barr DB, Bravo R (2006) Organic diets significantly lower children's dietary exposure to organophosphorus pesticides. Environ Health Perspect 114(2): 260–263
Maggio A, De Pascale S, Paradiso R, Barbieri G (2013) Quality and nutritional value of vegetables from organic and conventional farming. Sci Hortic 164:532–539
Mennan H, Ngouajio M (2012) Effect of brassica cover crops and hazelnut husk mulch on weed control in hazelnut orchards. Hortic Technol 22(1):99–105
Mirabelli C, Tullio M, Pierandrei F, Rea E (2007) Effect of arbuscular mycorrhizal fungi on micropropagated hazelnut (*Corylus avellana* L.) plants. In: III International symposium on acclimatization and establishment of micropropagated plants, vol 812, pp 467–472
Miraliakbari H, Shahidi F (2008) Lipid class compositions, tocopherols and sterols of tree nut oils extracted with different solvents. J Food Lipids 15:81–96
Moriya S, Shiga M, Adachi I (2003) Classical biological control of the chestnut gall wasp in Japan. In: Van Driesche RG (ed) Proceedings of the 1st international symposium on biological control of arthropods. USDA Forest Service, Forest Health Technology Enterprise Team, Washington, pp 407–441
Nadi M, Golchin A, Mozafari V, Saeidi T, Sedaghati E (2011) The effects of different vermicomposts on the growth and chemical composition of the pistachio seedlings. J Res Agric Sci 7(1):59–69
Nuts (2019) I.N.C & yearbook, D.F.S. 2020. https://www.nutfruit.org/files/tech/1587539172_INC_Statistical_Yearbook_2019-2020.pdf
Oerke EC, Dehne HW (2004) Safeguarding production – losses in major crops and the role of crop protection. Crop Protect J 23:275–285
Oliveira-Hofman C, Cottrell TE, Bock C, Mizell RF III, Wells L, Shapiro-Ilan DI (2021) Impact of a biorational pesticide on the pecan aphid complex and its natural enemies. Biol Control 161: 104709

Ozaktan H, Erdal M, Akkopru A, Aslan E (2012) Biological control of bacterial blight of walnut by antagonistic bacteria. J Plant Pathol 94. https://doi.org/10.4454/jpp.v94i1sup.009

Ozenc N, Caliskan N (2000) Effects of husk compost on hazelnut yield and quality. ISHS Acta Hortic 556:559–566

Ponder F Jr, Jones JE, Mueller R (2005) Using poultry litter in black walnut nutrient management. J Plant Nutr 28(8):1355–1364

Ram RM, Keswani C, Bisen K, Tripathi R, Singh SP, Singh HB (2018) Biocontrol technology: eco-friendly approaches for sustainable agriculture. In: Barh D, Azevedo VAC (eds) Omics technologies and bio-engineering. Elsevier, pp 177–190

Razavi Nasab A, Fotovat A, Astaraie A, Tajabadipour A (2019) Effect of organic waste and humic acid on some growth parameters and nutrient concentration of pistachio seedlings. Commun Soil Sci Plant Anal 50(3):254–264

Reganold JP, Glover PK, Andrews PK, Hinman HR (2001) Sustainability of three apple production systems. Nature 410:926–929

Research and Markets (2021) Organic fruit and nut farming global market report 2021: COVID-19 growth and change to 2030. https://www.researchandmarkets.com/content-images/187/187273/2/global-organic-fruit-and-nut-farming-market.png

Rivero ST, Moorillón VN, Borunda EO (2009) Growth, yield, and nutrient status of pecans fertilized with biosolids and inoculated with rizosphere fungi. Bioresour Technol 100(6): 1992–1998

Ros E (2010) Health benefits of nut consumption. Nutrients 2(7):652–682

Ryan SD, Dolatabadi N, Chan SF, Zhang XF, Akhtar MW, Parker J, Soldner F, Sunico CR, Nagar S, Talantova M, Lee B, Lopez K, Nutter A, Shan B, Molokanova E, Zhang YY, Han XM, Nakamura T, Masliah E, Yates JR, Nakaniski N, Andreyev AY, Okamoto S, Jaenisch R, Ambasudhan R, Lipton SA (2013) Isogenic human iPSC Parkinson's model shows nitrosative stress-induced dysfunction in MEF2-PGC1 alpha transcription. Cell 155:1652–1653

Salmani AF, Davarynejad GH, Abedi B (2020) Could effective microorganisms improve tolerance of UCB1 pistachio to salinity? Comunicata Scientiae 11:3435–3435

Salowi A (2010) Towards biological control of pistachio die-back. Adelaide, Australia, School of Agriculture, Food and Wine, University of Adelaide, p 116

Sathe SK, Kshirsagar HH, Roux KH (2005) Advances in seed protein research: a perspective on seed allergens. J Food Sci 70:93–120

Savino F, Maccario S, Guidi C, Castagno E, Farinasso D, Cresi F, Mussa GC (2006) Methemoglobinemia caused by the ingestion of courgette soup given in order to resolve constipation in two formula-fed infants. Ann Nutr Metab 50(4):368–371

Schroder JJ, Scholefield D, Cabral F, Hofman G (2004) The effects of nutrient losses from agriculture on ground and surface water quality: the position of science in developing indicators for regulation. Environ Sci Policy 7(1):15–23

Segura R, Javierre C, Lizarraga MA, Ros E (2006) Other relevant components of nuts: phytosterols, folate and minerals. Br J Nutr 96(2):36–44

Sexton SE, Lei Z, Zilberman D (2007) The economics of pesticides and pest control. Int Rev Environ Resour Econ 1(3):271–326

Shahidi F, Miraliakbari H (2005) Tree nut oils. In: Shahidi F (ed) Bailey's industrial oil and fat products, edible oil and fat products: specialty oils and oil products, vol 3, 6th edn. Wiley Interscience, Hoboken, NJ, pp 175–193

Shahidi-Bonjar GH, Barkhordar B, Pakgohar N, Aghighi S, Biglary S, Rashid Farrokhi P, Aghelizadeh A (2006) Biological control of Phytophthora drechsleri Tucker, the causal agent of pistachio gummosis, under greenhouse conditions by use of actinomycetes. Plant Pathol J 5(1):20–23

Shapiro-Ilan DI, Duncan LW, Lacey LA, Han R (2005) Orchard applications. In: Grewal PS, Ehlers R-U, Shapiro-Ilan DI (eds) Nematodes as biocontrol agents. CABI Publishing, Wallingford, UK, pp 215–229

Soares DJ, Vasconcelos PHMD, Camelo ALM, Longhinotti E, Sousa PHMD, Figueiredo RWD (2013) Prevalent fatty acids in cashew nuts obtained from conventional and organic cultivation in different stages of processing. Food Sci Technol 33:265–270

Stahl JM, Wilson H, Straser RK, Maccaro JJ, Daane KM (2021) Irrigated trap crops impact key hemipteran pests in organic pistachio orchard. Arthropod Plant Interact 15(6):949–959

Tapsell LC, Gillen LJ, Patch CS, Batterham M, Owen A, Baré M, Kennedy M (2004) Including walnuts in a low-fat/modified-fat diet improves HDL cholesterol-to-total cholesterol ratios in patients with type 2 diabetes. Diabetes Care 27(12):2777–2783

Taylor MD, Klaine SJ, Carvalho FP, Barcelo D, Everaarts J (eds) (2002) Pesticide residues in coastal tropical ecosystems: distribution, fate and effects. CRC Press

Tian G, Li Y (2022) Lignocellulose mulch increases the economic benefit of Chinese chestnut by suppressing weed and ameliorating soil properties. Sci Hortic 291:110576

Torabi A, Bonjar GHS, Abdolshahi R, Pournamdari M, Saadoun I, Barka EA (2019) Biological control of Paecilomyces formosus, the causal agent of dieback and canker diseases of pistachio by two strains of Streptomyces misionensis. Biol Control 137:104029

Van Sambeek JW, Garrett HE (2004) Ground cover management in walnut and other hardwood plantings, vol 243. United States Department of Agriculture Forest Service General Technical Report NC, p 85

Wang T, Zhang P, Ma C, Yasir Ali M, Gao G, Lu Z, Zalucki MP (2021) Is Orius sauteri Poppius a promising biological control agent for walnut aphids? An assessment from the laboratory to field. Insects 12(1):25

Willer H, Granatstein D, Kirby E (2010) The global extent and expansion of organic horticulture production. Acta Hortic 933:23–34

Willett W, Rockström J, Loken B, Springmann M, Lang T, Vermeulen S, Garnett T, Tilman D, DeClerck F, Wood A, Jonell M, Clark M, Gordon LJ, Fanzo J, Hawkes C, Zurayk R, Rivera JA, De Vries W, Sibanda LM, Afshin A, Chaudhary A, Herrero M, Agustina R, Branca F, Lartey A, Fan S, Crona B, Fox E, Bignet V, Troell M, Lindahl T, Singh S, Cornell SE, Srinath Reddy K, Narain S, Nishtar S, Murray CJ (2019) Food in the Anthropocene: the EAT–Lancet Commission on healthy diets from sustainable food systems. Lancet 393(10170):447–492

Zarei Jalalabadi A, Mohammadi S, Moradi M (2015) Study of the effect of different isolates of *Trichoderma spp.* on pistachio trees die-back disease in vitro. 1st National agriculture, environment and food safety conference. University of Jiroft, Jiroft, Kerman, Iran

Zhang Z (2009) Study approaches on the chestnut gall wasp, *Dryocosmus kuriphilus* Yasumatsu in China. Acta Hortic 844:425–432

Zhang Y, Liu J, Niu S, Kong M, Zhang J, Lu Y, Yao Y (2020) Animal wastes as fertilizers enhance growth of young walnut trees under soil drought conditions. J Sci Food Agric 100(8): 3445–3455

Chapter 12
Shelf Life Enhancement of Temperate Nuts

Shabir Ahmad Mir, Manzoor Ahamd Shah, Saqib Farooq, Kappat Valiyapeediyekkal Sunooj, Mohammad Maqbool Mir, Umar Iqbal, Sehrish Jan, Shabnam Ahad, and Tajamul F. Wani

Abstract Temperate nuts are one of the important crops which are consumed worldwide. The nuts can have a longer shelf life if postharvest operations are done properly. Nuts are harvested at an optimum maturity and then hulled, cleaned and dried to safe moisture level. Nuts may be marketed as unshelled or shelled. Also, the unshelled nuts have generally longer shelf life than shelled nuts and thus can be stored for longer period of time to meet the consumer demand. Many factors such as moisture content, packaging and storage conditions considerably affect the shelf life of nuts. Packaging material influences the microbial contamination and other biochemical changes such as oxidation and rancidity in nuts due to permeability to moisture and gases. Storage factors including temperature, oxygen concentration and light have the significant effect on the shelf life of nuts.

Keywords Moisture · Drying · De-shelling · Storage temperature · Shelf life · Oxygen concentration

---

S. A. Mir (✉)
Department of Food Science and Technology, Government College for Women, Srinagar, Jammu and Kashmir, India

M. A. Shah
Department of Food Science and Technology, Government Degree College for Women, Anantnag, Anantnag, Jammu and Kashmir, India

S. Farooq
Department of Food Technology, Islamic University of Science and Technology, Awantipora, Jammu and Kashmir, India

K. V. Sunooj
Department of Food Science and Technology, Pondicherry University, Puducherry, India

M. M. Mir · U. Iqbal · S. Jan · S. Ahad · T. F. Wani
Division of Fruit Science, FOH, Sher-e-Kashmir University of Agricultural Sciences and Technology of Kashmir, Srinagar, Jammu and Kashmir, India

M. M. Mir et al. (eds.), *Temperate Nuts*,
https://doi.org/10.1007/978-981-19-9497-5_12

1 Introduction

Temperate nuts have grained increasing popularity worldwide and have an important influence on the economy of several countries. There has been a gradual increase in area, production and productivity in the temperate nuts including chestnut, walnut, hazelnut pecan, pistachio, etc. Nuts have high nutrient contents and are consumed as healthy diet by the people of all age groups. Nuts provide valuable amounts of macronutrients, micronutrients and bioactive compounds. They are also rich sources of essential amino acids and tocopherols which are important for human health and reduce the risk of many diseases (Mir et al. 2022). The temperate nuts are harvested in a shorter harvesting season and are available in the market throughout the year. The freshly harvested nuts have high moisture percentage which makes them vulnerable to spoilage and quality deterioration.

Temperate nuts have usually longer shelf life when the postharvest operations are done properly. The improper postharvest operations may cause severe damage to nuts, qualitatively and quantitatively. The improper packaging and storage conditions lead to the microbial contamination, insect infestation and rancidity of temperate nuts. Nuts are dried immediately after harvest to prolong their shelf life. Nuts are low moisture foods and stored for longer period of time. The storage conditions after harvest are important as these affect the shelf life of nuts. Many studies have been conducted to maintain and increase the postharvest quality of temperate nuts (Chen and Pan 2022). The appropriate postharvest handling and storage enables the nut crops to be available throughout the year and throughout the world. Nuts are commonly stored at low temperature, low relative humidity and low oxygen concentration. Furthermore, other factors such as variety, harvest time and packaging also affect the quality and shelf-life stability of temperate nuts (Gama et al. 2018).

2 Postharvest Handling

Shelf life of temperate nuts are influenced by preharvest and postharvest factors. Preharvest factors include orchard management, climatic conditions, irrigation facility, fertiliser dose, nut cultivar, etc. (Mir et al. 2018). These factors have considerable effect on the quality and shelf life of nuts. Proper harvesting and handling of nuts are important to get the good quality of nuts and increase their marketability (Fig. 12.1).

Nuts are harvested at optimum maturation to avoid the quality loss and minimise the contamination associated with the microbial attacks and infestation with the insects (Kader 2013). The temperate nuts are generally harvest by shaken off trees to orchard ground at full maturity. The management of orchard floor is important for the harvest of nuts that are knocked to the ground during harvest to avoid the incorporation of debris which when the nuts are swept off the ground. The nuts should be collected from the orchard flour immediately and taken for further processing. The picked nuts are then stockpiled outdoor storage with aeration

Fig. 12.1 Nuts (**a**) pecan nuts, (**b**) walnuts, (**c**) chestnut

which favours the easy dehulling and cleaning. The nuts are dehulled manually or by abrasive huller, cleaned and immediately dried till the moisture content is reduced taking advantage of hot weather during the harvesting season. Drying of temperate nuts is one of the unit operations which can affect the shelf life of nuts. Natural drying takes a long time and may open the door for many pathogens which can limit their shelf life. At industrial scale, nuts are dried by using different types of industrial driers. However, orchardists dry the nuts by using the sunlight. Therefore, after harvesting, nuts should be dried immediately to the desired moisture content. The improper drying techniques considerably affect quality of nuts. The insufficient drying also affects the colour and sensory attributes of nuts (Chen and Pan 2022).

3 De-Shelling

The temperate nut kernels are covered by shells which have significant effect on their shelf life. Temperate nuts are supplied to the market shelled or unshelled. Shell is an effective natural package which protect the nut kernel from physical damage. Shells of nuts are cracked and removed manually or mechanically. The unshelled temperate nuts have usually twice the storage life as compared to shelled nuts. Shell prevents the entry of moisture and air which deteriorate the quality of nuts. Temperature has pronounced effect on the quality of shelled and unshelled nuts. The shelled nuts develop relatively darker colour than unshelled nuts at room temperature. However, unshelled nuts take more space during storage and also incur the higher cost for the transportation (Prabhakar et al. 2020).

4 Microbial Contamination

Mycotoxin producing fungi is the major concern in the food industry as it may lead to serious health issues. Temperate nuts are widely cultivated crops and may be contaminated with mycotoxins at any storage during production, processing, storage and distribution and can expose consumers to the risk of contamination (Kluczkovski 2019). The mycotoxin production depends on many factors such as moisture, relative humidity and temperature. Temperate nuts are generally contaminated with *Aspergillus* species, producing several types of aflatoxins. *Aspergillus flavus* and *Aspergillus parasiticus* are the common fungi observed in temperate nuts (Mir et al. 2022). Nut crops are also contaminated by *Penicillium* species, *Escherichia coli, Listeria monocytogenes* and *Salmonella* spp. Contamination of nuts with fungi and mycotoxins is difficult to avoid. Monitoring of the temperate nuts at different stages is important to ensure their safety. Furthermore, various techniques such as treatment with chemicals, ozone, irradiation and radiofrequency are used in nut industry for inactivating the mycotoxin producing fungi. These techniques have shown the promising results with the least effect on the qualitative properties of nuts. Therefore, it is important to improve the shelf life of nuts by employing different methods and maintain the quality and marketability during storage.

5 Packaging

Packaging is an important unit operation used for preserving the food quality. Temperate nuts are packed in various types of packaging materials which affect the quality and enhance their shelf life (Fig. 12.2). Selection of packaging material is important as it affects the product quality and overall nut quality and safety. Nuts are bulk or retail packed according to the demand or specifications. Packaging material considerably influences the microbial contamination during storage and distribution due to permeability of moisture and gas. Nuts are low moisture foods which require

Fig. 12.2 Packaging of nuts

the protection from moisture and oxygen. The unsuitable packaging material negatively affects the quality and act as obstacle for their exportation. The packaging materials which have high permeability increase the oxygen concentration in nuts which accelerates the oxidative reactions. The use of packing material with low oxygen permeability is usually recommended for the packaging of nuts (Gama et al. 2018).

The types of packaging material have considerable effect on the quality and shelf life of walnuts (Mexis et al. 2009). The walnuts retain the acceptable quality up to 2 months in polyethylene terephthalate-air, 4–5 months in polyethylene terephthalate/polyethylene-N_2 and 12 months in polyethylene terephthalate-silicon dioxide/polyethylene-N_2 filled pouches at 20 °C. The packaging of nuts in low density polyethylene is prone to contamination of aflatoxins as compared to the packaging material of polyethylene terephthalate (Macri et al. 2020). The packaging material having high permeability increases the moisture percentage of nuts which favours the microbial growth. Mohammad et al. (2019) reported that high oxygen permeability packaging material increases the oxygen percentage in the kernels which favours to the oxidative rancidity. Shakerardekani and Karim (2013) used the different types of plastic packaging material to pack pistachios nuts stored at ambient temperature. The best packaging material for packing of pistachios were polyethylene terephthalate followed by nylon, polyamide/polypropylene and polyvinyl chloride. The polyethylene terephthalate film significantly increased the shelf life of nuts up to 5 months at ambient storage. Furthermore, the packaging of pistachios in eight-layered plastic films using vacuum conditions showed effective resistance against the humidity and oxygen and prevent the production of aflatoxin producing fungi during storage.

6 Factors Affecting the Shelf Life

Contamination of nuts is the major problem which is associated with the improper postharvest practices. Most nuts have a long shelf life if kept in optimised storage conditions. However, if nuts are not stored properly, their shelf life is limited and become inedible due to rancidity which is related with the high fat content. Shelf life of harvested temperate nuts is considerably affected by many factors including moisture percentage during storage, storage temperature, relative humidity, concentration of oxygen and light, and packaging (Fig. 12.3).

6.1 Moisture of Nuts

Moisture is one of the critical factors which affects the storage stability. The freshly harvested nuts have high moisture percentage, and reducing the moisture quickly is important to retard the changes that affect the nut quality. High moisture during

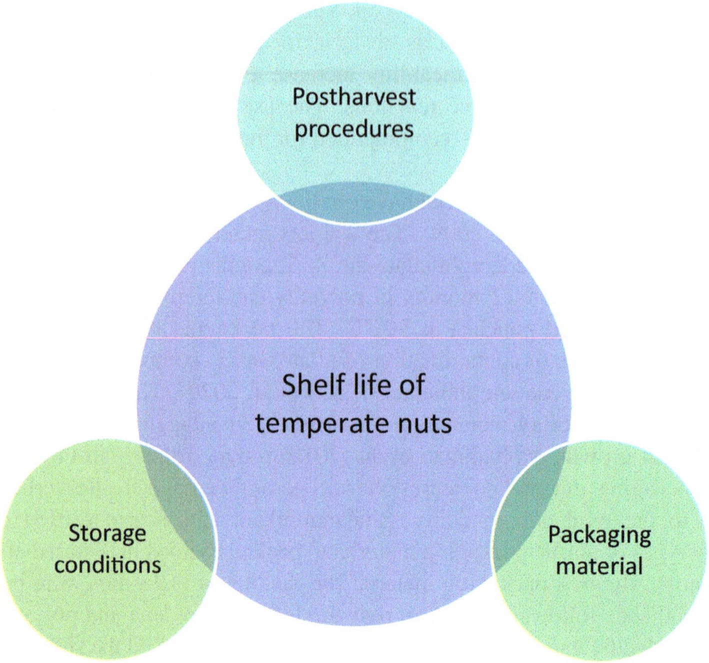

Fig. 12.3 Factors affecting shelf life of temperate nuts

storage lead to the microbial contamination, hydrolytic and oxidative rancidity. High moisture also has negative effect on the texture quality of nuts which ultimately leads to consumer unacceptability (Khir et al. 2013).

The moisture availability during storage favours fungal growth. The packaging material which is having high permeability to the water increases the moisture content of packaged nuts and ultimately leads to the fungal growth. Sometimes, dried nuts also promote absorption of moisture from the atmosphere and provide the favourable environment for fungi growth and mycotoxins. Different types of temperate nuts including pecans, walnut, pistachios having water activity of 0.75 develop the moulds after 16 weeks of storage, while as kernels of the same species with water activity of 0.9 develop the mould growth after 4 weeks of storage (Venkatachalam and Sathe 2006). High moisture at the beginning of storage also favours the rancidity and ultimately leads to brown colour and off flavour. Therefore, the moisture content of temperate nuts is reduced immediately after harvest to increase their shelf life and market value.

6.2 Storage Temperature

The temperature during storage of temperate nuts has a dramatic effect on the shelf life of nuts. The temperature during storage influences the rancidity and metabolic reactions involved in the quality degradation of nuts. Oxidation reactions and rancidity of fats are temperature dependent and have been found to increase with increasing temperature. The peroxide value of temperate nuts is also increased with increase of storage temperature and time. The almonds, walnuts and pistachios stored at 20 °C or above have higher peroxide value as compared to the nuts stored at 4 °C (Lin et al. 2012). The increase of peroxide value at higher storage temperature is due to the increased lipoxygenase and lipase enzyme activity at higher storage temperature (Ozturk et al. 2016). The storing of nuts or kernel at low storage temperature helps to minimise the oxidative reactions.

Ghirardello et al. (2013) studied the in-shell storage of hazelnuts stored at ambient room temperature and compared with the refrigerated storage at 4 °C and relative humidity (55%). The storage of hazelnuts at ambient temperature preserves the nuts below threshold limits of oxidative degradation up to 8 months, however, refrigeration maintains the quality of nuts up to 12 months. The acidity of hazelnuts (0.47% oleic acid) at ambient temperature after 1 year of storage was higher as compared to refrigerated storage (0.13% oleic acid). The authors recommended the modified atmosphere storage of nuts for longer storage. Johnson et al. (2009) demonstrated that hazelnuts stored at ambient temperature at high nitrogen concentration (99%) shown the comparable results to the low temperature storage conditions (3–6 °C) and 60% relative humidity.

6.3 Storage Gas Composition

The storage gas composition of temperate nuts is important factor which determines the shelf life because it facilitates the lipid oxidation. Oxygen is one of the critical factors which affects the quality of nuts during storage, and it accelerates the oxidation of lipids. The increased temperature and oxygen availability deteriorates the quality of nuts due to lipid oxidation and negatively affects the sensory properties. Storage of walnuts and pistachios at high oxygen produce more hexanal as compared stored at low oxygen concentration (Bakkalbaşı et al. 2012). Maté et al. (1996) reported that walnuts stored at ambient atmospheric concentration (21%) for 10 weeks has higher hexanal and peroxide value as compared to walnuts stored at low oxygen concentration (2.5%). The oxygen concentration during storage has more pronounced effect as compared to relative humidity on the production of hexanal and hydroperoxides in walnuts (Bakkalbaşı et al. 2012). Similar trend was observed in pistachio nuts which were stored at different oxygen levels (Sedaghat 2010). The pistachio nuts stored at oxygen concentration of 8% have shorter shelf life as compared to nuts stored at $<2\%$ oxygen concentration.

Hazelnuts were stored at different oxygen concentrations from 1 to 20% at temperature 7 and 25 °C (San Martin et al. 2001). The study revealed that after 1 year of storage, hazelnuts did not show any significant change in quality or rancidity. Furthermore, the authors recommended the low oxygen storage conditions and low temperature to delay the autoxidation and lipid rancidity for longer periods of storage. The controlled atmospheric conditions with low oxygen and high carbon dioxide concentrations, and vacuum packaging at low temperature were found effective for enhancing the shelf life of pecan nuts (Ribeiro et al. 2020).

6.4 Relative Humidity

Relative humidity has a significant effect on the storage stability of temperate nuts. High humidity affects the quality properties of nuts and favours the growth of mycotoxin producing fungi. High relative humidity increases the moisture percentage of nuts due to the absorption of moisture by dried kernels. The packaging material used for packaging influences the quality of nuts at high relative humidity. High permeable packaging material will allow the absorption of moisture from the environment and significantly affect the quality of nuts. The fungal growth is not observed in nuts when stored below 10 °C and the relative humidity of less than 40% (Shakerardekani and Karim 2013). The optimal relative humidity of walnuts should be less than 60%, while as in almond the humidity should be below 75% during storage (Lopez et al. 1995). The high relative humidity of nuts also leads to the higher free fatty acids, iodine and peroxide value as compared to nuts stored at low humidity (Christopoulos and Tsantili 2015).

6.5 Light

Light considerably affects the quality of temperate nuts. Light results in significant oxidative changes in walnut at 21 °C with the development of rancid flavour, while as storage at 5 °C in a dark does not lead to any oxidative change after 25 weeks of storage (Jensen et al. 2001). The pecans were exposed to different light sources such as sunlight, fluorescent and cook white light in flexible pouches (Prabhakar et al. 2020). The pecans stored in sunlight shown the significant darkening of nuts, while the pecans kept in cool white light and fluorescent experience the less colour change. The ultraviolet radiations of sunlight degrade the pigments of nuts and impart the dark colour to pecans. Mexis et al. (2009) also observed that walnuts retain the higher quality when stored in dark as compared to exposed in light.

7 Conclusion

Temperate nuts have generally long shelf life. Efficient postharvest handling is essential to preserve and quality and safety of temperate nuts. The quality of nuts is considerably affected by the drying and storage conditions. The degraded nut quality reduces the consumer acceptability and profitability. Many critical factors such as temperature, oxygen availability affect the quality of nuts during storage. The storage of nuts below 10 °C is usually recommended to prolong the shelf life. Nut quality should be monitored regularly during postharvest handling and storage. Optimised packaged condition are also required to maximise the quality of stored temperate nuts for longer time. Furthermore, the unshelled nuts have generally long shelf life than shelled nuts or kernels. It is recommended to store the unshelled nuts for longer period of time to increases consumer demand and maintain a continuous supply in the market.

References

Bakkalbaşı E, Yılmaz ÖM, Javidipour I, Artık N (2012) Effects of packaging materials, storage conditions and variety on oxidative stability of shelled walnuts. LWT Food Sci Technol 46(1): 203–209

Chen C, Pan Z (2022) Postharvest processing of tree nuts: current status and future prospects—a comprehensive review. Compr Rev Food Sci Food Saf 21(2):1702–1731

Christopoulos MV, Tsantili E (2015) Oil composition in stored walnut cultivars—quality and nutritional value. Eur J Lipid Sci Technol 117(3):338–348

Gama T, Wallace HM, Trueman SJ, Hosseini-Bai S (2018) Quality and shelf life of tree nuts: a review. Sci Hortic 242:116–126

Ghirardello D, Contessa C, Valentini N, Zeppa G, Rolle L, Gerbi V, Botta R (2013) Effect of storage conditions on chemical and physical characteristics of hazelnut (*Corylus avellana* L.). Postharvest Biol Technol 81:37–43

Jensen PN, Sørensen G, Engelsen SB, Bertelsen G (2001) Evaluation of quality changes in walnut kernels (Juglans regia L.) by Vis/NIR spectroscopy. J Agric Food Chem 49(12):5790–5796

Johnson JA, Yahia EM, Brandl DG (2009) Dried fruits and tree nuts. In: Yahia EM, Raton B (eds) Modified and controlled atmospheres for the storage, transportation, and packaging of horticultural commodities. CRC Press, Boca Raton, pp 507–526

Kader AA (2013) Impact of nut postharvest handling, de-shelling, drying and storage on quality. In: Improving the safety and quality of nuts. Woodhead Publishing, Cambridge, pp 22–34

Khir R, Pan Z, Atungulu GG, Thompson JF, Shao D (2013) Size and moisture distribution characteristics of walnuts and their components. Food Bioproc Tech 6(3):771–782

Kluczkovski AM (2019) Fungal and mycotoxin problems in the nut industry. Curr Opin Food Sci 29:56–63

Lin X, Wu J, Zhu R, Chen P, Huang G, Li Y, Ruan R (2012) California almond shelf life: lipid deterioration during storage. J Food Sci 77(6):C583–C593

Lopez A, Pique MT, Romero A, Aleta N (1995) Influence of cold-storage conditions on the quality of unshelled walnuts. Int J Refrig 18(8):544–549

Macri AM, Pop I, Simeanu D, Toma D, Sandu I, Pavel LL, Mintas OS (2020) The occurrence of aflatoxins in nuts and dry nuts packed in four different plastic packaging from the Romanian market. Microorganisms 9(1):61

Maté JI, Saltveit ME, Krochta JM (1996) Peanut and walnut rancidity: effects of oxygen concentration and relative humidity. J Food Sci 61(2):465–469

Mexis SF, Badeka AV, Riganakos KA, Karakostas KX, Kontominas MG (2009) Effect of packaging and storage conditions on quality of shelled walnuts. Food Control 20(8):743–751

Mir SA, Shah MA, Mir MM (eds) (2018) Postharvest biology and technology of temperate fruits. Springer International Publishing, Cham

Mir SA, Shah MA, Mir MM, Sidiq T, Sunooj KV, Siddiqui MW, Mousavi Khaneghah A (2022) Recent developments for controlling microbial contamination of nuts. Crit Rev Food Sci Nutr 2022:2038077. https://doi.org/10.1080/10408398.2022.2038077

Mohammad ZH, Murano EA, Moreira RG, Castillo A (2019) Effect of air-and vacuum-packaged atmospheres on the reduction of Salmonella on almonds by electron beam irradiation. LWT Food Sci Technol 116:108389

Ozturk I, Sagdic O, Yalcin H, Capar TD, Asyali MH (2016) The effects of packaging type on the quality characteristics of fresh raw pistachios (*Pistacia vera* L.) during the storage. LWT- Food Sci Technol 65:457–463

Prabhakar H, Sharma S, Kong F (2020) Effects of postharvest handling and storage on pecan quality. Food Rev Int 2020:1817066. https://doi.org/10.1080/87559129.2020.1817066

Ribeiro SR, Ribeiro QM, Klein B, dos Santos ID, Forgiarini S, Hamann JJ, Wagner R (2020) Effect of low oxygen on quality attributes of 'Barton' pecan nuts after long-term storage at different temperatures. Sci Hortic 263:109098

San Martin MB, Fernández-García T, Romero A, Lopez A (2001) Effect of modified atmosphere storage on hazelnut quality. J Food Process Preserv 25(5):309–321

Sedaghat N (2010) Pistachio nuts shelf life based on sensory evaluation. Am Eurasian J Agric Environ Sci 8(5):570–575

Shakerardekani A, Karim R (2013) Effect of different types of plastic packaging films on the moisture and aflatoxin contents of pistachio nuts during storage. J Food Sci Technol 50(2): 409–411

Venkatachalam M, Sathe SK (2006) Chemical composition of selected edible nut seeds. J Agric Food Chem 54(13):4705–4714

Chapter 13
Package and Storage of Temperate Nuts

K. Rama Krishna, M. P. Ellampirai, and T. J. Archana

Abstract Nuts are considered as a healthy snack because they are a good source of protein, fiber, oil, vitamins and minerals. These nuts are packed with energy and nutritional benefits, but they are also more vulnerable to moisture, oxygen and light. The postharvest management of temperate nuts is of utmost importance as it is the only possible way to enhance their shelf life. Thus, the packaging and storage of temperate nuts play a major role in determining the quality and longevity of these nuts. In this chapter, we try to provide information on the packaging and storage of temperate nuts (chestnut, hazelnut, pecan nut, walnut, pistachio nut, chilgoza). Different types of packaging material and MAP technology is of immense help in improving shelf life and maintain postharvest quality of temperate nuts.

Keywords Storage · Nuts · Packaging · MAP · Shelf-life

1 Introduction

Temperate nuts such as pecan nut, walnut, pistachio nut, hazelnut, chestnut, etc., are relished worldwide not only for their delicious taste but also for the nutrition and health benefits associated with them. According to the U.S. Food and Drug Administration, people consuming temperate nuts as a part of their diet (low in saturated fat and cholesterol) have a lower risk of getting heart diseases (Bruhn et al. 2010). This is because nuts are a high source of dietary fiber, protein and unsaturated fat. Temperate nuts like all other nuts and unlike other temperate, subtropical and tropical fruits crops need special care while packing and storage. This is basically because nuts are more vulnerable to moisture, oxygen and light. Often it is seen that

K. R. Krishna (✉) · M. P. Ellampirai
Department of Horticulture, School of Life Sciences, Central University of Tamil Nadu, Thiruvarur, TN, India

T. J. Archana
Division of Food Science and Postharvest Technology, ICAR-Indian Agriculture Research Institute, New Delhi, India

M. M. Mir et al. (eds.), *Temperate Nuts*,
https://doi.org/10.1007/978-981-19-9497-5_13

even if one of these conditions are not controlled properly, the quality of nuts is severely affected and might make them unfit for consumption. The aflatoxins produced by pathogenic bacteria (*Salmonella* sps., *Listeria* sps.) and fungi pose serious food safety concerns. The exposure to moisture and temperature than the recommended conditions will aggravate the situation further. Apart from these, the loss of quality of nuts due to rancidity, brittleness, darkening, insect infestation, mould growth and stale flavour will also deteriorate the consumer eating experience. So, this necessitates the use of proper packing and storage conditions for nut crops based on longevity and consumer preference.

1.1 Packaging of Temperate Nuts

Packaging is one of the important aspects of postharvest processing. Packaging provides easy handling operations, in the case of temperate nuts. As temperate nuts are high-value products the packaging material should be attractive as well apart from protecting the product physically and preventing spoilage. Several types of packaging materials are available in the market for the packing nuts such as net bags, paper bags, trays, jute bags, polyethylene/polypropylene cups, polycrystalline containers, metal boxes, cardboard boxes, glass jars, vacuum packaging, etc. However, all the packing materials are not suitable for packaging temperate nuts. This is because of the physicochemical nature associated (high oil, high protein, low moisture) with the nuts making them more susceptible to the environmental condition (air, moisture and temperature) in which they are stored. The package permeability towards air and moisture should be negligible so that the nuts can be stored for a longer time. Before packaging and storage, the moisture levels in the nuts must be brought down to the recommended levels which vary from crop to crop. Nano packaging is an emerging technology of packing which is gaining importance in nut packaging. This packaging technique helps to reduce microbial growth in nut crops. It also helps to increase the shelf life of the product.

1.2 Storage of Temperate Nuts

Temperate nuts such as pecan, pistachio, hazelnut and walnut are classified as low perishables except for chestnut as they are characterised by low water content and very low metabolic activity. However, the storability (the period for which the product remains good under storage conditions) of the nuts depends on the storage temperature and humidity. The nuts which are stored in room temperature conditions tend to have lower storage life as they are affected by both external and internal factors. When nuts are stored in room temperature, they are prone to insect infestation, moisture absorption, rancidity and microbial toxin development. This leads to the unacceptability of nuts for consumption. Higher room temperatures aggravate the rancidity process. Although rancid nuts are not unsafe for consumption, they tend to develop sharp flavour which is not preferred by consumers. Refrigeration or cold storage or freezing temperature storage have proven to provide longer storage life than ambient/room condition storage.

2 Packaging and Storage of Temperate Nuts

*2.1 Pecan Nut (***Carya illinoinensis;** *Juglandaceae)*

2.1.1 General Packaging

Pecan nut are semi-perishable in nature. Depending upon the climatic conditions and time of harvest, moisture content of the pecan nut fluctuates. The optimum storage moisture content of the pecan nut is 3.5–4.5%. Mechanical harvesting of nuts is followed by dehulling, washing, shelling, sorting, grading and packing. The packaging material should be based on physical attributes of nuts such as oxygen transmission rate, water vapour transmission rate as they are often susceptible to oxidation and aflatoxin contamination during storage. In the retail market, the polyethylene and low-density polyethylene (LDPE) package are most used for pecan nut packaging. Glass jar are best packaging material for pecans because of their impervious barrier properties helps to extend the storage life of pecans. Some of the packaging materials such as polyethylene, cellophane, nanocellulose, polyvinyl chloride, ethyl vinyl acetate are used for packing pecan nuts.

2.1.2 Modified Atmospheric Packaging (Map)

Pecan nuts have longer storage life when proper packaging with modified atmosphere was used. The packaging material made by laminating nylon and ethyl vinyl acetate and flushed with CO_2 gave pecans a storage life of 6 months at ambient conditions. The packaged pecans showed better flavour when compared to the pecans stored at refrigerated conditions. The pecan kernels when packed in cellophane packaging films coated with polyvinylidene chloride showed acceptability even after 6 months of storage at 24 °C and 60% RH. The recommended O_2 exchange rate for cellophane packaging was >0.08 mL, per 100 cm per h. The vacuum packaging of 'Barton' pecan nuts (shelled) in polypropylene plastic packaging (0.15 mm) at 20 °C maintained the storage life for 6 months with low contamination levels by *Aspergillus* sp. and no aflatoxin producing sp. (Ribeiro et al. 2021). For shipping, the pecan nuts single walled corrugated craft fiberboard with dimensions 38.7 × 28.6 × 25.7 cm can be used.

2.1.3 Storage

Pecans are considered as semi-perishable commodities and are prone to damage by moulds forming microorganisms and insects and may tend to develop kernel discoloration and poor flavour if proper storage conditions are not provided. The important factors affecting storage life of pecan nuts are temperature and humidity. Whole kernels have a longer storage life than broken pieces as whole kernels have low exposed surface area than broken pieces. The in-shell pecans can be stored for 2 years or more at 0 °F (−17.8 °C) and for 1 year at 32–45 °F (0–7.2 °C), while shelled pecans can be store for 2 years or more at 0 °F (−17.8 °C) and 1 year at 32 °F (0 °C) (Perry and Sibbett 1998). 'Barton' pecan nut when stored at controlled atmosphere storage conditions of 2 kPa of O_2 and 15 kPa of CO_2, able to ensure the absence/low contamination of microbial storage by *Aspergillus* sp., *Cladosporium* sp., *Penicillium* sp. and no aflatoxin producing sp. for 4 months at 10 °C and 20 °C storage temperatures (Ribeiro et al. 2021).

2.2 *Chestnut (*Castanea *Sp; Fagaceae)*

2.2.1 Packaging

Chestnuts are bulk packed and transported in boxes, bags and drums. As chestnuts are sensitive to impact, proper care should be given while handling. The packaging material should protect the chestnut from mould growth during storage. The commonly used packaging material in retail conditions are cardboard and paper boxes with or without bioplastic lining, PET lib boxes, PET trays with plastic films, PET plastic LLDPE seal pouches, plastic PE or PE nets, etc.

2.2.2 Modified Atmospheric Packaging

Modified atmospheric packaging is cheaper because it needs only plastic bags. MAP is sealing the products with different polymeric film packages which modify both the O_2 and CO_2 levels in the package atmosphere. The modification of air inside the package helps to decrease the rate of microbes and increases the shelf life and quality of the product (Mangaraj et al. 2009). Chestnut hybrid 'Bouche de betizac' and 'Garrone Nero' were MA packed with enriched CO_2 of 80% in polyethylene film (120 μm) and starch-based biodegradable film (180 μm) stored at 1 °C and 90–95% RH. The chestnut had a storage life of 90 days with good eating quality (Peano et al. 2014). Vacuum packaging (VAC) also helps to increase the product shelf life. Two packing materials were used for storage of chestnuts, one with MAP and other with vacuum packaging. The MA packed (Dupont Dura Fresh™ trays) and vacuum packed (Cryovac® Simple Steps™ trays) chestnuts stored at 4 °C were able to have storage life of 26 days with good eating quality and low microbial

Table 13.1 Modified atmospheric packaging of some chestnut cultivars

Cultivar	Packaging material	O_2	CO_2	Storage requirements	Storage period	References
Colossal	Plastic bowls covered with stretch films	–	–	0–2 °C and 85% RH	90 days	Ertan et al. (2015)
–	LDPE	1 kPa	–	0 °C	5 months	Talasila et al. (1995)
				−2 °C	6 months	
–	Microperforated PET/PE (12 μm/40 μm) micro perforated films	10.5%	10.9%	0 °C	110 days	Panagou et al. (2006)
		8.3%	12%	8 °C		

contamination (Bhisanbut et al. 2008). The quality of chestnut packed in single and two layers of polyethylene bags was good for 2 months. It has been found that two layers polyethylene bags facilitated decrease in O_2 and increase in CO_2 content (Homma et al. 2008). The chestnuts packed in vacuum packaging (polyamide and polyethylene) and MAP (polyester, 32% CO_2 and 0.3% O_2) proved to extend the storage life under 0 °C and 90% RH for 3 months (Fernandes et al. 2020). The chestnuts showed acceptable microbiological and physiochemical quality (Table 13.1).

2.2.3 Storage

The storage of chestnuts is different from those of other nuts. Chestnuts are highly perishable, with high moisture, high starch and low-fat content. Storage of chestnut at room temperature is not preferred as they lose moisture very quickly, become dry and hard, and develop mold within 2 weeks. Proper storage will avoid mold growth and retain quality for long time. When the chestnuts are stored at 0–2 °C and 90–98% RH within 1–2 days of harvest, they remain fresh for 4–8 weeks and roast properly (Kenneth et al. 2016). The in-shell chestnut can be stored for 1 month at the refrigerated condition of 0–7.2 °C and for 1 year at −17.8 °C. While shelled and dried chestnuts can be stored for 1 year at 0–7.2 °C, they tend to lose flavour and texture and do not taste great when roasted.

2.2.4 Controlled Atmospheric Storage

The chestnut cv. 'Marrone Florentino' were water cured (at 15 °C for 5 days followed by air drying for 4 h) and stored in CA condition of 15.20 kPa CO_2 and 3.04 kPa O_2 at −1 °C was able to preserve the organoleptic characteristics with reduced incidence of rot for 2 months (Cecchini et al. 2011). Chestnut cv. 'Catoto' and 'Platella' when stored under CA conditions of 2% O_2 and 20% CO_2 at 1 °C were able to store in good condition for 105 days (Mignani and Vercesi 2003). Similar reports are also witnessed when chestnut var. 'Marigoule' and 'Bouche de Betizac' were stored under CA conditions of 2% O_2 and 5% CO_2 at 1 °C showing no water loss and mold growth was significantly reduced (Rouves and Prunet 2002).

2.3 *Walnut (*Juglans regia, *Juglandaceae)*

2.3.1 Packaging

Walnuts are rich source of essential amino acids, Vit. E and polyunsaturated fatty acids, however, fresh walnuts are prone to spoilage due to high moisture content and respiratory activity (Ma et al. 2013). Hence, walnuts are usually stored and sold in dried form after removing hulls. The nuts are dried below 8% moisture using hot air temperature without exceeding 43 °C to avoid kernel rancidity (Labavitch 2004). The harvested nuts contain 35% of moisture. Packaging plays a crucial role in the prolonged storage life of walnuts. The packaging of walnuts is done in three ways, in-shell, retail, and bulk packed. As the walnuts undergo rancidity while storage, the infusion of packaging material with nitrogen gas and removal of air has been recommended practice. The vacuum packaging is the most preferred packaging for walnuts. The bulk packaging is done for in-shell nuts using large sacks with a capacity of 25–50 kg, while the peeled nuts are generally packed in plastic bags of half to 1 kg.

2.3.2 Modified Atmospheric Packaging

The most successful practical approach for storing walnuts has been determined as modified atmospheric storage (Ma et al. 2013). The walnuts under MAS could be stored up to 3 months (Wang et al. 2017). Shelled walnut packed in a 50 μm PE/PET bag with CO_2 composition lasted for 3 months (Javanmard 2017). Walnut kernels packed in polyethylenel polyamide pouches flushed with dry air, N_2, or CO_2 stored for 12 months at 1 °C. Low O_2 and temperature help to reduce the colour variation and increase the acceptable quality of walnut. For long-term storage of kernels, the lowest process of oxidation of lipids and hydrolysis reactions becomes necessary. The multilayer packaging act as a barrier for further spoilage.

2.3.3 Storage

The main objective of storage is to maintain low water content, must be attained after drying which helps to suppress enzyme activity, reduce the microbial contamination and retain texture. The optimum storage temperature is 0–10 °C (32–50 °F) and 50–65% RH.

2.3.4 Controlled Atmospheric Storage

The storage can be extended by using <1% O_2; CO_2 >80% in air and is effective in controlling insects. CAS is known to inhibit mold contamination of walnuts. According to Ma et al. (2020), the walnuts could be stored for 135 days under CAS (2% O_2 + 25% CO_2) at 0 ± 1 °C. The CA (5% O_2, 7.5% CO_2) storage along with 1-MCP treatment increase walnut shelf life up to 100 days (Ye et al. 2021).

2.4 *Pistachio Nut (*Pistacia vera, *Anacardiaceae)*

2.4.1 Packaging and Storage

The common storage problems of pistachios are mold contamination, storage pest, and rancidity. The longer storage of pistachio depends on its moisture absorption rate. Higher levels of moisture content (35%) of pistachio after soft hull splitting may lead to *Aspergillus* contamination. Hence suitable packaging material should be used to maintain the moisture content below 7% and to avoid exposure to atmospheric air (Ghadarijani and Javanshah 2006). The packaging of roasted pistachio nuts with five layers of packaging material such as modified polypropylene, compound film and metalised plastic flushed with N_2/CO_2 stored at 40 °C maintained the quality of nuts for 12 months (Raei et al. 2010). The PET packaging material can extend the shelf life of pistachios for 2–5 months (Shakerardekani and Karim 2013). The nuts packed in laminated pouches will last for 12 weeks at 50 °C (Tavakoli et al. 2019). The active modified atmospheric packaging (5% O_2 + 45% CO_2 + 50% N_2) in polypropylene trays preserved the better postharvest quality of pistachio nut at 4 ± 1 °C and 90 ± 5% RH (Sheikhi et al. 2019b).

The optimum storage temperature ranges from 0 to 10 °C. Fresh in hull pistachio nuts retained high firmness and sensory attributes when they are packed in polypropylene bags under MAP conditions of 100% CO_2 at 5 °C for 42 days (Shayanfar et al. 2011). MAP showed positive results when fresh raw pistachios were MA packed in polyvinyl chloride/polyethylene trays flushed with 30% CO_2, 4 ppm O_2 gas and 70% N_2, sealed and stored at 4 °C. The total mesophilic bacteria count and free fatty acid levels were low (Ozturk et al. 2016). De-hulled pistachio nuts cv. 'Kerman' was stored with up to 105 days with good physio-chemical properties when storage conditions (0 °C and 90% RH) coupled with passive MAP conditions. The passive MAP condition was able to maintain gas composition of 23.17–29.82% CO_2 and 0.95–3.35% O_2 in the package (Sheikhi et al. 2019a).

2.5 *Hazelnut* (**Corylus avellana**, ***Betulaceae***)

2.5.1 Packaging and Storage

Hazelnut can store better when the water content of kernels remains not more than 6% if shelled and 7% if in-shell. At any point of time the total water content of unshelled hazelnuts should not exceed 10–12% (Kenneth et al. 2016). As hazelnuts are normally marketed at ambient temperature, the use of low-O_2 MAP or vacuum packaging is recommended, while bulk containers can be used for marketing of in-shell hazelnut taking into account avoiding exposure to moisture and high RH. Roasted kernels when stored below 10 °C had a storage life of 6 months without the development of rancidity. The combination of reduced temperature and vacuum packaging can extend roasted kernel shelf-life up to 1 year. Low temperature and modified gaseous atmosphere techniques help to control enzymatic and chemical peroxidation of hazelnuts (San Martin et al. 2001). Hazelnuts both shelled and unshelled had a shelf life of 6 months when stored at 25 °C using vacuum packaging and raffia bags, but the vacuum-packed hazelnuts showed higher oxidative stability than raffia bags (Correia et al. 2020). The storage of hazelnuts at low temperatures has proven to have longer storage life. The in-shell hazelnuts were able to protect kernels from oxidative degradation for 8 months, while the refrigeration of hazelnut kernels at 4 °C and 55% RH had longer storage life up to 1 year. When modified atmospheric storage (1% O_2 and 99% nitrogen) was used in combination with low temperature, the nuts had higher quality and storage over 1 year (Ghirardello et al. 2013).

3 Conclusion

Packaging is one of the important unit operations in the food industry. Packaging material and storage conditions influence the quality and shelf life of temperate nuts. Packaging material protects the food from physical damage, prevents spoilage and attracts consumers. The improper packaging and storage conditions lead to microbial contamination, insect infestation and rancidity of temperate nuts. Nuts are commonly stored at low temperature, low relative humidity and low oxygen concentration.

References

Bhisanbut A, Shin J, Harte J, Fulbright D, Dolan K, Harte B (2008) The extension of chestnut product quality using modified atmosphere packaging and vacuum skin packaging. In: 16th IAPRI world conference on packaging, 8–12th June, Bangkok, Thailand

Bruhn C, Harris LJ, Giovanni M, Metz D (2010) Nuts: safe methods for consumers to handle, store and enjoy almonds, chestnuts, pecans, pistachios, and walnuts. UC/ANR Publication, Berkeley. http://ucfoodsafety.ucdavis.edu/files/44384.pdf

Cecchini M, Contini M, Massantini R, Monarca D, Moscetti R (2011) Effects of controlled atmospheres and low temperature on storability of chestnuts manually and mechanically harvested. Postharvest Biol Technol 61(2-3):131–136

Correia P, Filipe A, Ramalhosa E, Guiné R (2020) Hazelnut storage at controlled conditions with different packaging materials. In: 11th International conference on biotechnology and food science (ICBFS 2020)

Ertan E, Erdal E, Alkan G, Algül BE (2015) Effects of different postharvest storage methods on the quality parameters of chestnuts (*Castanea sativa* mill.). HortScience 50(4):577–581

Fernandes L, Pereira EL, Fidalgo MDC, Gomes A, Ramalhosa E (2020) Effect of modified atmosphere, vacuum and polyethylene packaging on physicochemical and microbial quality of chestnuts (*Castanea sativa*) during storage. Int J Fruit Sci 20(2):785–801

Ghadarijani MM, Javanshah A (2006) Distribution of aflatoxin in processed pistachio nut terminals. Acta Hortic 726:431–436

Ghirardello D, Contessa C, Valentini N, Zeppa G, Rolle L, Gerbi V, Botta R (2013) Effect of storage conditions on chemical and physical characteristics of hazelnut (*Corylus avellana* L.). Postharvest Biol Technol 81:37–43

Homma T, Inoue E, Matsuda T, Hara H (2008) Changes in fruit quality factors in Japanese chestnut (*Castanea crenata* Siebold and Zucc.) during long-term storage. Hortic Res (Japan) 7:591–598

Javanmard M (2017) Effect of modified atmosphere packaging and storage temperatures on quality of shelled raw walnuts. Int J Nutr Food Eng 11(7):510–514

Kenneth CG, Chien YW, Mikal S (2016) The commercial storage of fruits, vegetables, and florist and nursery stocks. US Department of Agriculture, Washington, DC. https://www.ars.usda.gov/arsuserfiles/oc/np/commercialstorage/commercialstorage.pdf

Labavitch JM (2004) Walnut. In: The commercial storage of fruits, vegetables, and florist and nursery stocks. US Department of Agriculture, Washington, DC. http://www.ba.ars.usda.gov/hb66/contents.html

Ma Y, Lu X, Liu X, Ma H (2013) Effect of 60Coγ-irradiation doses on nutrients and sensory quality of fresh walnuts during storage. Postharvest Biol Technol 84:36–42. https://doi.org/10.1016/j.postharvbio.2013.04.001

Ma Y, Li P, Watkins CB, Ye N, Jing N, Ma H, Zhang T (2020) Chlorine dioxide and sodium diacetate treatments in controlled atmospheres retard mold incidence and maintain quality of fresh walnuts during cold storage. Postharvest Biol Technol 161:111063. https://doi.org/10.1016/j.postharvbio.2019.111063

Mangaraj S, Goswami TK, Mahajan PV (2009) Applications of plastic films for modified atmosphere packaging of fruits and vegetables: a review. Food Eng Rev 1(2):133–158

Mignani I, Vercesi A (2003) Effects of postharvest treatments and storage conditions on chestnut quality. Acta Hortic 600:781–785

Ozturk I, Sagdic O, Yalcin H, Capar TD, Asyali MH (2016) The effects of packaging type on the quality characteristics of fresh raw pistachios (*Pistacia vera* L.) during the storage. LWT Food Sci Technol 65:457–463

Panagou EZ, Mallidis C, Vekiari SA (2006) The effect of modified atmosphere packaging of chestnuts in suppressing fungal growth and related physicochemical changes during storage in retail packages at 0 and 8 °C. Adv Hortic Sci 20:1000–1008

Peano C, Baudino C, Giuggioli NR, Girgenti V (2014) The use of a modified atmosphere for the storage of chestnut fruits. Ital J Food Sci 26(1):74–80

Perry E, Sibbett GS (1998) Harvesting and storing your home orchard's nut crop: almonds, walnuts, pecans, pistachios, and chestnuts. University of California, Berkeley. https://doi.org/10.3733/ucanr.8005

Raei M, Mortazavi A, Pourazarang H (2010) Effects of packaging materials, modified atmospheric conditions, and storage temperature on physicochemical properties of roasted pistachio nut. Food Anal Methods 3(2):129–132
Ribeiro SR, Garcia MV, Copetti MV, Brackmann A, Both V, Wagner R (2021) Effect of controlled atmosphere, vacuum packaging and different temperatures on the growth of spoilage fungi in shelled pecan nuts during storage. Food Control 128:108173
Rouves M, Prunet JP (2002) New technique for chestnut storage: effects of controlled atmosphere. Infos Ctifl (France) 186:33–35
San Martin MB, Fernández-García T, Romero A, Lopez A (2001) Effect of modified atmosphere storage on hazelnut quality. J Food Process Preserv 25(5):309–321
Shakerardekani A, Karim R (2013) Effect of different types of plastic packaging films on the moisture and aflatoxin contents of pistachio nuts during storage. J Food Sci Technol 50(2): 409–411
Shayanfar S, Kashaninejad M, Khomeiri M, Emam DZ, Mostofi Y (2011) Effect of MAP and different atmospheric conditions on the sensory attributes and shelf life characteristics of fresh pistachio nuts. Int J Nuts Relat Sci 2(3):47–57
Sheikhi A, Mirdehghan SH, Ferguson L (2019a) Extending storage potential of de-hulled fresh pistachios in passive-modified atmosphere. J Sci Food Agric 99(7):3426–3433
Sheikhi A, Mirdehghan SH, Karimi HR, Ferguson L (2019b) Effects of passive-and active-modified atmosphere packaging on physio-chemical and quality attributes of fresh in-hull pistachios (*Pistacia vera* L. cv. Badami). Foods 8(11):564. https://doi.org/10.3390/foods8110564
Talasila PC, Cameron AC, Taylor LJ (1995) Storage and modified-atmosphere packaging of Chinese chestnuts (*Castanea mollissima*). HortScience 30(4):815E
Tavakoli J, Sedaghat N, Mousavi Khaneghah A (2019) Effects of packaging and storage conditions on Iranian wild pistachio kernels and assessment of oxidative stability of edible extracted oil. J Food Process Preserv 43(4):e13911
Wang J, Li P, Gong B, Li S, Ma H (2017) Phenol metabolism and preservation of fresh in-hull walnut stored in modified atmosphere packaging. J Sci Food Agric 97(15):5335–5342
Ye N, Zhang P, Wang Y, Ma H, Zhang T (2021) Effects of controlled atmosphere on browning, redox metabolism and kernel quality of fresh in-hull walnut (*Juglans regia* L.). Hortic Environ Biotechnol 62(3):397–409

Chapter 14
Physiological Disorders

Sunil Kumar, Satyabrata Pradhan, Naveen Kumar Maurya, and Ashok Yadav

Abstract Many physiological disorders in temperate nuts affect both quality and storage life in all growing regions of the world. While the cause and control of some of these disorders have been found, others have eluded an answer so far. The extremes of environmental variables such as temperature, moisture, light, aeration and nutritional imbalances result in disturbances in the plant metabolic activities leading to these disorders. While the symptoms may appear disease-like, they can usually be prevented by altering environmental conditions. This chapter presents a critical review on cause and characteristics of physiological disorders in important temperate nut fruit crops besides providing an insight into the gaps and researchable issues.

Keywords Temperate nuts · Disorders · Oil rancidity · Mouse ear · Storage

1 Introduction

Nuts are botanically defined as a fruit in which the carpel wall is hard or bony in texture (Ros 2010). Fruit is derived from a hypogynous flower (filbert) or an epigynous one (walnut) and is enclosed in dry involucres (husk). It is only one seeded but in most cases derived from two carpels, i.e. English or Persian walnut (*Juglans regia*), almond (*Prunus amygdalus*), chestnut (*Castanea sativa*), hazelnut also known as filbert (*Corylus avellana*) and pecan nut (*Carya illinoensis*). Dry fruits are not juicy or succulent when mature and ripe. When dry, they may split open and discharge their seeds (called dehiscent fruits) or retain their seeds (called indehiscent

S. Kumar (✉)
ICAR-National Research Centre on Litchi, Muzaffarpur, India

S. Pradhan · N. K. Maurya
Division of Fruits and Horticultural Technology, ICAR-Indian Agricultural Research Institute, New Delhi, India

A. Yadav
ICAR-Central Agroforestry Research Institute, Jhansi, India

M. M. Mir et al. (eds.), *Temperate Nuts*,
https://doi.org/10.1007/978-981-19-9497-5_14

fruits). Nut crops are grown for their high calorie, nutritious seeds that humans have consumed over millennia, beginning with the gathering of wild nuts. Today, nuts, their oils, flours, and other edible products are popular worldwide. Cultivation of these nut crops contributes greatly to local and national economies, and they sell well because they are tasty and are an important part of a healthy diet (Preece and Aradhya 2019). The regular, moderate consumption of nuts is associated with a reduced risk of chronic degenerative diseases, in the context of a healthy diet and lifestyle (Dufoo-Hurtado et al. 2021). Temperate nuts are mainly produced in the middle latitudes ranging from 30° to 50° N and S. Their cultivation may extend to lower latitudes (15°–30° N and S) at higher altitudes and to higher latitudes where the congenial climate is available. Cool temperatures in the winter are essential for fulfilling their chilling requirements to ensure homogeneous flowering and fruit set and generate economically sufficient yields (Rai et al. 2015). In India, the temperate nuts cultivation is mainly confined to the union territory of Jammu and Kashmir and states of Himachal Pradesh, Uttarakhand and parts of Arunachal Pradesh. Most of the nut crops are grown in high hills of cold dessert areas (Dhillon and Rana 2003). Because of an ideal climate, this region offers tremendous opportunity for the production of high-quality temperate nuts (walnut, almond, chestnut, pecan nut and hazelnut) (Verma et al. 2010). In the changing climate scenario, temperatures are expected to increase in most parts of the world, with minimum temperatures rising most rapidly. This development may compromise the ability of many growers of temperate nuts to successfully produce the same array of crops as in the past (Luedeling et al. 2011). Apart from climate change, the productivity, as well as the quality of nut crops is affected to a greater extent due to various physiological disorders.

Physiological disorders are deviations from the normal physiological functioning and are often the result of imbalances in fruit metabolism induced by pre or postharvest environments (Shivashankar 2014). Physiological disorders are also commonly referred to as abiotic (i.e., non-living), non-pathogenic, nonparasitic or non-infectious diseases. Physiological disorders can occur in either or both preharvest and postharvest phases. Often a disorder is caused by the combined effects of both environmental conditions and management practices (Thornton et al. 2020) such as temperature, moisture, unbalanced soil nutrients, inadequate or excessive soil minerals, extremes of soil pH and poor drainage (Kumar and Kumar 2016). Postharvest storage conditions such as temperature, oxygen and carbon dioxide levels, as well as packaging, are contributing factors to the occurrence of many postharvest disorders. Many physiological disorders not only affect nuts production and productivity but may also affect fruit quality and its technical characteristics and commercial acceptability. This chapter highlights the important physiological disorders of temperate nuts.

2 Almond

2.1 *Double Fruit/Cleft Sutures*

Double or twin fruit is frequently seen in orchards. In many cases, the extent of the problem is minor, and these fruits can be easily eliminated during hand thinning. But in some orchards and on some cultivars, the problem can be of major importance. Whether a fruit will be double or not is determined the summer before fruiting when the flower buds are going through their initial development. The production of double kernelled fruits is frequent in some varieties, although the percentage depends on the year and large variations occur. It is known that conditions which favour pollination and fruit set give rise to a higher percentage of double kernelled fruits, although nothing is known about the influence of pre-blossom climatological conditions on double kernelled fruits (Egea and Burgos 1995).

2.2 *Buttons*

Buttons are fruits that, although they initially set, do not develop into full-sized fruit. These fruits generally have poorly developed or dead embryos because of incomplete fertilisation. This can be due to insufficient chilling received, a frost during bloom, or wet, cool weather during bloom. These fruits are hard to identify at the thinning time and may cause a grower to thin too heavily. Also, they provide a place for pests and diseases to survive and overwinter in a low fruit year.

3 Chestnut

Chestnut is not susceptible to any specific disorder. It is not chilling sensitive, but it can freeze when the temperature goes below −5 °C for a few days. Chestnut is not CO_2 sensitive, but it ferments in the absence of oxygen for several days. Thus, it is sensitive but only in extreme environmental conditions. Internal browning can appear after cutting the seed flesh at the end of the storage period which is increased after cooking. This might be due to the soluble sugar accumulation during senescence and the beginning of sprouting (Mencarelli and Vannini 2019).

4 Hazelnut

4.1 Kernel Black Tips

Blacktip on kernels appears to be associated with nuts having split or weak sutures. It appears to be caused by an oxidation process that occurs on the pellicle only. It deteriorates the nut quality due to which it fetches fewer returns.

4.2 Blank Nuts or Seedless Nuts

Blank means a hazelnut containing no kernel or a kernel filling less than the one-fifth capacity of the shell (Fig. 14.3). Most species of hazelnut are largely self-incompatible and self-incompatibility was often associated with a higher frequency of blanks (Beyhan and Marangoz 2007). One of the more intriguing aspects of the reproductive biology of hazelnuts is the temporal separation of pollination and fertilisation. At the time of pollination, the ovary is not formed and grows only if the flower is pollinated. Embryo abortion, rather than failed pollination or fertilisation, has been implicated in blank-fruit formation (Liu et al. 2012).

4.3 Brown Spots in Kernel Cavity

It is a commercially important physiological disorder particularly found in Spanish hazelnut cultivars. Several factors are associated with this disorder in the kernel cavity in the last stages of kernel development. Brown spots in the kernel cavity can affect 7–97% of the total production, depending on the year. A direct correlation is observed between the percentage of fruit affected and the altitude of the growing area. Although the definite cause of the brown spots in kernel cavity disorders is not known, it seems that orchards at higher elevation have a lack of accumulated heat units during the period of kernel development which occurs during the end of June to mid-July, which is correlated with an increase in the incidence of brown spots in kernel cavity. Therefore, it seems that the disorder could be related to some enzymatic process that could be affected by temperature (Tous et al. 2000).

4.4 Brown Stain

Brown stain is often described as a physiological disorder that causes distorted shells and leads to an increase in the incidence of blanks and poorly filled nuts. The main symptom is the presence of brownish liquid that soaks the side or end of the nut

(Fig. 14.4). Staining begins when nuts are about half grown. The cause is still unknown, but in severe attacks the shell becomes distorted, and the kernel destroyed (Silvestri et al. 2021).

5 Pecan Nut

5.1 Rosette and Little Leaf

The complex of symptoms known as 'rosette' and 'little leaf' in pecan is characterised by very short internodes (the portion of stems between nodes or leaves) and small, often chlorotic leaflets with wavy margins. The leaf finally shows a rosette appearance due to extensive crinkling. When the incidence is severe, twig and branch dieback occurs, resulting in retarded growth and development of the trees. The affected trees are devoid of nut production (Fig. 14.1). This is a nutritional disorder of pecan nut resulting from zinc deficiency. So, it can be corrected with a foliar spray of zinc sulphate (0.5%). Its application to the soil should be done at 900–1000 g/tree in sandy soils and 2.25 kg and 4.5 kg/tree in sandy-loam and heavier soils, respectively. Foliar Zn applications should be made before the rosette symptoms arise because these symptoms cannot be reversed within a given season (Heerema et al. 2010).

5.2 Mouse-Ear

Mouse-ear is a potentially severe anomalous growth disorder affecting pecan trees. It is characterised by small, roundish leaflets. The leaflets ultimately become round and

Fig. 14.1 Rosette and little leaf in peacan nut (Source: https://pubs.nmsu.edu/_h/H657/index.html)

Fig. 14.2 Mouse ear in peacan nut (Source: https://pubs.nmsu.edu/_h/H657/index.html)

Fig. 14.3 Blank nuts in hazel nut

wrinkled forming a cap. The leaflets develop into a mouse ear shape and the entire leaf is smaller than the normal size (Fig. 14.2). It has been reported that mouse ear in pecan is due to nickel (Ni) deficiency at bud break. Foliar application of Ni to mouse ear prone trees in mid-October or soon after bud break is an effective means of preventing or minimising mouse-ear disorder (Wood et al. 2004).

5.3 Premature Nut Drop

As the nuts develop during the season, pecan growers often observe three or four waves of premature nut drop. Weak and unpollinated flowers are shed by pecan trees during the first two waves of nut drop, which occur in May and June. Good orchard

Fig. 14.4 Brown stain in hazel nut

management practices (pruning, irrigation, nutrition, and pest control) year after year to minimise tree stresses and proper placement of compatible polliniser trees may help to reduce loss of crop during these early waves of nut drop. The severity of the third ('July drop') and fourth ('August drop') waves can be decreased by providing adequate irrigation and mineral nutrition at kernel fill stages of nut development (Heerema et al. 2010).

6 Walnut

6.1 Oil Rancidity

Although all nut species are prone to rancidity, walnuts are particularly susceptible, due to their elevated oil content (65–70%) and a high proportion of polyunsaturated fatty acids (PUFAs) (Martínez et al. 2010). The problem appears to be caused by poor seed storage conditions; elevated temperature and relative humidity, failure to use a controlled atmosphere and imbalance in O_2, concentration. The seed kernels are enclosed in a brown seed coat that contains antioxidants. The antioxidants protect the oil-rich seed from atmospheric oxygen so preventing rancidity (oxidative rancidity). Walnuts have the best colour and flavour when their water content is 2–8%. Higher water contents reduce storage life and increase the risk of rancidity. Rancidity caused by oxidative fat cleavage is particularly noticeable in the case of shelled walnuts, because of exposure to atmospheric oxygen. It is therefore essential to store walnuts in the dark and to protect them from oxygen since otherwise they become brown-coloured and develops a rancid odour and taste. Controlled atmospheric storage experiments under different O_2 partial pressures showed a beneficial effect of oxygen equal to or lower than 3 kPa to reduce deterioration of 'Howard' kernels (Ortiz et al. 2019).

6.2 *Winter Sunscald*

This type of injury occurs when the sun warms tree bark during the day and then the bark rapidly cools after sunset. These abrupt fluctuations may kill the inner bark. Young trees are most susceptible to winter sunscald. Wrapping trunks of susceptible trees with protective 'Tree wrap' is the most effective way to minimise this type of winter injury.

7 Conclusion

Lack of understanding of the causes of many physiological disorders in temperate nuts is a serious constraint in production of high-quality fruit. Detailed knowledge about the different causes and management practices of different physiological disorders in temperate nuts will not only aid the quality production to nut growers, but also it will be useful for researchers to generate innovative ideas to control these disorders through biotechnological interventions, breeding strategies or by understanding a physiological basis to overcome it.

There is an urgent need to understand the factors responsible for the occurrence of these disorders at physiological level which are hindering the quality production and export potential of our country and also a need to follow a different management approach as mentioned in this chapter to manage a particular disorder.

References

Beyhan N, Marangoz D (2007) An investigation of the relationship between reproductive growth and yield loss in hazelnut. Sci Hortic 113(2):208–215

Dhillon BS, Rana JC (2003) Temperate fruits genetic resources management in India-issues and strategies. Acta Hortic 662:39–146

Dufoo-Hurtado E, Luzardo-Ocampo I, Ceballos-Duque SM, Oomah BD, Maldonado-Celis ME, Campos-Vega R (2021) Nuts by-products: the Latin American contribution. In: Valorization of agri-food wastes and by-products. Academic Press, Cambridge, pp 289–315

Egea J, Burgos L (1995) Double kerneled fruits in almond (*Prunus dulcis* Mill.) as related to pre-blossom temperatures. Ann Appl Biol 126(1):163–168

Heerema R, Goldberg NP, Thomas S (2010) Diseases and other disorders of pecan in New Mexico. New Mexico State University, Las Cruces

Kumar R, Kumar V (2016) Physiological disorders in perennial woody tropical and subtropical fruit crops—a review. Indian J Agric Sci 86(6):703–717

Liu JF, Cheng YQ, Yan K, Liu Q (2012) An investigation on mechanisms of blanked nut formation of hazelnut (*Corylus heterophylla* fisch). Afr J Biotechnol 11(30):7670–7675

Luedeling E, Girvetz EH, Semenov MA, Brown PH (2011) Climate change affects winter chill for temperate fruit and nut trees. PLoS One 6(5):e20155

Martínez ML, Labuckas DO, Lamarque AL, Maestri DM (2010) Walnut (*Juglans regia* L.): genetic resources, chemistry, by-products. J Sci Food Agric 90(12):1959–1967

Mencarelli F, Vannini A (2019) Postharvest handling. In: The chestnut handbook: crop and forest management. CRC Press, Boca Raton, pp 255–274

Ortiz CM, Vicente AR, Fields RP, Grilo F, Labavitch JM, Donis-Gonzalez I, Crisosto CH (2019) Walnut (*Juglans regia* L.) kernel postharvest deterioration as affected by pellicle integrity, cultivar and oxygen concentration. Postharvest Biol Technol 156:110948

Preece JE, Aradhya M (2019) Temperate nut crops: chestnut, hazelnut, pecan, pistachio, and walnut. In: North American crop wild relatives, vol 2. Springer, Cham, pp 417–449

Rai R, Joshi S, Roy S, Singh O, Samir M, Chandra A (2015) Implications of changing climate on productivity of temperate fruit crops with special reference to apple. J Hortic 2(2):135–141

Ros E (2010) Health benefits of nut consumption. Nutrients 2(7):652–682

Shivashankar S (2014) Physiological disorders of mango fruit. Hortic Rev 42:313–348

Silvestri C, Bacchetta L, Bellincontro A, Cristofori V (2021) Advances in cultivar choice, hazelnut orchard management, and nut storage to enhance product quality and safety: an overview. J Sci Food Agric 101(1):27–43

Thornton M, Olsen N, Liang X (2020) Physiological disorders. In: Potato production systems. Springer, Cham, pp 447–478

Tous J, Ferrán X, Rius M, Sentis X, Plana J, Romero A (2000) The brown spots in kernel cavity disorder of hazelnut. Acta Hortic 556:397–402

Verma MK, Ahmed N, Singh AK, Awasthi OP (2010) Temperate tree fruits and nuts in India. Chron Hortic 50(4):43–48

Wood BW, Reilly CC, Nyczepir AP (2004) Mouse-ear of pecan: a nickel deficiency. HortScience 39(6):1238–1242

Chapter 15
Diseases of Temperate Nuts

Amir Mirzadi Gohari and Angela Feechan

Abstract Plant disease can significantly reduce the market size of edible nuts in the regions where these nut trees are growing. This chapter covers economically essential diseases of almond, pistachio, pecan nut, hazelnut, walnut, and chestnut crops. Here, we address the biotic agents limiting the cultivation of nut fruits worldwide and imposing huge economic yield losses in the nut industry. We summarise some important diseases in terms of developed symptoms on the infected tree, pathogen description, epidemiology, and recommended measures in managing nut disease. We discuss the environmental conditions, facilitating disease development, and, eventually, suggest appropriate measures to control nut diseases. Nut crops are common snacks and come in many fruit crops, which include almond, pistachio, hazelnut, walnut, chestnut, pecannut, and chilgoza. In 2020, the edible nuts market size in the U.S. was calculated at US$24 billion, and this country accounts for a 27.06% share in the global market (Anonymous, Edible nuts-global market trajectory and analytics, Global Industry Analysts, Inc., San Jose, 2021). A plethora of biotic and abiotic agents account for imposing significant economic losses on the nut industry. In this chapter, we aim to address the investigation of epidemiology and biology of the phytopathogens infecting nut crops and imposing economic losses on growers. Additionally, we summarise the remaining nut crops disease in a separate table.

Keywords Diseases · Temperate nuts · Symptomatology · Pathogen · Blight · Anthracnose

A. Mirzadi Gohari (✉)
Department of Plant Protection, University College of Agriculture and Natural Resources, University of Tehran, Karaj, Iran
e-mail: mirzadighohari@ut.ac.ir

A. Feechan
School of Agriculture and Food Science, University College Dublin, Dublin, Ireland

Present Address: Institute for Life and Earth Sciences, School of Energy, Geoscience, Infrastructure and Society, Heriot-Watt University, Edinburgh, UK

M. M. Mir et al. (eds.), *Temperate Nuts*,
https://doi.org/10.1007/978-981-19-9497-5_15

1 Almond

1.1 Almond Anthracnose

Anthracnose almond (AA) is one of the most economically important diseases of almonds, causing yield losses in the main areas of almond growing. This disease is found in Australia, France, Greece, Italy, the USA, South Africa, and Italy (Adaskaveg and Hartin 1997; López-Moral et al. 2020).

1.1.1 Symptomatology

The main parts affected by this disease are fruits. Infected green fruits become depressed, round, shriveled, and orange or brown. Lesions that are 5–12 mm in diameter and develop on the surface of the fruit during spring–summer. Under high humidity conditions, abundant whitish mycelium or orange spore masses are formed on the surface of attacked fruits, appearing as visible droplets. Acervuli, asexual fruiting bodies of the causal agent, containing numerous spores, are produced in the centre of lesions. Subsequently, many of the infected fruits become mummified and remain on the tree during the autumn and winter that are the major inoculum sources in the following seasons (López-Moral et al. 2017). In addition, some of the mummified fruits fall prematurely to the soil. Leaf infections in seriously attacked plants have also been reported, where water-soaked spots start to develop at the tip of leaves. Subsequently, these lesions extended to the entire leaf blade. Twig, shoot blight, and branch dieback happens on branches carrying the diseased fruits (Förster and Adaskaveg 1999; Peres et al. 2005).

1.1.2 Causal Organism

Colletotrichum acutatum and *C. godetiae* have been documented as two dominant species, causing almond anthracnose throughout the world. Both fungal species can be grown radially with concentric circles and extensive aerial mycelium on potato dextrose agar (PDA) under conditions of 23 ± 2 °C with a 12-h photoperiod. *C. acutatum* develops pink-orange colonies while the latter species produces grey colonies. The colony colour of both species on almond fruits artificially inoculated by *C. acutatum* sensu stricto (s.s.) and *C. godetiae* are similar to that observed on a PDA medium. The conidial characteristic feature of both species is not a helpful indicator to distinguish them between species, and both have a unicellular, hyaline conidium with two sharp ends (López-Moral et al. 2017). More recently, phylogenetic analysis coupled with the pathogenicity assay demonstrated that *C. acutatum* s.s. was the most common fungi in Australia, causing AA along with the *C. simmondsii* which was documented for the first time as the common causal agent of AA elsewhere in the world (de Silva et al. 2021).

1.1.3 Disease Cycle and Epidemiology

Anthracnose almond is depicted as a polycyclic disease. The pathogen survives on the tree in mummified fruits, infected branches or probably lives as an endophytic fungus on alternative host plants such as pepper and eggplant (Peres et al. 2005). These are primary inoculum establishing the primary infection during late winter to early spring once the first rain events occur. Wet weather and mild temperatures (10–25 °C) are central factors to stimulate fungal sporulation, infection, and disease development. Rain splash is the major player to disseminate the conidia produced from acervuli. These serve as inoculum for secondary and continuous infection, happening during spring until the rain stops. *C. acutatum* possess a vast host range, but the roles of an alternative host as a source of inoculum is unclear (Förster and Adaskaveg 1999).

1.1.4 Control

In an Integrated Pest Management (IPM) programme to efficiently manage anthracnose almond, preventive measures, including cultural practices, resistant cultivars, and biological control, should be included. The most effective strategy to cure anthracnose almond is fungicide applications, but it is necessary to couple these treatments with the cultural practice providing the best control. Demethylation inhibitors, quinone outside inhibitor, and succinate dehydrogenase inhibitor are the most efficient fungicide classes to manage anthracnose almond (López-Moral et al. 2020).

1.2 Red Leaf Blotch

Red leaf blotch (RLB) is a damaging foliar disease of almonds in the Mediterranean basin and Middle East regions. However, to date, there is no report indicating the occurrence of RLB in the USA and Australia (Miarnau et al. 2021).

1.2.1 Symptomatology

Lesions formed on both sides of infected leaves are circular or irregularly shaped spots and are initially pale green, turning successively yellowish-orange, red, brown, and finally black. These spots develop into larger blotches ranging from 10 to 22 mm in diameter and are composed of stromata of the causal pathogen, *Polystigma amygdalinum*. As the disease progresses, leaves curl and become necrotic. RLB can culminate in premature defoliation of the infected leaves, thus reducing the plants photosynthetic capacity. Consequently, this results in decreasing the almond yield quantitatively (Banihashemei 1990).

1.2.2 Causal Organism

Almond Red leaf blotch is caused by the fungus *Polystigma amygdalinum* PF Cannon (previously known as *Polystigma ochraceum*), producing perithecial stromata within living and infected leaves on the trees. Following falling infected leaves, perithecia grow and mature, they are flask-shaped with thick walls. Mature asci are clavate, short-stalked, very thin-walled with an apical ring (74–97 × 18–23 mm) containing eight ascospores that are hyaline, unicellular, smooth-surfaced, arranged biseriately, and thin-walled (13–16.5 × 5.5–6.5 mm). During the early stage of stromata development, the anamorph *Polystigmina* Sacc., initiates with the generation of internal conidiomata developed in the red lesions on infected leaves. This process leads to producing the globose pycnidia (80 × 120 mm), containing abundant conidia (22–34 × 0.75–1 mm) that are slender, hyaline, curved, and filiform.

1.2.3 Disease Cycle and Epidemiology

P. amygdalinum overwinters as a saprophyte on the infected fallen leaves where this fungus produces sexual reproductive structures. Primary infection establishes in early spring, coinciding with ascospore discharge from mature perithecia. Subsequently, these sexual spores infect the young leaves. The period of discharging primary inoculum depends mainly on temperature and rainfall during late winter and early spring. In Iranian conditions, ascospore discharge starts at the flowering stage in late April and lasts for 6 weeks until fruit set (mid-May), whereas under Spanish conditions this takes place from January to June. The time between inoculation and symptoms expression (incubation period) is approximately 30–40 days under conditions in the Maharlou region in Iran compared to 35–40 days in the Halat region of Lebanon. The earliest symptoms appear in May and continue to develop and change colour during late spring and summer. Pycnidia are produced in June, but their function in the disease cycle is unclear (Banihashemei 1990; Saad and Masannat 1997).

1.2.4 Control

Triforine, a systemic fungicide, at the lowest concentrations of 100 mg mL^{-1} provides a significant reduction of red leaf blotch when applied after the flowering stage. Chemical control starts at petal fall, followed by two or three subsequent sprays. This process depends on the duration of ascospore release from the perithecia (Banihashemei 1990). It is highly recommended to de-composite the leaf litter by applying urea on fallen leaves in Autumn or to bury the fallen leaves from the trees in the previous year to reduce the overwintering inoculum (Cannon 1996). To date, no resistant cultivar against *P. amygdalinum* has been identified (Miarnau et al. 2021).

1.3 Shot Hole

Shot hole disease is the most widespread fungal disease of stone fruit trees worldwide, including almond growing regions (Highberg 1986a; Adaskaveg et al. 1990).

1.3.1 Symptomatology

Localised circular spots occur on leaves and fruits as tiny purplish areas, developing into chlorotic and, subsequently, necrotic lesions, 3–10 mm in diameter (Fig. 15.1). Multiple infections and coalescing lesions on older and larger leaves may result in leaf defoliation, but this symptom is rare. Older infected leaves retain their lesions while lesions on young leaves drop out under warm and dry environmental conditions, leaving numerous circular holes, from which, the disease derives its name. Under cool and wet conditions, the fruiting structures appear as a small tan to dark speck (sporodochium) that frequently develop in the centre of the lesion. On fruits, lesions commonly occur on the upper side and are usually smaller (1–2 mm in diameter) with purplish margins, raised, and corky in appearance (Fig. 15.1). Gumming occurs if fruit lesions abscise (Highberg 1986a, b).

Fig. 15.1 Initial symptoms of almond shot hole disease occur on leaf (**a**) and fruits (**b**) as localised circular spots. (Photo credit: David doll: https://thealmonddoctor.com/shot-hole-fungus-on-almond)

1.3.2 Causal Organism

Shot hole disease of almond is caused by the fungus *Wilsonomyces carpophilus* (Lev.) Adaskaveg, Ogawa & Butler (syn. *Stigmina carpophila*) belonging to the Dothideomycetes. This fungus is characterised by forming sporodochial conidiomata and producing sympodial conidiogenous cells. The conidia are solitary, initially hyaline, and later turn golden brown, ellipsoidal to fusiform with thick double walls, and three to five transverse septa. *Ascospora beijerinckii* Vuillemin was introduced as the teleomorph, but this has never been demonstrated (Adaskaveg et al. 1990). Leaves and fruit lesions are ideal parts to isolate *W. carpophilus*, and can be cultured on potato-dextrose agar (PDA). Mycelial growth on PDA medium is slow and happens at 4–30 °C, with an optimum temperature of 15–20 °C (Ahmadpour et al. 2012).

1.3.3 Disease Cycle and Epidemiology

During the winter period, *W. carpophilus* survives as conidia for several months, infecting the dormant buds and twigs of almond trees. In spring, conidia are dispersed through rain splash to developing blossoms and leaves. Following spore landing on the leaves or fruit surface, germinated conidia elongate and subsequently penetrate indirectly through stomata or directly via forming an appressorium covered by a gelatinous matrix (Highberg 1986b). Leaf infection needs 8–12 h of moisture at 20–25 °C and symptom expression occurs within 4–15 days post-infection (Shaw 1990). In California, epidemics of shot hole frequently occur in early spring and late fall, once ample rainfall is common. Secondary infection initiates on leaves in fall when the weather is warm and wet since the leaf abscission is postponed and fungicide application is not regularly applied after harvest.

1.3.4 Control

Fungicide applications are a primary measure to manage shot hole disease, and one to three treatments are required from leaf emergence until 5 weeks after petal fall. Phthalimide, dicarboximide, dithiocarbamate, captan, captafol, and ziram are effective agents to control this disease. However, compounds belonging to the demethylation inhibitor (DMI) fungicide proved to be less efficient. Applying zinc sulphate foliar fertiliser to hasten leaf fall on the trees where the inoculum may increase is recommended. Sprinkler irrigation system leading to the wetness of leaves and fruits should be avoided. No commercial almond cultivars resistance towards shot hole are currently available (Teviotdale et al. 1989; Connell 2002).

1.4 Bacterial Canker

Bacterial gummosis, bud blight, and sour sap are other names to describe the bacterial canker afflicting *Prunus* spp. in all major production regions of the world, where nut fruits are growing. Almond trees, between 2 and 6 years old, are susceptible to bacterial canker causing significant economic losses, while those 5–6 years old are rarely affected by this destructive disease (Kennelly et al. 2007).

1.4.1 Symptomatology

Elliptical lesions are amber-coloured gumballs with undefined margins produce in winter and early spring on the trunks or scaffolds of trees ending at the graft union. Developed cankers do not extend into the soils and profusely exude gum in late winter and early spring. Under the bark, water-soaked, reddish-brown streaks or flecks develop into the phloem above and below the canker. Developed cankers on the branches around small twigs and spurs are diamond-shaped, and inactive scars are formed on them in the following years. A young tree attacked by this destructive disease frequently dies before budbreak. The disease is active in winter and inactive in spring. The healthy rootstock produces suckers at the base of the tree, below the dead scion in the following season. Infected trees emit a sour odour and fermented scent, which promoted the name sour sap disease (Hirano and Upper 1990).

1.4.2 Causal Organism

The causal agent of almond bacterial canker is *Pseudomonas syringae* pv. *syringae* (*Pss*) van Hall, a rod-shaped gram-negative with a size of 0.7–1.2 × 1.5–3 μm and possessing one to several polar flagella. On King's medium B, bacterial cells produce green fluorescent, water-soluble pigments emitting blue light under UV light. This organism is a non-produced oxidase and a strict aerobe enabling the consumption of several sources of energy. Tyrosinase activity and tartrate utilisation of most isolates measured by cultivating them on agar medium supplemented with L-tyrosine and on Simmons basal medium amended by sodium tartrate are negative, while that of gelatin liquefaction as well as aesculin hydrolysis is positive. *Pss* can induce ice-nucleating event, which promotes infection and disease development caused by this pathogen. This biotic agent also generates several toxins, including syringomycin, killing plant cells by disrupting cellular membranes. This event eventually leads to the formation of cankers on infected trees. The ideal temperature for the growth of species is determined to be 25–30 °C, nevertheless, growth occurring at 4 °C was also reported (Xin et al. 2018). Genetic analysis of the *Pss* strains recovered from diverse stone fruits demonstrated that these strains belong to a distinct clade separate from most of the strains isolated from other hosts (Little et al. 1998).

1.4.3 Disease Cycle and Epidemiology

Pss is an epiphytic bacterium that resides on the plant surface but does not inhabit soils. Bacterial canker isolates have been recovered occasionally from weeds, but the maximum population of bacterial cells occurs in developing buds, indicating that the inoculum presented on the trees causes disease. Bacteria are moved into infection sites through rain and fogs during late fall and winter. Bacterial cells are thought to penetrate mainly through natural openings such as leaf scars, stomata, and lenticels, but pruning wounds and other injuries for example from ice damage could function as entry points for infection. Freezing temperature and moisture are conducive conditions to develop canker so that the disease is likely to emerge in the lower parts of the trees where cold air accumulates. Nitrogen-deficient trees and those grown in sandy, heavy clay or low pH soil, and soil with high populations of ring nematodes creating wounded points are prone to bacterial canker (Hirano and Upper 1990; Kennelly et al. 2007).

1.4.4 Control

Chemical controls are an ineffective measure to manage bacterial canker. However, soil fumigation by applying nematicides to reduce the ring nematode populations and site selection play major roles in controlling the disease. Maintaining the proper level of nutrition, particularly nitrogen, plays an essential role in decreasing the disease incidence and severity (Cao et al. 2013). Most orchardists believe that postponing the pruning by the end of winter or early spring reduces bacterial canker, but this claim has never been experimentally corroborated.

2 Pistachio

2.1 Alternaria Late Blight

Alternaria late blight is one of the most common fungal diseases of pistachio. This disease causes shell staining, leaf loss, and the causal agent will move into the fruit. This event results in the reduction of nut quality, yield, and weakening of the tree. The disease has been reported in Egypt, the USA, Italy, and Australia (Pryor and Michailides 2002).

2.1.1 Symptomatology

On leaves, angular or circular lesions (3–7 mm in diameter) that are dark brown to black are formed on the leaf blade during mid-summer. Later in the season, lesions expand and merge, leading to the formation of large necrotic lesions (2.5–3 cm in diameter). Subsequently, the blotches turn black due to dense sporulation occurring particularly in the lesion centre and are the lesion is encircled by a typical chlorotic margin. Cleaning the surface of lesions with a finger will blacken the finger while that of a leaf lesion caused by the fungal pathogen *Botryosphaeria dothidea* does not darken the finger, since no spore is produced on the surface of these lesions. On fruits, small black specks (1 mm in diameter) associated with lenticels are developed on the hull of immature nuts. As the fruit grows, hull and black lesions, 1–5 mm in diameter, are produced on the mature fruit. These lesions are characteristically surrounded by a reddish-purple halo (Pryor and Michailides 2002).

2.1.2 Causal Organism

The causative organism of Alternaria late blight (ALB) has been documented as *Alternaria alternata* (Fr.) Keisel. Fungal colonies are filamentous, greenish-black or rusty brown and grow fast on potato dextrose agar (PDA) or malt extract agar (MEA). Conidiophores are solitary or arranged in small groups, erect or curved that are 3–6 nm × 20–50 nm in diameter. Conidia are multi-celled, obclavate or obpiriform, light brown to black, and are 9–18 nm × 20–63 nm in diameter. Conidia are multi-celled, obclavate, or obpiriform, light brown to black, and are 9–-18 nm × 20–63 nm in diameter. Conidia possess longitudinal and transverse cross-walls accompanied with a short conical beak (2–5 nm in diameter). This part is less than one-quarter of the conidial length (Troncoso-Rojas and Tiznado-Hernández 2014). *A. arborescens* E. Simmons and *A. tenuissima* (Kunze: Fr.) have been recovered from attacked pistachio and demonstrated experimentally to cause Alternaria late blight whereas, *A. infectoria* was weakly pathogenic on pistachio. Phylogenetic investigations demonstrated that isolates belonging to the alternata, tenuissima, and arborescens species composed a monophyletic clade in such a way these groups could not be further resolved (Pryor and Michailides 2002).

2.1.3 Disease Cycle and Epidemiology

Crop debris, fallen pistachio leaves, and weeds serve as overwintering sources for *Alternaria* spp., causing Alternaria late blight. The fungus in favourable conditions, which includes high relative humidity and dew, sporulates on the surface of attacked plant tissue, resulting in the abundant generation of spores. The conidia are readily dispersed to susceptible plant tissue via rain and wind. The necrotic areas on leaves due to Peter's scorch, a noninfectious disorder usually found only in male pistachio trees, and early splitting of fruit are commonly colonised by *Alternaria* spp. This disease becomes a problematic issue in orchards established close to the river or irrigated by sprinklers or via flooding due to the high humidity (Pryor and Michailides 2002).

2.1.4 Control

Management of Alternaria late blight can be achieved through a combination of proper cultural measures and multiple fungicide sprays. Cultural practices involve employing an irrigation system, allowing the ground surface to dry quickly, and pruning foliage to facilitate air movement and reduce humidity in the orchard. Additionally, harvesting on time to lower the fruit infection and the shell staining is recommended. Fungicides registered for controlling Alternaria late blight involve sterol demethylation inhibitors (DMIs), benzimidazoles, and strobilurins (Michailides and Morgan 1993). Due to the emergence of resistant Alternaria isolates, fungicides with diverse modes of action in a spray programme, either on a rotating schedule or in mixtures, is highly recommended. Resistance against fungicides such as azoxystrobin (Abound®), pyraclostrobin, and boscalid in the field populations of *Alternaria* spp. causing Alternaria late blight has been reported (Ma and Michailides 2004; Avenot et al. 2008).

2.2 *Panicle and Shoot Blight*

Botryosphaeria panicle and shoot blight is an economically major disease of pistachio in California, (USA) that currently threatens the industry, causing significant losses up to 40–100% because of the death of fruit panicles. This disease was first reported in 1984 in the USA and then subsequently reported in Italy, Greece, and South Africa (Michailides 2002).

2.2.1 Symptomatology

This disease affects shoots, fruits, rachis and leaves of pistachio (Fig. 15.2). Initial symptoms include black lesions (1–2 mm in diameter) on the infected parts that appear during mid-spring. Shoots emerging from either heavily or partially infected buds grow at a short distance, turn blank and eventually die. At the end of spring, the infected leaves become dry and shriveled during 3–5 days, and the infected shoots and leaves are easily distinguishable among the dark-healthy foliage. Infection of leaflet and leaves commonly leads to defoliation at the end of summer and can impose severe yield losses. As the summer progresses, large necrotic blotches with chlorotic margins (up to 15 mm in diameter) covered by asexual fruiting bodies (pycnidia) develop on the centre of infected leaves. On fruits, tiny, pin-sized, and black spots develop initially, and subsequently, they expand rapidly on one to several fruits in a panicle causing fruits to turn black and the infection moves through the peduncle to the rachis. The infection eventually reaches the shoot where the fruit cluster is produced, causing a sunken elliptical canker. Such cankers will be formed around the attacked leaves and bud scars (1–10 cm in length) and do not expand

Fig. 15.2 Infection of fruits caused by *Botryosphaeria dothidea* characterised by the formation of round and black spots on the leaves (**a**) and fruit surface (**b**); infections of leaves and fruits develop as large and irregular to round lesions, up to 25 mm in diameter, surrounded by chlorotic margins. (Picture is taken from Michailides and Morgan 2004)

further in the following years. The necrotic black rachises remain on the tree for 3–4 years is a typical symptom to diagnose this disease readily (Michailides and Morgan 2004).

2.2.2 Causal Organism

Panicle and shoot blight of pistachio is caused by the fungal pathogen *Botryosphaeria dothidea* (Moug.:Fr.) Ces. & De Not. (synonym *B. ribis* Gross. & Duggar). The pycnidial stage was initially reported on almonds and avocados as *Dothiorella* species but is currently determined as a species of *Fusicoccum*. This fungus produces black and asymmetrical pycnidia arranged solitary or in groups of 5–8 and contains an apical ostiole with white to creamy contents. Conidia are unicellular, hyaline, fusiform with rounded ends and are 15–29 × 5–8 μm in length. In culture, colonies are white and olivaceous, turning dark grey and black in reverse as the colony ages. Fungal isolates derived from pistachio grow fast and well on acidified potato dextrose agar at 20–36 °C with an optimum temperature of 27–30 ° C. At a higher temperature of 36–39 °C, pycnidiospores failed to form colonies on PDA. The teleomorph of *B. dothidea* was discovered in blackberry, firethorn and olive, either growing next to pistachio or at a distance of several miles away from pistachio (Michailides and Morgan 2004). Phylogenetic studies of 304 isolates recovered from pistachio displaying the symptoms of panicle and shoot blight demonstrated that 262 (86.2%) of the examined isolates are *Neofusicoccum mediterraneum*, whereas only 17 (5.6%) isolates were determined as *B. dothidea*. Interestingly, this study indicated that the colony growth and conidial features of *Neo. mediterraneum* and *B. dothidea* are very similar, proposing that the principal causal agent of pistachio panicle and shoot blight is *Neof. mediterraneum*. Hence, it is assumed that the *B. dothidea* discovered two decades ago according to morphological signatures and reported the biotic agent causing Botryosphaeria panicle and shoot blight might be *Neof. mediterraneum* based on molecular investigations (Chen et al. 2014). The sexual stage of *Neof. mediterraneum* was described recently in southern Spain, but this has not yet been reported on Californian pistachio (Moral et al. 2015).

2.2.3 Disease Cycle and Epidemiology

B. dothidea overwinters as pycnidia which had developed on the last year's blighted buds, shoots, rachises, petioles, and fruit. Additionally, old cankers caused by *B. dothidea* could produce conidia for at least 6 years, contributing to establishing the primary infection. In wet conditions, cirrhi containing abundant pycnidiospores are discharged from pycnidia. The released spores are dispersed mainly through the rain, but insects, birds, and water from sprinkler irrigation play a role in disseminating the disease (Michailides 1993). Rain occurred anytime during the growing season transfers inoculum to developing tissues resulting in initiating the infection. A period of at least 9–12 h continuous wetness is required to develop symptoms on the infected pistachio leaves. Disease development on fruit occurs at high temperatures (27–33 °C) so that this disease becomes severe in late spring to summer when the temperature rises (Turechek and Stevenson 1998). Nutritional deficiency and drought are considered as triggering factors to predispose trees to the infection caused by *B. dothidea*. The trapped airborne ascospores in some pistachio orchards suggested their role is to function as a source of inoculum and they probably play a central function in long distance dissemination.

2.2.4 Control

When panicle blight disease is established in a pistachio orchard, disease management is difficult to achieve. Intense pruning and eliminating the infected parts are strongly recommended as essential cultural practices to reduce the inoculum level. Applying a sprinkler with a low trajectory angle (12°) plays a key role in reducing disease intensity since panicle and foliage does not expose to a prolonged wetness condition (Michailides 1992). Shortening the duration of irrigation from 48 to 24 h is an effective measure to lower the disease incidence, and irrigating merely during daytime for 12 + 12 h on two successive days is also recommended. Application of strobilurins fungicides, including azoxystrobin, pyraclostrobin, and trifloxystrobin, to prevent or to reduce the disease incidence during spring and later in summer to prevent disease development is an effective practice to manage this destructive disease (Ma et al. 2002).

3 Pecan Nut

3.1 Pecan Scab

Pecan scab is the most serious fungal disease of pecan that can remarkedly reduce the quality of pecan produced in a season. This disease is commonly prevalent in the southeastern United States. Nevertheless, this disease has been documented in other parts of the USA, and Mexico, Canada, Central America, New Zealand, and South America.

3.1.1 Symptomatology

This disease is characterised by the formation of small (1–5 mm in diameter), roughly circular, olive to black spots on the leaves, shucks, and twigs of the infected plant (Fig. 15.3). These specks may fuse together to develop larger blackened regions. As the conidia are generated on the surface of spots, they are initially velvety, but they become hard, dry, crack and fall out as the disease progresses. Lesions on leaves, fruits, and shoots are similar, but those formed on nut shucks commonly develop a visible stroma that is in a black, velvety and cushion-like mass (Demaree 1924). These structures provide the major source of overwintering, resulting in establishing the primary infection in the following spring. Severe fruit infection could culminate in defoliation, reducing the size and quality of the nuts, and cause, huge economic loss to the pecan nut industry. The surface of heavily infected and old nuts is a suitable site for the development of the pink rot caused by the fungus *Cephalothecium roseum* (Terabe et al. 2008).

Fig. 15.3 Typical symptoms of pecan scab characterized by the formation of small (1–5 mm in diameter), roughly circular, olive to black spots on the leaves (**a**), nut shucks (**b**), and young shoots (**c**). (Pictures are derived from Hoefnagels and Mason 2016)

3.1.2 Causal Organism

Venturia effusa (G. Winter) Rossman & W.C. Allen (formerly known as: *Fusicladium effusum* G. Winter) is the causal agent of the pecan nut scab, reproducing both asexually and sexually, but it is mainly the anamorph that has been noticed in nature (Charlton et al. 2020). *F. effusum* produces conidiophores that are solitary, semi-macronematеous, pale to dark brown, mostly erect, septate, and 22–130 × 4–6 μm in size. Asexual spores (conidia) are borne in commonly branched chains, unicellular and 10–24 × 5–10 μm in size. They are in olive-brown to brown, pyriform, subcylindrical, and ellipsoid to fusiform in shape with one or more protuberant scars on their surfaces and are formed on the surfaces of the leaves, fruits, and shoots of the host (Schubert et al. 2013). Certain media containing peptone, dextrose, and yeast extract provide a maximum growth of *F. effusum* in vitro for 21 days at 24–30 °C (Barnes 1964).

3.1.3 Disease Cycle and Epidemiology

F. effusum survives on lesions developed on the leaves, twigs, and shucks that are overwintering sites to produce the primary inoculum in the following spring. Conidiophores carrying the conidia are produced in the overwintering stromata in the following spring. After rainfall, conidia are discharged in air currents and dispersed mainly through wind and rain splash. Following landing on developing nuts and/or an emerging leaf, conidia germinate under conducive environmental conditions. These include the period in which the temperature is 20–30 °C, and leaf wetness continues for 24–48 h (Turechek and Stevenson 1998). By 36 h after infection (hai), the developed germ tube penetrates the cuticle by aid of an appressorium and colonises laterally within the host tissue. About 7–9 days post-infection, symptoms on the infected organs emerge. Conidiophores grow upward and rupture the cuticle, resulting in disease symptoms (lesions) on the tissue surface. These conidiophores generate new conidia acting as inoculum to establish secondary infection. Thus, the pecan scab is considered a polycyclic disease (Gottwald and Bertrand 1982).

3.1.4 Control

Use of resistant cultivars is the most efficient measure to control pecan scab. The selection of these cultivars depends on the regions since regional strains exist and high genetic diversity has been reported in the natural population of *F. effusum*. Resistant cultivars such as Elliott and Kanza are recommended as ideal options in the commercial orchards of Georgia, whereas the cultivar named Gloria Grande is not suggested for most regions. Multiple applications of preventive fungicide are the most effective practice to minimise scab development throughout a growing season.

The timing of spraying plays a major role in providing successful control since the treatment should be applied before infection. Seven to ten fungicidal applications at intervals of 10–14 days from bud break to pollination, and 14–21 days from pollination to shell hardening are recommended. However, the interval between treatments may be decreased to 7–10 days leading to an increased fungicidal application in each growing season. A variety of chemical compounds to manage scab pecan have been recommended. These include the methyl benzimidazole carbamates (MBCs), demethylation inhibitors (DMIs), succinate dehydrogenase inhibitors (SDHIs) and quinone outside inhibitors (QoIs). Cultural practices such as tree spacing and orientation play an essential role in rendering enough exposure to sunlight and adequate airflow circulation. Finally, sanitary measures such as cleaning the garden are proposed to reduce the amount of primary inoculum that may cause infection (Bock et al. 2017; Standish et al. 2021).

3.2 Anthracnose

Pecan anthracnose caused by the *Colletotrichum gloeosporioides* species complex is a widespread fungal disease found wherever the pecan trees are grown. This disease is responsible for significant yield losses since the causal agent damages the nuts remarkedly. The highly infected trees show leaf defoliation in late autumn. This event results in lowering the quality and quantity of the yield. Pecan anthracnose is reported in the USA, Argentina, New Zealand, Korea, and China (Brenneman 1989; Mantz et al. 2010; Zhang et al. 2019; Oh et al. 2021).

3.2.1 Symptomatology

The disease mainly attacks pecan leaves and nuts. The initial symptoms appear as brown and irregular-shaped flecks developed along the edges of a leaflet, spreading quickly over the entire leaf. This event eventually results in causing leaf defoliation. A characteristic necrotic line separating the healthy tissue from the diseased ones can be easily observed on the infected leaves. Initial nut infection occurs as sunken brown spots that are 1–3 cm in diameter. These lesions usually develop at the basal end of the shuck, covering a large part of the shuck's surface. Nut infection occurs in the mid-summer, leading to the formation of smaller nuts in size and inhibiting typical shuck split (Brenneman 1989). A study showed that the leaves placed at the bottom of the tree are more commonly infected compared to those located in higher parts (Oh et al. 2021). Nut mummification, rotting and dropping with the rate of 30–50% can be observed in the pecan tree attacked by anthracnose (Zhang et al. 2019).

3.2.2 Causal Organism

Pecan anthracnose is caused by the ascomycete *Glomerella cingulata* (Stoneman) Spauld. & Schrenk that its anamorph belongs to the *C. gloeosporioides* species complex. *C. gloeosporioides* is reported in Argentina as the causal agent of pecan anthracnose, producing conidia that are oblong with rounded ends (21–15 × 9–6 μm) and the size of appressoria measured 18–6 × 9–6 μm (Mantz et al. 2010). In China, *C. nymphaeae* was documented as a fungal pathogen to cause pecan anthracnose. This fungus produces conidiophores that are hyaline, septate, and branched. Conidia are fusiform to cylindric with one end round, aseptate, hyaline (12.3 × 4.1 μm in diameter). The appressoria were described as elliptical, clavate, or irregular in outline with a size of 8.1 × 5.2 μm in diameter (Zhang et al. 2019). In Korea, *C. siamense* was described as the causal agent of pecan anthracnose. This fungus generates asexual spores that are cylindrical, hyaline, and solitary (15.5–17.7 × 3.4–4.8 μm) (Oh et al. 2021).

3.2.3 Disease Cycle and Epidemiology

C. gloeosporioides species complex survives as the sexual phase (perithecia) mainly on dead tissues or as the asexual stage (conidia) on living or dead tissues. The overwintering sites include the nut shucks or peduncles generated during the previous season. Rain and wind play major roles in the movement of the conidia to the infection entry points. *C. gloeosporioides* penetrates the epidermal cells using appressoria and colonises the attacked cell biotrophically without inducing apparent symptoms and damage to living cells (de Silva et al. 2017). The optimum temperature to establish the infection of pecan leaflets tree is between 20 and 25 °C. Pecan anthracnose is a polycyclic disease in such a way that the disease cycle is repeated several times in a growing season. The disease has a long latent period, and the conducive environmental conditions favouring the causal agent include cool and wet weather (Brenneman 1989).

3.2.4 Control

Planting resistant varieties and destroying infected plant debris such as dead shucks and leaves, that are overwintering sites are recommended to avoid running into anthracnose in pecan groves. Fungicide application to avoid disease breakout is an effective measure to manage the disease. Almost all tested chemicals against pecan scab are effective toward the pecan anthracnose. The tested fungicide chemistries include triazoles, guanidines, benzimidazoles, and strobilurins (Littrell 1981).

4 Hazelnuts

4.1 Bacterial Blight

Bacterial blight is the most economically important and predominant disease of hazelnuts, reported in nearly all hazelnut-growing countries. In the last decade, ten reports have been documented from nine different countries, describing outbreaks of this destructive disease. This disease has been reported from countries, including the USA, Chile, Iran and Germany demonstrating that the causal agent disperses through the propagation materials (Lamichhane and Varvaro 2014).

4.1.1 Symptomatology

This disease creates symptoms on leaves, buds, trunks and occasionally on husks, and nuts. Developed symptoms varied in orchards and nurseries due to the different applied growing systems. In spring, small angular or irregularly circular, pale yellow-green, and water-soaked specks (2–3 mm in diameter) initially appeared on leaves when the stage of leaf development was ended, and the temperature was conducive for leaf infection. These symptoms turn reddish-brown lesions over time. Oily polygonal spots merged to cause general chlorosis of the lamina and browning the leaf margin frequently occur in the infected leaves. During summer, black green, water-soaked areas in the bark resulting in the formation of 10–20 cm long cankers with longitudinally cracking appearance occurs in the bark. Once stems are surrounded by lesions, leaf necrosis and death can happen far away from the stem. Infection of branch and trunk occur commonly on young plants (1–4 years old), frequently resulting in tree death. If the bark of the infected limb is removed, the cortical tissues beneath the bark can be observed. On the fruit, small and circular lesions, ranging from 0.5 to 1.5 mm in diameter, brown or black, surrounded by a water-soaked area are developed. These are superficial lesions, penetrating the shell by just a few mm.

4.1.2 Causal Organism

The bacterial blight of hazelnuts is caused by the *Xanthomonas arboricola* pv. *corylina* (*Xac*), which is a gram-negative rod bacterium with a single polar flagellum. Culturing this bacterium on NAS, YDCA, GYCA, and YPGA media at 35 °C for 72 h leads to the formation of yellow-mucoid and domed colonies. Cellobiose, D-galactose α-D-glucose, D-mannose, sucrose and trehalose can be used as sole carbon source by this bacterium but not adonitol, D-mannitol, D-sorbitol, dulcitol, erythritol, inulin, L-rhamnose, or L-xylose. *Xac* isolates are negative for oxidase, indole, lecithinase, urease, tyrosinase and nitrate reduction, but they are positive for hydrolysis of gelatin, esculin, arbutin and starch. This bacterium produces a yellow carotenoid pigment in the culturing medium similar to other *Xanthomonas* bacteria (Lamichhane et al. 2013; Webber et al. 2020).

4.1.3 Disease Cycle and Epidemiology

Xac survives as bacterial cells in cankers formed on the larger branches and trunks of trees that are the main sources of primary inoculum in the following season. A sticky ooze containing abundant bacterial cells exudes out of cankers during rainy weather. The bacterial ooze disseminates through rain splash and infects other host parts such as buds and young leaves. This event results in the initiation of a new infection. *Xac* can also be dispersed mechanically through contaminated pruning tools and naturally through rain and wind. Following landing on the host, *Xac* penetrates the tissues through natural openings such as stomata and wounds caused by cultural practices such as pruning. Wetness of leaves and a temperature of 20 °C are favourable conditions to stimulate infection by *Xac*. Stress caused by cultural practice and pedo-climatic factors plays a pivotal role in the occurrence and the spread of the hazelnut bacterial blight. Higher values of rainfall, thermal shock, and soil nitrogen are pedo-climatic conditions, that positively influence disease incidence caused by *Xac* (Lamichhane et al. 2013; Lamichhane and Varvaro 2014).

4.1.4 Control

Copper-based compounds such as the Guardsman copper oxychloride 50% are specifically applied to control hazelnut bacterial blight. These chemical agents provide satisfactory results by reducing the epiphytic bacterial populations, but their efficacy is reduced remarkedly when the bacterium exists internally inside dormant buds and cankers. Nevertheless, accumulating copper compounds in soils and bacterial strains developing resistance against these agents are become problematic issues in the agricultural ecosystem (Mantovi et al. 2003). Once the disease establishes in a production region, the only measure to limit the damage is to reduce pathogen inoculation. Eliminating the infected plant materials, sealing the wounds, and spraying copper-based compounds are recommended measures to manage the disease (Lamichhane and Varvaro 2014).

4.2 Eastern Filbert Blight

Eastern filbert blight (EFB) is a damaging fungal disease of hazelnuts (filberts), which is endemic to a vast region in eastern North America where the American hazelnut (*Corylus americana*), a potential source of EFB, is cultivating. The disease is a key agent limiting the cultivation of the commercial European hazelnuts

(*C. avellana*) in such a way to cause significant damage on these cultivars. The disease kills European hazelnuts within 5–10 years (Pinkerton 1993).

4.2.1 Symptomatology

Typical symptoms of eastern filbert blight include the formation of dead or collapsed areas of bark (sunken cankers) on branches or main trunks. Infected branches may die rapidly, and dead leaves remain attached to them. Cankers initially develop on new twigs and enlarge over time. The cambium below the formed pumps on infected limb become necrotic. Oval to football-shape stromata usually develop within cankers in rows of two. The surface of stromata are initially white and later become black as perithecia develop. Cankers extend horizontally along the branches of susceptible cultivars, ranging from 30 to 90 cm. The canopy of the infected tree is destroyed completely within 7–10 years after initial infection. Depending on the type of cultivar grown, the tree productivity declines sharply 3–10 years after infection. Eastern filbert blight does not infect leaves, fruits or nuts (Gottwald and Cameron 1980a, b).

4.2.2 Causal Organism

Eastern filbert blight is caused by the fungus *Anisogramma anomala* (Peck) E. Müller (Ascomycota: *Diaporthales*). The anamorph, the conidial stage has never been observed in this fungus but *A. anomala* produces a dark-walled perithecium, which are ovate to pear-shaped and are 50–830 × 1040–2160 μm in diameter. Mature asci are mainly clavate and unitunicate (10–15 × 45–65 μm) with a long, thread-like stipe. The asci contain eight ascospore that are arranged uniseriate to sub-biseriate with little uniformity. The ascospores are unequally two-celled, and hyaline. The small and cap-like cell are 1.1–1.4 × 1.1 μm in diameter while the large and biguttulate cell measures 8–12 × 4–5 μm when mature (Gottwald and Cameron 1979).

4.2.3 Disease Cycle and Epidemiology

Anisogramma anomala overwinters as the sexual stage borne in stromata on cankered branches. This fungus apparently is biotrophic requiring 2 years to complete its lifecycle with a symptomless period of 13–16 months. In late winter and early spring, the rereleased spores move a short distance to the infection sites by water splash and over longer distances by wind. It is demonstrated that only immature tissues near the apical meristem of actively growing shoots (budbreak stage) are susceptible to attack by the germinating ascospores (Johnson et al. 1994). The fungus penetrates the epidermal cell directly without requiring wounding or natural openings as entry sites and produces a vesicle-like structure within the infected cells

(Pinkerton et al. 1995). Following penetration, *A. anomala* colonises the phloem, cambium and the outermost layer of the xylem. The tissue colonisation by the *A. anomala* leads to the generating of the sunken cankers on disease branches and twigs 12–14 months after initial infection. Typical stromata borne 50–100 perithecia, containing around 10^3 asci, which reach maturity in autumn. A broad range of temperatures (8–25 °C) and a period of 24–72 h high humidity are conducive factors for initiating infection by *A. anomala* (Stone 1992).

4.2.4 Control

It is necessary to apply an IPM programme to control **Eastern filbert blight.** Scouting orchards to search for dying branches in late summer and for the canker and stromata in late fall and winter are recommended. Prune out cankered limbs or any diseased branches around two to three feet beyond the infection site and burn diseased wood is a practical measure to limit the spread of the disease in the orchards (Johnson et al. 1996). Application of resistant cultivars, including Lewis and Clark with quantitative resistance traits and Dorris and Jefferson, carrying a single dominant resistance gene is an efficient measure to combat this notorious disease (Muehlbauer et al. 2014). Multiple fungicide applications, initiating at bud break and continuing at 2-week intervals is the best way to control Eastern filbert blight. Evaluation of multiple chemical agents demonstrated that picoxystrobin, pyraclostrobin, and trifloxystrobin provide significant control compared to non-treated trees, of between 64 and 74% (Pscheidt et al. 2017).

5 Walnut

5.1 Anthracnose

One of the most widespread and damaging foliar diseases of both black (*Juglans nigra* L.) and Persian (*J. regia* L.) walnuts is anthracnose. This destructive disease leads to premature defoliation, reduction of plant development, and reduction of the quality and quantity of nuts. This disease has a worldwide distribution and has been reported from almost all walnut-growing areas, including Europe, Canada, Asia, North and South America. Yield losses of 60–80% in infected walnut plants are also reported (Saremi and Amiri 2010).

5.1.1 Symptomatology

The disease attack leaves, twigs, fruits, and occasionally shoots of the current season's growth. Initial symptoms appear on the infected leaves as black and circular to irregular circular spots. These specks become gradually numerous and eventually merge into large necrotic areas surrounded by a halo margin. Attacked leaves and leaflets drop prematurely, but some infected leaflets are retained attached to the tree during the growing season. Sunken and necrotic spots that are smaller than those of infected leaves appear on the husks of infected nuts. Nuts developed from diseased trees often possess dark and shriveled meats that are unattractive for marketing. Additionally, necrotic lesions that are oval to irregular circles surrounded by a reddish-brown margin are rarely formed on the shoots. On the surface of infected organs, solitary and dark acervuli, the asexual fruiting bodies, are formed (Woeste and Beineke 2001; Yang et al. 2021).

5.1.2 Causal Organism

The walnut anthracnose is caused by the fungal pathogen *Ophiognomonia leptostyla* (Fr.) Ces. et de Not. (anamorph: *Marssonina juglandis* (Lib.) Magn.). Conidiophores are hyaline, septate, branched irregularly, with one celled packed together in a small layer producing conidia at the tips. The conidia are ovoid, falcate or with only one end rounded and the other pointed, bi-celled being unequal with prominent oil globules. They are hyaline of 15–26 × 2–5 μm in size and diameter. Brown perithecia generated on fallen leaves are single, amphigenous, globose, and reddish-brown with an elongated cylindric beak. The beak is 140–170 × 25–40 μm in diameter. The developed asci are hyaline aparaphysate (56–62 × 14–16 μm in diameter) and contain eight ascospores. They are hyaline, furisode, straight to slightly curved septate that are 15–19 × 4–5 μm in diameter. The oatmeal agar provides the fastest growth condition for *O. leptostyla* in vitro, where this fungus can cover half of the petri dish incubated at 21 °C with a 12-h photoperiod and 30% relative humidity after 96 h and the entire plate in 2 weeks. The optimum temperature for ascospore and conidial germination of *O. leptostyla* are 26 °C and 24 °C, respectively (Sogonov et al. 2008).

5.1.3 Disease Cycle and Epidemiology

O. leptostyla survives primarily as perithecia on fallen leaven but also overwinters as mycelium on fruit and twig lesions. Perithecia are developed during wintertime at a temperature between 7–10 °C. The temperature of 15–30 °C, frequent precipitation, humidity over 65%, and wet weather are conducive conditions resulting in discharging the ascospore developed on the dropped leaves under the walnut trees to cause primary infection. Infected leaves are covered by black small asexual fruiting bodies (acervuli), producing abundant conidia. Asexual spores are released from the new lesions formed on leaves, stems, and fruits, initiating a secondary cycle of infection (Vonica 1970). This event only occurs when rains continue in the spring and when there is sufficient rainfall in the fall. Both ascospores and conidia penetrate directly into plant tissues, and there is no need for a wound to establish infection. The optimal temperature to initiate infection is around 21 °C and the incubation period is about 2–3 weeks post-inoculation.

5.1.4 Control

Cultural and chemical practices are recommended to manage walnut anthracnose. Burying fallen leaves (overwintering sites) during autumn and winter time, to a depth of 10–15 cm is an efficient way to reduce the primary inoculum. Raking and eliminating the infected twigs, leaves, and fruits play a major role in reducing the future infection. The application of nitrogen fertilisers on the soil surface in spring provides plant growth promotion and declining disease severity (Neely 1981). No commercial resistant cultivars against walnut anthracnose are available. Application of fungicides such as Syllit (Dodine) and Bordeaux mixture are recommended as an efficient practice when the walnut leaves size is around half of their final size, followed by three additional treatments every 2 weeks (Zamani et al. 2011).

5.2 Blackline

Blackline is widespread and the main viral disease of English walnut trees (*Juglans regia*) propagated on seedlings of Northern California black walnut (*J. hindsii*) and natural hybrid Paradox (*J. hindsii* × *J. regia*). The disease was initially discovered in the USA in 1924 and was attributed to an abiotic agent(s). The disease is currently distributed in several countries of Europe, North America, Chile, New Zealand, Australia, and China, Turkey, and Japan. Blackline culminates in inducing a programmed cell death at the rootstock-scion union, leading to the decline and killing of the English walnut scion (Büttner et al. 2011).

5.2.1 Symptomatology

Blackline rarely occurs in trees less than 10 years old but commonly develops in trees 15–25 years old. The initial symptoms include weakly terminal growth, yellowing, and the premature falling of leaves from certain branches. As the disease progress, the infected trees show dieback of terminal shoots and decline usually accompanied by production of abundant suckers from the rootstock. The typical feature of the blackline is the presence of a narrow necrotic stripe of necrotic cambium and phloem tissues at the union between rootstock and scion. At the early stage of disease, the blackline is not continuous around the union, but the disease gradually extends around it, resulting in a complete girdling. The disease kills the tissue cells responsible for transporting nutrients and water between the rootstock and scion. Infected trees commonly die within 4–6 years after noticing the first symptoms, but some walnut cultivars stand longer (Mircetich 1980; Brooks and Bruening 1995).

5.2.2 Causal Organism

Blackline is caused by the *Cherry leaf roll virus* (CLRV), possessing a wide host range that encompasses 17 genera of annual and woody plants. CLRV belongs to the Nepovirus in the *Comoviridae* family, subfamily *Comovirinae* (Sanfaçon et al. 2009). This virus spreads mainly through graft, pollen, or seed and has no nematode vector. The CLRV genome is bipartite and composed of single-stranded positive-sense RNA-1 (~8 kb) and RNA-2 (~7 kb), encoding a polyprotein. A study demonstrated that the 3′ terminal non-coding regions (1.2–1.6 kb) of the genomic RNAs of CLRV are nearly identical in sequence (Borja et al. 1995).

5.2.3 Disease Cycle and Epidemiology

The primary way of blackline spreading is through the production and dissemination of pollen from the infected trees. Subsequently, the inoculum is transferred to the female flowers of neighboring healthy English walnut trees. Diseased pollen can move several miles through wind, therefore, enhancing the introduction of the blackline to the healthy neighboring orchards. Diseased pollen can move several miles through the wind. This event enhances the introduction of the blackline to healthy neighboring orchards. Initial CLRV infection leads to the expression of mild and conspicuous symptoms on susceptible English walnut trees, while systematic infection and CLRV movement to the graft union takes several years so that the infected trees serve as a source of inoculum to establish the subsequent infection (Mircetich et al. 1981; Mircetich 1984).

5.2.4 Control

There is no cure for a tree infected by CLRV, and there is no practical method to detect blackline before symptoms express. Nevertheless, few practices are suggested to reduce the disease spreading in orchards. Planting Persian walnut varieties that have been grafted on Persian walnut rootstocks are a preventive measure to avoid the blackline. A strong type of Persian seedlings such as Manregian is recommended as rootstocks. Eliminating any visible diseased or killed tree is suggested as the only practical measure to reduce the spread of the disease (Nafesa 2015).

6 Chestnut

6.1 *Blight*

Chestnut blight is one of the most destructive diseases restricting chestnut cultivation in the USA and as a result is a central factor, when determining which chestnut species are grown in commercial orchards in North America. American chestnut (*Castanea dentata*) is highly susceptible to the disease, while Chinese chestnut (*C. mollissima*) is a resistant cultivar toward chestnut blight. The disease was accidentally imported from Asia through plant materials (Japanese chestnut trees) to the USA. The disease is an essential factor in killing four billion American chestnut trees and imposing serious economic impacts on the nut industries (Anagnostakis 1987).

6.1.1 Symptomatology

The disease infects the above-ground tree parts such as stems, branches, twigs, and limbs. The typical symptoms include the formation of necrotic reddish-brown patches (so-called cankers) on the bark that can kill small branches in a short period of a few months. Within the bark, brown mycelial fans that are a typical features of chestnut infection are formed. In severe infection, bark cankers grow quickly and girdle stems, resulting in the disruption of nutrient transportation. This event leads to the death of all plant materials beyond the cankers. Leaves on such branches turn brown and wither but remain attached to the infected necrotic branches, resulting in a situation so-called a flag (Prospero and Rigling 2013).

6.1.2 Causal Organism

Chestnut blight is caused by a sordariomycete fungus, *Cryphonectria parasitica* (Murrill) Barr (previously known as *Endothia parasitica* [Murrill] Anderson & Anderson). This fungus produces a subglobose perithecium, 300–400 μm in diameter, in small stromata, which are white to brown with a long cylindrical neck. Stromata usually contain 15–30 perithecia formed on the surfaces of the stroma as several black ostiolate necks. Asci (30–60 × 7–9 μm in diameter) are unitunicate, ellipsoid to subclavate containing eight ascospores. These spores are forcibly released and carried in air currents. Ascospores (7–12 × 3–5.5 μm in diameter) are two-celled separated via a median septum, hyaline, and ellipsoid or ovoid covered by a gelatinous envelope. Conidia (3–5 × 1–2 μm in diameter) are unicellular, hyaline, ellipsoidal to bacilliform in shape. Conidia ooze is extruded in mucilaginous spore tendrils during rainy weather. They are initially yellowish but become coral red once they become older (Rigling and Prospero 2018).

6.1.3 Disease Cycle and Epidemiology

C. parasitica survives as mycelium in the lesions and infected barks. In early spring, asexual fruiting bodies termed as pycnidia develop on the surfaces of the lesions, and conidia ooze containing abundant spores may be observed on the bark. Several weeks to months later, dark sexual fruiting bodies, perithecia, also emerged in the colonised tissue (Prospero et al. 2006). Rain-splash, wind, birds, insects, or animals are the main players to carry both spore types to the healthy tissues, resulting in the establishment of the infection. Following landing, spores germinate and penetrate the bark through wounds made by insects (Meyer et al. 2015). The developing cankers cause the tree to wilt and lose its capacity to tolerate the infection. The tissue beyond the cankers is girdled and dies. Finally, the cankers spread and gradually kill the tress leaving just the root networks alive. The fungus may overwinter as a saprobe in dead chestnut tissues for at least 2 years (Prospero et al. 2006). Air temperature is a key factor impacting the canker development and potentially the spread of the *C. parasitica*. It was demonstrated that cankers expand by 1 mm per day in North America at an average daily temperature of 20 °C (Anagnostakis and Aylor 1984).

6.1.4 Control

Quarantine and eradication, especially removing and burning of the infected limbs, are two measures to prevent the introduction and the spreading of *C. parasitica*. Chemical control is an impractical practice for forest conditions. However, epoxiconazole provided satisfactory efficacy to control chestnut blight under controlled conditions such as nurseries or an individual tree (Trapiello et al. 2015). Hypovirulence refers to a viral disease infecting the *C. parasitica* and slowing down the canker progression. This event has been relatively effective in Europe but failed in eastern North America (Rigling and Prospero 2018). Breeding for resistance by backcrossing the resistant Chinese chestnut into the susceptible American chestnut genome is another approach to manage the disease (Bauman et al. 2014).

6.2 Phytophthora Root Rot

Phytophthora root rot, known as ink disease, is one of the most destructive diseases of American chestnut (*C. dentata*), resulting in the killing of chestnut trees from lower-elevation forests in the south-eastern United States before introducing the *C. parasitica*. Ink disease was initially recorded in Portugal in 1838 and then spread largely on sweet chestnut (*C. sativa*) in Europe and on American chestnut in the USA (Crandall et al. 1945). Additionally, the disease is a major factor, limiting the re-introduction of American chestnut trees with improved resistance against chestnut blight (Jacobs 2007). Ink disease also attacks the seedling of chestnut nurseries, leading to rapid or gradual wilting (Jung et al. 2016).

6.2.1 Symptomatology

Infected chestnut trees display small-sized and chlorotic leaves becoming yellow or wilted. The disease kills the entire crown as it progresses, resulting in extensive dieback and tree death. Dry leaves and fruits remain attached to the dead tree over the winter. These symptoms are attributed to extensive damage to the root systems and irregular wedge-shaped necrotic streaks originating from the primary roots into the collar (Fig. 15.4). Lesions on the roots showed a blue to black (inky blue) exudate, staining the surrounding soil near the roots so that Phytophthora root rot was named ink disease. Infected trees lose their ability to resprout since the disease destroys the root networks (Crandall et al. 1945; Grandall 1950).

6.2.2 Causal Organism

Phytophthora root rot is caused by the soilborne oomycete *Phytophthora cinnamomi* Rands, and this is the only biotic agent causing ink disease on American chestnuts. Recently, *P. cambivora*, *P. cryptogea*, and *P. heveae* were recovered from American and backcross hybrid chestnut seedlings. Infection assays demonstrated that these Phytophthora species cause necrotic lesions on roots of American chestnut under greenhouse conditions (Sharpe 2017). Three Phytophthora species, including

Fig. 15.4 Typical irregular wedge-shaped necrotic streaks of the ink disease originating from the primary roots into the collar of a young (**a**), and mature tree (**b**). (Picture is from Jung et al. 2018)

P. cambivora, *P. citricola*, and *P. cactorum*, were experimentally found associated with the occurrence of ink disease outbreaks in central Italy (Vettraino et al. 2001). In a subsequent study, seven Phytophthora species associated with the ink disease occurring on sweet chestnut in five European countries in 35 regions were isolated. *P. cambivora* and *P. cinnamomi* were the only species isolated from trees showing ink disease symptoms (Vettraino et al. 2005). *P. cinnamomi* produces particular coralloid hyphal on several isolation media and generates botryose and spherical hyphal swellings. These structures are involved in the formation of the resting spore structure named chlamydospores. The sporangia of *P. cinnamomi* are ellipsoid to ovoid, 57 × 33 μm in diameter, non-papillate, and non-deciduous (Sharpe 2017).

6.2.3 Disease Cycle and Epidemiology

Phytophthora cinnamomi is a soilborne pathogen persisting in the soils in infected roots as hyphae and chlamydospores. Soil moisture and temperature >12 °C are conducive agents in stimulating sporangia production and the release of zoospores. These spores can swim through the soil or by surface flowing water to reach the root of a chestnut tree. They encyst and germinate to produce germ tubes, which penetrate the root tips. Under favourable conditions, generation of sporangia and zoospores release occurs again on the host surface within 3–5 days, leading to an explosive epidemic (Rhoades et al. 2003; Sharpe 2017). Transferring soils containing infective structures such as motile zoospores or chlamydospores play a role in the dissemination of the ink disease (Ristaino and Gumpertz 2000). In locations where the minimum and maximum temperatures were below 1.4 °C and above 28 °C, respectively, the *P. cinnamomi* was not recovered (Vettraino et al. 2005).

6.2.4 Control

There is no chemical cure to manage ink disease since these treatments are costly and ineffective towards soil-borne diseases. Additionally, they impose deleterious impacts on human health and ground water (Branzanti et al. 1999). However, the chemical agent fosetyl-aluminum is able to reduce the mycelial growth of *P. cinnamomi* in vitro and disease severity of Phytophthora oak root disease in vivo (González et al. 2017). The generation of resistant cultivars carrying resistance traits towards the ink and chestnut blight diseases are underway to restore the American chestnut to its native range (Westbrook et al. 2019). The most effective practice to manage the *P. cinnamomi* is application of aggressive spot eradication, including the removal of infected trees, fumigation and fungicide application to eradicate this destructive pathogen from localised infestations and limit its further spread (Sena et al. 2018). Additionally, we summarise the remaining nut crops disease in Table 15.1.

Table 15.1 Some important diseases of temperate nuts

Almond disease			
Diseases caused by fungi			
No.	Common name	Causal agent	Symptoms
1	Alternaria leaf spot	*Alternaria alternate*	Fairly large *brown spots* on leaves
2	Armillaria root disease	*Armillaria mellea*	Infected trees develop pale foliage with small leaves, a lack of new growth
3	Ceratocystis canker	*Ceratocystis fimbriata*	Amber-coloured gum is found at the canker margins. Infected tissue turns brown
4	Verticillium wilt	*Verticillium dahlia*	Leaves on one or more branches, often on only one side of the tree, will turn yellow or *wilt* early in the growing season
5	Leucostoma canker	*Leucostoma persoonii and Leucostoma cinctum*	Cankers located on the main trunk, branch crotches, scaffold limbs, and older branches
6	Phomopsis canker and fruit rot	*Phomopsis amygdali*	Brown, sunken, elongated, necrotic lesions were observed around the buds
7	Phytophthora diseases	*Phytophthora* spp.	Root rot, crown rot and trunk cankers
8	Powdery mildew	*Podosphaera* spp.	Russeting on almond hulls
9	Rosellinia root rot	*Rosellinia* necatrix	Yellow foliage, shriveled fruit, and little or no new growth
10	Rust	*Tranzschelia discolor*	Small, yellow spots on the upper surface of leaves
Diseases caused by bacteria			
No.	Common name	Causal agent	Symptoms
1	Almond leaf scorch	*Xylella fastidiosa*	Marginal scorching of leaves
2	Bacterial canker	*Pseudomonas syringae*	Oozing of gum (gummosis) at infection sites
3	Crown Gall	*Agrobacterium tumefaciens*	Rough, abnormal *galls* on roots or trunk
4	Bacterial spot	*Xanthomonas arboricola* pv. *pruni*	Sunken brown corky lesions on fruit, often oozing amber coloured gum
Diseases caused by viruses			
No.	Common name	Causal agent	Symptoms
1	Yellow bud mosaic	*Tomato ringspot virus* (TmRsV)	Crinkled or distorted leaves, necrotic spots may develop and then will drop off leaving a tattered appearance
Diseases caused by nematodes			
No.	Common name	Causal agent	Symptoms

(continued)

Table 15.1 (continued)

Almond disease			
1	Dagger nematodes	*Xiphinema* spp.	The feeding at the meristematic root-tips destroys root cells and reduces root volume
2	Ring nematodes	*Mesocriconema* spp.	Infestation will result in stunted roots, which sometimes proliferate and form dense mats
3	Root-knot nematodes	*Meloidogyne* spp.	Infected root systems show characteristic knots or galls
4	Root-lesion nematodes	*Pratylenchus* spp.	Lesion nematodes feed and migrate inside roots causing black lesions
Hazelnut diseases			
Diseases caused by fungi			
No.	Common name	Causal agent	Symptoms
1	Armillaria root disease	***Armillaria* spp.**	Usually poor growth of the shoots together with premature dropping of the leaves
2	Anthracnose	*Gloeosporium coryli*	The initial symptoms on leaves are small specks of necrotic lesions that expand with time and become necrotic blotches
3	Powdery mildew	*Phyllactinia guttata*	White powdery growth appears on the underside of leaves but often not until late in the season
Diseases caused by bacteria			
No	Common name	Causal agent	Symptoms
1	Bacterial canker	*Pseudomonas syringae* pv. *avellanae*	Infected catkins release only sparse amounts of viable pollen and often completely wilt
2	Crown Gall	*Agrobacterium tumefaciens*	Rough-surfaced galls, usually at or near the soil line, on a graft site or bud union, or on roots and lower stems
Pecan diseases			
Diseases caused by fungi			
No.	Common name	Causal agent	Symptoms
1	Armillaria root diseases	*Armillaria* spp.	Decline or death of a tree
2	Brown leaf spot	*Ragnhildiana diffusa*	Circular, reddish *brown spots* form on the upper and lower *leaf* surface
3	Phytophthora root and crown rot	*Phytophthora* spp.	Reddish brown Phytophthora crown rot canker with zonate margin
4	Powdery mildew	*Microsphaera penicillata*	**White or gray spots or blotches on leaves, buds, and stems**

(continued)

Table 15.1 (continued)

Pecan diseases			
Diseases caused by bacteria			
No.	Common name	Causal agent	Symptoms
1	Pecan bacterial leaf scorch	***Xylella fastidiosa* subsp. *multiplex***	The **leaflets turn tan to brown at the tips with discoloration and tissue desiccation**
Pistachio diseases			
No.	Common name	Causal agent	Symptoms
1	Armillaria root disease	*Armillaria mellea*	Decline or death of a tree
2	Aspergillus fruit rot	*Aspergillus niger* and other *Aspergillus* spp.	Aflatoxin contamination and a blight that turns the hulls light beige to yellow
3	Phytophthora diseases	*Phytophthora* spp.	Diseased trees show reduced growth, thinned canopy and early defoliation for several years
4	Powdery mildew	*Phyllactinia guttata*	Small powdery white patches on leaves and fruit which can expand to cover the entire leaf or fruit surface
5	Rust	*Uromyces terebinthi*	Round or irregularly shaped red-brown pustules on leaves, flowers, pedicels and/or fruit
6	Stigmatomycosis	*Nematospora coryli*	Internal symptoms were brown necrotic areas and malformation of the cotyledons
7	Verticillium wilt	*Verticillium dahlia*	Interveinal patches of yellowing or scorching of the leaves on affected branches
Diseases caused by bacteria			
No.	Common name	Causal agent	Symptoms
1	Bacterial dieback	*Xanthomonas translucens*	Dieback of twigs and branches; tree producing an excessive amount of resin
Walnut diseases			
Diseases caused by fungi			
No.	Common name	Causal agent	Symptoms
1	Armillaria root disease	*Armillaria mellea*	Small, discolored leaves which drop early; death of branches; death of plant
2	Walnut cankers	*Botryosphearia* spp. *Phomopsis* spp.	Fungi cause canker on affected tree
3	Downy leaf spot	*Phyllactinia guttata*	Small, powdery white spots on leaves and fruit; spots spread to cover entire leaf
Diseases caused by bacteria			
No.	Common name	Causal agent	Symptoms
1	Crown rot	*Xanthomonas arboricola* pv. *Juglandis*	Numerous tiny, angular, *brown spots* on the leaves
2	Walnut blight	*Xanthomonas campestris*	Small, water-soaked spots on immature fruit which darken and rapidly enlarge
3	Crown gall	*Agrobacterium tumefaciens*	Galls of various sizes on roots and root crown below the soil line

7 Conclusion

According to published reports (Anonymous 2021), there is a sizable demand for edible nuts, hence it is important to implement effective management stategies to prevent biotic agents from damaging nut plants. Basically, nut crops offer a unique challenge to apply management strategies since these trees have a long life and a particular growth habit. Having a deep knowledge of the interaction between nut crops and their corresponding pathogens is required to design novel management approaches. The new investigation, employing the transcriptomics and metabolomics techniques, will result in discovering the candidate genes involved in mounting or triggering the defence responses in resistant cultivars. It is inevitable to perform this type of research nowadays, culminating in offering novel and effective measures to manage nut crops disease. Another issue is that researchers need to focus on developing rapid molecular tools such as LAMP (Loop-mediated isothermal amplification) technology enabling growers to detect pathogen infection in nurseries. Some nut trees generated in nurseries with scion cultivars propagated to their rootstocks, and latent infections in these places are frequent.

References

Adaskaveg JE, Hartin RJ (1997) Characterization of *Colletotrichum acutatum* isolates causing anthracnose of almond and peach in California. Phytopathology 87(9):979–987

Adaskaveg JE, Ogawa JM, Butler EE (1990) Morphology and ontogeny of conidia in *Wilsonomyces carpophilus*, gen. nov. and comb. nov., causal pathogen of shot hole disease of *Prunus* species. Mycotaxon 37:275–290

Ahmadpour A, Ghosta Y, Javan-Nikkhah M, Ghazanfari K, Fatahi R (2012) Study on morphology, pathogenicity and genetic diversity of *Wilsonomyces carpophilus* isolates, the causal agent of shot hole of stone fruit trees based on RAPD-PCR in Iran. Arch Phytopathol Plant Protect 45(17):2076–2086

Anagnostakis SL (1987) Chestnut blight: the classical problem of an introduced pathogen. Mycologia 79(1):23–37

Anagnostakis SL, Aylor DE (1984) The effect of temperature on growth of *Endothia* (*Cryphonectria*) *parasitica* in vitro and in vivo. Mycologia 76(3):387–397

Anonymous (2021) Edible nuts-global market trajectory and analytics. Global Industry Analysts, Inc., San Jose. https://www.researchandmarkets.com/reports/5030789/edible-nuts-global-market-trajectory-and

Avenot H, Morgan DP, Michailides TJ (2008) Resistance to pyraclostrobin, boscalid and multiple resistance to Pristine® (pyraclostrobin + boscalid) fungicide in *Alternaria alternata* causing alternaria late blight of pistachios in California. Plant Pathol 57(1):135–140

Banihashemei Z (1990) Biology and control of *Polystigma ochraceum*, the cause of almond red leaf blotch. Plant Pathol 39(2):309–315

Barnes G (1964) Growth responses of the pecan scab fungus,'*Fusicladium effusum*', on various common and exotic agar media. Oklahoma State University, Stillwater

Bauman JM, Keiffer CH, McCarthy BC (2014) Growth performance and chestnut blight incidence (*Cryphonectria parasitica*) of backcrossed chestnut seedlings in surface mine restoration. New For 45:813–828

Bock CH, Brenneman TB, Wood BW, Stevenson KL (2017) Challenges of managing disease in tall orchard trees—pecan scab, a case study. CAB Rev 12(8):1–18

Borja M, Sánchez F, Rowhani A, Bruening G, Ponz F (1995) Long, nearly identical untranslated sequences at the 3′ terminal regions of the genomic RNAs of *cherry leafroll virus* (walnut strain). Virus Genes 10(3):245–252

Branzanti MB, Rocca E, Pisi A (1999) Effect of ectomycorrhizal fungi on chestnut ink disease. Mycorrhiza 9(2):103–109

Brenneman TB (1989) Recent occurrence of pecan anthracnose caused by *Glomerella cingulata*. Plant Dis 73(9):775

Brooks M, Bruening G (1995) Transient gene expression of antisense RNA and coat protein-encoding sequences reduced accumulation of *cherry leafroll virus* in tobacco protoplasts. Virology 208(1):132–141

Büttner C, Von Bargen S, Bandte M, Myrta A (2011) Chapter 24: *Cherry leaf roll virus*. In: Virus and virus-like diseases of pome and stone fruits. The American Phytopathological Society, St. Paul, pp 119–125

Cannon PF (1996) Systematics and diversity of the *phyllachoraceae* associated with *rosaceae*, with a monograph of *Polystigma*. Mycol Res 100(12):1409–1427

Cao TA, Duncan RC, Kirkpatrick BA, Shackel KM, Dejong T (2013) Effect of calcium and nitrogen fertilization on bacterial canker susceptibility in stone fruits. Fruits 68(3):245–254

Charlton ND, Yi M, Bock CH, Zhang M, Young CA (2020) First description of the sexual stage of *Venturia effusa*, causal agent of pecan scab. Mycologia 112(4):711–721

Chen SF, Morgan DP, Michailides TJ (2014) *Botryosphaeriaceae* and *Diaporthaceae* associated with panicle and shoot blight of pistachio in California, USA. Fungal Divers 67(1):157–179

Connell JH (2002) Leading edge of plant protection for almond. HortTechnology 12(4):619–622

Crandall BS, Gravatt GF, Ryan MM (1945) Root disease of *Castanea* species and some coniferous and broadleaf nursery stocks, caused by *Phytophthora cinnamomi*. Phytopathology 35:162–180

de Silva DD, Crous PW, Ades PK, Hyde KD, Taylor PWJ (2017) Life styles of *Colletotrichum* species and implications for plant biosecurity. Fungal Biol Rev 31(3):155–168

de Silva DD, Mann RC, Kaur J, Ekanayake PN, Sawbridge TI, McKay S, Taylor PWJ, Edwards J (2021) Revisiting the *Colletotrichum* species causing anthracnose of almond in Australia. Australas Plant Pathol 50(3):267–279

Demaree J (1924) Pecan scab with special reference to sources of the early spring infections. J Agric Res 28:321–330

Förster H, Adaskaveg JE (1999) Identification of subpopulations of *Colletotrichum acutatum* and epidemiology of almond anthracnose in California. Phytopathology 89(11):1056–1065

González M, Caetano P, Sánchez ME (2017) Testing systemic fungicides for control of *Phytophthora* oak root disease. For Pathol 47(4):e12343

Gottwald TR, Bertrand PF (1982) Patterns of diurnal and seasonal airborne spore concentrations of *Fusicladium effusum* and its impact on a pecan scab epidemic. Phytopathology 72(3):330–335

Gottwald TR, Cameron HR (1979) Studies in the morphology and life history of *Anisogramma Anomala*. Mycologia 71(6):1107–1126

Gottwald TR, Cameron HR (1980a) Infection site, infection period, and latent period of canker caused by *Anisogramma anomala* in European Filbert. Phytopathology 70(11):1083–1087

Gottwald TR, Cameron HR (1980b) Disease increase and the dynamics of spread of canker caused by *Anisogramma anomala* in European Filbert in the Pacific Northwest. Phytopathology 70(11): 1087–1092

Grandall BS (1950) The distribution and significance of the chestnut root rot Phytophthoras, *P. cinnamomi* and *P. cambivora*. Plant Dis Report 34(6):194–196

Highberg LM (1986a) Yield reduction in almond related to incidence of shot-hole disease. Plant Dis 70(9):825–828

Highberg LM (1986b) Survival of shot-hole inoculum in association with dormant almond buds. Plant Dis 70(9):828–831

Hirano SS, Upper CD (1990) Population biology and epidemiology of *pseudomonas syringae*. Annu Rev Phytopathol 28(1):155–177

Hoefnagels MH, Mason B (2016) Pecan scab. In: The plant health instructor. The American Phytopathological Society, St. Paul. https://doi.org/10.1094/PHI-I-2016-0620-01

Jacobs DF (2007) Toward development of silvical strategies for forest restoration of American chestnut (*Castanea dentata*) using blight-resistant hybrids. Biol Conserv 137(4):497–506

Johnson KB, Pinkerton JN, Gaudreault SM, Stone JK (1994) Infection of European hazelnut by *Anisogramma anomala*: site of infection and effect of host developmental stage. Phytopathology 84(12):1465–1470

Johnson KB, Mehlenbacher SA, Stone JK, Pscheidt JW, Pinkerton JN (1996) Eastern filbert blight of European hazelnut: It's becoming a manageable disease. Plant Dis 80(12):1308–1316

Jung T, Orlikowski L, Henricot B, Abad-Campos P, Aday AG, Aguín Casal O, Bakonyi J, Cacciola SO, Cech T, Chavarriaga D, Corcobado T, Cravador A, Decourcelle T, Denton G, Diamandis S, Dogmus-Lehtijärvi HT, Franceschini A, Ginetti B, Glavendekic M, Hantula J, Hartmann G, Herrero M, Ivic D, Horta Jung M, Lilja A, Keca N, Kramarets V, Lyubenova A, Machado H, di San M, Lio G, Mansilla Vázquez PJ, Marçais B, Matsiakh I, Milenkovic I, Moricca S, Nagy ZÁ, Nechwatal J, Olsson C, Oszako T, Pane A, Paplomatas EJ, Pintos Varela C, Prospero S, Rial Martínez C, Rigling D, Robin C, Rytkönen A, Sánchez ME, Scanu B, Schlenzig A, Schumacher J, Slavov S, Solla A, Sousa E, Stenlid J, Talgø V, Tomic Z, Tsopelas P, Vannini A, Vettraino AM, Wenneker M, Woodward S, Peréz-Sierra A (2016) Widespread *Phytophthora infestations* in European nurseries put forest, semi-natural and horticultural ecosystems at high risk of *Phytophthora* diseases. For Pathol 46(2):134–163

Jung T, Pérez-Sierra A, Durán A, Jung MH, Balci Y, Scanu B (2018) Canker and decline diseases caused by soil- and airborne Phytophthora species in forests and woodlands. Persoonia: Molecular Phylogeny and Evolution of Fungi 40:182–220

Kennelly MM, Cazorla FM, De Vicente A, Ramos C, Sundin GW (2007) *Pseudomonas syringae* diseases of fruit trees: Progress toward understanding and control. Plant Dis 91(1):4–17

Lamichhane JR, Varvaro L (2014) *Xanthomonas arboricola* disease of hazelnut: current status and future perspectives for its management. Plant Pathol 63(2):243–254

Lamichhane JR, Fabi A, Ridolfi R, Varvaro L (2013) Epidemiological study of hazelnut bacterial blight in Central Italy by using laboratory analysis and geostatistics. PLoS One 8(2):e56298

Little EL, Bostock RM, Kirkpatrick BC (1998) Genetic characterization of *pseudomonas syringae* pv. *Syringae* strains from stone fruits in California. Appl Environ Microbiol 64(10):3818–3823

Littrell RH (1981) Management of pecan fruit and foliar diseases with fungicides. Plant Dis 65:769

López-Moral A, Raya-Ortega MC, Agustí-Brisach C, Roca L, Lov era, M., Luque, F., Arquero, Q. and Trapero, A. (2017) Morphological, pathogenic, and molecular characterization of *Colletotrichum acutatum* isolates causing almond anthracnose in Spain. Plant Dis 101(12): 2034–2045

López-Moral A, Agustí-Brisach C, Lovera M, Arquero O, Trapero A (2020) Almond anthracnose: current knowledge and future perspectives. Plan Theory 9(8):945

Ma Z, Michailides TJ (2004) A real-time PCR assay for the detection of azoxystrobin-resistant *Alternaria* populations from pistachio orchards in California. Crop Prot 23(12):1259–1263

Ma Z, Morgan DP, Felts D, Michailides TJ (2002) Sensitivity of *Botryosphaeria dothidea* from California pistachio to tebuconazole. Crop Prot 21(9):829–835

Mantovi P, Bonazzi G, Maestri E, Marmiroli N (2003) Accumulation of copper and zinc from liquid manure in agricultural soils and crop plants. Plant Soil 250(2):249–257

Mantz G, Minhot R, Morrelli G, Maiale S (2010) First report of *Colletotrichum gloeosporioides* causing pecan anthracnose in Argentina. J Plant Pathol 92(2):544

Meyer JB, Gallien L, Prospero S (2015) Interaction between two invasive organisms on the European chestnut: does the chestnut blight fungus benefit from the presence of the gall wasp? FEMS Microbiol Ecol 91(11):1–10

Miarnau X, Zazurca L, Torguet L, Zuniga E, Batlle I, Alegre S, Luque J (2021) Cultivar susceptibility and environmental parameters affecting symptom expression of red leaf blotch of almond in Spain. Plant Dis 105(4):940–947

Michailides TJ (1992) Effects of temperature and wetness duration on infection of pistachio by *Botryosphaeria dothidea* and management of disease by reducing duration of irrigation. Phytopathology 82(12):1399–1406

Michailides TJ (1993) Spore release by *Botryosphaeria dothidea* in pistachio orchards and disease control by altering the trajectory angle of sprinklers. Phytopathology 83(2):145–152

Michailides TJ (2002) Panicle and shoot blight. In: Teviotdale BL, Michailides TJ, Pscheidt JW (eds) Compendium of nut crop diseases in temperate zones. The American Phytopathological Society, St. Paul, pp 68–69

Michailides TJ, Morgan DP (1993) Principles for the control of *Alternaria* late blight of pistachio caused by *Alternaria alternata* in California. In: Caruso T, Barone E, Sottile F (eds) Proceeding of the GREMPA meeting–Pistachio, 9th. Bront, Sciacca, Italy. Springer, New York, pp 56–63

Michailides T, Morgan DP (2004) Panicle and shoot blight of pistachio: a major threat to the California pistachio industry. The American Phytopathological Society, St. Paul. https://doi.org/10.1094/APSnetFeature-2004-0104

Mircetich SM (1980) Natural spread, graft-transmission, and possible etiology of walnut blackline disease. Phytopathology 70(10):962–968

Mircetich SM (1984) The relationship of *Cherry Leafroll Virus* and blackline disease of English walnut trees. Phytopathology 74(4):423–428

Mircetich SM, Dezoeten GA, Lauritis JA (1981) Etiology and natural spread of blackline disease of English walnut trees. Acta Hortic 94:147–152

Moral J, Pérez-Rodríguez M, Michailides TJ, Trapero A (2015) First report of the teleomorph of *Neofusicoccum mediterraneum*, a pathogen of olive. Phytopathology 105(11):97–98

Muehlbauer MF, Honig JA, Capik JM, Vaiciunas JN, Molnar TJ (2014) Characterization of eastern filbert blight-resistant hazelnut germplasm using microsatellite markers. J Am Soc Hortic Sci 139(4):399–432

Nafesa L (2015) *Cherry leaf roll virus* and blackline disease in walnut. University of California, Davis

Neely D (1981) Application of nitrogen fertilizer to control anthracnose of black walnut. Plant Dis 65(7):580–581

Oh JY, Heo JI, Lee DH (2021) First report of anthracnose on pecan (*Carya illinoensis*) caused by *Colletotrichum siamense* in Korea. Plant Dis 105(10):3296. https://doi.org/10.1094/pdis-11-20-2458-pdn

Peres NA, Timmer LW, Adaskaveg JE, Correll JC (2005) Lifestyles of *Colletotrichum acutatum*. Plant Dis 89(8):784–796

Pinkerton JN (1993) Susceptibility of European hazelnut clones to eastern filbert blight. Plant Dis 77(3):261–266

Pinkerton JN, Stone JK, Nelson SJ, Johnson KB (1995) Infection of European hazelnut by *Anisogramma anomala*: ascospore adhesion, mode of penetration of immature shoots, and host response. Phytopathology 85(10):1260–1268

Prospero S, Rigling D (2013) Chestnut blight. In: Gonthier P, Nicolotti G (eds) Infectious forest diseases. CABI, Wallingford, pp 318–338

Prospero S, Conedera M, Heiniger U, Rigling D (2006) Saprophytic activity and sporulation of *Cryphonectria parasitica* on dead chestnut wood in forests with naturally established hypovirulence. Phytopathology 96(12):1337–1344

Pryor BM, Michailides TJ (2002) Morphological, pathogenic, and molecular characterization of *Alternaria* isolates associated with alternaria late blight of pistachio. Phytopathology 92(4): 406–416

Pscheidt JW, Heckert S, Cluskey SA (2017) Evaluation of quinone outside and succinate dehydrogenase inhibitors for effectiveness against eastern filbert blight of hazelnut. Plant Dis 101(11): 1868–1873

Rhoades CC, Brosi SL, Dattilo AJ, Vincelli P (2003) Effect of soil compaction and moisture on incidence of *phytophthora* root rot on American chestnut (*Castanea dentata*) seedlings. For Ecol Manag 184(1-3):47–54

Rigling D, Prospero S (2018) *Cryphonectria parasitica*, the causal agent of chestnut blight: invasion history, population biology and disease control. Mol Plant Pathol 19(1):7–20

Ristaino JB, Gumpertz ML (2000) New frontiers in the study of dispersal and spatial analysis of epidemics caused by species in the genus *Phytophthora*. Annu Rev Phytopathol 38(1):541–576

Saad AT, Masannat K (1997) Economic importance and cycle of *Polystigma ochraceum*, causing red leaf blotch disease of almond in Lebanon. EPPO Bull 27(4):481–485

Sanfaçon H, Wellink J, Le Gall O, Karasev A, Van Der Vlugt R, Wetzel T (2009) Secoviridae: a proposed family of plant viruses within the order *Picornavirales* that combines the families *Sequiviridae* and *Comoviridae*, the unassigned genera *Cheravirus* and *Sadwavirus*, and the proposed genus Torradovirus. Arch Virol 154(5):899–907

Saremi H, Amiri ME (2010) Evaluation of resistance to anthracnose (*Marssonina juglandis*) among diverse Iranian clones of walnut (*Juglans regia* L.). J Food Agric Environ 8(2):375–378

Schubert K, Rischel A, Braun DU (2013) A monograph of *Fusicladium* s.lat. (Hyphomycetes). Schlechtendalia 9:1–132

Sena K, Crocker E, Vincelli P, Barton C (2018) *Phytophthora cinnamomi* as a driver of forest change: implications for conservation and management. For Ecol Manag 409:799–807

Sharpe SR (2017) *Phytophthora* species associated with American, Chinese, and backcross hybrid chestnut seedlings in field sites in the Southeastern United States. Clemson University, Clemson

Shaw DA (1990) Influence of wetness period and temperature on infection and development of shot-hole disease of almond caused by *Wilsonomyces carpophilus*. Phytopathology 80(8): 749–756

Sogonov MV, Castlebury LA, Rossman AY, Mejía LC, White JF (2008) Leaf-inhabiting genera of the *Gnomoniaceae, Diaporthales*. Stud Mycol 62:1–77

Standish JR, Brenneman TB, Bock CH, Stevenson KL (2021) Fungicide resistance in *Venturia effusa*, cause of pecan scab: current status and practical implications. Phytopathology 111(2): 244–252

Stone JK (1992) Natural infection period and susceptibility of vegetative seedlings of European hazelnut to *Anisogramma anomala*. Plant Dis 76(4):348–352

Terabe NI, Martins CM, Homechin M (2008) Microorganisms associated with fruits of different cultivars of pecan nut. Ciencia Agrotecnol 32(2):659–662

Teviotdale BL, Sibbett GS, Viveros M, Mark CI (1989) Effect of fungicides on shot hole disease of almonds. Calif Agric 43(3):21–23

Trapiello E, González-Varela G, González AJ (2015) Chestnut blight control by agrochemicals in *Castanea sativa* under managed conditions. J Plant Dis Protect 122(3):120–124

Troncoso-Rojas R, Tiznado-Hernández ME (2014) *Alternaria alternata* (black rot, black spot). In: Postharvest decay: control strategies. Elsevier Inc., Amsterdam, pp 147–187

Turechek WW, Stevenson KL (1998) Effects of host resistance, temperature, leaf wetness, and leaf age on infection and lesion development of pecan scab. Phytopathology 88(12):1294–1301

Vettraino AM, Natili G, Anselmi N, Vannini A (2001) Recovery and pathogenicity of *Phytophthora* species associated with a resurgence of ink disease in *Castanea sativa* in Italy. Plant Pathol 50(1):90–96

Vettraino AM, Morel O, Perlerou C, Robin C, Diamandis S, Vannini A (2005) Occurrence and distribution of *Phytophthora* species in European chestnut stands, and their association with ink disease and crown decline. Eur J Plant Pathol 111(2):169–180

Vonica I (1970) Biology and control of *Gnomonia leptostyla* and its conidial state *Marssonina juglandis*, causing anthracnose of walnut. Analele Institutului de Cercetari pentru Protectia Plantelor 8:69–80

Webber JB, Putnam M, Serdani M, Pscheidt JW, Wiman NG, Stockwell VO (2020) Characterization of isolates of *Xanthomonas arboricola* pv. *corylina*, the causal agent of bacterial blight, from Oregon hazelnut orchards. J Plant Pathol 102(3):799–812

Westbrook JW, James JB, Sisco PH, Frampton J, Lucas S, Jeffers SN (2019) Resistance to *Phytophthora cinnamomi* in American chestnut (*Castanea dentata*) backcross populations that descended from two Chinese chestnut (*Castanea mollissima*) sources of resistance. Plant Dis 103(7):1631–1641

Woeste KE, Beineke WF (2001) An efficient method for evaluating black walnut for resistance to walnut anthracnose in field plots and the identification of resistant genotypes. Plant Breed 120(5):454–456

Xin XF, Kvitko B, He SY (2018) *Pseudomonas syringae*: what it takes to be a pathogen. Nat Rev Microbiol 16(5):316–328

Yang H, Cao G, Jiang S, Han S, Yang C, Wan X, Zhang F, Chen L, Xiao J, Zhu P, Zhang D, He F, Xing W (2021) Identification of the anthracnose fungus of walnut (*Juglans* spp.) and resistance evaluation through physiological responses of resistant vs. susceptible hosts. Plant Pathol 70(5): 1219–1229

Zamani AR, Imani A, Mirza Aghayan M, Mohammadi R (2011) A study and comparison of control methods of anthracnose disease in walnut trees of roodbar region. Int J Nuts Relat Sci 2(4): 75–81

Zhang YB, Meng K, Shu JP, Zhang W, Wang HJ (2019) First report of anthracnose on pecan (*Carya illinoensis*) caused by *Colletotrichum nymphaeae* in China. Plant Dis 103(6):1432

Chapter 16
Integrated Pest Management of Temperate Nuts

Bashir Ahmad Rather, Jamasb Nozari, Mohammad Maqbool Mir, and Umar Iqbal

Abstract All temperate nuts (almond, hazelnut, walnut, pecan, chestnut and chilgoza) are widely planted and support a substantial portion of total agricultural income in the world. They have economic, medicinal and nutritional benefits and are consumed throughout the world. They are vulnerable to stress from different biotic factors including insect pests such as aphids, borers, leaf rollers, scales, etc. that cause direct as well as indirect considerable damage to various plant parts, thereby lowering productivity and product quality. However, they are safeguarded from the insect pest damage to some extent due to the protection of the kernel/seed from the environment by shell and/or hull. Successful insect pest management of these crops involves combination of orchard sanitation, monitoring of the orchards for incidence and potential damage due to major insect pests and selection of the proper combination of the best pest management strategies. Ecologically sustainable pest management strategies play a significant role in reducing the pest load on nut crops by making the crop environment less favourable for these pests. Relying on chemicals for insect pest management along with their indiscriminate and injudicious use has resulted in several adverse effects such as environmental pollution, ecological imbalance, pesticide resistance, pest resurgence, secondary pest outbreaks, etc. on these crops.

Keywords Nuts · Crop · Pest management · Biotic · Aphids · Damage · Shell and hull

B. A. Rather (✉)
Mountain Research Centre for Field Crops (MRCFC), Sher-e-Kashmir University of Agricultural Sciences and Technology of Kashmir, Khudwani, Jammu and Kashmir, India

J. Nozari
Department of Plant Protection, College of Agriculture and Natural Resources, University of Tehran, Tehran, Iran

M. M. Mir · U. Iqbal
Division of Fruit Science, FOH, Sher-e-Kashmir University of Agricultural Sciences and Technology of Kashmir, Srinagar, Jammu and Kashmir, India

M. M. Mir et al. (eds.), *Temperate Nuts*,
https://doi.org/10.1007/978-981-19-9497-5_16

1 Introduction

Nuts are generally defined as fruits of plants which can be sold with or without a shell. Nut is botanically defined as a fruit from angiosperms which is dry, indehiscent and seeded with a hard pericarp. It means that nuts do not automatically crack open to release the seed (Jackson and McNeil 1999).

They are regarded as a dietary source of abundant calories, unsaturated fatty acids and oils, proteins, vitamins, minerals, and fibre, as well as some antioxidants. They are also notable as a meat-production alternative. The immense health benefits and their increasing trend in worldwide consumption, the nuts support a sustainable growth for the market in the world (Maestri et al. 2018). All these nuts are susceptible to a range of insect pests, diseases which result in low yield and low-quality produce. Pest management in the nut crops is an ongoing challenge for all farmers from large scale commercial to small producers. Crop yield in these fruits is significantly impacted by disease outbreaks, invading insect pest species, and the changing insect pest environment. Plant pathogens, adverse weather conditions, weeds and various pests are the main causes of crop yield reduction or destruction of plants.

2 Description of insect/mite pests

The key insect pests currently associated with these crops, and which pose a serious threat to the nuts, are summarised as under:

S. no.	Name of the nut crop	Names of the key insect pest	Scientific name
A.	**Almond** *(Prunus amygdalus)*	Aphids	*Hyalopterus amygdalinus, H. pruni, Myzys persicae*
1.		Almond stone wasp	*Eurytoma amygdali*
2.		Navel orange worm	*Amyelois transitella*
3.		San Jose Scale	*Quadraspidiotus perniciosus*
4.		Peach Twig Borer	*Anarsia lineatella*
5.		Bark beetle	*Scolytus amygdali*
6.		Spider mites	*Panonychus ulmi and Tetranychus urticae*
7.		Almond mealy bug	*Drosicha dalbergiae*
B	**Hazelnut** *(Corylus avellana)*		
1.		**Aphids** Filbert aphid Hazelnut aphid	*Myzocallis coryli* *Corylobium avellanae*
2.		**Stink bugs** Brown Marmorated Stink Bug True bug	*Halyomorpha halys* *Phylus coryli*

(continued)

S. no.	Name of the nut crop	Names of the key insect pest	Scientific name
3.		Nut weevil	*Culculio nucum*
4.		**Leaf rollers and other leaf eating caterpillars** Filbert leaf roller Oblique banded leaf roller	*Archips rosana* *Choristoneura rosaceana*
5.		Filbert bud mite	*Phytoptus avellanae*, *Cecidophyopsis vermiformis*
6.		Filbertworm	*Cydia latiferreana*
C.	**Walnut** *(Juglans regia)*		
1.		Walnut weevil	*Alcides porrectirostris*
2.		Chaffer beetles	*Adoretus* sp., *Brahmina* sp., *Heteronychus* sp
3.		Walnut huskfly	*Rhagoletis completa*
4.		San Jose scale	*Quadraspidiotus perniciosus*
5.		Codling moth	*Cydia pomonella*
6		Hairy caterpillar	*Lymantria obfuscata*
7.		Walnut blister mite	*Aceria erineus*
8.		Stem borer	*Acolesthes sarta*
9.		Pinhole borer	*Scolytis nitidus*
10.		Defoliator	*Chaetoprocta* sp.
11.		Walnut aphid	*Chromaphis juglandicola* and *Panaphis juglandis*
D	**Pecan** *Carya illinoensis)*	Pecan weevil	*Curculio caryae*
1.		**Aphids** Black margined aphid Black pecan aphid Yellow pecan aphid	*Monellia caryella* *Monelliopsis pecanis* *Meranocailis caryaefolia*
2.		Pecan nut casebearer	*Acrobasis nuxvorella*
3.		Hickory shuckworm	*Cydia caryana*
4.		Pecan leaf phylloxera	*Phylloxera devastratrix*
5.		Shothole borer	*Scolytus rugulosus*
E	**Chestnut** *(Castanea sativa)*		
1.		**Chestnut weevils** Lesser chestnut weevil, Larger chestnut weevil	*Curculio sayi* *Curculio caryatypes*
2.		Shothole borer	*Scolytus rugulosus.*
F	**Chilgoza** *(Pinus gerardiana)*	Indian meal moth	(*Plodia interpunctella*)
		Cone worms	*Dioryctria abietella*
		Two seed bugs	*Leptoglossus corculus* and *Tetyra bipunctata*

2.1 *Almond*

Almond tree is considered unique within agricultural ecosystems enabling it to grow and survive in areas with as little as 180 mm annual rainfall. It is native to western Asia, but the highest production in the world has been recorded in the USA. Almond flowers earlier than other *Prunus* spp. but they are more susceptible to late spring frosts and a number of diseases and insect pests. Several insect pest species attack the different plant parts of almonds including kernels, thereby reducing orchard vigour and yield. The major insect pests attacking the tree are different types of aphids, borers, scales and mites. Besides some minor pests are also inflicting damage during the growing season.

2.1.1 Aphids

Aphids are considered as one of the major pests of almonds. Almonds are infested by three different species of aphids: *Hyalopterus amygdali* (Blanchard), *H. pruni* (Geoffroy), and *Myzus persicae* (Sulzer). The most prevalent aphid is *Myzus persicae* Sulz., followed by *Hyalopterus amygdali* (Blanchard), which feeds on the leaves, blooms and fruitlets, causing leaf curling and finally fruit drop (Plates 16.1 and 16.2).

These aphids suck the plant sap and infested leaves roll up and drop prematurely and new growth is stunted as a result of feeding by the aphids. The young shoots are infested by colonies of aphids as a result of which the terminal shoots become bunchy, malformed and sticky. These aphids also produce honeydew as excrement which causes sooty moulds to develop on the surface of leaves. Aphids reproduce fast by parthenogenesis and numerous generations are completed in a year. The population of aphids has been seen to increase with increased nitrogenous fertiliser application and irrigation, which has been linked to shoot flushing and leaf formation (Boulif 2022). There are a number of natural enemies which either feed or breed on

Plate 16.1 Green peach aphid (Photo from sarafraz-hezarmasjed.ir)

Plate 16.2 *Hyalopterus amygdali* on the host leaf

aphids. Different species of Braconidae, Eulophidae, Encytridae and Pteromalidae are parasitoids of aphids and many species of Chrysopidae, Coccinellidae, Lygaeidae, Miridae, Anthocoridae, Syrphidae, etc. are general predators of aphids (Almatni and Khalil 2022).

2.1.2 San Jose Scale

San Jose scale although considered a major pest of apple, pear and stone fruits, but it can also impact almonds and is active from March to December. Scale insects have piercing and sucking type of mouth parts which suck plant sap from twigs and branches and render them unfruitful and even death of small branches under severe conditions. The damage is caused by nymphs and female scales on fruits also and adversely affect the quality and render them unmarketable. Additionally, scales also excrete a lot of honeydew, which promotes the development of black sooty mould. When they are young, scale nymphs or crawlers are small, flat, oval, and mobile. They move around easily on the surface of the plant and can spread to new plants by irrigation, wind, rain, and human and mechanical activity. With the passage of time, the crawlers become established on the plant surface and begin sucking the plant sap. A new generation of crawlers is produced when adults form into a distinctive scale and start laying eggs/nymphs beneath it (Bessin 2003) (Plate 16.3).

Plate 16.3 San Jose scale infested shoot (Courtesy of Catholic University, Piacenza, Italy)

2.1.3 Almond Stone Wasp

It is also an important pest that affects almonds and attacks the kernels of the developing fruits. The pest, which is univoltine, overwinters as a fully mature larva inside the stones of the fruit that has fallen from the trees beneath them. The adults emerge from the mummified fruits that have fallen with the arrival of spring, usually in late April to early May, and after mating, the females lay their eggs within the freshly developed almond fruits. Before the stone formation the egg is placed into the endosperm of fruit with incubation period ranging from 20 to 22 days. The hatching process starts around the time when the embryo of the seed becomes visible (Arnaudov et al. 2020).

2.1.4 Navel Orangeworm

It is one of the main pests of almonds which significantly damage the nut crop throughout the growing season. The larvae of the pest directly damage the nuts by penetrating the kernels and polluting them with frass and webbing, while the adults are able to introduce a fungus during oviposition that result in aflatoxin, a pollutant that is toxic and carcinogenic (Plate 16.4).

This not only lowers crop output and quality but also increases the difficulty in sorting and processing the crop after harvest. There is very little tolerance or economic threshold limit for the infestation of this pest which is hardly less than 2% of crop damage. The larvae either pupate within the infested nut or outside of the shell. The pupae are dark brown in colour and are typically encased within silk cocoons. Adults are small and grey 9–10 mm moths with a wingspan of 19–20 mm (Wilson et al. 2020).

Egg Larva Adult

Plate 16.4 Egg larva adult

2.1.5 Peach Twig Borer

The peach twig borer also causes damage to almonds, in addition to being a significant pest of peach and apricot fruit. It is active in almond orchards and attacks the crop during fruit development. The larvae move into the growing shoots where they feed from inside out. The first-generation larvae mainly cause harm to the flowers and shoots whereas the larvae of the subsequent generations primarily feed on fruits. Any form of damage could cause significant economic damage. Furthermore, the larvae damage the nuts by eating developing nuts, leaves, and shoots directly, as well as by encouraging damage due to navel orange worm. The females of this worm are drawn to almonds whose hulls have been prematurely split by *Anarsia* for oviposition (Baspinar et al. 2018).

2.1.6 Spider Mites

Spider mites are considered indirect pests in almonds as they do not feed directly on the fruits/nuts. The plant damage is caused by sucking the cell contents out of the leaves. Early in the season, they are often located in the lower to centre parts of the tree, and as temperatures rise, they begin to spread. Common signs of mite damage include leaf speckling, yellowing, and premature leaf fall. High mite populations can be detected by webbing on leaves and tree terminals. In almonds significant mite injury has been observed to cause economic damage in subsequent seasons in the form of reduced vegetative tree growth and crop reduction. The excessive leaf drop also interferes with harvest operations and nut drying in the current season.

The overwintering mites are found under rough almond bark, in ground litter, and on winter weeds. They are reddish orange in colour. The colour changes during the season from yellow to green to black depending on age and host food. Egg laying takes place on leaves. During favourable conditions, mites develop within 7 days, and undergo moulting thrice to form a mature adult and eight to ten generations are completed in a season. Spider mites are frequently a problem in orchards that are under water stress, and they can also become a problem when broad-spectrum insecticides particularly pyrethroids which disrupts the natural enemy complex and kills a variety of predatory insects and mites. Almond trees can endure mild mite populations without suffering economic damage. Presence and relative abundance of predators is crucial in the management of mites. High predator-to-pest mite ratio in orchards need not to be treated.

2.1.7 Almond Mealy Bug

Mealy bug is an important pest of almonds which is identified by the presence of insects covered with white cottony mass below the collar region of plants. The old and weak plantations are typically attacked, which leads to nutrient deficiencies, a sickly appearance, and ultimately low yield. The insects weaken the host plant by sucking the sap from its delicate roots, branches, and stem. Typically, feeding is accompanied by the secretion of honey dew, which makes the plant sticky and promotes the development of sooty moulds. The most obvious symptoms are plant wilting and stunted growth. Strong infestation causes nursery plants to dry up (Anonymous 2018).

2.2 Hazelnut

2.2.1 Aphids

The filbert aphid (*Myzocallis coryli*) and the hazelnut aphid (*Corylobium avellanae*) are the two main species of aphids that pose a serious threat to hazelnut trees. Most often, the underside of hazelnut leaves will contain aphids. They cause leaf yellowing, wilting, and distortion by sucking plant cell fluids. Excreting honeydew leads to the development of black sooty moulds on the surface of leaves (Plates 16.5 and 16.6).

The filbert aphid overwinters as eggs in crevices on bark and around bud scales. When the eggs hatch in the spring the young aphids begin feeding on the underside of leaves. The aphid population increases more rapidly in the summer because they

Plate 16.5 Myzocallis coryli adult (Infojardin.com)

Plate 16.6 *Corylobium avellanae* adult and nymphs (biodiversidadvirtual.Or)

reproduce more frequently and reach maturity more quickly. Aphid colonies with all life stages present are typically seen on the underside of leaves. In the summer, several generations are completed. The aphids mate and begin to lay their winter eggs in late summer and early fall (Anonymous 2022a).

2.2.2 Stink or Shield Bugs

Brown marmorated stink bug is the most common bug causing damage to the hazelnuts. It is a serious pest that attacks various plant species, including ornamental plants, nuts, tree fruits, and grapes. The consumption of nuts by adults and nymphs results in corked and blank nuts. The adults congregate in the fall in and around buildings in search of warm places to spend the winter, which is a nuisance. The adult is shield-shaped, about the size of a coin, and has a brown marbled appearance. On the edge of the abdomen, brown and white markings alternate. The distinguishing characteristic from other stink bugs is the presence of distinctive white bands on the antennae. Bright orange, black, or brown are the colours of the immature stages, while the pear-shaped mature stages have white markings on their legs and antennae (Wiman and Chernoli 2022) (Plates 16.7 and 16.8).

2.2.3 True Bugs

True bug (*Phylus coryli*) is also one of the main insect pests that attacks hazelnut leaves, and the damage can be seen readily on the leaves throughout the growing season. It is omnivorous as it feeds on some aphids and is native to Europe. It is biovoltine and completes at least two generations per year. The black adults, yellow eggs and nymphs are present during the season (Wiman and Chernoli 2022) (Plate 16.9).

Plate 16.7 An adult brown marmorated stink bug

Plate 16.8 Brown marmorated stink bug eggs (Source: DISAFA Entomology, University of Torino)

2.2.4 Leaf Rollers and Other Leaf-Eating Caterpillars

The nut crops are being damaged by a variety of leaf rollers and other leaf-eating caterpillars, such as the European/Filbert leaf roller, winter moth, and Bruce spanworm. Numerous fruits, ornamental trees, and shrubs are other hosts of these pests. The larvae consume buds and leaves. Some of these leaf rollers eat inside rolled leaves while others skeletonise leaf surfaces. These pests cause harm all through the growing season, with bud damage occurring in the spring and leaf damage typically occurring in the spring and summer. Complete defoliation occurs in small and young trees under severe infestations. These pests have a variety of life cycles, but the majority overwinter as eggs or larvae. With the arrival of spring, the overwintering

Plate 16.9 Feeding damage to hazelnut leaf from *Phylus coryli*

Plate 16.10 Bruce spanworm: early spring bud feeder

eggs hatch and the larvae become active. The larvae then begin feeding on the green tissues that emerge initially (Wiman and Chernoli 2022) (Plates 16.10 and 16.11).

Leaf rolling caterpillars use silken webbing to roll up leaves as they expand and hide during the day. Winter moth and Bruce spanworm are univoltine and the larvae feed on buds between March and May while the larvae of the European or filbert leaf roller only feed during the middle of the summer. The bivoltine oblique banded leaf roller cause damage to the leaves and buds in both the spring and the middle of the summer. The feeding damage can be monitored by observing the bud leaves for larvae. Some species of moths, such as the European/Filbert leaf roller and the oblique banded leaf roller in summer, can also be monitored by using pheromone-baited traps to determine population size and time of flight (Wiman and Chernoli 2022).

Plate 16.11 Leaf roller: spring and summer bud and leaf feeder

Plate 16.12 Filbertworm larva

2.2.5 Filbert Worm

The filbert worm is considered one of the important pests in some hazelnut growing areas. It is a caterpillar that feeds in hazelnuts and oaks. Adult moths have golden streaks across each forewing and are grey to reddish in colour. After mating in the spring, from June through the fall, the female moths lay their eggs singly close to growing nuts. The larvae enter a nut to feed after the eggs hatch, where they stay for up to 4 weeks until they are about 1 cm long. The larvae have a brown head and a pale body. The boring signs of larvae on the harvested nuts include frass and webbing in addition to escape holes. Mature larvae spend the winter in a cocoon about 2–5 cm deep on the orchard floor (Plates 16.12 and 16.13).

Plate 16.13 Adult filbertworm moth [*Photo courtesy of Ken Gray Insect Image (Collection, Oregon State University, CC BY 4.0)*]

2.2.6 Bud Mites

The world's hazelnut growing regions are all host to the dangerous hazelnut pest known as the Filbert Big Bud Mite (FBBM). The filbert bud mite, also known as *Phytoptus or Phytocoptella avellanae*, and its vagrant variant (*Cecidophyopsis vermiformis*) are the two species of this pest that cause damage to the nut. Both species pose a serious threat to the health of crop, thus they need to be adequately handled and monitored.

The gall form can be identified by the presence of distinctive galls or by the misshaped hazelnut buds that result from the gall mite feeding on them. Bud mites feed inside flower and leaf buds, causing the smaller buds to enlarge and become blasted or 'big' buds.

These damaged buds dry up and fall off, and mites move to new buds. Buds may open fully or partially, but will be distorted, rigid, and brittle and catkins produced have little or no pollen. These galls or big buds are formed due to distortions induced by chemicals arising from mite infestation causing buds to become swollen, fleshy, deformed and pink. Different hazelnut varieties differ in susceptibility to bud mite infestation. These tiny sausage shaped mites overwinter within the buds. In summer they move from old buds to newly formed buds and cause them to expand from late summer through winter. Mostly the mites remain protected within buds and can be detected only by the damage caused by them. When the damage due to the pest is observed, start management which includes pruning and destroying of infested buds wherever possible (Anonymous 2022a).

Mites move to new buds because of the drying up and falling off of damaged buds. Although buds may open fully or partially, they will be distorted, rigid and brittle and the catkins they generate will contain little or no pollen. These galls or big buds develop from chemical aberrations brought on by mite infection, which causes buds to swell, flesh out, deform, and turn pink. The susceptibility of different hazelnut cultivars varies to bud mite infestation varies. These smaller sausage-like mites hibernate inside the buds. They expand from late summer through winter as a result of their movement from old to freshly created buds in the summer (Plates 16.14 and 16.15).

Plate 16.14 Galls on a hazelnut plant

Plate 16.15 Dense colonies inside buds infected by the FBBM (Source: DISAFA, Entomology, University of Torino)

The big buds produced by the pest becomes susceptible to other species of mites. The adult FBBM is white, elongated, about 0.2–0.3 mm long, with two pairs of legs. The gall form has a simple life cycle with a single nymph form, which resembles the adult, while the vagrant form has a more complex life cycle, in which it undergoes two nymph forms.

The FBBM overwinter in the core tissue of swollen buds and has a multivoltine life cycle that involves at least six generations in a single year. Nymphs depart from their winter buds in the spring and spread to other buds, where they eat, mature, and reproduce. FBBM migration typically takes place during the second or third leaf stage of a shoot and lasts for about 30–50 days. Migration normally occurs when the daytime temperature is greater than 15–20 °C and does not occur at lower temperatures. Eggs are laid from late February to mid-April and mite population is at its highest from March to early April (Anonymous 2022a).

2.2.7 Nut Weevil

The Nut Weevil overwinters as mature larvae in the soil and has a life cycle much slower cycle than other major hazelnut pests. It is univoltine and normally only produces one generation per year, but it may remain dormant for up to 3 years. Adults emerge from the ground in the spring and move to trees bearing rich, pulpy fruit (such pears, cherries, and peaches), where they begin feeding. Then, between May and June, they relocate to hazelnut trees, where the females lay their eggs after using their snouts to bore a hole in the fruit. Within the nut, larvae grow while consuming the kernel and finish their development in about a month. When the larvae reach maturity, they make small, distinctive circular holes in the shell from which they emerge drop to the ground for overwintering. The damage is primarily caused by the larvae, which consume the nut kernel and render it useless, lowering crop yield. The extent of damage due to this pest is often highly dependent on the variety of tree and thickness of the shell. Adults after emergence feed on young nuts leading to premature fall of young nuts but not to the extent of economic damage (Anonymous 2022a) (Plates 16.16 and 16.17).

2.2.8 Ambrosia Beetles

Different species of *Ambrosia* beetles are causing damage to different varieties in hazelnut orchards. The nature of damage is somehow similar in different cultivars due to which the varieties are often confused on observing the symptoms of damage.

Plate 16.16 Adult nut weevils—*Curculio nucum*

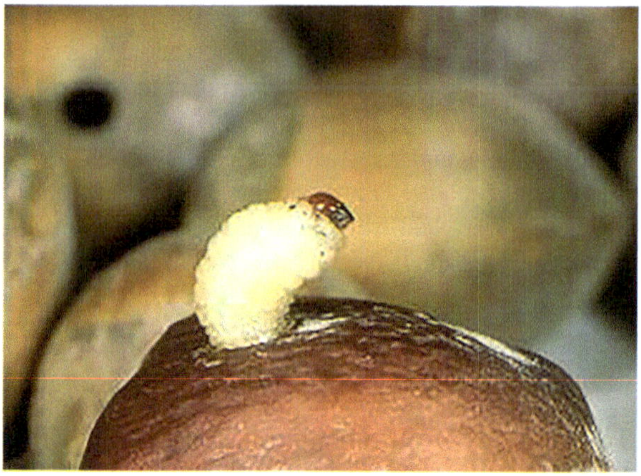

Plate 16.17 A mature larva dropping out from a nut. (Source: DISAFA Entomology, University of Torino)

Plate 16.18 An adult Ambrosia beetle

Adult *Ambrosia* beetles are small, around 2–4 mm long, and are dark brown to black in colour. They can attack a variety of fruit and forest trees and have a very varied diet. Beetle larvae are legless and have a creamy colour. Typically, ambrosia beetles produce one generation per year and overwinter as adults in channels drilled into trees, which is the main mechanism of nature of damage to crops. Females leave their overwintering sites in the spring when the temperature rises to about 18 °C. They fly to other trees where they bore through the wood to create a distinctive network of canals that are 2–5 mm in diameter. Females start depositing eggs in the tree after 10–15 days, and the larvae appear after 4–6 weeks. Larvae mature inside the host and stay there until the next spring. Ambrosia beetles attack trees in stressful situations including drought, wet soil, infestations, and illness. Young trees can die from sap flow obstruction caused by the pest damage, while the older crop may rapidly decline or suddenly dry up after blossoming. Pathogens may spread across plants because of Ambrosia beetle (Wiman and Chernoli 2022) (Plate 16.18).

2.3 *Walnut*

The oldest fruit in the world to have been cultivated is the walnut (*Juglans regia* L.), popularly known as the Persian walnut (Ozkan and Koyuncu 2005). The biochemical makeup of the walnut plant, which contains polyunsaturated fatty acids, makes it highly significant for human health in terms of nutrition, economy, and medicine. It is crucial for supplying high-quality timber, as well (Savage et al. 2001). To produce a marketable and respectable yield, walnut actually contends with a variety of insect and disease pests. Walnuts are vulnerable to a number of illnesses and insect pests, and even the majority of them pose a threat to plant life. They can also spoil a substantial amount of the edible nuts and the aesthetic appeal of the fruit.

Numerous insect pests cause damage to walnut trees directly, while others infest nuts and make them unfit for human consumption. Walnut weevil (*Alcides porrectirostris* Marsha), walnut chaffer beetle (*Anomala* sp. and *Holotrichia* sp.), San Jose scale (*Quadraspidiotus perniciosus*), walnut green aphid (*Chromaphis juglandicola* Kalt and *Panaphis juglandis*), stem borer (*Aeolesthes sarta*); pin hole borer (*Scolytis nitidus*); hairy caterpillar (*Lymantria obfuscate*); leaf roller (*Archips argyrospilus*); mealy bug (*Drosicha dalbergia*); grey weevil, *Mylocerus* sp.; defoliator (*Chaetoprota* sp.); and (*Chromaphis juglandicola*) (DPPQS 2003). Because it is challenging to spray the massive trees, pest control in the walnut is a tremendous issue. The most cutting-edge methods currently available for controlling insect pests in walnuts include integrated pest management programmes that use beneficial insects, mating disruption, insect growth regulators, improved monitoring techniques, and precise treatment timing based on the insect's life cycle (Shankar and Sharma 2012).

2.3.1 Walnut Weevil

Walnut weevil has been recorded in all the walnut growing regions throughout the world. Female weevil lays eggs in the pits excavated on the fruits. The grubs bore deeper on hatching and feed on the kernels and the kernels are reduced into a black mass. The grubs also cause extensive premature dropping of the fruits. The affected fruits also show dark brown spots which are dried resinous excretions. The adult weevil comes out by biting circular holes. The adult weevils feed on petioles, female floral buds, tender shoots and even on young fruits. Leaves, fruits are damaged by weevils during June–July, fall of nuts to the extent of 50–60% has been observed. Jet black weevils appearing in early May are noticed, which deposit eggs after mating on the green nutlets. The larvae after hatching bore the nuts and cause the damage (Shankar and Sharma 2012) (Plates 16.19).

Plate 16.19 Walnut weevil

2.3.2 Chaffer Beetles

Different genera and species of chaffer beetles are causing damage to the walnuts. The grubs and adults cause damage to roots and leaves, respectively. The entire tree is defoliated in case of heavy infestation of the pest. Grubs attack the plant throughout the year while the adult beetles are active during June–July (Shankar and Sharma 2012) (Plate 16.20).

2.3.3 San Jose Scale

These small insects suck the sap from leaves, tender shoots, branches, and trunk. The infested tree portions have a greyish crust covering them. Scales cause red stains on the bark and fruits, which decreases the market value of the fruits. Trees that have been severely affected lack vigour, and their foliage becomes sparse and spotted with yellow colour. Only growth is first halted, but as the infestation spreads, the affected tree dies from devitalisation. Often, only a few branches of a tree will first become infested and perish. Even after scale infestation on strongly infested trees and limbs is successfully controlled, the branches and buds have been so weakened by the scale infestation that no fruit is borne for one or two seasons (Shankar and Sharma 2012).

Plate 16.20 Chaffer beetle (Holotrichia sps.)

2.3.4 Walnut Aphids

Walnut aphids include the walnut green aphid (*Chromaphis juglandicola*) and Dusky-veined aphid (*Panaphis juglandis*). These are serious pests of nursery plantations, medium trees, and walnut production (Olson and Buchner 2002). Walnut aphids are smaller than dusky-veined aphids, pale yellow, and feed on the underside of leaves. Dusky-veined aphids feed along the mid vein on the top surface of leaves and have dark banded markings on their backs. Both the aphid species secrete honeydew after sucking plant juice. The midrib of the leaf turns black when the dusky-veined aphid feeds on it.

Walnut aphids include walnut aphid (*Chromaphis juglandicola*) and Dusky-veined aphid (*Callaphis juglandis*). This is a serious pest of nursery plantation and medium trees and are also affecting walnut production (Olson and Buchner 2002). Walnut aphids are pale yellow, much smaller than the dusky-veined aphid, and feed on the lower surface of leaves. Dusky-veined aphids have dark banded spots on their backs and are found feeding along the mid-vein on the top surface of leaves. Both aphids suck plant juice and deposit honeydew. Feeding of the dusky-veined aphid causes the leaf midrib to turn black. Some kinds of the husk become toxic from the honeydew, which also turns it black. High populations may cause leaf drop and a reduction in the production and quality of nuts. In the spring and early summer, when nuts are developing quickly, aphid populations are most destructive. The nymphs and adults of these aphid species twist and distort the leaves by sucking the plant sap from the petioles, flowers, and fruits. Fruits that have already formed wither away as well (Plates 16.21 and 16.22).

Plate 16.21 Green walnut aphid or small walnut aphid

Plate 16.22 Dusky-veined aphid or large walnut aphid (*Callaphis julandis*)

2.3.5 Walnut Tree Trunk Borer

The tree trunk borer is a pest that can cause significant damage to temperate fruits including apple, walnut, almond, cherry, etc. (Sheikh 1985). The pest poses a major threat to older trees throughout the year and is active year-round. The cuts or wounds in the tree bark are the places where the eggs are deposited. The grubs on hatching bore in the bark and sapwood and then tunnel downwards. The pest makes zigzag tunnels into trunk and main branches thereby reducing the life of trees and fruit yield (Bhat et al. 2010). The attack of this pest results in big holes measuring 1.50–2.00 inches in length in the main stem or old branches of trees from which saw dust comes out thereby causing drying of trees or branches (Shankar and Sharma 2012).

2.3.6 Shothole Borer

This pest is serious in semi dried trees where attack of San Jose scale has been noticed for 2–3 years. Pinholes are created as the adult and grubs bore through the sapwood and hardwood of the branches. The circular pin holes are seen on main trunk, branches and tender shoots of the trees containing small brown insects. The leaves on infested branches turn yellowish and wilt once their surfaces get punctured. The tree may succumb in extreme circumstances. The infested branches release a fermenting odour on breakage (Shankar and Sharma 2012).

2.3.7 Flat Headed Tree Borer

The infestation of this pest is very mild in stem and branches. The common symptoms of the pest include big galleries under bark, deep into cambium, dieback of branches and drying of bark and stem. Grubs are medium sized, elongate, bore on infested parts (Shankar and Sharma 2012).

2.3.8 Walnut Blister Mite

This tiny eriophyid mite infects leaves, causing blister-like swelling on the upper leaf surface and a concave yellow or brown pocket on the underside of the leaf. Typically, the trees are not severely damaged by these insects. This mite spends the winter under bud scales. The mites feed among the leaf hairs beneath the leaves when the weather is favourable. Throughout the summer, there are several generations that attack fresh foliage as soon as it sprouts. Predatory mites such as phytoseiid mites, try to keep these mites under check, if broad-spectrum insecticide applications are avoided. Cold temperatures and a lot of rain also reduce the amount of mites. Broad leaved weeds and excessive nitrogen application increase mite infestation (Shankar and Sharma 2012) (Plates 16.23 and 16.24).

2.3.9 Walnut Husk Fly

The most significant pest of walnuts is the walnut husk fly, which originated in the south-central United States but has since spread to all regions where walnuts are grown. Its wings have three distinct dark bands, one of which wraps around the wing to make a V-shape, and it is roughly the size of a housefly with a yellow patch below the places where the wings are linked to the body. The larvae or maggots are whitish up to 0.19-inch length that feed in bunches inside the husk. These larvae cause the nutshell to become stained thereby lowering the quality of the nuts. Huskfly infestation is characterised by dark and soft blotches but hard and dry blotches on mature husks are symptoms of a disease known as walnut blight. The damage is largely

Plate 16.23 Blister mite on walnut nut trees

Plate 16.24 Walnut husk fly (Photo from s3.wp.wsu.edu)

cosmetic staining on the nutshell and is primarily a concern for in-shell walnuts where a severe infestation can affect kernels. The adult huskfly should be the focus of management as the huskfly eggs and larvae are sheltered in the walnut husk and the insect spends a lot of time under the soil (Shankar and Sharma 2012).

The adults of walnut husk flies emerge in the early to mid-June after overwintering as pupae in the soil. Female flies lay their eggs under the husk in clusters of about 15 eggs each. The egg deposits appear as tiny black spots on the husk. The eggs hatch within 5 days. The black area grows as a result of the maggot feeding within the husk, but it nevertheless remains soft, smooth, and undamaged. In most cases, the outer surface of husk is undamaged, but the fleshy interior degrades and discolours the nutshell. After consuming the husk for 3–5 weeks, maggots reach maturity. When they are ready to pupate, they descend to the ground and dig several inches into the earth. Most emerge as adults in the following summer, but some remain in the soil for two or more years (Anonymous 2019).

2.4 *Pecan*

Pecan (*Carya illinoensis* (Wang) K. Koch is the most important nut crop and is native to the United States. There is a close relationship between the seasonal development of the pecan and the development of different insect pests which attack the fruit crop. The different insect pest problems associated with various developmental stages of the pecan include the yellow and black aphids, pecan casebearer, spider mites and pecan weevil, *Curcalio caryae* (Hom), hickory shuckworm, *Cydia caryana* (Fitch). The commercially most important foliar feeding pests are usually aphids, especially *Monellia caryella* (Fitch), *Meranocailis caryaefolia* (Davis) *Monelliopsis pecanis* (Bissel). There are three species of *Phylloxera devastratrix* which cause damage to Pecan, but only Pecan *phylloxera* causes economic damage.

2.4.1 Pecan *phylloxera*

It includes leaf phylloxerans, *P. notabilis* and *P. russelae* and stem phylloxera *P. devastratrix*. Phylloxerans attack foliage, shoots and fruits and results in severe damage. Leaf phylloxera generally causes minor damage, but stem phylloxera can be devastating as the name implies. It results in the formation of prominent galls on branches which results in deformation and weakening of branches and eventually dying (Harris and Jackman 1991). The pest overwinters as eggs which hatch as the bud break occurs in spring and young phylloxera move out and begin feeding at new leaves or terminals. A gall is soon formed which envelopes the insect. Stem phylloxera galls replace nut clusters and heavy infestations can eliminate production and reduce growth of new shoots (Plate 16.25).

Plate 16.25 Damage to foliage by *phylloxera* in Pecan. (Photo from extension.missouri.edu)

2.4.2 Pecan Weevil

It is one of the key pests of pecan nut although not present in all orchards. The weevil has a 1, 2 or 3 year cycle. Mostly development is completed in 2 years. Adults emerge in late summer about the time the nuts are emerging and fly or crawl into the tree canopy. Adult feeding causes younger nuts to fall and there is also loss of quality of nuts usually observed at the shelling plant after harvest. Once the maturing nuts reach the half-shell hardened stage, females begin laying eggs in nuts. Developing larvae feed on the kernel destroying it and emerge to drop to the ground. They dig in the soil and pupate emerging as adults in 1, 2 or 3 years.

2.4.3 Pecan Nut Casebearer

The pecan nut casebearer *Acrobasis nuxvorella* Nuenzig (Lepidoptera: Pyralidae) is monophagous and is regarded as a major and most destructive pest of pecan in the world (Harris et al. 1998). The larvae cause damage by tunneling into nutlets and begin feeding. Depending on the climate and geographic location, the pest completes 3–4 generations/year (Neunzig 1986). The newly generated larvae or first-generation larvae are more destructive to pecan crop due to their capability of consuming 3–4 nutlets in a cluster on which they feed and reduce the yield. The nature of damage is identified by the silk webbing and buildup of frass at the base of the tree (Ring and Harris 1984) (Plate 16.26).

Fewer nuts are required by the larvae of subsequent generations to complete development because these nuts are larger. Thus, the larvae of future generations are less damaging to the crop. But feeding by second-generation larvae can result in significant nut loss, especially if the first-generation larvae are not controlled. Later, the larvae enter into the shuck, because of the hardening of the nutshell. This damage is similar to that of hickory shuckworm, *Cydia caryana* (Fitch). The only difference is that the larvae of hickory shuckworm are dirty white while the larvae of pecan nut casebearer are olive grey (Knutson and Ree 2019).

Plate 16.26 Pecan nut casebearer moths captured in a pheromone trap. Frass and silk webbing on pecan nutlets (Photo: Bill Ree, Texas A&M University)

Plate 16.27 Hickory shuckworm adult moth larva damage (Source: www.isuagcenter.com)

2.4.4 Hickory Shuckworm

The pest got its name hickory shuckworm from the larvae which feed and develop inside the pecan shell. It is also considered one of the major pests of pecan. The shuckworm larvae have been observed to cause several kinds of pecan nut damage. First, the shuckworm larvae eat inside the nut before the shell hardens, their attack causes premature nut drop. Further, the shuck mining activities of the larvae after the shell hardens also result in the formation of poor kernels, shuck sticking, scarring, and discoloration of the shell, as well as a delay in nut maturity (Plate 16.27).

The pest overwinters as a fully developed larva inside the nut's shuck. In the shuck, pupation occurs in the late winter or early spring. The adult hickory shuckworm is a tiny, nocturnal moth that is 3/8 inches long, smoky black in colour, and has a wingspan of about ½ inches. Even during the day, it is rarely noticeable due to its modest size and discrete coloration. The hickory shuckworm larvae are between one-third and half an inch long. The larva measures between a third and a half of an inch in length, with a creamy to dirty white body and a reddish-brown head. It resembles nut curculio larvae in appearance; however, the nut curculio has legs, whereas the hickory shuckworm has not. After emerging in spring, the female shuckworm moth immediately begins to lay eggs on the nuts. All through the summer, up until the shuck-split stage, eggs are laid. The pest has three to four successful generations each year and is multivoltine. Pest mass emergence peaks have been seen in the middle of May, late June, the middle of August, and the early to middle of September (Hall 2013).

2.5 *Chestnut*

Chestnut, (*Castanea* spp.) is native to Asia and is popular by many names which include European, American, Japanese and Spanish chestnut. It is a deciduous tree cultivated for its tasty seeds (nuts) and belongs to Fagaceae family. The chestnut tree has an erect growth habit, a solid trunk coated with grey bark, and a maximum diameter of 2 m. The seeds of the chestnut tree are generated in clusters of one to three, while the tree produces flowers on long catkins. A substantial, spiky bur that is roughly 10 cm (4 in) in diameter surrounds and protects the seeds. There is a thin, dark brown shell covering the kernel inside. Chestnut yield and quality have been observed to be affected by a number of arthropod pests that attack the tree. These pests include Chestnut weevil (Lesser chestnut weevil, *Curculio sayi*, and larger chestnut weevil, *Curculio caryatypes*), Chestnut Filbertworm, *Cydia latiferrana* and Shothole borer *Scolytus rugulosus*. Moreover, chestnut is also attacked by a few minor pests (Cipollini 2017).

2.5.1 Chestnut Weevil

The pest is univoltine and mainly attacks chestnut where it feeds on the amydaceous content of the chestnut fruit. The chestnut weevil is yellowish grey in colour and has an elongated oval shape and a characteristic long rostrum. The rostrum in females slightly extends beyond the length of the body while in the male it is noticeably shorter. Larvae are apodal and creamy white grubs which bore into the nuts. The fallen nuts on ground have been observed with small circular holes drilled from the inside (Jackson and McNeil 1999). Females lay eggs by piercing the husk with their rostrum and inserting an egg into the hole. The newly emerged larvae penetrate into the underlying chestnut fruit and feeds on the amydaceous substrate of the kernel. As many as 18 larvae per fruit have been reported, although each fruit in general hosts not more than two or three larvae. Chestnuts should be collected every day and roasted for 30 min at 60 °C (140 °F) after curing to kill any larvae inside the nuts. The timely collection of nuts when they fall is seen to be an important cultural technique for managing them because it stops larvae from emerging from the nut and entering the soil (Cipollini 2017).

2.5.2 Mites

Chestnut trees are vulnerable to severe feeding damage from many spider mite species which include European red mite and two-spotted spider mite. European red mite has been observed as the predominant species in case of chestnut. The infested leaves initially appear mottled and speckled and under severe infestation become bronze coloured and brittle leading to premature defoliation and reduced photosynthetic activity. Reduced photosynthetic activity can lead to reduced nut size and affect the yield in subsequent years as well as increase sensitivity to winter injury (Anonymous 2022b).

European red mites hibernate as eggs in bud scales and in bark cracks. Eggs are small, approximately the size of a pinhead, and have a single stipe, or hair, protruding out of the top. Overwintering eggs hatch as the temperature rises, and nymphs crawl onto the newly sprouting leaves and begin feeding. The hair on adult ERM is red and gives them a spikey appearance. Feeding is primarily done by adults and nymphs on the upper surface of the leaves. The development of the first generation is slow and is completed in at least 4 weeks as it is completely temperature dependent.

Plate 16.28 Some harmful mites of the tetranychidae family on nut trees

Summer generations are completed quickly in at least 10 days (Wood 2003) (Plate 16.28).

2.5.3 Shothole Borer

Shothole borers are small brownish black beetles which are polyphagous and damage ornamental, forest trees, shrubs as well as nut trees. Primarily, shothole borer is a problem on injured or stressed or weaker plants, but healthy trees also get affected if they are growing adjacent to blocks of neglected trees. Chestnut trees growing adjacent to borer infested trees also get affected (Cipollini 2017) (Plate 16.29).

The larvae are around 0.17 inches long, whitish, and legless. Adults and larvae eat through the cambial and vascular tissues of trees, weakening them and causing individual stems and branches to wilt and die. All the branches along with tree trunk get fully covered in galleries. The shothole borer larvae overwinter under the

Plate 16.29 *Scolytus rugulosus* (Shothole borers)

bark of infested trees. They pupate there, then in the spring or early summer, the adults emerge, mate, and fly to trees that are vulnerable, where they feed at the base of leaves or small twigs. After that, they burrow into the tree, creating galleries that are parallel to the direction of the wood. Along the gallery, they deposit eggs (Chernoli and Wiman 2022).

2.6 Chilgoza

Chilgoza pine (*Pinus gerardiana*) is a conifer tree having short lateral branches. It is a delicious nut fruit with a high fat content. The nuts are eaten raw or after roasting them. The seeds are severely attacked by seed-eating insects or worms known as the Indian meal moth (*Plodia interpunctella* Hübner). Its larvae (caterpillars) are commonly known as 'wax worms'. Cones/berries and seeds are highly prone to insect attack, whether in field or in stored conditions. The study on the insect prevalence on this pine species revealed that seed borers are major insect-pests which attack the cones and berries in the natural forests as well as stored seeds. The two seed bugs, *Leptoglossus corculus* and *Tetyra bipunctata* and two species of coneworm *Dioryctria abietella and Cateremna cedrella* (=*Euzophera*) are most damaging. There are some other insect species which causes potential damage to the chilgoza pine which include pine seed worms (*Cydia* spp.), cone borers (*Eucosma* spp.), cone beetles (*Conophthorus* spp.), and tip moths (*Rhyacionia* spp.). The two species of cone borers, which belong to family Pyralidae of Lepidoptera, occur as serious pests of cones and seeds of conifers including chilgoza seeds. The adult moths lay eggs on the green scales of 1 year old cones. Young larvae bore into the scales and feed on the seeds and throw out their excreta through the entrance holes, a brownish matter mixed with resin which later on dries up. The larva cuts the hard shell inside the cones of the seeds which covers the edible portion. The damage to the seeds continues one after another. Study on the management of insect pests of chilgoza seeds reveals that the freezing treatments are very successful against the attack of insect pests on chilgoza seeds (Plate 16.30).

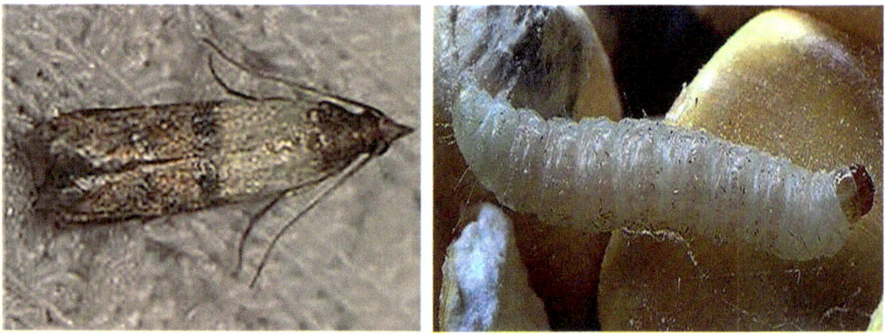

Plate 16.30 Indian meal moth *Plodia interpunctella* (lucidcentral.org)

2.7 *Integrated Pest Management (IPM)*

IPM technology is the management of pest population in such a manner that economic damage is avoided and adverse side effects of chemical pesticides in the environment are minimised. The IPM package includes various management strategies for containing the pest and disease problems. Pest monitoring is one of the important components of IPM in decision making, to manage the pest problem. It can be done through agro-ecosystem analysis (AESA), field scouting and installing different types of traps such as light, sticky and pheromone traps. If the population density is low, it will be managed by birds and other bio-agents. The determination of economic threshold level (ETL) against major pests helps in taking appropriate management strategies when pest population crosses ETL (DPPQS 2003). Pest control of trees can be provided by direct method, i.e. by using pesticide or by creating conditions that prevent pest attack, i.e. indirect. The use of either of these two methods depends significantly on a more complete and specific knowledge of the pest. The purpose of both methods is to reduce the amount of damage to plants. There are different cultural practices which help in minimising the damage due to insect pests in nut crops. Selection of planting site with well drained and slightly acidic soil is best suited for these nut crops. The best management strategy has been observed in keeping trees in healthy condition through different cultural practices viz., pruning and training, balanced fertiliser application and proper irrigation scheduling. Because healthy trees repel the beetles by plugging bore holes with sap and resins and have the compensation ability to overcome the damage. Removal and destruction of infested wood on the trees and severe pruning of infested branches and cutting of dried trees helps in minimising the pest attack (Olson and Buchner 2002). Different insecticides are recommended for the successful management of insect pests in nut fruits. Carbamates have been observed to cause immediate knockdown effect with residual toxicity of 7 days. Similarly, Organophosphates can act in 3 days but provide a residual toxicity to the target pests for 10–14 days. Pyrethroids have good immediate activity and provide 7–10 days of residual control but they are toxic to natural enemies. Neonicotinoids act initially as a contact poison for 2–5 days and have a longer residual period as antifeedants. Neem-based products which exhibit a residual toxicity of 1–2 days and kaolin clay which act as physical barrier and irritant are included among organic options. The integrated pest management strategies for different categories of insect pests in temperate nuts include the following preventive and curative measures:

- Maintain vigour of fruit trees by giving proper irrigation and fertilisation.
- Proper pruning and training methods and removal of infested shoots.
- Mealy bug infestation should be managed by placing slippery bands of alkathene sheets of 25–30 cm wide on grease for adhering at 0.5–1.0 m height on the trunk above the ground at fruitlet stage.
- Avoid planting aphid or mealy bug or scale infested plants.
- Secondary host plants of insect pests should be removed.

- Trunks of young trees should be whitewashed to prevent sunburn and reduce potential hazard of attack from borers. Yellow sticky traps can also be used for trapping these beetles.
- Collection and destruction of adult June and other chaffer beetles. Light traps and beating during dusk are used for trapping beetles of white grub and found promising in controlling white grub adults.
- A number of beneficial insects such as green lacewing, *Chrysoperla corne* (Stephens), ladybird beetles, syrphid flies and the mealybug destroyer *Cryptolaemus montrouzieri* (Mulsant) have been found to feed on aphids, scales and mealy bugs.
- Shade trees especially willows, poplars, etc. should not be planted in and around fruit orchards.
- Orchard sanitation greatly reduces the damage due to these pests.
- Dormant or delayed dormant spray of any horticulture mineral oil (HMO) at 2%, i.e. 2 L in 100 L of water has been observed to be effective against dormant stages of many insect pests including aphids, scales, etc. Mixing an organophosphorous insecticide such as Ethion 50EC at 0.05% can give more efficient results.
- Chemical treatment involves spraying of Dimethoate 30EC or Chloropyriphos 20EC or at 1 mL/L of water at fruit let stage, if curling of leaves due to aphids or other sucking pests is noticed.
- Application of Carbofuran 3CG granules at 100 g/tree under the tree canopy during the season if need arises is recommended for aphids, mealy bugs and other sucking pests including white grubs, beetles, etc.
- Some reduced risk insecticides and biopesticides such as Spinosad 45 SC at 0.5 mL/L, or Chlorontraniprole 18.5 SC at 0.5 mL/L or Imidaclopirid 17.8 SL at 0.25 mL/L or Lamda-cyhalothrin 5EC at 0.6 mL/L or Bifenthrin 10EC at 0.8 mL/L, Neem oil at 5 mL/L, etc. are effective for the management of different insect pests such as caterpillars, aphids, leaf rollers, beetles, borers, etc. in the nut crops.
- Collection and destruction of egg masses, larvae, pupae and adults of hairy caterpillars which are present from July to March on the tree trunk or scaffold branches and other hidden places. Different egg, larval and pupal parasitoids have been reported as natural regulating factors of Indian gypsy moth in the country (Rishi and Shah 1985).
- Burlapping of tree trunks with gunny bags dipped in Chloropyriphos 20EC at 1 mL/L of water after 20–30 days from fruit let stage for collection and destruction of defoliating nocturnal caterpillars/larvae.
- Application of Quinalphos 25 EC or Chloropyriphos 20EC at 1 mL/L of water at fruit let stage will give excellent results in managing aphids, scales, hairy caterpillars, leaf rollers and other defoliating caterpillars, beetles and weevils, and nut boring insects.
- Fall ploughing helps in reduction of chaffer and click beetles by exposing the grubs to predators.

- Light traps should also be installed for monitoring, collection and destruction of beetles and nocturnal lepidopteran pests.
- The bearing and non-bearing trees should be shaken during large scale emergence of defoliating beetles at dusk for collection and destruction thereof.
- Pest management is really a big challenge in the walnut and other nut trees because of the difficulty faced to spraying the giant trees. Therefore, an integrated approach which include monitoring techniques, habitat management practices, mechanical and physical control strategies, beneficial insects, mass trapping and mating disruption, insect growth regulators and accurate timing of treatment based on study of life cycle of the insect pest should be adopted for successful management of insect pests in walnut and other nut trees (Olson and Buchner 2002).
- The integrated pest management of navel orange worm involves an integrated approach which includes sanitation of orchards, accuracy in the timing of insecticide treatments and timely harvesting of the crop. Presently, novel tools such as mating disruption and sterile insect technique are being explored for the management.
- The integrated approach for weevil and huskfly infestation in walnut includes field sanitation with collection and destruction of fallen infested nuts. Yellow sticky traps should also be used for monitoring and timing of insecticide application because timely application of sprays is important to prevent adult flies from laying eggs. The traps should be placed in the upper half of the canopy in late May or early June. Smaller and medium sized plants can be sprayed with 0.02% Chlorpyriphos 20EC or 0.05% Dichlorvos 76EC during the month of April-May.
- Shothole borer infestation in case of nut crops is successfully managed by cutting and burning of infested branches. Tree trunk should be plastered with Chlorpyriphos 1.5 WP and soil in the ratio of 1:1. Also spray Chloropyriphos 20EC at 1 mL/L during May and July at the time of peak season of adult emergence of first and second generation of the pest, respectively.
- Stem borer infestation can be managed by cleaning and plugging of holes during May–September with cotton or pieces of cloth impregnated with Dichlorovos 76EC at 3 mL/L of water or petrol. Naphthene balls or Odonil plugging can also be used for plugging. Tree trunks should also be plastered with 10% Malathion dust and soil in the ratio of 1:6.
- The successful management of mite infestation involves monitoring of Spider mite infestation and their predators at least once every 2 weeks from March to early May and then monitor at least once a week. When treatments are required, selective miticides should be chosen that cause least harm to predators. The most effective spider mite management programmes are focused on integrated strategies involving habitat management practices, biocontrol and need based application acaricides. In case of mite infestations pruning and destruction of infested twigs/branches during dormancy. Spraying with HMO at 2% during delayed dormancy and spraying of Dicofol at 1 mL/L or Fenzaquin 10EC or Hexythiazox

5.45EC or Spiromesifen 22.9 SC at 40 mL/100 L of water or Fenpyroximate 5EC at 1 mL/L or Propergite 57EC at 1 mL/L of water during the season.

- For successful management of codling moth in walnut and other nut crops monitoring with pheromone and degree-day developmental models are considered very important tools for optimising the timing of IPM interventions particularly the timing of insecticide application. In addition, cultural practices such as bark scrapping, collection and destruction of fallen infested fruits and burlapping of tree trunks help in minimising the damage due to codling moth.
- The damage due to chestnut weevil can be prevented by harvesting burs from tree rather than waiting for nuts to fall or harvest frequently from the ground. The damage can also be prevented by postharvest treatment, i.e. after curing, the nuts should be immersed in 140 degrees F water bath for 30 minutes and then cooled.
- The management of Filbert Big Bud Mite (FBBM) involves monitoring the incidence from late winter to early spring. At least 200 buds per orchard of one hectare or all buds on four branches of 10% of plants/hectare should be inspected for the mite damage. The economic threshold for chemical treatment has been determined as if 15% of buds inspected show signs of galling. The accurate timing for management has been observed during migration of mites from infested to healthy buds which usually begins from late March to early April. At least two applications of sulphur products or lime sulphur are recommended for application during migration phase, first in late winter and second after an interval of 8–10 days. They will kill exposed mites as they move from old buds to new buds. Predatory mites may also help control these pest mites.
- Monitoring of population of stink bugs and nut weevil is important considering the severity risk of crops posed by these pests. Two different methods of monitoring include plant beating and traps. Use of pheromone traps is most popular method in hazelnut orchards. Placing of traps should be done between May and harvesting time, and should be checked weekly during the growing season. If more than ten adult specimens are caught/trap/week or more than five nymph specimens are caught per trap per week during the season or if nymph specimens are found over two consecutive weeks chemical treatment should be carried out immediately.
- The best management strategy against the *Ambrosia* beetles in chestnut is to upkeep the vigour of the plants by preventing water stress during dry spell and by application of balanced dose of fertilisers recommended for the crop. In healthy and vigorous trees, the development of the eggs and larvae is obstructed when the sap fills up the galleries. The attack of *Ambrosia* beetles is first identified by presence of entrance holes about 0.5–1 mm in diameter on the branches of a tree. Different types of traps can be utilised for monitoring and management purposes like in mass trapping in which 8–10 traps are recommended as a mass trapping strategy.
- The most effective management strategy for pecan nut casebearer, *A. nuxvorella,* is to target the hatching first instar larva to prevent its entry into the nut. It involves need based insecticide use and their timely application which is achieved

by monitoring this pest with sex pheromone baited traps and using a degree day prediction model (Stevenson et al. 2003) combined with a sequential sampling plan (Ring et al. 1989). This IPM strategic plan has significantly reduced the amount of pesticides used to control *A. nuxvorella* compared to when calendar sprays were used to control this pest (Stevenson et al. 2003). In addition, this strategy has caused a significant reduction in the target pest population, reduced pesticide use and increase crop quality and yield. Moreover, this has also conserved natural enemies and effectively managed other foliar pecan pests, such as aphids, mites and leaf miners, that occur later in the pecan growing season (Harris et al. 1998).

- Phylloxerans can be managed with Chloropyriphos, pyrethroides and nicotinyl insecticides. Timing of application is very critical to prevent yield loss. Application of insecticide at bud break is usually effective.
- Time of application of chemical intervention is also very important for management of hickory shuckworm because its emergence varies from orchard to orchard and year to year. Shuckworm activity can be monitored with black light traps and by inspecting aborted nuts for signs of nut entry by shuckworm larvae. If you do not have a black light trap, adequate control of hickory shuckworms usually can be obtained by making two insecticide applications. The first application should be made at about half-shell hardening, usually during mid-August. This should be followed by a second application about 2 weeks later. The timing of these applications coincides with the period of greatest shuckworm activity and the stage of nut development where the greatest amount of damage to the nut from the shuckworm occurs.

3 Conclusion

Nuts are a highly heterogeneous category of fruits with a greater diversity of insect pest spectrum which cause damage to these crops. The different categories of pests include sucking pests, foliage feeders, leaf rollers, defoliators, mites, fruit weevils, borers, etc. The successful insect pest management of these nut crops involves close monitoring for incidence and potential damage due to major insect pests. Monitoring provides knowledge about the current pests and crop situation and is also helpful in selecting the best possible combinations of the pest management strategies. There should be emphasis on biocontrol in IPM programmes with conservation and augmentation of natural enemies of pests such as insect predators, parasitoids, parasitic nematodes, fungi and bacteria which will reduce the pest load and outbreak of secondary pests. In addition to application of different methods of IPM, there should be accurate timing of chemical control strategy or organic option based on monitoring or need based use of insecticides and acaricides in nut crops. This will always manage the pest populations below economically damaging levels and increase the yield and quality of nut crops.

4 Future Strategies

The future strategies should begin with an overview of temperate nuts, followed by discussion of critical crop protection aspects including the basics of IPM in nut production. Brief description of each pest along with information about biology, ecology and population dynamics of the insect pests is necessary and needs to be studied. The management options should involve prevention, avoidance, monitoring and then suppression. The focus should be on long term prevention of pests or their damage through a combination of techniques such as habitat management, physical and mechanical methods, biocontrol, resistant varieties, etc. and need based use of pesticides. The IPM programmes should emphasise on the protection of beneficial organisms that occur naturally in the orchards. Pesticides should be used only if monitoring indicates that they are needed as per the established guidelines. The need of the hour is to evaluate the key insect pests causing damage to the nut crops and coordinate the integrated pest management strategies.

References

Almatni W, Khalil N (2022) A primary survey of aphid species on almonds and peach and natural enemies of *Brachycaudus amygdalinus* in As-Sweida, Syria. Ministry of Agriculture, Damascus, pp 109–115. https://www.orgprints.org/1654

Anonymous (2018) Package of practices. In: Temperate fruits, vol 1. Directorate of Extension Sher-e-Kashmir University of Agricultural Sciences and Technology of Kashmir Shalimar, Srinagar, p 199

Anonymous (2019) Chapter 14: Home and garden pest management guide for British Columbia. In: Pests of nut trees. Government of British Columbia, Victoria, pp 1–12

Anonymous (2022a) Integrated pest management for hazelnut crops. A practical handbook. ipm_guide_eng.pdf. USAID. Action against hunger. UNDP, New Delhi

Anonymous (2022b) Chestnut pests—Shothole borer (*Scolytus rugulosus*). In: Nut crops. Oregon State University, Corvallis. http://pnwhandbooks.org/node/7978

Arnaudov V, Davidova R, Vasilev V (2020) Biology, ecology and control of the plum seed wasp [*Eurytoma schreineri* Schrener (Hymenoptera: Eurytomidae)]. AGROFOR Int J 5(1):54–60

Baspinar H, Doll D, Rijal J (2018) Chapter 12. Pest management in organic almonds. University of California Cooperative Extension, Woodland. http://www.researchgate.net/publications322750268, pp 329–345

Bessin R (2003) San Jose scale. Extension entomologist. UK cooperative extension service. University of Kentucky, Lexington

Bhat MA, Mantoo MA, Zaki FA (2010) Management of tree trunk borer, *Aeolesthes sarta* (Solsky) infesting apple in Kashmir. J Insect Sci 23(2):124–112

Boulif M (2022) A review of research on disease and pests of almonds in Morocco. Options Méditerr Série A Sémin Méditerr 119:199–204

Chernoli E, Wiman N (2022) http.//pnwhandbooks.options for commercial and home use. In: Nut crops. Oregon State University, Corvallis, pp 1–26

Cipollini ML (2017) Insect pests of chestnuts with emphasis on *Ambrosia* beetles. In: TACF. 34th Annual meeting, Portland, ME. Berry College and Georgia Chapter of the American Chestnut Foundation, Portland

DPPQS (2003) Integrated pest management package for walnut. Directorate of Plant Protection, Quarantine and Storage, Faridabad

Hall MJ (2013) Hickory Shuckworm. Bug Biz. Pest management and insect identification series. Pecan Research and Extension Station. Louisiana State University Agricultural Center, Baton Rouge

Harris KM, Jackman JA (1991) Pecan arthropod management. Texas A and M University, College Station, pp 1–19

Harris M, Ree B, Cooper J, Jackman J, Young J, Lacewell R, Knutson A (1998) Economic impact of pecan integrated pest management implementation in Texas. J Econ Entomol 91:1011–1020

Jackson D, McNeil D (1999) Stone fruit Part II. In: Temperate and sub-tropical fruit production, 2nd edn. Cab International, Wallingford, p 321

Knutson AE, Ree B (2019) Managing insect and mite pests of commercial pecans in Texas. In: Texas agri life extension service bulletin Ento-048. Texas A & M, College Station, TX. https://www.agrilifebookstore.org/

Maestri E, Imperiale D, Marmiroli N (2018) Nut, nut products and other seeds. In: Food integrity handbook: a guide to food authenticity issues and analytical solutions. University of Parma, Parma, pp 2–11

Neunzig HH (1986) The moths of America north of Mexico: Pyraloidea, Pyralidae. The Wedge Entomological Research Foundation, Washington, DC

Olson WH, Buchner RP (2002) Leading edge of plant protection for walnuts. Hortic Technol 12(4): 615–618

Ozkan G, Koyuncu MA (2005) Physical and chemical composition of some walnut (*Juglans regia* L.) genotypes grown in Turkey. Grasas Aceites 56:141–146

Ring DR, Harris MK (1984) Nut entry by first summer generation pecan nut casebearer. Southwest Entomol 9:13–21

Ring DR, Harris MK, Payne JA (1989) Sequential sampling plan for integrated pest management of pecan nut casebearer (Lepidoptera: Pyralidae). J Econ Entomol 82:906–909

Rishi ND, Shah KA (1985) Survey and bioecological studies on the natural enemy complex of Indian gypsy moth, *Lymantria obfuscata* Walker (Lepidoptera: Lymantriidae). J Entomol Res 9(1):82–93

Savage GP, McNeil DL, Dutta PC (2001) Some nutritional advantages of walnuts. Acta Hortic 544: 557–563

Shankar U, Sharma RM (2012) Pest management in walnut: an overview. In: Abrol DP, Shankar U (eds) Ecologically based integrated pest management. New India Publishing Agency, New Delhi, pp 765–785

Sheikh AG (1985) Insect pests of temperate fruits and their management. In: Proceedings of national workshop-cum-seminar on temperate fruits. SKUAST, Kashmir, pp 95–98

Stevenson DE, Knutson AE, Ree W, Jackman JA, Dean A, Matis JH, McVay J, Nesbitt M, Mizell R, Dutcher J (2003) Pecan nut casebearer pheromone monitoring and degree-day model validation across the pecan belt. Southwest Entomol 27:57–73

Wilson H, Burks CS, Reger JE, Wenger JA (2020) Biology and management of navel orangeworm (Lepidoptera: Pyralidae) in California. J Integr Pest Manag 11(1):1–15

Wiman N, Chernoli E (2022) Hazelnut pests includes management options for commercial and home use. In: Nut crops. Oregon State University, Corvallis, pp 1–26

Wood BW (2003) Pecan production in North America. Southwest Entomol 21:1–19

The manufacturer's authorised representative in the EU is Springer Nature Customer Service Centre GmbH, Europaplatz 3, 69115 Heidelberg, Germany. If you have any concerns regarding our products, please contact ProductSafety@springernature.com

Printed and bound by CPI Group (UK) Ltd, Croydon, CR0 4YY
01/10/2026
02225868-0012